AF332957

Analysis and Control of Linear Systems

Analysis and Control of Linear Systems

Edited by
Philippe de Larminat

First published in France by Hermès Science/Lavoisier entitled "Analyse des systèmes linèaires"
and "Commande des systèmes linèaires"
First published in Great Britain and the United States in 2007 by ISTE Ltd

ISTE Ltd
6 Fitzroy Square
London W1T 5DX
UK
www.iste.co.uk

ISTE USA
4308 Patrice Road
Newport Beach, CA 92663
USA

© ISTE Ltd.
© LAVOISIER

First South Asian Edition 2007

Library of Congress Cataloging-in-Publication Data

[Analyse des systèmes linèaires/Commande des systèmes linèaires. eng] Analysis and control
of linear systems analysis and control of linear systems/edited by Philippe de Larminat.
  p. cm.
 ISBN 13: 978-1-905209-35-4
 ISBN 10: 1-905209-35-5
 1. Linear control systems. 2. Automatic control. I. Larminat, Philippe de.

TJ220.A5313 2006
629.8'32—dc22

2006033665

British Library Cataloguing-in-Publication Data
A CIP record for this book is available from the British Library
ISBN 10: 1-905209-35-5
ISBN 13: 978-1-905209-35-4

Printed in Brijbasi Art Press Ltd., I-72, Sector-9, Noida, U.P. India.

Table of Contents

Chapter 5. Signals: Deterministic and Statistical Models 141
Eric LE CARPENTIER

**Chapter 6. Kalman's Formalism for State Stabilization
and Estimation** 159
Gilles DUC

Chapter 14. Multi-variable Modal Control 445
Yann LE GORREC and Jean-Francois MAGNI

Chapter 15. Robust H_∞/LMI Control. 479
Gilles DUC

Preface

This book is about the theory of continuous-state automated systems whose inputs, outputs and internal variables (temperature, speed, tension, etc.) can vary in a continuous manner. This is contrary to discrete-state systems whose internal variables are often a combination of binary sizes (open/closed, present/absent, etc.).

The word "linear" requires some explanation. The automatic power control of continuous-state systems often happens through actions in relation to the gaps we are trying to control. Thus, it is possible to regulate cruise control by acting on the acceleration control proportionally to the gap observed in relation to a speed instruction. The word "proportional" precisely summons up a linear control law.

Some processes are actually almost never governed by laws of linear physics. The speed of a vehicle, even when constant, is certainly not proportional to the position of the accelerator pedal. However, if we consider closed loop control laws, the return will correct mistakes when they are related either to external disturbances or to gaps between the conception model and the actual product. This means that modeling using a linear model is generally sufficient to obtain efficient control laws. Limits to the automated systems performances generally come from the restricted power of motors, precision of captors and variability of the behavior of the processes, more than from their possible non-linearity.

It is necessary to know the basics of linear automated systems before learning about the theory of non-linear systems. That is why linear systems are a fundamental theory, and the problems linked to closed-loop control are a big part of it.

Input-output and the state representations, although closely linked, are explained in separate chapters (1 and 2). Discrete-time systems are, for more clarity, explained in Chapter 3. Chapter 4 explains the structural properties of linear systems. Chapter

5 looks into deterministic and statistical models of signals. Chapter 6 introduces us to two fundamental theoretical tools: state stabilization and estimation. These two notions are also covered in control-related chapters. Chapter 7 defines the elements of modeling and identification. All modern control theories rely on the availability of mathematical models of processes to control them.

Modeling is therefore upstream of the control engineer. However, pedagogically it is located downstream because the basic systems theory is needed before it can be developed. This same theory also constitutes the beginning of Chapter 8, which is about simulation techniques. These techniques form the basis of the control laws created by engineers.

Chapter 9 provides an analysis of the classic invariable techniques while Chapter 10 summarizes them. Based on the transfer function concept, Chapter 11 addresses pole placement control and Chapter 12 internal control. The three following chapters cover modern automation based on state representation. They highlight the necessary methodological aspects. H_2 optimization control is explained in Chapter 13, modal control in Chapter 14 and H_∞ control in Chapter 15. Chapter 16 covers linear time-variant systems.

Part 1

System Analysis

Chapter 1

Transfer Functions and Spectral Models

1.1. System representation

A system is an organized set of components, of concepts whose role is to perform one or more tasks. The point of view adopted in the characterization of systems is to deal only with the input-output relations, with their causes and effects, irrespective of the physical nature of the phenomena involved.

Hence, a system realizes an application of the input signal space, modeling magnitudes that affect the behavior of the system, into the space of output signals, modeling relevant magnitudes for this behavior.

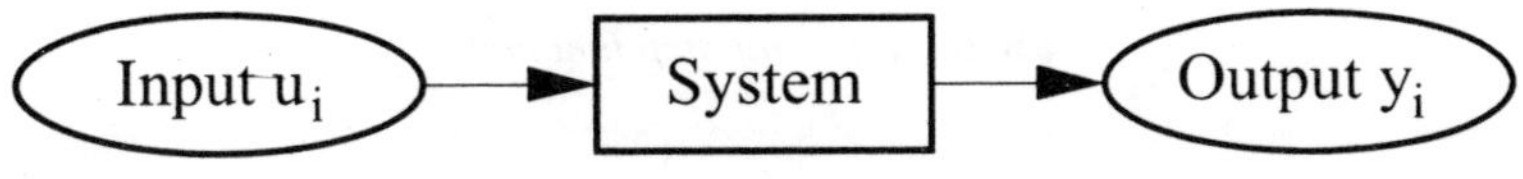

Figure 1.1. *System symbolics*

In what follows, we will consider mono-variable, analog or continuous systems which will have only one input and one output, modeled by continuous signals.

Chapter written by Dominique BEAUVOIS and Yves TANGUY.

1.2. Signal models

A continuous-time signal $(t \in R)$ is represented *a priori* through a function $x(t)$ defined on a bounded interval if its observation is necessarily of finite duration.

When signal mathematical models are built, the intention is to artificially extend this observation to an infinite duration, to introduce discontinuities or to generate Dirac impulses, as a derivative of a step function. The most general model of a continuous-time signal is thus a distribution that generalizes to some extent the concept of a digital function.

1.2.1. *Unit-step function or Heaviside step function U(t)*

This signal is constant, equal to 1 for the positive evolution variable and equal to 0 for the negative evolution variable.

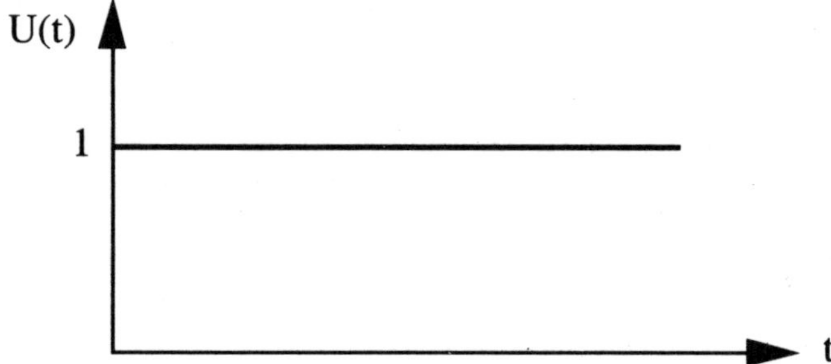

Figure 1.2. *Unit-step function*

This signal constitutes a simplified model for the operation of a device with a very low start-up time and very high running time.

1.2.2. *Impulse*

Physicists began considering shorter and more intense phenomena. For example, an electric loading $M\mu$ can be associated with a mass M evenly distributed according to an axis.

What density should be associated with a punctual mass concentrated in 0? This density can be considered as the bound (simple convergence) of densities $M\mu_n(\sigma)$ verifying:

$$\mu_n(\sigma) = \frac{n}{2} \quad -\frac{1}{n} \leq \sigma \leq \frac{1}{n} \qquad\qquad M\int_{-\frac{1}{n}}^{+\frac{1}{n}} \mu_n(\sigma)\,d\sigma = M$$

$$\mu_n(\sigma) = 0 \quad \text{elsewhere}$$

This bound is characterized, by the physicist, by a "function" $\delta(\sigma)$ as follows:

$$\delta(\sigma) = 0 \quad \sigma \neq 0$$

$$\text{with} \quad \int_{-\infty}^{+\infty} \delta(\sigma)\,d\sigma = 1$$

$$\delta(0) = +\infty$$

However, this definition does not make any sense; no integral convergence theorem is applicable.

Nevertheless, if we introduce an auxiliary function $\varphi(\sigma)$ continuous in 0, we will obtain the mean formula:

$$\lim_{n\to+\infty} \int_{-\frac{1}{n}}^{+\frac{1}{n}} \phi(\sigma)\,\mu_n(\sigma)\,d\sigma = \varphi(\eta) \quad \text{because} \quad -\frac{1}{n} \leq \eta \leq \frac{1}{n}$$

Hence, we get a functional definition, indirect of symbol δ. δ associates with any continuous function at the origin its origin value. Thus, it will be written in all cases:

$$\varphi(0) = \langle \delta, \varphi \rangle = \int_{-\infty}^{+\infty} \varphi(\sigma)\,\delta(\sigma)\,d\sigma$$

δ is called a *Dirac impulse* and it represents the most popular distribution. This impulse δ is also written $\delta(t)$.

For a time lag t_o, we will use the notations $\delta(t - t_o)$ or $\delta_{(t_o)}(t)$; the impulse is graphically "represented" by an arrow placed in $t = t_o$, with a height proportional to the impulse weight.

In general, the Dirac impulse is a very simplified model of any impulse phenomenon centered in $t = t_o$, with a shorter period than the time range of the systems in question and with an area S.

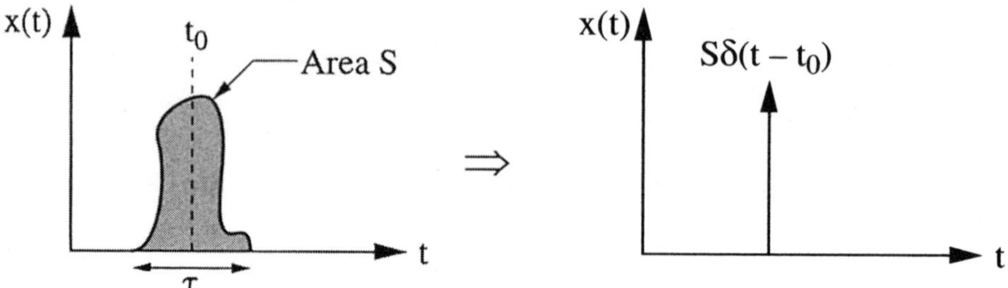

Figure 1.3. *Modeling of a short phenomenon*

We notice that in the model based on Dirac impulse, the "microscopic" look of the real signal disappears and only the information regarding the area is preserved.

Finally, we can imagine that the impulse models the derivative of a unit-step function. To be sure of this, let us consider the step function as the model of the real signal $u_o(t)$ represented in Figure 1.4, of derivative $u_o'(t)$. Based on what has been previously proposed, it is clear that $\lim_{\tau \to 0} u_o'(t) = \delta(t)$.

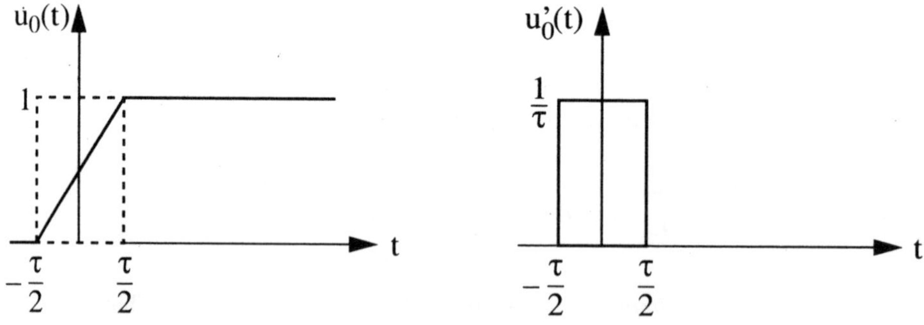

Figure 1.4. *Derivative of a step function*

1.2.3. *Sine-wave signal*

$x(t) = A\cos(2\pi f_o t + \varphi)$ or $x(t) = Ae^{j(2\pi f_o t + \varphi)}$ for its complex representation. f_o designates the frequency expressed in Hz, $\omega_o = 2\pi f_o$ designates the impulse expressed in rad/s and φ the phase expressed in rad.

A real value sine-wave signal is entirely characterized by f_o ($0 \leq f_o \leq +\infty$), by A ($t = t_o$), by φ ($-\pi \leq \varphi \leq +\pi$). On the other hand, a complex value sine-wave signal is characterized by a frequency f_o with $-\infty \leq f_o \leq +\infty$.

1.3. Characteristics of continuous systems

The input-output behavior of a system may be characterized by different relations with various degrees of complexity. In this work, we will deal only with *linear systems* that obey the physical principle of *superposition* and that we can define as follows: a system is *linear* if to any combination of input constant coefficients $\sum a_i x_i$ corresponds the same output linear combination, $\sum a_i y_i = \sum a_i G(x_i)$.

Obviously, in practice, no system is rigorously linear. In order to simplify the models, we often perform linearization around a point called an *operating* point of the system.

A system has an *instantaneous response* if, irrespective of input x, output y depends only on the input value at the instant considered. It is called *dynamic* if its response at a given instant depends on input values at other instants.

A system is called *causal* system if its response at a given instant depends only on input values at previous instants (possibly present). This characteristic of causality seems natural for real systems (the effect does not precede the cause), but, however, we have to consider the existence of systems which are not strictly causal in the case of delayed time processing (playback of a CD) or when the evolution variable is not time (image processing).

The pure delay system $\tau > 0$ characterized by $y(t) = x(t - \tau)$ is a dynamic system.

1.4. Modeling of linear time-invariant systems

We will call LTI such a system. The aim of this section is to show that the input-output relation in an LTI is modeled by a convolution operation.

1.4.1. *Temporal model, convolution, impulse response and unit-step response*

We will note by $h_\tau(t)$ the response of the real impulse system represented in Figure 1.5.

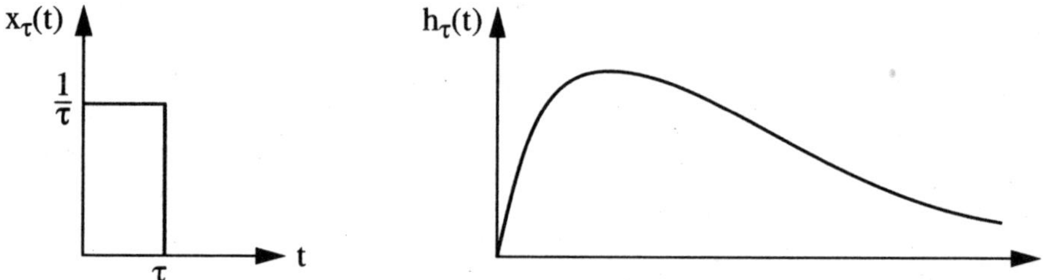

Figure 1.5. *Response to a basic impulse*

Let us approach any input $x(t)$ by a series of joint impulses of width τ and amplitude $x(k\tau)$.

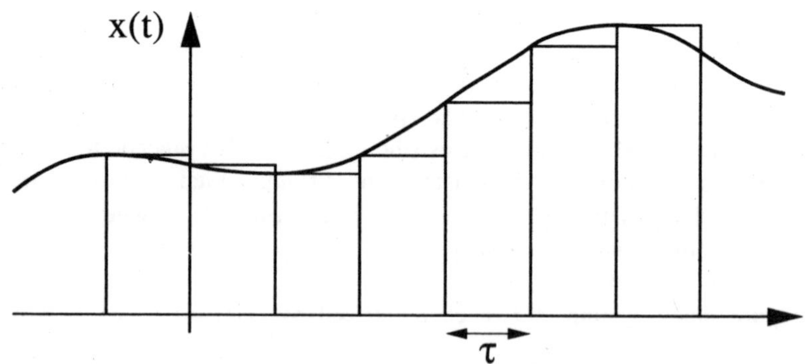

Figure 1.6. *Step approximation*

By applying the linearity and invariance hypotheses of the system, we can approximate the output at an instant t by the following amount, corresponding to the recombination of responses to different impulses that vary in time:

$$y(t) \cong \sum_{-\infty}^{+\infty} \tau\, x(k\tau) h_\tau (t - k\tau)$$

In order to obtain the output at instant t, we will make τ tend toward 0 so that our input approximation tends toward x. Hence:

$$\lim_{\tau \to 0} x_\tau (t) = \delta(t) \text{ and } \lim_{\tau \to 0} h_\tau (t) = h(t)$$

where $h(t)$, the response of the system to the Dirac impulse, is a characteristic of the system's behavior and is called an *impulse response*.

If we suppose that the system preserves the continuity of the input, i.e. for any convergent sequence $x_n(t)$ we have $G\left(\lim_{n \to \infty} x_n(t) \right) = \lim_{n \to \infty} G(x_n(t))$, we obtain:

$$y(t) = \int_{-\infty}^{+\infty} x(\theta) h(t - \theta) d\theta$$

or:

$$y(t) = \int_{-\infty}^{+\infty} h(\sigma) x(t - \sigma) d\sigma \text{ through } \sigma = t - \theta$$

which defines the *convolution integral* of functions x and h, noted by the asterisk:

$$y(t) = x * h(t) = h * x(t)$$

1.4.2. *Causality*

When the system is causal, the output at instant t depends only on the previous inputs and consequently function $h(t)$ is identically zero for $t < 0$. The impulse response, which considers the past in order to provide the present, is a causal function and the input-output relation has the following form:

$$y(t) = \int\limits_{0}^{+\infty} h(\theta)x(t-\theta)d\theta = \int\limits_{-\infty}^{t} x(\theta)h(t-\theta)d\theta$$

The output of a causal time-invariant linear system can be interpreted as a weighted mean of all the past inputs having excited it, a weighting characteristic for the system considered.

1.4.3. *Unit-step response*

The *unit-step response* of a system is its response $i(t)$ to a unit-step excitation. The use of the convolution relation leads us to conclude that the unit-step response is the integral of the impulse response:

$$i(t) = \int\limits_{0}^{t} h(\theta)d\theta$$

This response is generally characterized by:

– the rise time t_m, which is the time that separates the passage of the unit-step response from 10% to 90% of the final value;

– the response time t_r, also called establishment time, is the period at the end of which the response remains in the interval of the final value $\pm\,\alpha\%$. A current value of α is 5%. This time also corresponds to the period at the end of which the impulse response remains in the interval $\pm\,\alpha\%$; it characterizes the transient behavior of the system output when we start applying an excitation and it also reminds that a system has several inputs which have been applied before a given instant;

– the possible overflow defined as $\dfrac{y_{max} - y(\infty)}{y(\infty)}$ expressed in percentage.

1.4.4. *Stability*

1.4.4.1. *Definition*

The concept of stability is delicate to introduce since its definition is linked to the structures of the models studied. Intuitively, two ideas are outlined.

A system is labeled as stable around a point of balance if, after having been subjected to a low interference around that point, it does not move too far away from it. We talk of asymptotic stability if the system returns to the point of balance and of stability, in the broad sense of the word, if the system remains some place near that point. This concept, intrinsic to the system, which is illustrated in Figure 1.7 by a ball positioned on various surfaces, requires, in order to be used, a representation by equations of state.

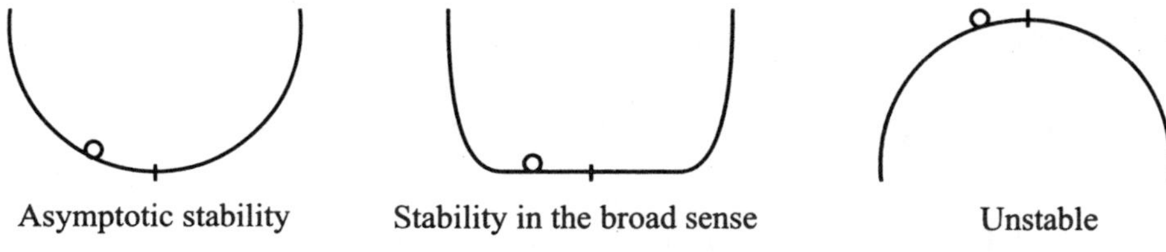

Asymptotic stability   Stability in the broad sense   Unstable

Figure 1.7. *Concepts of stability*

Another point of view can be adopted where the stability of a system can be defined simply in terms of an input-output criterion; a system will be called stable if its response to any bounded input is limited: we talk of *L(imited) I(nput) L(imited) R(esponse) stability*.

1.4.4.2. *Necessary and sufficient condition of stability*

An LTI is BIBO (bounded input, bounded output) if and only if its impulse response is positively integrable, i.e. if:

$$\int_{-\infty}^{+\infty} \left| h(\theta) \right| d\theta < +\infty$$

The sufficient condition is immediate if the impulse response is positively integrable and applying a bounded input to the system, $\forall t \; \left| x(t) \right| < M$, leads to a bounded output because:

$$\forall t \; \left| y(t) \right| \leq \int_{-\infty}^{+\infty} \left| x(t-\theta) \right| \left| h(\theta) \right| d\theta \leq M \int_{-\infty}^{+\infty} \left| h(\theta) \right| d\theta \leq +\infty$$

Let us justify the necessary condition: the system has a bounded output in response to any bounded excitation, then its impulse response is positively integrable.

To do this, let us demonstrate the opposite proposition: if the impulse response of the system is not absolutely integrable:

$$\forall K, \exists T \quad \int_{-\infty}^{+T} |h(\theta)|\, d\theta > K$$

there is a bounded input that makes the output diverge.

It is sufficient to choose input x such that:

$$x(T - \theta) = \operatorname{sgn}(h(\theta)) \quad \text{for} \quad \theta < T \quad \text{and} \quad x(T - \theta) = 0 \quad \text{for} \quad \theta > T$$

then $y(T) = \displaystyle\int_{-\infty}^{T} h(\theta)\,\operatorname{sgn}(h(\theta))\,d\theta > K \quad \forall K$ which means that y diverges.

1.4.5. *Transfer function*

Any LTI is modeled by a convolution operation, an operation that can be considered in the largest sense, i.e. the distribution sense. We know that if we transform this product through the proper transform (see section 1.4.1), we obtain a simple product.

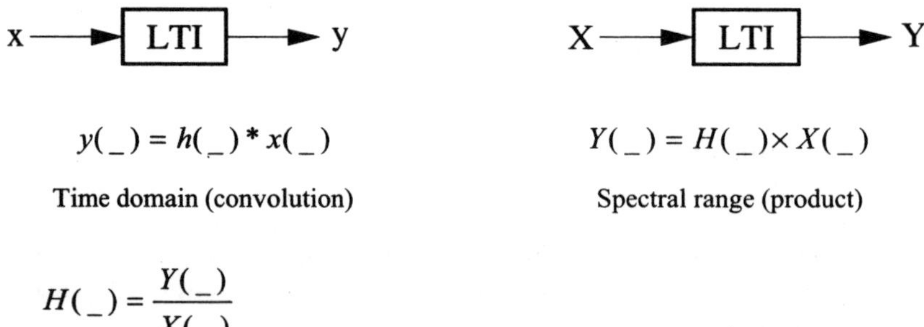

$$y(\_) = h(\_) * x(\_) \qquad\qquad Y(\_) = H(\_) \times X(\_)$$

Time domain (convolution)   Spectral range (product)

$$H(\_) = \frac{Y(\_)}{X(\_)}$$

This formally defined transform ratio is the transform of the impulse response and is called a *transfer function* of LTI.

The use of transfer functions has a considerable practical interest in the study of system association as shown in the examples below.

1.4.5.1. *Cascading (or serialization) of systems*

Let us consider the association of Figure 1.8.

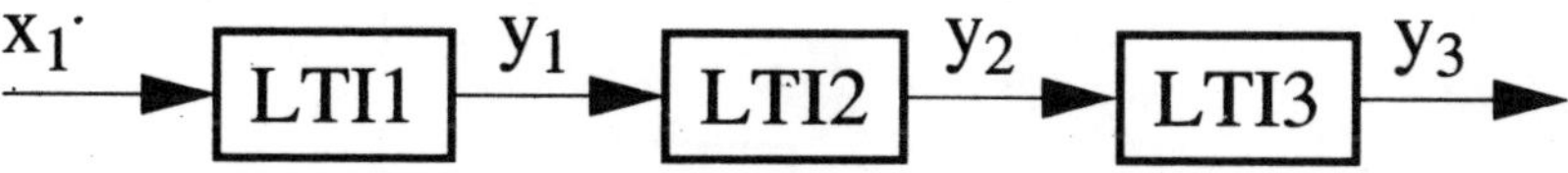

Figure 1.8. *Serial association*

Hence $y_3(\_) = h_3(\_) * \left(h_2(\_) * \left(h_1(\_) * x_1(\_)\right)\right)$. This leads, in general, to a rather complicated expression.

In terms of transfer function, we obtain:

$$H(\_) = H_1(\_) \times H_2(\_) \times H_3(\_)$$

i.e. the simple product of three basic transfer functions. The interest in this characteristic is that any processing or transmission chain basically consists of an association of "basic blocks".

1.4.5.2. *Other examples of system associations*

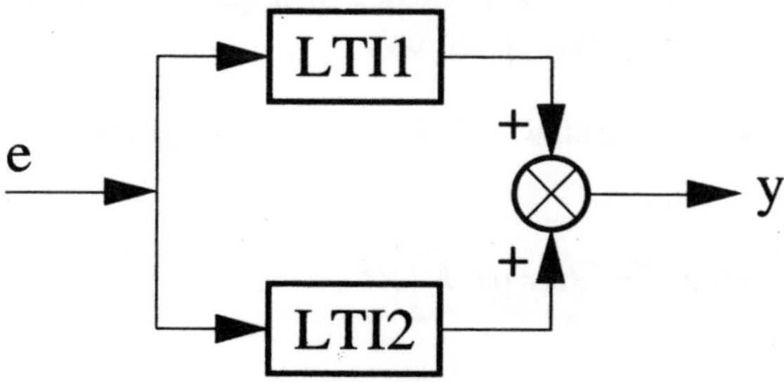

Figure 1.9. *Parallel association*

$$H(\_) = \frac{Y(\_)}{E(\_)} = H_1(\_) + H_2(\_)$$

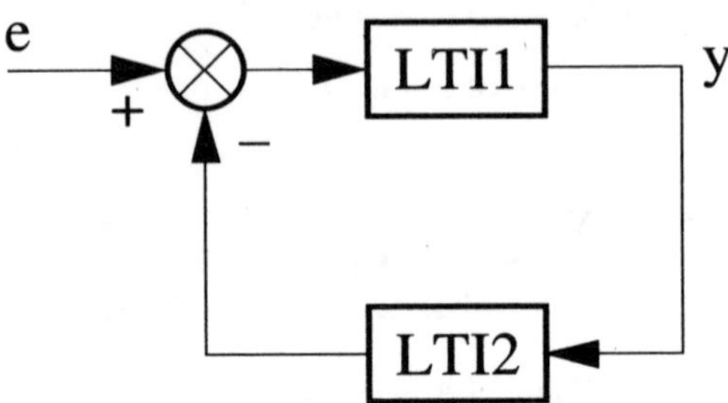

Figure 1.10. *Loop structure*

$$H(\_) = \frac{Y(\_)}{E(\_)} = \frac{H_1(\_)}{1 + H_1(\_) \times H_2(\_)}$$

The term $1 + H_1(\_) \times H_2(\_)$ corresponds to the return difference, which is defined by $1 -$ (product of loop transfers). The loop transfers, introduced in the structure considered here, are the sign minus the comparator, the transfers H_1 and H_2.

1.4.5.3. *Calculation of transfer functions of causal LTIs*

In this section, we suppose the existence of impulse response transforms while keeping in mind the convergence conditions.

Using the Fourier transform, we obtain the frequency response $H(f)$:

$$H(f) = \int_0^{+\infty} h(\theta) e^{-2\pi j f \theta} d\theta = |H(f)| e^{j\Phi(f)}$$

where $|H(f)|$ is the module or gain, $\Phi(f)$ is the phase or phase difference of the frequency response.

Through the Laplace transform, we obtain the transfer function of the system $H(p)$, which is often referred to as *isomorphic* transfer function:

$$H(p) = \int_0^{+\infty} h(\theta) e^{-p\theta} d\theta$$

The notations used present an ambiguity (same *H*) that should not affect the informed reader: when the impulse response is positively integrable, which corresponds to a stability hypothesis of the system considered, we know that the Laplace transform converges on the imaginary axis and that it is mistaken with Fourier transform through $p = 2\pi j f$. Hence, the improper notation (same *H*):

$$H(p)\big|_{p=2\pi jf} = H(f)$$

We recall that the transfer functions have been formally defined here and the convergence conditions have not been formulated. For the LTIs, which are system models that can be physically realized, the impulse responses are functions whose Laplace transform has always a sense within a domain of the complex plane to define.

On the other hand, the frequency responses, which are defined by the Fourier transform of the impulse response, even considered in the distribution sense, do not always exist. The stability hypothesis ensures the simultaneous existence of two transforms.

EXAMPLE 1.1.– it is easily verified whether an integrator has as an impulse response the Heaviside step function $h(t) = u(t)$ and hence:

$$H(p) = \frac{1}{p} \qquad\qquad H(f) = \frac{1}{2}\delta(f) + \frac{1}{2\pi j}\, Pf\, \frac{1}{f}$$

where $Pf\, \dfrac{1}{f}$ designates the pseudo-function distribution $\dfrac{1}{f}$.

An LTI with localized constants is represented through a differential equation with constant coefficients with $m < n$:

$$b_0\, y(t) + \ldots + b_n\, y^{(n)}(t) = a_0\, x(t) + \ldots + a_m\, x^{(m)}(t)$$

By supposing that $x(t)$ and $y(t)$ are continuous functions defined from $-\infty$ to $+\infty$, continuously differentiable of order m and n, by a two-sided Laplace transform we obtain the transfer function $H(p)$:

$$(b_0 + b_1 p + \ldots + b_n p^n)Y(p) = (a_0 + a_1 p + \ldots a_m p^m)X(p)$$

$$H(p) = \frac{a_0 + a_1 p + \ldots a_m p^m}{b_0 + b_1 p + \ldots + b_n p^n}$$

Such a transfer function is called rational in p. The coefficients of the numerator and denominator polynomials are real due to their physical importance in the initial differential equation. Hence, the numerator roots, called *zeros*, and the denominator roots, called *transfer function poles*, are conjugated real or complex numbers.

If $x(t)$ and $y(t)$ are causal functions, the Laplace transform of the differential equation entails terms based on initial input values $x(0)$, $x'(0)$, $x^{(m-1)}(0)$ and output values $y(0)$, $y'(0)$, $y^{(n-1)}(0)$; the concept of state will make it possible to overcome this dependence.

1.4.6. *Causality, stability and transfer function*

We have seen that the necessary and sufficient condition of stability of an SLI is for its impulse response to be absolutely integrable: $\int_{-\infty}^{+\infty}|h(\theta)|d\theta < +\infty$.

The consequence of the hypothesis of causality modifies this condition because we thus integrate from 0 to $+\infty$.

On the other hand, if we seek a necessary and sufficient condition of stability for the expression of transfer functions, the hypothesis of causality is determining.

Since the impulse response $h(\theta)$ is a causal function, the transfer function $H(p)$ is holomorphic (defined, continuous, derivable with respect to the complex number p) in a right half-plane defined by $\operatorname{Re}(p) > \sigma_o$. The absolute integrability of $h(\theta)$ entails the convergence of $H(p)$ on the imaginary axis.

A CNS of EBRB stability of a *causal LTI* is that its transfer function is holomorphic in the right half-plane defined by $Re(p) \geq 0$.

When:

$$H(p) = e^{-\tau p}\, \frac{N(p)}{D(p)}$$

where $N(p)$ and $D(p)$ are polynomials, it is the same as saying that *all the transfer function poles are negative real parts*, i.e. placed in the left half-plane.

We note that in this particular case, the impulse response of the system is a function that tends infinitely toward 0.

1.4.7. *Frequency response and harmonic analysis*

1.4.7.1. *Harmonic analysis*

Let us consider a *stable* LTI whose impulse response $h(\theta)$ is canceled after a period of time t_R. For the models of physical systems, this period of time t_R is in fact rejected infinitely; however, for reasons of clarity, let us suppose t_R as finite, corresponding to the response time to 1% of the system.

When this system is subject to a harmonic excitation $x(t) = Ae^{2\pi jf_0 t}$ from $t = 0$, we obtain:

$$y(t) = \int_0^t h(\theta)\, Ae^{2\pi jf_0(t-\theta)}\, d\theta = Ae^{2\pi jf_0 t} \int_0^t h(\theta)\, Ae^{-2\pi jf_0\theta}\, d\theta$$

For $t > t_R$, the impulse response being zero, we have:

$$\int_0^t h(\theta)\, e^{-2\pi jf_0\theta}\, d\theta = H(f_0) = \int_0^{+\infty} h(\theta)\, e^{-2\pi jf_0\theta}\, d\theta = |H(f_0)|\, e^{j\Phi(f_0)}$$

and hence for $t > t_R$, we obtain $y(t) = AH(f_o)e^{2\pi jf_0 t} = A|H(f_0)|e^{j(2\pi f_0 t + \Phi(f_0))}$.

This means that the system, excited by a sine-wave signal, has an output that tends, after a transient state, toward a sine-wave signal of same frequency. This signal, which is a characteristic of a steady (or permanent) state, is modified in amplitude by a multiplicative term equal to $|H(f_0)|$ and with a phase difference of $\Phi(f_0)$.

$$x(t) = A_x \sin(2\pi f_0 t) \quad \longrightarrow \boxed{\text{LTI}} \longrightarrow \quad y(t) = A_y \sin(2\pi f_0 t + \Phi)$$

$$\frac{A_y}{A_x} = |H(f_0)| \text{ module or gain} \qquad \Phi = \arg H(f_o) \text{ phase}$$

We note that $H(f)$ is nothing else but the Fourier transform of the impulse response, the frequency response of the system considered.

1.4.7.2. *Existence conditions of a frequency response*

The frequency response is the Fourier transform of the impulse response. It can be defined in the distribution sense for the divergent responses in $|t|^\alpha$ but not for exponentially divergent responses (e^{bt}). However, we shall note that this response is always defined under the hypothesis of stability; in this case and only in this case, we pass from transfer functions with complex variables to the frequency response by determining that $p = 2\pi j f$.

EXAMPLE 1.2.– let $h(t) = u(t)$ be the integrator system:

$$H(p) = \frac{1}{p}$$

$$H(f) = \frac{1}{2}\delta(f) + \frac{1}{2\pi j} Pf \frac{1}{f} \text{ and not } \frac{1}{2\pi j f} \text{ because the system is not EBRB}$$

stable.

$H(p) = TL(u(t))$ is defined according to the functions in the half-plane $Re(p) > 0$, whereas $H(f) = TF(u(t))$ is defined in the distribution sense.

Unstable filter of first order: $h(t) = e^t \quad t \geq 0$

$$H(p) = \frac{1}{p-1} \text{ defined for } Re(p) > 1, \quad H(f) \text{ is not defined, even in the}$$

distribution sense.

Hence, even if the system is unstable, we can always consider the complex number obtained by formally replacing p by $2\pi j f$ in the expression of the transfer function in p. The result obtained is not identified with the frequency response but may be taken as a harmonic analysis, averaging certain precautions as indicated in the example in Figure 1.11.

Let us consider the unstable causal system of transfer function $\dfrac{1}{p-1}$, inserted into the loop represented in Figure 1.11.

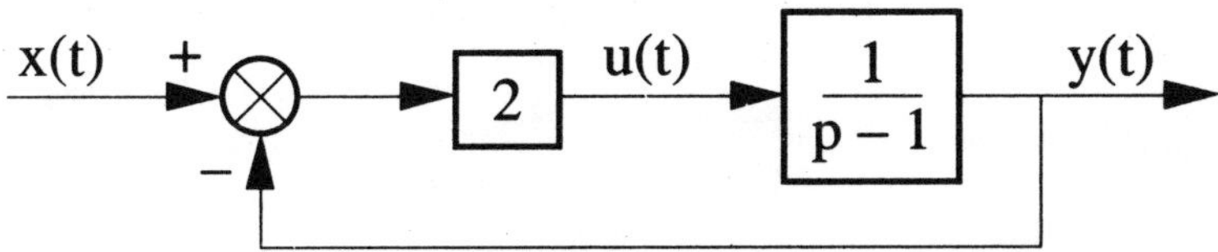

Figure 1.11. *Unstable system inserted into a loop*

The transfer function of the system is $\dfrac{2}{p+1}$. The looped system is stable and hence we can begin its harmonic analysis by placing an input sine-wave signal $x(t) = A_x \sin(2\pi f_0 t)$. During the stationary regime, $y(t)$ and $u(t)$ are equally sinusoidal, hence:

$$y(t) = A_y \sin(2\pi f_0 t + \Phi_y) \text{ with } \frac{A_y}{A_x} = \left| \frac{2}{2\pi j f_0 + 1} \right| \text{ and } \Phi_y = \arg\left(\frac{2}{2\pi j f_0 + 1} \right)$$

$$u(t) = A_u \sin(2\pi f_0 t + \Phi_u) \text{ with } \frac{A_u}{A_x} = \left| \frac{2(2\pi j f_0 - 1)}{2\pi j f_0 + 1} \right| , \Phi_u = \arg\left(\frac{2(2\pi j f_0 - 1)}{2\pi j f_0 + 1} \right)$$

Hence:

$$\frac{A_y}{A_u} = \left| \frac{1}{2\pi j f_0 - 1} \right| = H(p)\big|_{p=2\pi j f_0}$$

$$\Phi_y - \Phi_u = \arg\left(\frac{1}{2\pi j f_0 - 1} \right) = \arg H(p)\big|_{p=2\pi j f_0}$$

Table 1.1 sums up the features of a system's transfer function, the existence conditions of its frequency response and the possibility of performing a harmonic analysis based on the behavior of its impulse response.

1.4.7.3. *Diagrams*

$h(t)$	$H(p)$	$H(2\pi jf) = H(j\omega)$
$\displaystyle\int_{-\infty}^{+\infty}\left\|h(\theta)\right\|d\theta < +\infty$ EBRB stability	$H(p)$ has its poles on the left of the imaginary axis. Holomorphy	TF exists Possible direct analysis $TF = H(p)\big\|_{p=2\pi jf}$
$t \to +\infty\ \left\|h(t)\right\| \sim t^{n-1}$	$H(p)$ has a pole of order n at the origin. Holomorphy	Directly impossible harmonic analysis impossible directly except for a simple pole at the origin
$t \to +\infty\ \left\|h(t)\right\| \sim e^{kt}$ $k > 0$	$H(p)$ has poles on the right of the imaginary axis. Holomorphy	Possible analysis if the system is introduced in a stable looping and $H(j\omega) = H(p)\big\|_{p=j\omega}$
$t \to +\infty\ \left\|h(t)\right\| \sim e^{j\omega t}$	$H(p)$ has poles on the imaginary axis. Holomorphy	

Table 1.1. *Unit-step responses, transfer functions and existence conditions of the frequency response*

Frequency responses are generally characterized according to impulse $\omega = 2\pi j f$ and data $|H(j\omega)|$ and $\Phi(j\omega)$ grouped together as diagrams. The following are distinguished:

– *Nyquist diagram* where the system of coordinates adopts in abscissa the real part, and in ordinate the imaginary part $H(p)\big|_{p=j\omega}$;

– *Black diagram* where the system of coordinates adopts in ordinate the module expressed in decibels, like:

$$20\log_{10}\big(\big\|H(p)\big\|_{p=j\omega}\big) \text{ and in abscissa } \arg H(p)\big|_{p=j\omega} \text{ expressed in degree;}$$

– *Bode diagram* which consists of two graphs, the former representing the module expressed in decibels based on $\log_{10}(\omega)$ and the latter representing the phase according to $\log_{10}(\omega)$. Given the biunivocal nature of the logarithm function and in order to facilitate the interpretation of the diagram, the axes of the abscissas are graduated in ω.

1.5. Main models

1.5.1. *Integrator*

This system has for impulse response $h(t) = KU(t)$ and for transfer function in p:

$$H(p) = \frac{K}{p} \quad \mathrm{Re}\,(p) > 0$$

The unit-step response is a slope ramp K : $i(t) = KtU(t)$.

The frequency response, which is the Fourier transform of the impulse response, is defined only in the distribution sense:

$$H(f) = \frac{1}{2}\delta(f) + \frac{1}{2\pi j} Pf \frac{1}{f}$$

The evolution of $H(p)\big|_{p=j\omega}$ according to ω leads to the diagrams in Figure 1.12.

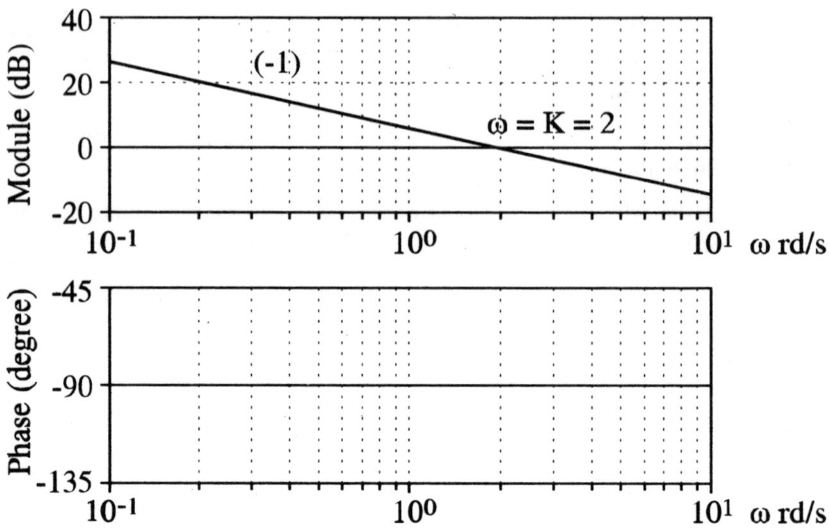

Figure 1.12. *Bode diagram*

The module is characterized by a straight line of slope (–1), –6 dB per octave (factor 2 between 2 impulses) or –20 dB per decade (factor 10 between two impulses), that crosses the axis 0dB in $\omega = K$.

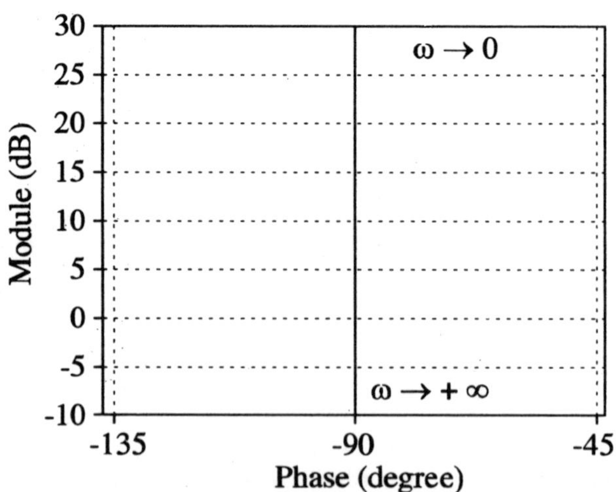

Figure 1.13. *Black diagram*

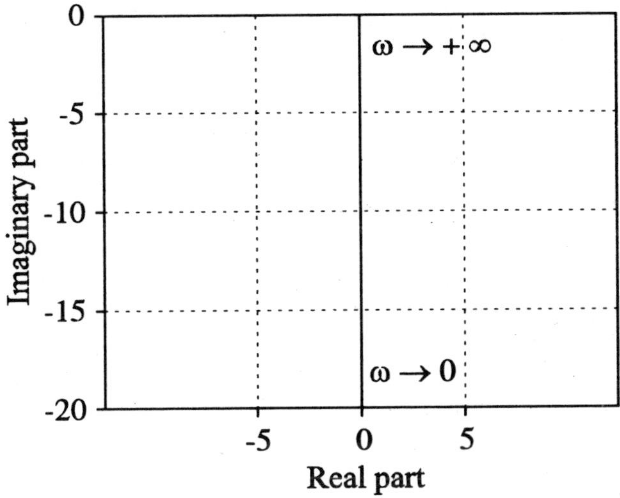

Figure 1.14. *Nyquist diagram*

1.5.2. *First order system*

This causal system, with an impulse response of $h(t) = \frac{K}{T} e^{-\frac{t}{T}} U(t)$, has a transfer function:

$$H(p) = \frac{K}{1+Tp} \quad \mathrm{Re}(p) > -\frac{1}{T}$$

The unit-step response admits as time expression and as Laplace transform the following functions:

$$i(t) = K\left(1 - e^{-\frac{t}{T}}\right) U(t) \qquad\qquad I(p) = \frac{K}{p(1+Tp)}$$

It has the following characteristics:

– the final value is equal to K, for an input unit-step function;

– the tangent at the origin reaches the final value of the response at the end of time T, which is called time constant of the system.

The response reaches $0.63\,K$ in T and $0.95\,K$ in 3 T.

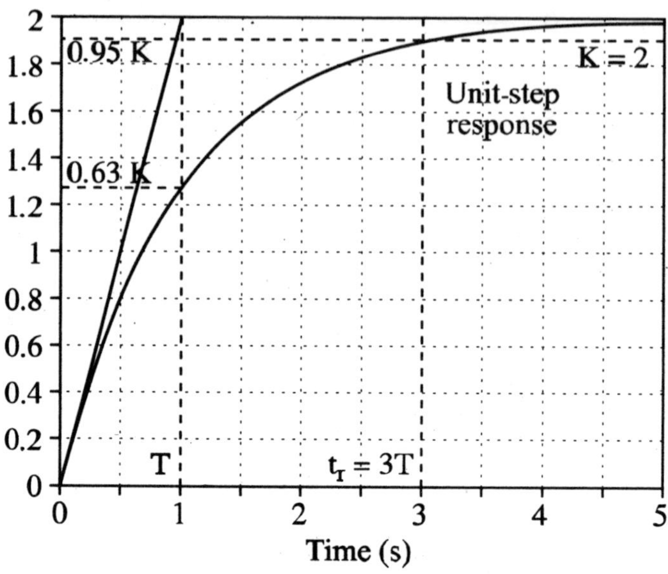

Figure 1.15. *Unit-step response of the first order model*

The frequency response is identified with the complex number $H(j\omega)$:

$$\left|H(j\omega)\right| = \frac{K}{\sqrt{1+\omega^2 T^2}} \qquad \arg(H(j\omega)) = -Arctg(\omega T)$$

In the Bode plane we will thus have:

$$20\log_{10}\frac{K}{\sqrt{1+\omega^2 T^2}} \quad \text{and} \quad -Arctg(\omega T) \text{ according to } \log_{10}(\omega)$$

The asymptotic behavior of the gain and phase curves is obtained as follows:

$$\omega T \ll 1 \qquad\qquad\qquad \omega T \gg 1$$

$$20\log_{10}|H| \cong 20\log_{10} K \qquad 20\log_{10}|H| \cong 20\log_{10}\left(\frac{K}{T}\right) - 20\log_{10}\omega$$

$$\arg H = 0\,rd \qquad\qquad\qquad \arg H = -\frac{\pi}{2}\,rd$$

These values help in building a polygonal approximation of the plot called Bode asymptotic plot:

– gain: two half-straight lines of slope (0) and –20 *dB/decade* noted by (–1);

– phase: two asymptotes at $0rd$ and $-\frac{\pi}{2}rd$.

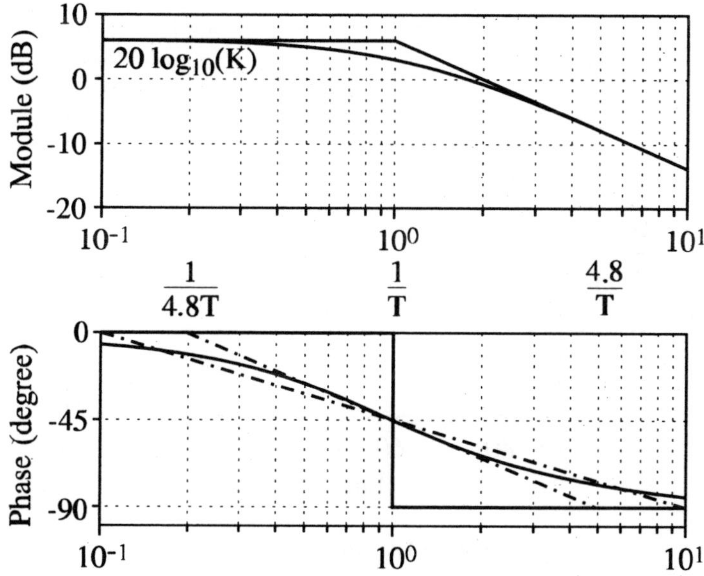

Figure 1.16. *Bode diagram of the first order system*

The gain curve is generally approximated by the asymptotic plot.

The plot of the phase is symmetric with respect to the point $(\omega = \frac{1}{T}, \phi = -45^{\circ})$. The tangent at the point of symmetry crosses the asymptote 0° at $\omega = \frac{1}{4.8T}$ and, by symmetry, the asymptote -90° at $\omega = \frac{4.8}{T}$.

The gaps δG and $\delta \phi$ between the real curves and the closest asymptotic plots are listed in the table of Figures 1.17 and 1.18.

ω	$\dfrac{1}{8T}$	$\dfrac{1}{4T}$	$\dfrac{1}{2T}$	$\dfrac{1}{T}$	$\dfrac{2}{T}$	$\dfrac{4}{T}$	$\dfrac{8}{T}$
δG	$\approx 0\ db$	$\approx 0\ db$	$-1\ db$	$-3\ db$	$-1\ db$	$\approx 0\ db$	$\approx 0\ db$
$\delta\phi$	$7°$	$14°$	$26.5°$	$45°$	$26.5°$	$14°$	$7°$

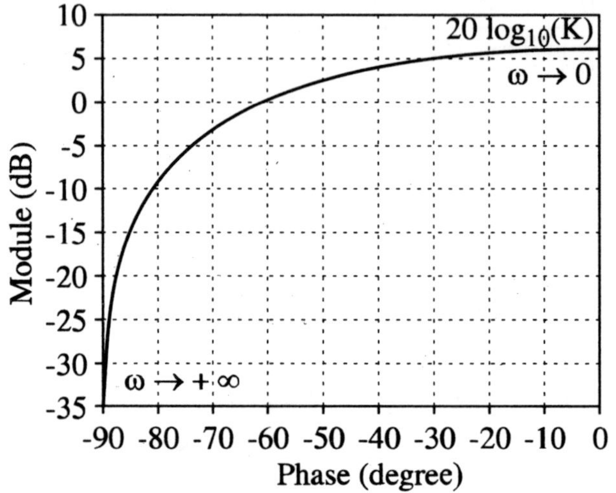

Figure 1.17. *Black diagram*

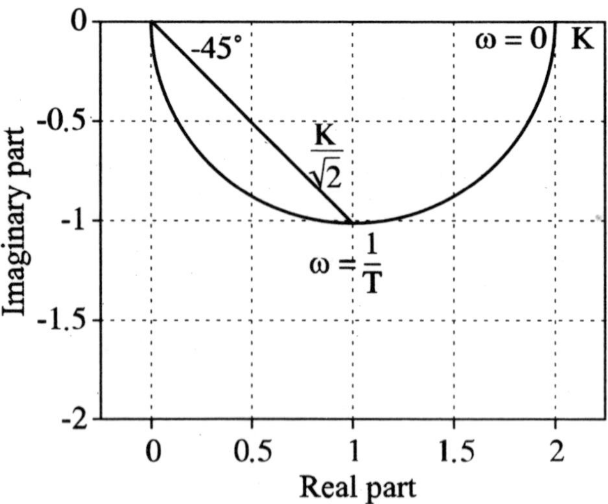

Figure 1.18. *Nyquist diagram*

1.5.3. *Second order system*

The second order system, of angular frequency $\omega_0 = \dfrac{1}{T_0}$, and of damping coefficient ξ, is defined by a transfer function such as:

$$H(p) = \frac{K}{1 + 2\dfrac{\xi}{\omega_0}p + \dfrac{p^2}{\omega_0^2}}$$

1.5.3.1. *Unit-step response*

The theorems of initial and final values make it possible to easily comprehend the asymptotic features of the unit-step response: zero initial value, final value equal to K, tangent at the origin with zero slope.

Based on the value of ξ with respect to 1, the transfer function poles have a real or complex nature and the unit-step response looks different.

$\xi > 1$: the transfer function poles are real and the unit-step response has an aperiodic look (without oscillation):

$$i(t) = K\left(1 - \frac{1}{2}\left((1-a)e^{-\left(\xi+\sqrt{\xi^2-1}\right)\frac{t}{T_0}} + (1+a)e^{-\left(\xi-\sqrt{\xi^2-1}\right)\frac{t}{T_0}}\right)\right)$$

where $a = \dfrac{\xi}{\sqrt{\xi^2-1}}$

The tangent at the origin is horizontal.

If $\xi \gg 1$, one of the poles prevails over the other and hence:

$$i(t) \approx K\left(1 - e^{-\frac{\omega_0 t}{2\xi}}\right)$$

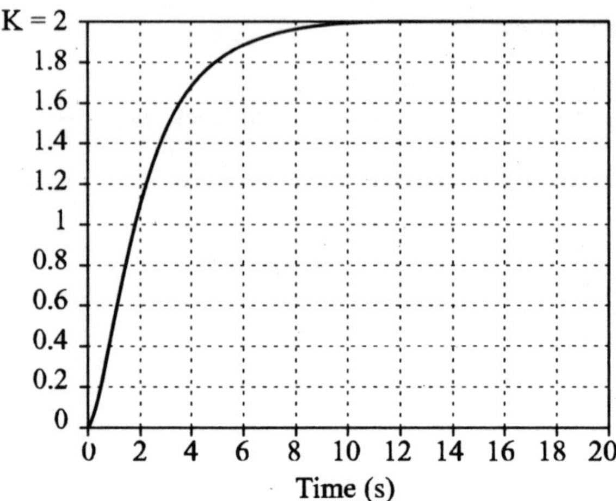

Figure 1.19. *Unit-step response of the second order system $\xi \geq 1$*

$\xi = 1$: critical state. The roots of the denominator of the transfer function are real and merged, and the unit-step response is:

$$i(t) = K\left(1 - \left(1 + \frac{t}{T_0}\right)e^{-\frac{t}{T_0}}\right)$$

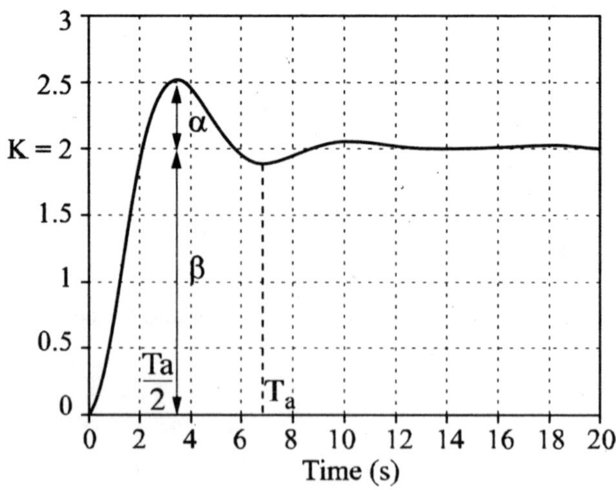

Figure 1.20. *Unit-step response of the second order system $\xi < 1$*

$\xi < 1$: oscillating state. The two poles of $H(p)$ are conjugated complex numbers and the unit-step response is:

$$i(t) = K\left(1 - \frac{e^{-\xi\omega_0 t}}{\sqrt{1-\xi^2}}\cos\left(\omega_0 t\sqrt{1-\xi^2} - \alpha\right)\right) \text{ where } tg\alpha = \frac{\xi}{\sqrt{1-\xi^2}}$$

$T_a = \dfrac{2\pi}{\omega_0\sqrt{1-\xi^2}}$ is the pseudo-period of the response.

The instant of the first maximum value is $t_m = \dfrac{T_a}{2}$.

The overflow is written $D = 100\dfrac{\alpha}{\beta} = e^{-\frac{\pi\xi}{\sqrt{1-\xi^2}}}$

The curves in Figure 1.21 provide the overflow and the terms $\omega_0 t_r$ (t_r is the establishment time at 5%) and $\omega_0 t_m$ according to the damping ξ.

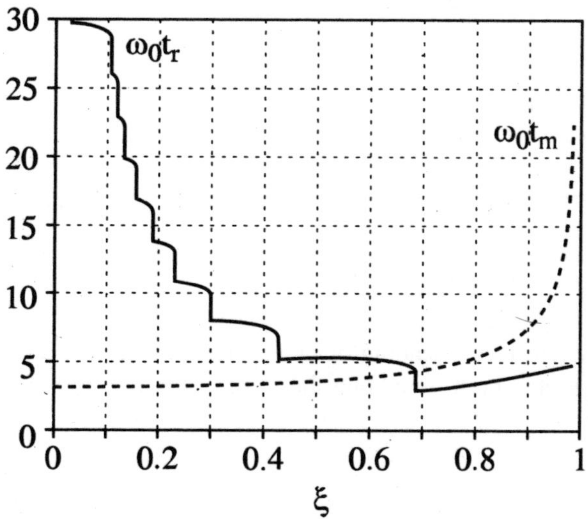

Figure 1.21. $\omega_0 t_r$ and $\omega_0 t_m$ according to the damping ξ

The alternation of slow and fast variations of product $\omega_0 t_r$ is explained because instant t_r is defined in reference to the last extremum of the unit-step response that exits the band at a final level of ±5%. When ξ increases, the numbers of extrema considered can remain constant (slow variation of t_r), or it can decrease (fast variation of t_r).

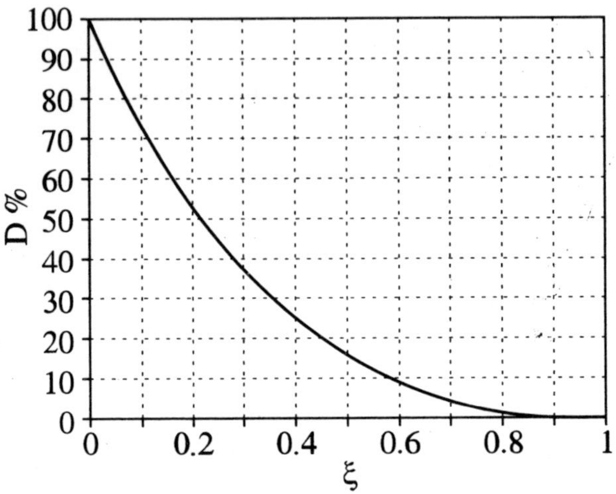

Figure 1.22. *Overflow according to the damping* ξ

1.5.3.2. Frequency response

$$H(j\omega) = \frac{K}{\left(1 - \dfrac{\omega^2}{\omega_0^2}\right) + 2j\xi\dfrac{\omega}{\omega_0}}$$

$\xi \geq 1$: the system is a cascading of two systems of the first order H_1 and H_2. The asymptotic plot is built by adding the plots of the two systems separately built (see Figure 1.23).

$\xi < 1$: the characteristics of the frequency response vary according to the value of ξ. Module and phase are obtained from the following expressions:

$$|H(j\omega)| = \frac{K}{\sqrt{\left(1 - \dfrac{\omega^2}{\omega_0^2}\right)^2 + 4\,\xi^2\,\dfrac{\omega^2}{\omega_0^2}}} \qquad \phi(\omega) = -Arctg(\frac{2\xi\omega\omega_o}{\omega_o^2 - \omega^2})$$

For $\xi < \dfrac{1}{\sqrt{2}}$, the module reaches a maximum $A_{\max} = \dfrac{K}{2\xi\sqrt{1 - \xi^2}}$ in an angular

frequency called *of resonance* $\omega_r = \omega_0\sqrt{1 - 2\xi^2}$.

We note that the smaller ξ, the more significant this extremum and the more the phase follows its asymptotes to undertake a sudden transition along ω_o.

Finally, for $\xi = 0$, the system becomes a pure oscillator with a infinite module in ω_o and a real phase mistaken for the asymptotic phase.

Figures 1.23, 1.24, 1.25 and 1.26 illustrate the diagrams presenting the aspect of the frequency response for a second order system with different values of ξ.

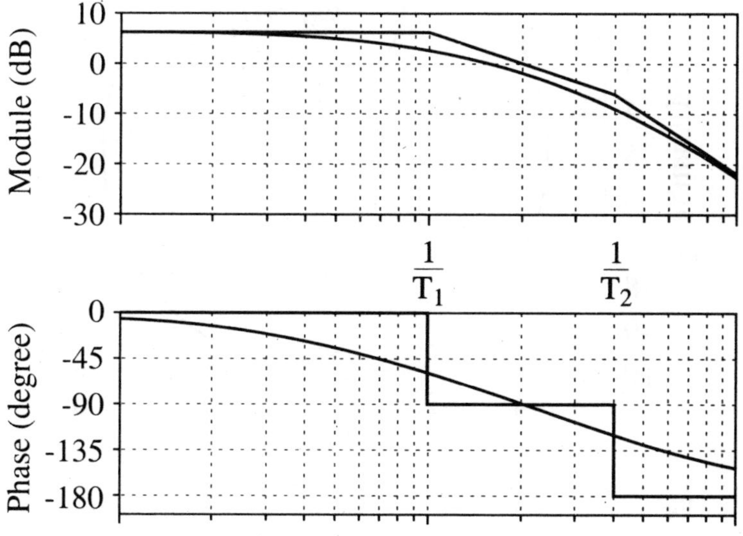

Figure 1.23. *Bode diagram of a second order system with $\xi \geq 1$*

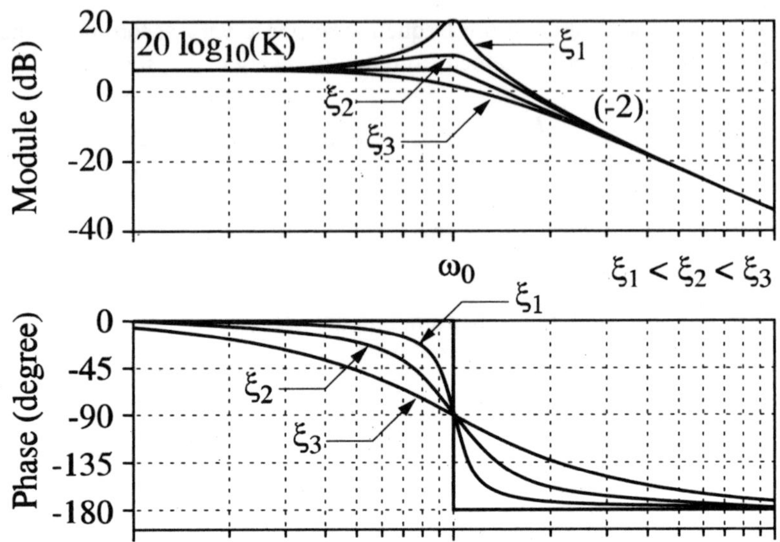

Figure 1.24. *Bode diagram of a second order system with ξ < 1*

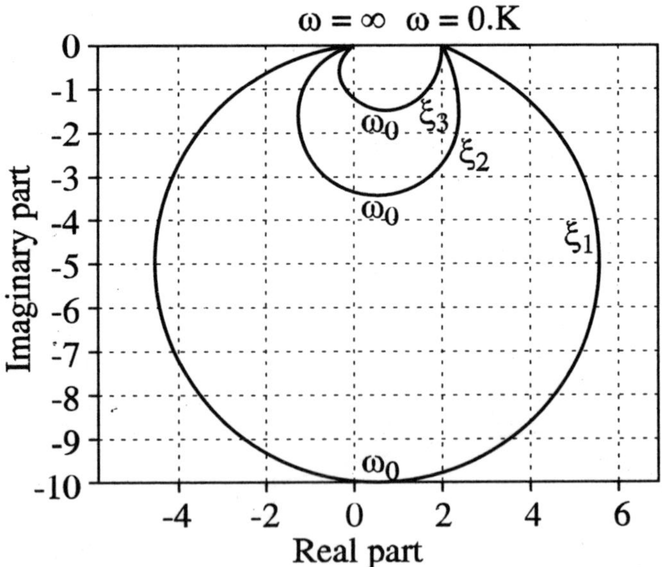

Figure 1.25. *Nyquist diagram of a second order system with ξ < 1*

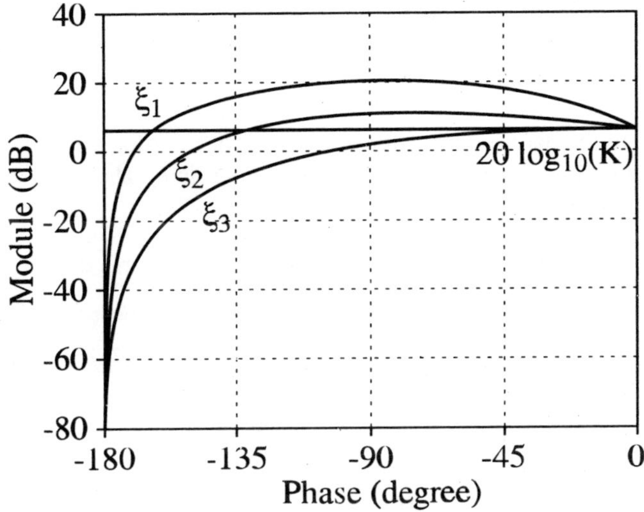

Figure 1.26. *Black diagram of a second order system with* $\xi < 1$

1.6. A few reminders on Fourier and Laplace transforms

1.6.1. *Fourier transform*

Any signal has a reality in time and frequency domains. Our ear is sensitive to amplitude (sound level) and frequency of a sound (low or high-pitched tone). These time and frequency domains, which are characterized by variables that are opposite to one another, are taken in the broad sense: if a magnitude evolves according to a distance (atmospheric pressure according to altitude), the concept of frequency will be homogenous, contrary to a length.

The Fourier transform is the mathematical tool that makes it possible to link these two domains. It is defined by:

$$X(f) = \int_{-\infty}^{+\infty} x(t)\,e^{-2\pi jft}\,dt = TF(x(t))$$

When we seek the value $X(f)$ for a value f_o of f that means that we seek in the whole history, past and future, of $x(t)$ which corresponds to frequency f_o. This corresponds to an infinitely selective filtering.

The energy exchanged between $x(t)$ and the harmonic signal of frequency f_o ($e^{2\pi jf_o t}$) can be finite. $X(f_o)$ is then finite, or infinite if $x(t)$ is also a harmonic signal and $X(f)$ is then characterized by a Dirac impulse $\delta_{(f_o)}(f)$.

According to the nature of the signal considered, by using various mathematical theories concerning the convergence of indefinite integrals, we can define the Fourier transform in the following cases:

– positively integrable signal: $\int |x(t)| \, dt \leq \infty$. The integral definition of the TF converges in absolute value. $X(f)$ is a function that tends toward 0 infinitely;

– integrable square signal or finite energy signal $\int x(t)^2 \, dt \leq \infty$. The integral definition of the TF exists in the sense of the convergence in root mean square:

$$\lim_{A \to \infty} \int \left| X(f) - \int_{-A}^{+A} x(t) e^{-2\pi jft} \, dt \right|^2 df = 0$$

– slightly ascending signal: $\exists A$ and k $|t| > A \Rightarrow |x(t)| < |t|^k$. The Fourier transform exists in the distribution sense. We also note the transforms in the sense of following traditional distributions:

$$TF(\delta_{(a)}) = e^{-2\pi jaf} \quad \text{and its reciprocal function} \quad TF(e^{2\pi jat}) = \delta_{(a)}(f)$$

$$TF\left(\sum_{k=-\infty}^{+\infty} \delta(t - kT) \right) = \sum_{k=-\infty}^{+\infty} e^{-2\pi jkT} = \frac{1}{T} \sum_{k=-\infty}^{+\infty} \delta\left(f - \frac{k}{T} \right)$$

1.6.2. *Laplace transform*

When the signal considered has an exponential divergence, irrespective of the mathematical theory considered, we cannot attribute any sense to the integral definition of the Fourier transform.

The idea is to add to the pure imaginary argument $2\pi jf$ a real part σ which is chosen in order to converge the integral considered:

$$\int_{-\infty}^{+\infty} x(t) e^{-(\sigma + 2\pi jf)t} \, dt$$

By determining that $p = \sigma + 2\pi jf$, we define a function of a complex variable, called the Laplace transform of $x(t)$, defined into a vertical band of the complex plane, which is determined by the conditions on σ ensuring the convergence of the integral:

$$X(p) = \int\limits_{-\infty}^{+\infty} x(t)e^{-pt}\,dt = TL(x(t))$$

The instability phenomena that can interfere in a linear system are characterized by exponential divergence signals; hence, we perceive the interest in the complex variable transformations for the analysis and synthesis of linear systems.

We note that the complex variable that characterizes the Laplace transform is noted by s.

Let us suppose that $x(t)$ is of exponential order, i.e. locally integrable and as it is in two positive real numbers A and B and in two real numbers α and β so that:

$$\forall t \geq t_1 \geq 0 \quad |x(t)| \leq Ae^{\alpha t}$$

$$\forall t \leq t_2 \leq 0 \quad |x(t)| \leq Be^{\beta t}$$

and $x(t)$ locally integrable $\Leftrightarrow \forall a,b$ finite $\int\limits_{a}^{b} |x(t)|\,dt < \infty$.

The Laplace transform (LT) exists if $|X(p)|$ exists. However:

$$|X(p)| \leq \left|\int\limits_{-\infty}^{t_2} x(t)e^{-pt}\,dt\right| + \left|\int\limits_{t_2}^{t_1} x(t)e^{-pt}\,dt\right| + \left|\int\limits_{t_1}^{+\infty} x(t)e^{-pt}\,dt\right|$$

$\left|\int\limits_{t_2}^{t_1} x(t)e^{-pt}\,dt\right|$ is bounded because $x(t)$ is locally integrable.

$\left|\int\limits_{-\infty}^{t_2} x(t)e^{-pt}\,dt\right| \leq B\int\limits_{-\infty}^{t_2} e^{(\beta-\sigma)t}\,dt$ which converges if $\sigma = \mathrm{Re}(p) < \beta$.

$$\left| \int_{t_1}^{+\infty} x(t)e^{-pt}\,dt \right| \leq A \int_{t_1}^{+\infty} e^{(\alpha-\sigma)t}\,dt \quad \text{which converges if } \sigma = \text{Re}(p) > \alpha.$$

The LT thus exists for $\alpha < \sigma = \text{Re}(p) < \beta$.

Let σ_1 and σ_2 be the values of α and β ensuring the tightest increases of the signal module for t tending toward $\pm\infty$. We will call the group consisting of *function* $X(p)$ and the *convergence band* $[\sigma_1,\sigma_2]$ *Laplace transform* (two-sided) which is sometimes noted by:

$$TL\{x(t)\} = \{X(p),[\sigma_1,\sigma_2]\}$$

We note that in the convergence band, the integral that defines the Laplace transform is absolutely convergent, hence the properties of holomorphy, continuity and derivability with respect to p, of e^{-tp} are carried over $X(p)$.

EXAMPLE 1.3. – consider the signal defined by
$$\begin{aligned} x(t) &= e^{at} & t \geq 0 \\ x(t) &= e^{bt} & t \leq 0 \end{aligned}.$$

Determining the transform of $x(t)$ supposes the following evaluations:

$$\int_{0}^{+\infty} e^{(a-p)t}\,dt = \left[-\frac{e^{(a-p)t}}{p-a} \right]_{0}^{+\infty} = \frac{1}{p-a} \quad \text{if } a < \text{Re}(p)$$

$$\int_{-\infty}^{0} e^{(b-p)t}\,dt = \left[-\frac{e^{(b-p)t}}{p-b} \right]_{-\infty}^{0} = -\frac{1}{p-b} \quad \text{if } \text{Re}(p) < b$$

Provided a is strictly less than b, so that there is a complex plane domain where the two integrals considered are convergent, $x(t)$ will admit for LT the function:

$$X(p) = \frac{1}{p-a} - \frac{1}{p-b}$$

Saying that $a < b$ means in the time domain:

– for $a > 0$, the causal part $(t \geq 0)$ exponentially diverges which implies, $b > 0$, that the anti-causal part $(t < 0)$ converges faster toward 0 than the causal part diverges;

– for $\cdot b < 0$, the anti-causal part $(t < 0)$ exponentially diverges which implies, $a < 0$, that the causal part $(t \geq 0)$ converges faster toward 0 than the anti-causal part diverges.

All the anti-causal signals, zero for $t > 0$, that have a Laplace transform are such that this transform is defined into a left half-plane (containing $Re(p) = -\infty$).

All the causal signals, zero for $t < 0$, that have a Laplace transform are such that this transform is defined into a right half-plane, (containing $Re(p) = +\infty$). The transform of such signals is again called *one-sided transform*.

For a positively integrable (causal or anti-causal) signal:

$$X(p) \text{ for } Re(p) = 0 \text{ is increased in module by } \int_{-\infty}^{+\infty} |x(t)| \, dt \, .$$

Its Laplace transform always exists, the associated convergence band containing the imaginary axis. Hence, we notice the identity between the Laplace transform and the Fourier transform because on the imaginary axis:

$$X(p)\Big|_{p=2\pi jf} = \int_{-\infty}^{+\infty} x(t) e^{-2\pi jft} \, dt = TF(x)$$

Finally, we note that the concept of Laplace transform can be generalized in the case where the signals considered are modeled by distributions. We recall from what was previously discussed in this chapter that the popular Dirac impulse admits the constant function equal to 1 as Laplace transform:

$$TL(\delta(t)) = \int_{-\infty}^{\infty} \delta(t) e^{-pt} \, dt = [e^{-pt}]\Big|_{t=0} = 1$$

$$TL(\delta_{(a)}(t)) = \int_{-\infty}^{\infty} \delta(t-a)e^{-pt}\,dt = [e^{-pt}]\Big|_{t=a} = e^{-ap}$$

1.6.3. *Properties*

As we have already seen, the Fourier and Laplace transforms reveal the same concept adapted to the type of signal considered. Thus, these transforms have similar properties and we will sum up the main ones in the following table.

We recall that $U(t)$ designates the unit-step function.

<table>
<tr><th>Fourier transform</th><th>Laplace transform</th></tr>
<tr><td>

Linearity

$$TF(\lambda x + \mu y) = \lambda TF(x) + \mu TF(y)$$

</td><td>

$$TL(\lambda x + \mu y) = \lambda TL(x) + \mu TL(y)$$

The convergence domain is the intersection of each domain of basic transforms.

$$x(t) \xrightarrow{\quad TL \quad} \begin{cases} X(p) \\ \sigma_1 < Re(p) < \sigma_2 \end{cases}$$

</td></tr>
<tr><td>

$$x(t) \xrightarrow{\quad TF \quad} X(f)$$

</td><td></td></tr>
<tr><td>

Delay

$$x(t-\tau) \xrightarrow{\quad TF \quad} X(f)e^{-2\pi j f\tau}$$

</td><td>

$$x(t-\tau) \xrightarrow{\quad TL \quad} \begin{cases} X(p)e^{-\tau p} \\ \sigma_1 < Re(p) < \sigma_2 \end{cases}$$

</td></tr>
<tr><td>

Time reverse

$$x(-t) \xrightarrow{\quad TF \quad} X(-f) = X^*(f)$$

</td><td>

$$x(-t) \xrightarrow{\quad TL \quad} \begin{cases} X(-p) \\ -\sigma_2 < Re(p) < -\sigma_1 \end{cases}$$

</td></tr>
<tr><td>

Signal derivation

$$\frac{dx}{dt} \xrightarrow{\quad TF \quad} (2\pi j f)X(f)$$

This property verifies that the signal is modeled by a function or a distribution.

</td><td>

The signal is modeled by a continuous function:

$$\frac{dx}{dt} \xrightarrow{\quad TL \quad} \begin{cases} pX(p) \\ \sigma_1 < Re(p) < \sigma_2 \end{cases}$$

The signal is modeled by a function that has a discontinuity of the first kind in t_o:

$$(x(t_o+0) - x(t_o-0)) = S_o$$

$$\frac{dx}{dt} \xrightarrow{\quad TL \quad} \begin{cases} pX(p) - S_o\,e^{-t_o p} \\ \sigma_1 < Re(p) < \sigma_2 \end{cases}$$

</td></tr>
</table>

Case of a causal signal:

$$x(t) = U(t)x(t)$$

$$S_o = x(0^+) - 0$$

$$\frac{dx}{dt}u(t) \xrightarrow{\;TL\;} \begin{cases} pX(p) - x(0^+) \\ \sigma_1 < Re(p) < \sigma_2 \end{cases}$$

Transform of a convolution

Through a simple calculation of double integral, it is easily shown that:

$$TF(x * y) = X(f)\,Y(f)$$

$$TL(x * y) = X(p)\,Y(p)$$

defined in the intersection of the convergence domains of $X(p)$ and $Y(p)$

The Laplace transform also makes it possible to determine the behavior at the time limits of a causal signal with the help of the following two theorems.

THEOREM OF THE INITIAL VALUE.– *provided the limits exist, we have*:

$$\lim_{t \to 0^+} x(t) = \lim_{Re(p) \to +\infty} pX(p)$$

THEOREM OF THE FINAL VALUE.– *provided the limits exist, we have:*

$$\lim_{t \to +\infty} x(t) = \lim_{Re(p) \to 0} pX(p)$$

The convergence domain of $X(p)$ is the right half-plane, bounded on the left by the real part of the pole which is at the most right (convergence abscissa σ_0) because signals are causal.

1.6.4. *Laplace transforms of ordinary causal signals*

$x(t)$	$X(p)$	σ_0 convergence abscissa
$\delta_{(0)}$	1	$-\infty$
$\delta_{(0)}^{(n)}$	p^n	$-\infty$
$U(t) = 1 \quad t \geq 0$ $U(t) = 0 \quad t < 0$	$\dfrac{1}{p}$	0
$U(t)e^{-at}$	$\dfrac{1}{p+a}$	$-a$
$U(t)t^n$	$\dfrac{n!}{p^{n+1}}$	0
$U(t)t^n e^{-at}$	$\dfrac{n!}{(p+a)^{n+1}}$	$-a$
$U(t)\sin(\omega t)$	$\dfrac{\omega}{p^2 + \omega^2}$	0
$U(t)\cos(\omega t)$	$\dfrac{p}{p^2 + \omega^2}$	0
$U(t)e^{-at}\sin(\omega t)$	$\dfrac{\omega}{(p+a)^2 + \omega^2}$	$-a$
$U(t)e^{-at}\cos(\omega t)$	$\dfrac{p+a}{(p+a)^2 + \omega^2}$	$-a$

1.6.5. *Ordinary Fourier transforms*

$x(t)$	$TF(x)(t)$
1	$\delta_{(0)}$
t^n	$\left(-\dfrac{1}{2\pi j}\right)^n \delta_{(0)}^{(n)}$
$e^{2\pi j f_0 t}$	$\delta_{(f_0)}$
$\sin(2\pi f_0 t)$	$\dfrac{1}{2j}\delta_{(f_0)} - \dfrac{1}{2j}\delta_{(-f_0)}$
$\cos(2\pi f_0 t)$	$\dfrac{1}{2}\delta_{(f_0)} + \dfrac{1}{2}\delta_{(-f_0)}$
$sgn(t)$	$\dfrac{1}{\pi j} Pf \dfrac{1}{f}$
$U(t)$	$\dfrac{1}{2}\delta_{(0)} + \dfrac{1}{2\pi j} Pf \dfrac{1}{f}$
$\delta_{(0)}$	1
$\delta_{(t_0)}$	$e^{-2\pi j f t_0}$
$\displaystyle\sum_{n=-\infty}^{+\infty} \delta_{(nT)}$	$\dfrac{1}{T}\displaystyle\sum_{n=-\infty}^{+\infty} \delta_{\left(n/T\right)}$
$x(t) = 1 \quad t \in [0, T_0]$ $= 0 \quad elsewhere$	$T_0\, e^{-\pi j f T_0} \dfrac{\sin(\pi f T_0)}{\pi f T_0}$

1.7. Bibliography

[AZZ 88] D'ÀZZO J.J., HOUPIS C.H., *Linear Control System Analysis and Design*, McGraw-Hill, 1988 (3rd edition).

[BAS 71] BASS J., *Cours de mathématiques*, vol. III, Masson, Paris, 1971.

[LAR 96] DE LARMINAT P., *Automatique*, Hermès, Paris, 1996 (2nd edition).

[ROD 71] RODDIER F., *Distributions et Transformation de Fourier*, Ediscience, 1971.

[ROU 90] ROUBINE E., *Distributions et signal*, Eyrolles, Paris, 1990.

Chapter 2

State Space Representation

Control techniques based on spectral representation demonstrated their performances though numerous industrial implementations, but they also revealed their limitations for certain applications.

The objective of this chapter is to provide the basis for a more general representation than the one adopted for the frequency approach and to offer the necessary elements to comprehend time control through the state approach.

Time control, which is largely used in space and aeronautic applications but also in industrial applications such as servomechanisms, is based on the representation of systems through state variables. According to the matrix structure adopted, this modeling is also currently used during the synthesis of control laws, irrespective of the method chosen.

This internal representation, which is richer and more global than the input-output representation, is enabling the representation, in the form of a matrix, of any system: invariant or non-invariant, linear or non-linear, mono-variable or multi-variable, continuous or discrete. This will be presented in this chapter, along with a few fundamental properties such as stability, controllability and observability.

Chapter written by Patrick BOUCHER and Patrick TURELLE.

2.1. Reminders on the systems

A physical system receives external excitations, stores information and energy, and delivers them according to the traditional diagram represented in Figure 2.1.

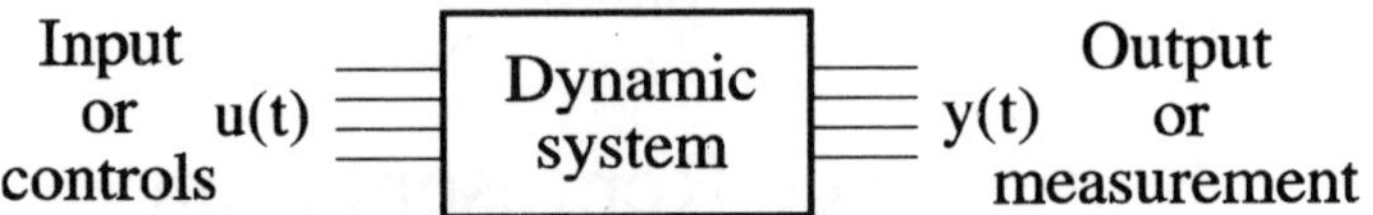

Figure 2.1. *Dynamic system*

The controls and outputs are generally multiple and their evolution is as a function of time. In the determinist case which we are dealing with here, knowing the controls $\mathbf{u}(t)$ from $-\infty$ makes it possible to know the outputs $\mathbf{y}(t)$ at instant t.

2.1.1. *Internal representation of determinist systems: the concept of state*

A system can be represented externally through the relations that link the inputs to the outputs, which are described in the vector form:

$$\mathbf{y}(t) = h\{\mathbf{u}(\tau)\} \quad \tau \in [0, \infty[$$

Hence, linear and invariant systems (LIS) are traditionally represented, in the mono-variable case, by the convolution equation:

$$y(t) = h(t) * u(t) = \int_{-\infty}^{\infty} h(\tau) u(t - \tau) d\tau = \int_{-\infty}^{\infty} h(t - \tau) u(\tau) \, d\tau$$

This representation is linked to the concept of transfer function by Laplacian transformation of the convolution equation:

$$Y(p) = H(p) U(p)$$

However, we are soon limited at the level of these representations by the non-linearities and non-stationarity of systems. Hence, it is interesting to consider what we call the state of a system in order to obtain a so-called "internal" representation.

The state is a vector quantity $\mathbf{x}(t)$ defined at any instant which represents the set of information and energies stored by the system at instant t.

EXAMPLE 2.1.– let us consider the system consisting of a body in freefall in the field of gravity (supposed to be constant). The principle of energy conservation leads to the relation:

$$E_{\text{total}} = E_{\text{potential}} + E_{\text{kinetic}} = mgz + \frac{1}{2}mv^2 = \text{constant}$$

Knowing v (speed of the body) and z (altitude of the body) is sufficient in order to characterize the evolution of this mechanical system. Hence, vector $\mathbf{x}(t) = \begin{bmatrix} z(t) \\ v(t) \end{bmatrix}$ makes it possible to describe this evolution.

Thus, knowing the state of a system at instant $t' < t$ and the controls $\mathbf{u}(t)$ applied to the system between instants t' and t, the system output is written as:

$$\mathbf{y}(t) = h_{t,t'}[\mathbf{x}(t'), \mathbf{u}(\tau)] \quad \text{for} \quad t' \leq \tau \leq t$$

Similarly, the evolution of the state will be expressed by the relation:

$$\mathbf{x}(t) = \varphi_{t,t'}[\mathbf{x}(t'), \mathbf{u}(\tau)] \quad \text{for} \quad t' \leq \tau \leq t$$

We note that $\mathbf{x}(t')$ can be expressed from $\mathbf{x}(t'')$ $(t'' < t')$ and the controls $\mathbf{u}(\tau')$ applied to the system between instants t'' and t':

$$\mathbf{x}(t') = \varphi_{t',t''}[\mathbf{x}(t''), \mathbf{u}(\tau')] \quad \text{for} \quad t'' \leq \tau' \leq t'$$

which leads to:

$$\mathbf{x}(t) = \varphi_{t,t'}[\varphi_{t',t''}[\mathbf{x}(t''), \mathbf{u}(\tau')], \mathbf{u}(\tau)] \quad \text{for} \quad t'' \leq \tau' \leq t' \leq \tau \leq t$$

Between t'' and t, we have:

$$\mathbf{x}(t) = \varphi_{t,t''}[\mathbf{x}(t''), \mathbf{u}(\tau'')] \quad \text{for} \quad t'' \leq \tau'' \leq t$$

The comparison between these two results leads to the property of transition, which is essential for the systems that we are analyzing here:

$$\varphi_{t,\,t''}[\mathbf{x}(t''),\,\mathbf{u}(\tau'')] = \varphi_{t,\,t'}[\varphi_{t',\,t''}[\mathbf{x}(t''),\,\mathbf{u}(\tau')],\,\mathbf{u}(\tau)]$$

$$t'' \le \tau'' \le t \qquad\qquad t'' \le \tau' \le t' \le \tau \le t$$

and which characterizes the transition of a state $\mathbf{x}(t'')$ to a state $\mathbf{x}(t)$ by going through $\mathbf{x}(t')$.

The state of a system sums up its past entirely.

2.1.2. *Equations of state and equations of measurement for continuous systems*

Knowing state $\mathbf{x}(t)$ at instant t and the controls applied for $t \le \tau \le t + \Delta t$, we have the relations:

$$\mathbf{x}(t + \Delta t) = \varphi_{t+\Delta t.t}[\mathbf{x}(t), \mathbf{u}(\tau)]$$

$$\frac{\mathbf{x}(t + \Delta t) - \mathbf{x}(t)}{\Delta t} = \frac{\varphi_{t+\Delta t,t}[\mathbf{x}(t), \mathbf{u}(\tau)] - \varphi_{t,t}[\mathbf{x}(t),\ \mathbf{u}(t)]}{\Delta t}$$

The equation of state is then obtained by going at the bound $(\Delta t \to 0)$ in this relation. Hence, we obtain the matrix differential equation:

$$\frac{d\mathbf{x}}{dt} = f[\mathbf{x}(t), \mathbf{u}(t), t]$$

in which the control vector $\mathbf{u}(t)$ has m components, and the state vector $\mathbf{x}(t)$ is characterized by n components called state variables.

The vector of measurements or of observations has l components and the equation of measurement has the following form:

$$\mathbf{y}(t) = h[\mathbf{x}(t), \mathbf{u}(t), t]$$

The evolution of a dynamic system subjected to inputs $\mathbf{u}(t)$ and delivering information $\mathbf{y}(t)$ is entirely characterized by a vector $\mathbf{x}(t)$ of size n linked to the inputs and outputs by the relations:

$$\dot{\mathbf{x}}(t) = f[\mathbf{x}(t), \mathbf{u}(t), t]$$

$$\mathbf{y}(t) = h[\mathbf{x}(t), \mathbf{u}(t), t]$$

$$\text{with: } \mathbf{x}(t) = \begin{bmatrix} x_1(t) \\ x_2(t) \\ \vdots \\ x_n(t) \end{bmatrix} \; ; \mathbf{u}(t) = \begin{bmatrix} u_1(t) \\ u_2(t) \\ \vdots \\ u_m(t) \end{bmatrix} \text{ and } \mathbf{y}(t) = \begin{bmatrix} y_1(t) \\ y_2(t) \\ \vdots \\ y_l(t) \end{bmatrix}$$

2.1.3. *Case of linear systems*

When the system is linear (principle of superposition of causes and effects), the equations of state and measurement have the following form:

$$\dot{\mathbf{x}}(t) = \mathbf{A}(t)\,\mathbf{x}(t) + \mathbf{B}(t)\,\mathbf{u}(t)$$

$$\mathbf{y}(t) = \mathbf{C}(t)\,\mathbf{x}(t) + \mathbf{D}(t)\,\mathbf{u}(t)$$

in which:

- $\mathbf{A}(t)$ is the evolution matrix $(\dim n \times n)$;
- $\mathbf{B}(t)$ is the control matrix $(\dim n \times m)$;
- $\mathbf{C}(t)$ is the observation matrix $(\dim l \times n)$;
- $\mathbf{D}(t)$ is the direct transmission matrix $(\dim l \times m)$.

We note that state representation is not unique and in any case we would have to talk of a state representation (see section 2.3.2).

2.1.4. *Case of continuous and invariant linear systems*

The four state representation matrices are constant, irrespective of time:

$$\dot{\mathbf{x}}(t) = \mathbf{A}\,\mathbf{x}(t) + \mathbf{B}\,\mathbf{u}(t)$$

$$\mathbf{y}(t) = \mathbf{C}\,\mathbf{x}(t) + \mathbf{D}\,\mathbf{u}(t)$$

2.2. Resolving the equation of state

Firstly, let us recall the nature of the magnitudes used:

– state vector: $\mathbf{x}(t)$ column vector $n \times 1$;

– control vector: $\mathbf{u}(t)$ column vector $m \times 1$;

– observation vector: $\mathbf{y}(t)$ column vector $l \times 1$;

– equation of state: $\dot{\mathbf{x}}(t) = \mathbf{A}(t)\,\mathbf{x}(t) + \mathbf{B}(t)\mathbf{u}(t)$;   [2.1]

– equation of measurement: $\mathbf{y}(t) = \mathbf{C}(t)\,\mathbf{x}(t) + \mathbf{D}(t)\,\mathbf{u}(t)$.   [2.2]

We will approach the resolution of the equation of state in two instances:

– free state $\mathbf{u}(t) = 0$;

– forced state $\mathbf{u}(t) \neq 0$.

2.2.1. *Free state*

This refers to solving the equation $\dot{\mathbf{x}}(t) = \mathbf{A}(t)\,\mathbf{x}(t)$ from the initial condition $\mathbf{x}(t_0) = \mathbf{x}_0$. Since the equation is linear, the solution $\mathbf{x}(t)$ is expressed linearly according to $\mathbf{x}(t_0)$ as follows:

$$\mathbf{x}(t) = \Phi(t, t_0)\,\mathbf{x}(t_0)$$

The matrix $\Phi(t, t_0)$ of size $n \times n$ is called a system transition matrix. It has the following properties:

$$\Phi(t_2,\ t_0) = \Phi(t_2, t_1)\,\Phi(t_1,\ t_0)\ ;\ \text{(transition property previously mentioned)}$$

$$\Phi(t, t) = \mathbf{I}\ ;\ \left(\mathbf{x}(t) = \Phi(t, t)\,\mathbf{x}(t)\right)$$

$$\Phi(t_1, t_2) = \Phi(t_2, t_1)^{-1}$$

$$\left(\mathbf{x}(t_2) = \Phi(t_2, t_1)\mathbf{x}(t_1) \Rightarrow \mathbf{x}(t_1) = \Phi(t_1, t_2)\mathbf{x}(t_2)\right)$$

$$\frac{d\Phi}{dt}(t, t_0) = \mathbf{A}(t)\Phi(t, t_0)\,\forall t_0 \,;\, \left(\dot{\mathbf{x}}(t) = \mathbf{A}(t)\ \mathbf{x}(t) \text{ and } \dot{\mathbf{x}}(t) = \dot\Phi\left(t,\ t_0\right)\ \mathbf{x}(t_0)\right)$$

2.2.2. *Forced state*

This refers to solving the equation $\dot{\mathbf{x}}(t) = \mathbf{A}(t)\,\mathbf{x}(t) + \mathbf{B}(t)\,\mathbf{u}(t)$ for $\mathbf{u}(t) \neq 0$, knowing the general solution of the homogenous equation $\left(\mathbf{x}(t) = \Phi(t,t_0)\,\mathbf{x}(t_0)\right)$. Then, a particular solution of the complete equation is searched for as $\mathbf{x}_p(t) = \Phi(t,t_0)\mathbf{z}(t)$, where function $\mathbf{z}(t)$ is the unknown factor obtained from the initial condition $\mathbf{x}_p(t_0) = 0 \quad (\mathbf{z}(t_0) = 0)$.

By deriving $\mathbf{x}_p(t)$ and by transferring in [2.1] we have:

$$\dot{\mathbf{x}}_p(t) = \frac{d\Phi}{dt}(t, t_0)\,\mathbf{z}(t) + \Phi(t, t_0)\,\dot{\mathbf{z}}(t) = \mathbf{A}(t)\Phi\ (t, t_0)\,\mathbf{z}(t) + \mathbf{B}(t)\,\mathbf{u}(t)$$

However:

$$\frac{d\Phi}{dt}(t, t_0) = \mathbf{A}(t)\Phi(t, t_0) \Rightarrow \Phi(t, t_0)\,\dot{\mathbf{z}}(t) = \mathbf{B}(t)\,\mathbf{u}(t) \Leftrightarrow$$

$$\dot{\mathbf{z}}(t) = \Phi\left(t, t_0\right)^{-1}\mathbf{B}(t)\,\mathbf{u}(t)$$

or by considering the initial condition on $\mathbf{z}$:

$$\mathbf{z}(t) = \int_{t_0}^{t}\Phi(\tau, t_0)^{-1}\mathbf{B}(\tau)\mathbf{u}(\tau)d\tau \Rightarrow \mathbf{x}_p(t) = \Phi(t, t_0)\int_{t_0}^{t}\Phi(\tau, t_0)^{-1}\mathbf{B}(\tau)\,\mathbf{u}(\tau)\,d\tau$$

and by using the properties of the transition matrix mentioned above:

$$\mathbf{x}_p(t) = \int_{t_0}^{t}\Phi(t, \tau)\,\mathbf{B}(\tau)\,d\tau$$

The general solution of the complete equation is then expressed by:

$$\mathbf{x}(t) = \Phi(t,t_0)\,\mathbf{x}(t_0) + \int_{t_0}^{t} \Phi(t,\tau)\,\mathbf{B}(\tau)\,\mathbf{u}(\tau)\,d\tau$$

and the output $\mathbf{y}(t)$ is then simply calculated from $\mathbf{x}(t)$:

$$\mathbf{y}(t) = \mathbf{C}(t)\left[\Phi(t,t_0)\,\mathbf{x}(t_0) + \int_{t_0}^{t} \Phi(t,\tau)\,\mathbf{B}(\tau)\mathbf{u}(\tau)\,d\tau\right] + \mathbf{D}(t)\,\mathbf{u}(t)$$

The only difficulty of this solving method is the calculation of the transition matrix. This method often calls upon numeric solving techniques.

2.2.3. *Particular case of linear and invariant systems*

When matrices $\mathbf{A}$, $\mathbf{B}$, $\mathbf{C}$ and $\mathbf{D}$ are independent of time, the transition matrix $\Phi(t,t_0)$ takes a simple particular form because Φ depends only on the difference $t-t_0$. Indeed, equation $\dfrac{d\Phi}{dt}(t,t_0) = \mathbf{A}\Phi(t,t_0)$, with $\mathbf{A}$ constant, is resolved by analogy with the scalar case by searching for a solution as the sum of an entire matrix sequence.

$$
\begin{aligned}
\Phi(t,\ t_0) \ \ &= \mathbf{A}_0 \ \ \ +\mathbf{A}_1(t-t_0)+\mathbf{A}_2(t-t_0)^2 +\cdots+\mathbf{A}_k(t-t_0)^k +\cdots \\
\dot{\Phi}(t,t_0) \ \ &= 0 \ \ \ \ \ \ +\mathbf{A}_1 + 2\mathbf{A}_2(t-t_0)+\cdots+k\ \ \mathbf{A}_k(t-t_0)^{k-1} +\cdots \\
&= \ \ \ \ \ \ \ \ \mathbf{A}[\mathbf{A}_0 + \mathbf{A}_1(t-t_0 + \mathbf{A}_2(t-t_0)^2 +\cdots+\mathbf{A}_k(t-t_0)^k +\cdots]
\end{aligned}
$$

By identifying term to term these two developments, we successively obtain:

$$\mathbf{A}_1 = \mathbf{A}\ \ \mathbf{A}_0,\ \ \mathbf{A}_2 = \frac{1}{2}\ \ \mathbf{A}\ \ \mathbf{A}_1,\ \cdots,\ \ \mathbf{A}_k = \frac{1}{k}\mathbf{A}\ \ \mathbf{A}_{k-1},\ \cdots$$

By considering the fact that $\Phi(t_0,\ t_0) = \mathbf{A}_0 = \mathbf{I}$, for $\Phi(t,t_0)$ we obtain the following development:

$$\Phi(t,\ t_0) = \mathbf{I} + \mathbf{A}(t-t_0) + \frac{1}{2!}\mathbf{A}^2(t-t_0)^2 + \cdots + \frac{1}{k!}\mathbf{A}^k(t-t_0)^k + \ldots$$

which is not that of the exponential function in the scalar case. Hence, we symbolically mark this development:

$$\Phi(t,t_0) = e^{\mathbf{A}(t-t_0)}$$

and therefore:

$$\mathbf{x}(t) = e^{\mathbf{A}(t-t_0)}\mathbf{x}(t_0) + \int_{t_0}^{t} e^{\mathbf{A}(t-\tau)}\mathbf{B}(\tau)\,\mathbf{u}(\tau)\,d\tau$$

In this particular case of linear and invariant systems, the transition matrix can be calculated analytically from, among others, the values of matrix $\mathbf{A}$. We will see some of these methods in the following section.

2.2.4. Calculation method of the transition matrix $e^{A(t-t_0)}$

Use of serial development

When the power calculation of matrix $\mathbf{A}$ is done simply (in particular through simple recurrences), we can use the definition of the matrix exponential function as an entire sequence:

$$e^{\mathbf{A}(t-t_0)} = \mathbf{I} + \mathbf{A}(t-t_0) + \frac{1}{2!}\mathbf{A}^2(t-t_0)^2 + \cdots + \frac{1}{k!}\mathbf{A}^k(t-t_0)^k + \ldots$$

EXAMPLE 2.2.

$$\mathbf{A} = \begin{bmatrix} 0 & 1 \\ 0 & 0 \end{bmatrix} \Rightarrow \mathbf{A}^2 = \begin{bmatrix} 0 & 0 \\ 0 & 0 \end{bmatrix}$$

$$\Rightarrow e^{\mathbf{A}(t-t_0)} = \mathbf{I} + \mathbf{A}(t-t_0) = \begin{bmatrix} 1 & t-t_0 \\ 0 & 1 \end{bmatrix}$$

$$\mathbf{A} = \begin{bmatrix} 0 & 1 \\ 0 & -a \end{bmatrix} \Rightarrow \mathbf{A}^2 = \begin{bmatrix} 0 & -a \\ 0 & a^2 \end{bmatrix}$$

$$\Rightarrow \mathbf{A}^3 = \begin{bmatrix} 0 & a^2 \\ 0 & -a^3 \end{bmatrix} \cdots \Rightarrow \mathbf{A}^k = \begin{bmatrix} 0 & (-a)^{k-1} \\ 0 & (-a)^k \end{bmatrix}$$

$$e^{\mathbf{A}}(t-t_0) = \mathbf{I} + \mathbf{A}(t-t_0) + \frac{1}{2}\mathbf{A}^2(t-t_0)^2 + \cdots + \frac{1}{k!}\mathbf{A}^k(t-t_0)^k + \cdots = \begin{bmatrix} 1 & \dfrac{1-e^{a(t-t_0)}}{a} \\ 0 & e^{-a(t-t_0)} \end{bmatrix}$$

Sylvester formula (distinct eigenvalues of A)

If $\lambda_1, \lambda_2 \cdots, \lambda_n$ are the n distinct eigenvalues of $\mathbf{A}$ (solutions of the equation: $\det(\mathbf{A} - \lambda\mathbf{I}) = 0$), the transition matrix $e^{\mathbf{A}(t-t_0)}$ is expressed by:

$$e^{\mathbf{A}(t-t_0)} = \sum_{i=1}^{n} e^{\lambda_i(t-t_0)} \prod_{\substack{j=1 \\ j \neq i}}^{n} \left[\frac{\mathbf{A} - \lambda_j \mathbf{I}}{\lambda_i - \lambda_j} \right]$$

EXAMPLE 2.3.– $\mathbf{A} = \begin{bmatrix} 0 & 1 \\ 0 & -a \end{bmatrix}$ $\lambda_1 = 0, \lambda_2 = -a$ distinct   eigenvalues   (the eigenvalues of a triangular matrix are presented in its diagonal).

$$\sum_{i=1}^{n} e^{\lambda i(t-t_0)} \prod_{\substack{j=1 \\ j \neq i}}^{n} \left[\frac{\mathbf{A} - \lambda_j \mathbf{I}}{\lambda_i - \lambda_j} \right] \;=\; e^{\lambda_1(t-t_0)} \left[\frac{\mathbf{A} - \lambda_2 \mathbf{I}}{\lambda_1 - \lambda_2} \right] + e^{\lambda_2(t-t_0)} \left[\frac{\mathbf{A} - \lambda_1 \mathbf{I}}{\lambda_2 - \lambda_1} \right]$$

$$= \; 1 \times \left[\frac{\mathbf{A} + a\mathbf{I}}{0+a} \right] + e^{-a(t-t_0)} \left[\frac{\mathbf{A} - 0 \times \mathbf{I}}{-a-0} \right]$$

$$= \; \frac{1}{a}\mathbf{A} + \mathbf{I} - \frac{1}{a}\mathbf{A}\; e^{-a(t-t_0)} = \frac{1-e^{-a(t-t_0)}}{a}\mathbf{A} + \mathbf{I}$$

$$= \; \begin{bmatrix} 0 & \dfrac{1-e^{a(t-t_0)}}{a} \\ 0 & e^{-a(t-t_0)} - 1 \end{bmatrix} + \begin{bmatrix} 1 & 0 \\ 0 & 1 \end{bmatrix} = \begin{bmatrix} 1 & \dfrac{1-e^{-a(t-t_0)}}{a} \\ 0 & e^{-a(t-t_0)} \end{bmatrix}$$

Sylvester interpolation method

When the eigenvalues of $\mathbf{A}$ are not distinct, we cannot use the previous formula. Hence, we will do as follows: we will suppose that λ_1 is a simple eigenvalue and that λ_2 is a double eigenvalue. Thus, we build a matrix in which the rows of simple eigenvalues are of the same type:

$$\begin{bmatrix} 1 & \lambda_1 & \lambda_1^2 & \cdot & \lambda_1^{n-1} & e^{\lambda_1(t-t_0)} \\ 1 & \lambda_2 & \lambda_2^2 & \cdot & \lambda_2^{n-1} & e^{\lambda_2(t-t_0)} \\ 0 & 1 & 2\lambda_2 & \cdot & (n-1)\lambda_2^{n-2} & (t-t_0)e^{\lambda_2(t-t_0)} \\ 1 & \lambda_3 & \lambda_3^2 & \cdot & \lambda_3^{n-1} & e^{\lambda_3(t-t_0)} \\ \cdot & \cdot & \cdot & \cdot\,\cdot & & \cdot \\ \mathbf{I} & \mathbf{A} & \mathbf{A}^2 & \cdot & \mathbf{A}^{n-1} & e^{\mathbf{A}(t-t_0)} \end{bmatrix}$$

The first row corresponding to a multiple eigenvalue is done in the same way, but the following row is the derivative of the current row with respect to the multiple eigenvalue (here λ_2). If the eigenvalue is double, the next row has a standard form with the next eigenvalue. If the eigenvalue is triple, we repeat the procedure and we derivate again the second row corresponding to this triple eigenvalue and so on. Hence we obtain a matrix of n rows and $n+1$ columns which we complete by a last row built from the successive powers of matrix $\mathbf{A}$ and completed by the transition matrix $e^{\mathbf{A}(t-t_0)}$.

Hence, Sylvester's method consists of formally calculating the determinant of this matrix and of extracting the transition matrix from this calculation by writing that this *determinant*, formally developed with respect to its last row, is *zero*.

EXAMPLE 2.4.– if we take the same example: $\mathbf{A} = \begin{bmatrix} 0 & 1 \\ 0 & -a \end{bmatrix}$ and build the matrix which in our example is limited to:

$$\begin{bmatrix} 1 & \lambda_1 & e^{\lambda_1(t-t_0)} \\ 1 & \lambda_2 & e^{\lambda_2(t-t_0)} \\ \mathbf{I} & \mathbf{A} & e^{\mathbf{A}(t-t_0)} \end{bmatrix} = \begin{bmatrix} 1 & 0 & 1 \\ 1 & -a & e^{-a(t-t_0)} \\ \mathbf{I} & \mathbf{A} & e^{\mathbf{A}(t-t_0)} \end{bmatrix}$$

The determinant of this matrix developed in relation with the last row is:

$$\Delta = \mathbf{I} \times \begin{vmatrix} 0 & 1 \\ -a & e^{-a(t-t_0)} \end{vmatrix} - \mathbf{A} \times \begin{vmatrix} 1 & 1 \\ 1 & e^{-a(t-t_0)} \end{vmatrix} + e^{\mathbf{A}(t-t_0)} \times \begin{vmatrix} 1 & 0 \\ 1 & -a \end{vmatrix}$$

$$\Delta = a\mathbf{I} - (e^{-a(t-t_0)} - 1)\mathbf{A} - ae^{\mathbf{A}(t-t_0)}$$

$$\Delta = 0 \Rightarrow \quad e^{\mathbf{A}(t-t_0)} = \mathbf{I} + \frac{1 - e^{-a(t-t_0)}}{a}\mathbf{A} = \begin{bmatrix} 1 & \dfrac{1 - e^{-a(t-t_0)}}{a} \\ 0 & e^{-a(t-t_0)} \end{bmatrix}$$

NOTE 2.1.– regarding these last two techniques, we note that the transition matrix $e^{\mathbf{A}(t-t_0)}$ is expressed by a finite degree polynomial $(n-1)$ of matrix $\mathbf{A}$. This result is due to the fact that any matrix verifies its characteristic equation, $\det(\mathbf{A} - \lambda\mathbf{I})$ (theorem of Cayley Hamilton) and thus all powers of $\mathbf{A}$ of a degree more than n are expressed by a linear combination of first powers $n - 1$.

Method of modes

When matrix $\mathbf{A}$ is diagonalizable (it is at least the case when the eigenvalues of $\mathbf{A}$ are distinct), we can calculate the transition matrix by using the diagonal form of $\mathbf{A}$.

$$\mathbf{A} = \mathbf{M}\Lambda\mathbf{M}^{-1} \Rightarrow \mathbf{A}^2 = \mathbf{M}\Lambda\mathbf{M}^{-1}\mathbf{M}\Lambda\mathbf{M}^{-1} = \mathbf{M}\Lambda^2\mathbf{M}^{-1} \dots \mathbf{A}^k = \mathbf{M}\Lambda^k\mathbf{M}^{-1}$$

$$e^{\mathbf{A}(t-t_0)} = \mathbf{I} + \mathbf{A}(t-t_0) + \frac{1}{2!}\mathbf{A}^2(t-t_0)^2 + \cdots + \frac{1}{k!}\mathbf{A}^k(t-t_0)^k + \cdots =$$

$$\mathbf{M}\left[\mathbf{I} + \Lambda(t-t_0) + \frac{1}{2!}\Lambda^2(t-t_0)^2 + \cdots + \frac{1}{k!}\Lambda^k(t-t_0)^k + \cdots\right]\mathbf{M}^{-1} = \mathbf{M}e^{\Lambda(t-t_0)}\mathbf{M}^{-1}$$

However, matrix $e^{\Lambda}(t-t_0)$, by build, is the diagonal matrix whose diagonal elements are the scalar exponential functions $e^{\lambda_i(t-t_0)}$. The calculation of the transition matrix is then done by determining a system of eigenvectors of $\mathbf{A}(\{\mathbf{x}_1, \mathbf{x}_2, \cdots, \mathbf{x}_n\})$ in order to define the basic change matrix $\mathbf{M}$.

EXAMPLE 2.5.

$$\Rightarrow \mathbf{A} = \begin{bmatrix} 0 & 0 \\ 0 & -a \end{bmatrix} \qquad \text{eigenvalues } \lambda_1 = 0, \ \lambda_2 = -a$$

$$\Rightarrow \Lambda = \begin{bmatrix} 0 & 0 \\ 0 & -a \end{bmatrix} \Rightarrow \ e^{\Lambda(t-t_0)} = \begin{bmatrix} 1 & 0 \\ 0 & e^{-a(t-t_0)} \end{bmatrix}$$

Determining $\mathbf{M}$:

$$\mathbf{A}\,\mathbf{x}_1 = \lambda_1\,\mathbf{x}_1 = 0 \Rightarrow \mathbf{x}_1 = \begin{bmatrix} 1 \\ 0 \end{bmatrix} \ \text{(for example)}$$

$$\mathbf{A}\,\mathbf{x}_2 = \lambda_2\,\mathbf{x}_2 = -a\,\mathbf{x}_2 \Rightarrow \mathbf{x}_2 = \begin{bmatrix} 1 \\ -a \end{bmatrix} \text{(for example)}$$

$$\mathbf{M} = [\mathbf{x}_1 | \mathbf{x}_2] = \begin{bmatrix} 1 & 1 \\ 0 & -a \end{bmatrix} \Rightarrow \mathbf{M}^{-1} = \begin{bmatrix} 1 & \dfrac{1}{a} \\ 0 & -\dfrac{1}{a} \end{bmatrix}$$

$$\Rightarrow e^{\mathbf{A}(t-t_0)} = \begin{bmatrix} 1 & 1 \\ 0 & -a \end{bmatrix}\begin{bmatrix} 1 & 0 \\ 0 & e^{-a(t-t_0)} \end{bmatrix}\begin{bmatrix} 1 & \dfrac{1}{a} \\ 0 & -\dfrac{1}{a} \end{bmatrix}$$

$$e^{\mathbf{A}(t-t_0)} = \begin{bmatrix} 1 & \dfrac{1-e^{-a(t-t_0)}}{a} \\ 0 & e^{-a(t-t_0)} \end{bmatrix}$$

2.2.5. *Application to the modeling of linear discrete systems*

Let us consider the equation providing the evolution of a state of a system between two instants t_0 and t:

$$\mathbf{x}(t) = \Phi(t,t_0)\,\mathbf{x}(t_0) + \int_{t_0}^{t}\Phi(t,\tau)\,\mathbf{B}(\tau)\,\mathbf{u}(\tau)\,d\tau$$

This equation is valid for any pair of instants $t_0 < t$.

Hence, let us suppose that between two instants t_k and t_{k+1} the control $\mathbf{u}(t)$ applied to the system is constant and equal to $\mathbf{u}_k = \mathbf{u}(t_k)$. The evolution of the state between these two instants can be expressed as follows:

$$\mathbf{x}(t_{k+1}) = \Phi(t_{k+1}, t_k)\mathbf{x}(t_k) + \left[\int_{t_k}^{t_{k+1}} \Phi(t_{k+1}, \tau)\mathbf{B}(\tau)d\tau \right] \mathbf{u}(t_k)$$

where: $\Phi(t_{k+1}, t_k) = \mathbf{F}(k)$ and $\int_{t_k}^{t_{k+1}} \Phi(t_{k+1}, \tau)\mathbf{B}(\tau)\,d\tau = \mathbf{G}(k)$ are two functions of k.

When instants t_k are multiples of a sampling period $T(t_k = kT)$ and when between the two sampling instants the control applied to the system is constant, the evolution of this state is:

$$\mathbf{x}(k+1) = \mathbf{F}(k)\mathbf{x}(k) + \mathbf{G}(k)\mathbf{u}(k)$$

which represents the discrete model of the continuous system of the equation of state:

$$\dot{\mathbf{x}}(t) = \mathbf{A}(t)\,\mathbf{x}(t) + \mathbf{B}(t)\,\mathbf{u}(t) \text{ operated by a control:}$$

$$\mathbf{u}(t) = \mathbf{u}(kT) \text{ for } kT \leq t < (k+1)T.$$

Moreover, when the system is invariant, $\dot{\mathbf{x}}(t) = \mathbf{A}\mathbf{x}(t) + \mathbf{B}\mathbf{u}(t)$, then we obtain:

$$\mathbf{F}(k) = e^{\mathbf{A}T} = \mathbf{F}$$

and: $$\mathbf{G}(k) = \int_{t_k}^{t_{k+1}} \Phi(t_{k+1}, \tau)\ \mathbf{B}\ d\tau = \int_{t_k}^{t_{k+1}} e^{\mathbf{A}(t_{k+1}-\tau)}\mathbf{B}\ d\tau = \int_0^T e^{\mathbf{A}(\theta)}\mathbf{B}\ d\theta = \mathbf{G}$$

These two constant matrices are independent of k and the discrete model becomes:

$$\mathbf{x}(k+1) = \mathbf{F}\,\mathbf{x}(k) + \mathbf{G}\,\mathbf{u}(k)$$

In both cases, the equation of measurement is added to the equation of state in order to complete the model.

2.3. Scalar representation of linear and invariant systems

2.3.1. *State passage $\rightarrow$ transfer*

Linear and invariant systems are characterized by an external representation that has the form of a transfer matrix linking the controls to the outputs:

$$\mathbf{Y}(p) = \mathbf{H}(p)\ \mathbf{U}(p)$$

For mono-variable systems (y and u scalar), we define the transfer function of the system by the ratio: $H(p) = \dfrac{Y(p)}{U(p)}$.

In the case of multi-variable systems, the transfer is characterized by matrix $\mathbf{H}(p)$ whose size is linked to the size of output vectors $\mathbf{y}$ and control vectors $\mathbf{u}$. If $\mathbf{y}$ has a size l and $\mathbf{u}$ has a size m, transfer matrix $\mathbf{H}(p)$ has the size $l \times m$. The problem that arises is the passage from an internal representation:

$$\dot{\mathbf{x}}(t) = \mathbf{A}\,\mathbf{x}(t) + \mathbf{B}\,\mathbf{u}(t)$$

$$\mathbf{y}(t) = \mathbf{C}\,\mathbf{x}(t) + \mathbf{D}\,\mathbf{u}(t)$$

to the external representation $\mathbf{H}(p)$.

This passage will be done, in the case of the linear and invariant systems we are dealing with, by using the Laplace transform on the equations of state and measurement:

$$p\,\mathbf{X}(p) - \mathbf{x}(t_0) = \mathbf{A}\,\mathbf{X}(p) + \mathbf{B}\,\mathbf{U}(p) \quad \Rightarrow \quad [p\,\mathbf{I} - \mathbf{A}]\,\mathbf{X}(p) = \mathbf{x}(t_0) + \mathbf{B}\,\mathbf{U}(p)$$

$$\Rightarrow \quad \mathbf{X}(p) = [p\mathbf{I} - \mathbf{A}]^{-1}\mathbf{x}(t_0) + [p\mathbf{I} - \mathbf{A}]^{-1}\mathbf{B}\,\mathbf{U}(p)$$

From the analogy between this transform and the expression of time response:

$$\mathbf{x}(t) = e^{\mathbf{A}(t-t_0)}\mathbf{x}(t_0) + \int_{t_0}^{t} e^{\mathbf{A}(t-\tau)}\,\mathbf{B}\,\mathbf{u}(\tau)\,d\tau$$

we obtain the following relations:

$$[p\ \ \mathbf{I} - \mathbf{A}]^{-1} = L\{e^{\mathbf{A}(t-t_0)}\}$$

$$[p\ \ \mathbf{I} - \mathbf{A}]^{-1}\mathbf{B}\ \ \mathbf{U}(p) = L\{e^{\mathbf{A}(t-t_0)} * \mathbf{B}\mathbf{u}(t)\} = L\left\{\int_{t_0}^{t} e^{\mathbf{A}(t-\tau)}\mathbf{B}\mathbf{u}(\tau)\ \ d\tau\right\}$$

which provide a new calculation technique of the transition matrix:

$$e^{\mathbf{A}(t-t_0)} = L^{-1}\left\{[p\mathbf{I} - \mathbf{A}]^{-1}\right\}$$

If the initial state is zero (hypothesis used in order to define the transfer function of a system), we have:

$$\mathbf{X}(p) = [p\mathbf{I} - \mathbf{A}]^{-1}\,\mathbf{B}\mathbf{U}(p)\quad\text{and}\quad \mathbf{Y}(p) = [\mathbf{C}[p\ \ \mathbf{I} - \mathbf{A}]^{-1}\,\mathbf{B} + \mathbf{D}]\,\mathbf{U}(p),$$

hence we obtain the transfer matrix:

$$\mathbf{H}(p) = [\mathbf{C}[p\ \ \mathbf{I} - \mathbf{A}]^{-1}\ \ \mathbf{B} + \mathbf{D}]$$

Element $H_{ij}(p)$ of matrix $\mathbf{H}(p)$ represents the transfer between the control $u_j(t)$ and an output $y_i(t)$, when the other controls are at zero. Thus, we can obtain from the matrix:

$$\mathbf{Y}(p) = \mathbf{H}(p) \times \mathbf{U}(p) \Rightarrow \begin{bmatrix} Y_1(p) \\ Y_2(p) \\ \vdots \\ Y_l(p) \end{bmatrix} = \begin{bmatrix} \ddots & \cdots & \cdots & \cdots \\ \cdots & H_{ij}(p) & \cdots & \cdots \\ \cdots & & \ddots & \cdots \\ \cdots & \cdots & \cdots & \ddots \end{bmatrix} \times \begin{bmatrix} U_1(p) \\ U_2(p) \\ \vdots \\ U_m(p) \end{bmatrix}$$

$$H_{ij}(p) = \frac{Y_i(p)}{U_j(p)} \quad \text{to} \quad U_k(p) = 0 \quad \text{for } k \neq j$$

NOTE 2.2.– there are systematic and recursive methods of calculating matrix $[p\mathbf{I} - \mathbf{A}]^{-1}$ such as the Leverrier-Souriau algorithm; firstly, matrix $[p\mathbf{I} - \mathbf{A}]^{-1}$ is written as:

$$[p\mathbf{I} - \mathbf{A}]^{-1} = \frac{\mathbf{B}_0 p^{n-1} + \mathbf{B}_1\ p^{n-2} + \cdots + \mathbf{B}_{n-1}}{d_0\ p^n + d_1\ p^{n-1} + \cdots d_n}$$

where:

$$d_0\, p^n + d_1\, p^{n-1} + \cdots + d_n = \det\left[p\,\mathbf{I} - \mathbf{A}\right]$$

which is an expression where the idea is to determine the n square matrices $\mathbf{B}_j$ $n \times n$ and the $n+1$ scalar coefficients d_i.

Hence, we make the following iterative calculations:

$$d_0 = 1$$
$$\mathbf{B}_0 = \mathbf{I} \qquad d_1 = -\text{trace}\{\mathbf{B}_0\mathbf{A}\} \cdot$$
$$\mathbf{B}_1 = \mathbf{B}_0\mathbf{A} + d_1\mathbf{I} \qquad d_2 = -\frac{1}{2}\,\text{trace}\{\mathbf{B}_1\mathbf{A}\}$$
$$\mathbf{B}_2 = \mathbf{B}_1\mathbf{A} + d_2\mathbf{I} \qquad d_3 = -\frac{1}{3}\,\text{trace}\{\mathbf{B}_2\mathbf{A}\}$$

$$\mathbf{B}_k = \mathbf{B}_{k-1}\ \mathbf{A} + d_k\mathbf{I} \qquad d_{k+1} = -\frac{1}{k+1}\,\text{trace}\{\mathbf{B}_k\ \mathbf{A}\}$$
$$\vdots$$
$$\mathbf{B}_{n-1} = \mathbf{B}_{n-2}\mathbf{A} + d_{n-1}\mathbf{I} \qquad d_n = -\frac{1}{n}\,\text{trace}\{\mathbf{B}_{n-1}\,\mathbf{A}\}$$

and the last relation that represents the verification of the calculation and must give:

$$\mathbf{B}_n = \mathbf{B}_{n-1}\,\mathbf{A} + d_n\,\mathbf{I} = 0$$

which ends the calculation of terms defining $[p\mathbf{I} - \mathbf{A}]^{-1}$.

2.3.2. *Change of basis in the state space*

As we have already mentioned, the representation of a state of a system is not unique. Only the controls applied to the system and the resulting outputs are the physical magnitudes of the systems. Hence, there is an infinite number of internal representations of a system which depend on the state vector chosen.

We will verify it by performing a change of basis in the state space. Let us assume that $\mathbf{x}(t) = \mathbf{M}\,\tilde{\mathbf{x}}(t)$ with $\mathbf{M}$ constant and $\tilde{\mathbf{x}}(t)$ state of a system in the new basis:

$$\dot{\mathbf{x}}(t) = \mathbf{M}\,\dot{\tilde{\mathbf{x}}}(x) = \mathbf{A}\,\mathbf{x}(t) + \mathbf{B}\,\mathbf{u}(t) = \mathbf{A}\,\mathbf{M}\,\tilde{\mathbf{x}}(t) + \mathbf{B}\,\mathbf{u}(t) \Rightarrow \dot{\tilde{\mathbf{x}}} = \mathbf{M}^{-1}\mathbf{A}\mathbf{M}\,\tilde{\mathbf{x}}(t) + \mathbf{M}^{-1}\,\mathbf{B}\,\mathbf{u}(t)$$

$$\dot{\tilde{\mathbf{x}}}(t) = \tilde{\mathbf{A}}\,\tilde{\mathbf{x}}(t) + \tilde{\mathbf{B}}\,\mathbf{u}(t)$$

$$\mathbf{y}(t) = \mathbf{C}\,\dot{\mathbf{x}}(t) = \mathbf{A}\,\mathbf{x}(t) + \mathbf{B}\,\mathbf{u}(t) = \mathbf{A}\,\mathbf{M}\tilde{\mathbf{x}}(t) + \mathbf{D}\,\mathbf{u}(t) \qquad \mathbf{y}(t) = \tilde{\mathbf{C}}\,\tilde{\mathbf{x}}(t) + \tilde{\mathbf{D}}\,\mathbf{u}(t)$$

In the new basis, $\tilde{\mathbf{x}}(t) = \mathbf{M}^{-1}\,\mathbf{x}(t)$ and:

$$\dot{\tilde{\mathbf{x}}}(t) = \tilde{\mathbf{A}}\,\tilde{\mathbf{x}}(t) + \tilde{\mathbf{B}}\,\mathbf{u}(t)$$

$$\mathbf{y}(t) = \tilde{\mathbf{C}}\,\tilde{\mathbf{x}}(t) + \tilde{\mathbf{D}}\,\mathbf{u}(t)$$

with: $\tilde{\mathbf{A}} = \mathbf{M}^{-1}\,\mathbf{A}\mathbf{M}$ $\tilde{\mathbf{B}} = \mathbf{M}^{-1}\mathbf{B}$ $\tilde{\mathbf{C}} = \mathbf{C}\mathbf{M}$ and $\tilde{\mathbf{D}} = \mathbf{D}$.

Hence, let us calculate:

$$\begin{aligned}
\tilde{\mathbf{H}}(p) = \tilde{\mathbf{C}}[p\,\mathbf{I} - \tilde{\mathbf{A}}]^{-1}\,\tilde{\mathbf{B}} + \tilde{\mathbf{D}} \;&=\; \mathbf{C}\mathbf{M}[p\mathbf{I} - \mathbf{M}^{-1}\mathbf{A}\mathbf{M}]^{-1}\mathbf{M}^{-1}\mathbf{B} + \mathbf{D} \\
&=\; \mathbf{C}[\mathbf{M}[p\mathbf{I} - \mathbf{M}^{-1}\mathbf{A}\mathbf{M}]\,\mathbf{M}^{-1}]^{-1}\mathbf{B} + \mathbf{D} \\
&\;\; \mathbf{C}[p\mathbf{I} - \mathbf{A}]^{-1}\mathbf{B} + \mathbf{D} = \mathbf{H}(p)
\end{aligned}$$

The change of basis did not modify the transfer matrix of the system.

2.3.3. *Transfer passage $\rightarrow$ state*

We will deal here only with mono-variable systems. Generalizing of multi-variable systems is done easily for each scalar transfer function $H_{ij}(p)$.

Hence, let us consider a transfer function:

$$H(p) = \frac{Y(p)}{U(p)} \text{ with: } H(p) = \frac{a_0 + a_1 p + a_2 p^2 + \cdots + a_{n-1} p^{n-1}}{b_0 + b_1 p + b_2 p^2 + \cdots + p^n} = \frac{N(p)}{D(p)}$$

We note that the coefficient of the highest degree term of the denominator is standardized at $b_n = 1$. We also note that the degree of the numerator is at least one unit lower than the one of the denominator. If it is in any other way (same degree maximum for a physically conceivable system), the rational fraction $\frac{N(p)}{D(p)}$ is decomposed into a full part d_0 and a rational fraction of the type shown above. This full part d_0 characterizes the direct transmission of the control and hence represents an element of matrix $\mathbf{D}$ of the state model.

Among all the forms of transfer passage $\rightarrow$ state, we will describe two particular forms adapted to problems of control and estimation.

Companion form for the control

We will start from the expression: $Y(p) = \dfrac{N(p)}{D(p)} U(p)$.

Let us assume that:

$$Y(p) = N(p)X_1(p) \text{ with } X_1(p) = \frac{U(p)}{D(p)}$$

where $X_1(p)$ represents the Laplace transform of $x_1(t)$, the first of n state variables constituting $\mathbf{x}(t)$. It is thus possible to write the two following polynomial relations:

$$D(p)\ X_1(p) = (b_0 + b_1 p + b_2 p^2 + \cdots + p^n)\ X_1(p) = U(p)$$

$$Y(p) = N(p)\ X_1(p) = (a_0 + a_1 p + a_2 p^2 + \cdots + a_{n-1} p^{n-1})X_1(p)$$

In the time domain, the previous equations lead to:

$$\frac{d^n x_1}{dt^n}(t) + \cdots + b_1 \frac{dx_1}{dt}(t) + b_0 \, x_1(t) = u(t)$$

$$y(t) = a_{n-1} \frac{d^{n-1} x_1}{dt^{n-1}}(t) + \cdots + a_1 \frac{dx_1}{dt}(t) + a_0 \, x_1(t)$$

Thus, by choosing as state variables $x_1(t)$ and its $(n-1)$ first derivatives, the equations of state and measurement are written:

$$\dot{x}_1(t) = x_2(t)$$

$$\dot{x}_2(t) = x_3(t)$$

$$\vdots$$

$$\dot{x}_{n-1}(t) = x_n(t)$$

$$\dot{x}_n(t) = -b_0 \, x_1(t) - b_1 \, x_2(t) - \cdots - b_{n-1} \, x_n(t) + u(t)$$

$$y(t) = a_0 \, x_1(t) + a_1 \, x_2(t) + \cdots + a_{n-1} \, x_n(t)$$

which lead to the matrix form called *companion form for the control* whose block diagram is given in Figure 2.2.

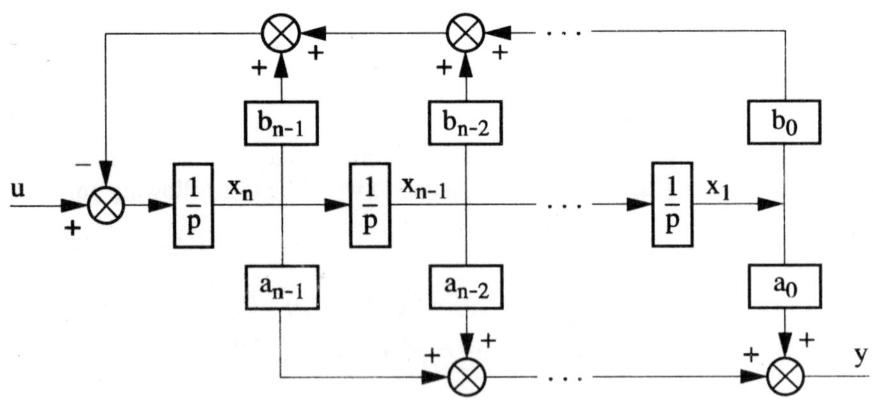

Figure 2.2. *Companion form for the control*

This representation, whose state variables corresponding to this form are currently called *phase variables* (phase plane for $n = 2$), is summed up by the following equation of state and equation of measurement:

$$\dot{\mathbf{x}}(t) = \begin{bmatrix} 0 & 1 & 0 & . & . & 0 \\ 0 & 0 & 1 & 0 & . & 0 \\ . & . & 0 & 1 & 0 & . \\ . & . & . & . & . & . \\ . & . & . & . & 0 & 1 \\ -b_0 & -b_1 & -b_2 & . & . & -b_{n-1} \end{bmatrix} \mathbf{x}(t) + \begin{bmatrix} 0 \\ 0 \\ . \\ . \\ 0 \\ . \\ 1 \end{bmatrix} u(t)$$

$$y(t) = [a_0 \quad a_1 \cdots a_{n-1}] \, \mathbf{x}(t)$$

Companion form for the observation

Let us bring back the fraction form:

$$\frac{Y(p)}{U(p)} = \frac{a_0 + a_1 p + a_2 p^2 + \cdots + a_{n-1} p^{n-1}}{b_0 + b_1 p + b_2 p^2 + \cdots + p^n} = \frac{N(p)}{D(p)}$$

By dividing $N(p)$ and $D(p)$ by p^n, we obtain:

$$\frac{Y(p)}{U(p)} = \frac{a_0 \dfrac{1}{p^n} + a_1 \dfrac{1}{p^{n-1}} + a_2 \dfrac{1}{p^{n-2}} + \cdots + a_{n-1} \dfrac{1}{p}}{b_0 \dfrac{1}{p^n} + b_1 \dfrac{1}{p^{n-1}} + b_2 \dfrac{1}{p^{n-2}} + \cdots + b_{n-1} \dfrac{1}{p} + 1}$$

which leads to:

$$Y(p) = \left[a_{n-1} \frac{1}{p} + \cdots + a_1 \frac{1}{p^{n-1}} + a_0 \frac{1}{p^n} \right] U(p)$$
$$- \left[b_{n-1} \frac{1}{p} + \cdots + b_1 \frac{1}{p^{n-1}} + b_0 \frac{1}{p^n} \right] Y(p)$$

$$Y(p) = \frac{1}{p} \left[a_{n-1} \; U(p) - b_{n-1} \; Y(p) + \frac{1}{p} \left[\begin{matrix} a_{n-2} \; U(p) - b_{n-2} \; Y(p) \\ + \dfrac{1}{p} \left[\cdots + \dfrac{1}{p} [a_0 \; U(p) - b_0 \; Y(p)] \cdots \right] \end{matrix} \right] \right]$$

Hence, through $X_n(p) = L\{x_n(t)\} = Y(p)$, we obtain:

$$X_n(p) = \frac{1}{p}[a_{n-1}\ U(p) - b_{n-1}\ X_n(p) + X_{n-1}(p)]$$

$$X_{n-1}(p) = \frac{1}{p}[a_{n-2}\ U(p) - b_{n-2}\ X_n(p) + X_{n-2}(p)]$$

$$\vdots$$

$$X_1(p) = \frac{1}{p}[a_0\ U(p) - b_0\ X_n(p)]$$

These n equations are directly transcribed into the time domain as follows:

$$\dot{x}_1(t) = a_0\, u(t) - b_0\, x_n(t)$$

$$\dot{x}_2(t) = a_1\, u(t) - b_1\, x_n(t) + x_1(t)$$

$$\dot{x}_3(t) = a_2\, u(t) - b_2\, x_n(t) + x_2(t)$$

$$\vdots$$

$$\dot{x}_n(t) = a_{n-1}\, u(t) - b_{n-1}\, x_n(t) + x_{n-1}(t)$$

Grouped as a matrix, they represent the *companion form for the observation*:

$$\dot{\mathbf{x}}(t) = \begin{bmatrix} 0 & 0 & \cdots & & -b_0 \\ 1 & 0 & 0 & \cdots & -b_1 \\ 0 & 1 & 0 & \cdots & -b_2 \\ \cdot & \cdot & \cdot & \cdot & \cdot \\ \cdot & \cdot & 0 & 1 & 0 & -b_{n-2} \\ \cdot & \cdot & \cdot & 0 & 1 & -b_{n-1} \end{bmatrix} \mathbf{x}(t) + \begin{bmatrix} a_0 \\ a_1 \\ \cdot \\ \cdot \\ \cdot \\ \cdot \\ \cdot \\ a_{n-1} \end{bmatrix} u(t)$$

the equation of measurement being reduced to:

$$y(t) = [0\quad 0\quad 0\quad \cdots\quad 0\quad 1]\, \mathbf{x}(t)$$

The block diagram of this representation is given in Figure 2.3.

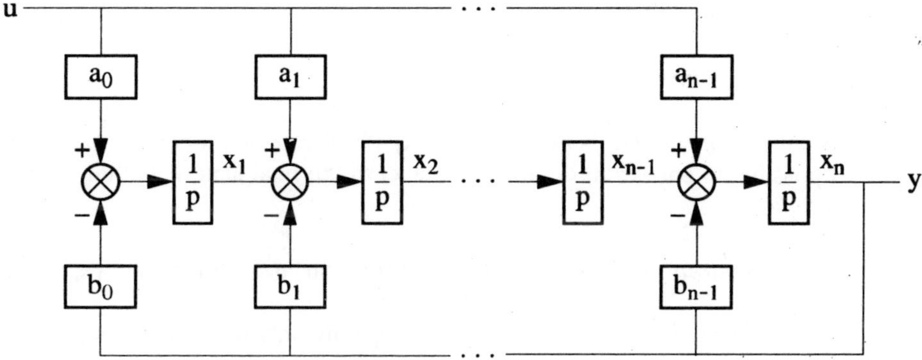

Figure 2.3. *Companion form for the observation*

2.3.4. *Scalar representation of invariant and linear discrete systems*

In section 2.2.5 we have seen the modeling of discrete linear and invariant systems as:

$$\mathbf{x}(k+1) = \mathbf{F}\,\mathbf{x}(k) + \mathbf{G}\,\mathbf{u}(k)$$
$$\mathbf{y}(k) \quad = \mathbf{H}\,\mathbf{x}(k) + \mathbf{J}\,\mathbf{u}(k)$$

The scalar representation of such systems is obtained by using the transformation in z on the equation of state (zero initial conditions):

$$z\ \mathbf{X}(z) = \mathbf{F}\mathbf{X}(z) + \mathbf{G}\mathbf{U}(z) \quad \Rightarrow \quad [z\mathbf{I} - \mathbf{F}]\,\mathbf{X}(z) = \mathbf{G}\mathbf{U}(z)$$
$$\Rightarrow \quad \mathbf{X}(z) = [z\mathbf{I} - \mathbf{F}]^{-1}\ \mathbf{G}\mathbf{U}(z)$$
$$\mathbf{Y}(z) = \mathbf{H}\mathbf{X}(z) + \mathbf{G}\mathbf{U}(z) \quad \Rightarrow \quad \mathbf{Y}(z) = [\mathbf{H}[z\mathbf{I} - \mathbf{F}]^{-1}\ \mathbf{G} + \mathbf{J}]\,\mathbf{U}(z) = \mathbf{T}(z)\,\mathbf{U}(z)$$

This last relation makes it possible to define the transfer matrix of the discrete system:

$$\mathbf{T}(z) = [\mathbf{H}[z\mathbf{I} - \mathbf{F}]^{-1}\mathbf{G} + \mathbf{J}]$$

The calculation of matrix $[z\ \mathbf{I} - \mathbf{F}]^{-1}$ can be done by using the Leverrier-Souriau algorithm which has been described for the calculation of matrix $[p\ \mathbf{I} - \mathbf{A}]^{-1}$ in section 2.3.1.

2.4. Controllability of systems

2.4.1. *General definitions*

A system is *controllable* between $\mathbf{x}_0$ and $\mathbf{x}_1$ if we can find a control $\mathbf{u}(t)$ that makes it possible to pass from the initial state $\mathbf{x}_0$ to the final state $\mathbf{x}_1$ within a *finite period of time*.

A system is *entirely controllable* if there is a solution $\mathbf{u}(t)$ for any pair $\{\mathbf{x}_0, \mathbf{x}_1\}$.

A system is *partially controllable* if we can operate only on certain components of the state of a system between $\mathbf{x}_0$ and $\mathbf{x}_1$.

EXAMPLE 2.6.– let us consider the size 4 mono-variable system defined by the equation of state:

$$\dot{\mathbf{x}}(t) = \begin{bmatrix} a_1 & 0 & 0 & 0 \\ 0 & a_2 & 0 & 0 \\ 0 & 0 & a_3 & 0 \\ 0 & 0 & 0 & a_4 \end{bmatrix} \mathbf{x}(t) + \begin{bmatrix} b_1 \\ b_2 \\ 0 \\ 0 \end{bmatrix} u(t)$$

It is clear that only variables $x_1(t)$ and $x_2(t)$ are controllable by $u(t)$. The evolution of variables $x_3(t)$ and $x_4(t)$ remains independent of the control.

2.4.2. *Controllability of linear and invariant systems*

Firstly, we will limit ourselves to the case of a mono-variable system whose control $u(t)$ is reduced to a scalar term. Hence, we have to define the necessary conditions (i.e. the sufficient conditions) so that the system described by the equation of state· $\dot{\mathbf{x}}(t) = \mathbf{A}\,\mathbf{x}(t) + \mathbf{B}\,u(t)$ is controllable. The linearity of the equation makes it possible to calculate for a zero initial state $(\mathbf{x}(t_0) = 0)$.

If we consider $\mathbf{x}_1 \in R^n$, there is a control $u(t)$ so that, for finite t_1, we have:

$$\mathbf{x}(t_1) = e^{\mathbf{A}(t_1 - t_0)}\,\mathbf{x}(t_0) + \int_{t_0}^{t_1} e^{\mathbf{A}(t_1 - \tau)}\mathbf{B}\;u(\tau)\;d\tau = \int_{t_0}^{t_1} e^{\mathbf{A}(t_1 - \tau)}\mathbf{B}\;u(\tau)\quad (\text{si } \mathbf{x}(t_0) = 0)$$

For any instant t, we have $\mathbf{x}(t) = \displaystyle\int_{t_0}^{t} e^{\mathbf{A}(t-\tau)}\mathbf{B}\ u(\tau)d\tau$.

However, according to the Cayley-Hamilton theorem, $e^{\mathbf{A}(t-\tau)}$ can develop as a matrix polynomial of degree $n-1$:

$$e^{\mathbf{A}(t-\tau)} = \alpha_0(\tau,\ t)\ \mathbf{I} + \alpha_1(\tau,\ t)\mathbf{A} + \cdots + \alpha_{n-1}(\tau,\ t)\mathbf{A}^{n-1} = \sum_{i=0}^{n-1}\alpha_i(\tau,\ t)\mathbf{A}^i$$

because, as we have seen during the calculation of the matrix exponential function, $\mathbf{A}^n$ and all powers more than $\mathbf{A}$ are linear combinations of $(n-1)$ first powers.

Thus:

$$\mathbf{x}(t) = \int_{t_0}^{t}\left[\sum_{i=0}^{n-1}\alpha_i(\tau,\ t)\mathbf{A}^i\right]\mathbf{B}u(\tau)\ d\tau = \sum_{i=0}^{n-1}\mathbf{A}^i\mathbf{B}\int_{t_0}^{t}\alpha_i(\tau,t)\ u(\tau)\ d\tau$$

By supposing that $\rho_i = \displaystyle\int_{t_0}^{t}\alpha_i(\tau,t)u(\tau)\,d\tau$, the previous equation becomes:

$$\mathbf{x}(t) = \mathbf{B}\,\rho_0 + \mathbf{AB}\,\rho_1 + \cdots + \mathbf{A}^{n-1}\,\mathbf{B}\,\rho_{n-1}$$

an expression in which matrices $\mathbf{A}^i\mathbf{B}$ are column vectors (like $\mathbf{B}$ – mono-variable control) and functions $\rho_i(t)$ are scalar.

By using a vector notation: $\rho(t) = \begin{bmatrix}\rho_0(t)\\ \rho_1(t)\\ .\\ .\\ \rho_{n-1}(t)\end{bmatrix}$, we can sum up the previous equation as follows: $\mathbf{x}(t) = [\mathbf{B}|\mathbf{AB}|\cdots|\mathbf{A}^{n-1}\mathbf{B}]\rho(t) = \mathbf{Q_G}\rho(t)$, where matrix $\mathbf{Q_G}$ is the square matrix $n\times n$ built by juxtaposing n column vectors $\mathbf{B}, \mathbf{AB},\cdots,\mathbf{A}^{n-1}\mathbf{B}$.

This result shows that irrespective of control $\mathbf{u}(t)$, $\mathbf{x}(t)$ remains within the vector sub-space generated by vectors $\mathbf{A}^i\mathbf{B}$, for $i = 0,1,\cdots,n-1$.

To reach any point $\mathbf{x}(t)$ of the state space, the n $\mathbf{A}^i\mathbf{B}$ columns must be linearly independent. In other words, it is necessary that:

$$\operatorname{rank}[\mathbf{Q_G}] = n$$

If this condition is satisfied, we have $\rho(t) = \mathbf{Q_G}^{-1}\,\tilde{\mathbf{x}}(t)$.

The solution in $\rho(t)$ of this problem is unique but there is an infinite number of $u(t)$ solutions satisfying:

$$\rho_i(t) = \int_{t_0}^{t} \alpha_i(\tau,t)\, u(\tau)\, d\tau$$

The control $u(t)$ is defined only by its n "projections" on the functions $\alpha_i(\tau,t)$.

To conclude, a mono-variable linear and invariant system characterized by its equation of state:

$$\dot{\mathbf{x}}(t) = \mathbf{A}\,\mathbf{x}(t) + \mathbf{B}\,u(t)$$

is entirely controllable by control $u(t)$ if and only if the controllability matrix:

$$\mathbf{Q_G} = [\mathbf{B} \mid \mathbf{AB} \mid \cdots \mid \mathbf{A}^{n-1}\mathbf{B}]$$

is of rank n (and thus reversible or regular).

This result can be generalized to the multi-variable case with the following adjustments: if $\dim \mathbf{B} = n \times m$, $\rho_i(t)$ is no longer a scalar function but a vector of size m (that of $u(t)$). Consequently, $\rho(t)$ is a vector of size $n \times m$. Considering these adjustments, the expression:

$$\mathbf{x}(t) = [\mathbf{B} \mid \mathbf{AB} \mid \cdots \mid \mathbf{A}^{n-1}\mathbf{B}]\,\rho(t) = \mathbf{Q}_G\rho(t)$$

remains valid, but matrix $\mathbf{Q_G}$ is of size $n \times (n \times m)$.

The system will be entirely controllable if matrix $\mathbf{Q_G}$ is of rank n, which means that we can extract n linearly independent columns from:

$$\mathbf{Q_G} = [\mathbf{B} \mid \mathbf{AB} \mid \cdots \mid \mathbf{A}^{n-1}\mathbf{B}]$$

2.4.3. *Canonic representation of partially controllable systems*

When rank $[\mathbf{Q_G}] = q < n$, the relation $\mathbf{x}(t) = \mathbf{B}\,\rho_0 + \mathbf{AB}\,\rho_1 + \cdots + \mathbf{A}^{n-1}\mathbf{B}\rho_{n-1}$ limits the state evolution $\mathbf{x}(t)$ of the system to a vectorial sub-space of size q generated by the independent columns of matrix $\mathbf{Q_G}$. Hence, we will seek a basis of the state space where the equation of state makes it possible to split the state vector into an entirely controllable part of size q and a non-controllable part of size $n - q$. Thus, let us choose a basis of the vector sub-space consisting of q linear independent combinations of q independent vectors of $\mathbf{Q_G}$.

Let $\mathbf{T}_1$ of size $n \times q$ be the matrix built from these q column vectors.

Let us build the basis change matrix $\mathbf{T}$ (so that $\mathbf{x}(t) = \mathbf{T}\,\tilde{\mathbf{x}}(t)$) by completing block $\mathbf{T}_1$ by a block $\mathbf{T}_2$ of size $n \times (n - q)$ so that:

$$\mathbf{T} = \; [\mathbf{T}_1 \mid \mathbf{T}_2] \downarrow n \; \text{ is reversible}$$
$$\rightarrow \rightarrow$$
$$q \quad n-q$$

In section 2.3.2 we have seen that in a basis change the new equations of state and measurement of the system are:

$$\dot{\tilde{\mathbf{x}}}(t) = \mathbf{T}^{-1}\mathbf{A}\,\mathbf{T}\,\tilde{\mathbf{x}}(t) + \mathbf{T}^{-1}\mathbf{B}\,\mathbf{u}(t) = \tilde{\mathbf{A}}\,\tilde{\mathbf{x}}(t) + \tilde{\mathbf{B}}\,\mathbf{u}(t)$$

$$\mathbf{y}(t) = \mathbf{C}\,\mathbf{T}\,\tilde{\mathbf{x}}(t) + \mathbf{D}\,\mathbf{u}(t) = \tilde{\mathbf{C}}\,\tilde{\mathbf{x}}(t) + \tilde{\mathbf{D}}\,\mathbf{u}(t)$$

In order to calculate matrices $\tilde{\mathbf{A}}, \tilde{\mathbf{B}}$ and $\tilde{\mathbf{C}}$, we will use the following property: let $\mathbf{M} = \underbrace{[\mathbf{M}_1 \mid \mathbf{M}_2 \mid \cdots \mid \mathbf{M}_r]}_{r \text{ columns } (r \leq q)} \downarrow n\,rows$ be a matrix extracted from $\mathbf{Q_G}$ by linear combinations of the independent columns of $\mathbf{Q_G}$. By construction, columns $\mathbf{M}_i$ of $\mathbf{M}$ are linked to columns $\mathbf{T}_{1j}$ of $\mathbf{T}_1$ through a relation of the type:

$$M_i = \sum_{j=1}^{q} k_{ij}\, T_{1j} = k_{ij}\, T_{11} + k_{i2}\, T_{12} + \cdots + k_{iq}\, T_{1q} \quad \text{for}: \quad i = 1, 2, \cdots, r \; (r \leq q)$$

This expression translated as a matrix becomes:

$$M_i = [T_{11} \mid T_{12} \mid \cdots \mid T_{1q}] \begin{bmatrix} k_{i1} \\ k_{i2} \\ \cdot \\ \cdot \\ k_{iq} \end{bmatrix}$$

By grouping the M_i in order to build matrix M, we obtain:

$$M = T_1 K = \begin{bmatrix} M_1 \mid M_2 \mid \cdots \mid M_r \end{bmatrix}$$

To conclude, if $M(\dim n \times r)$ is built from linear combinations of the columns of matrix Q_G, then there is a matrix $K\,(\dim q \times r)$ such that:

$$\underset{n \times r}{M} = \underset{n \times q}{T_1}\; \underset{q \times r}{K}$$

We will use this property to calculate matrices $\tilde{A}, \tilde{B}$ and $\tilde{C}$.

$$\tilde{A} = T^{-1}\, AT = T^{-1} A\, [T_1 \mid T_2] = T^{-1}\, [\, AT_1 \mid AT_2]$$

However, $AT_1\,(n \times q)$ is, by construction, extracted from $Q_G \Rightarrow \exists K_1\,(q \times q)$ so that $AT_1 = T_1\, K_1$, hence:

$$T^{-1}A\; T = T^{-1}[T_1\, K_1 \mid AT_2] = [\, T^{-1}T_1\, K_1 \mid T^{-1}AT_2]$$

$$\underset{q}{\xrightarrow{\hspace{1cm}}} \quad \underset{n-q}{\xrightarrow{\hspace{1cm}}} \qquad \underset{q}{\xrightarrow{\hspace{1cm}}} \quad \underset{n-q}{\xrightarrow{\hspace{1cm}}}$$

but: $\quad T^{-1}\; T = T^{-1}[T_1 \mid T_2] = [\, T^{-1}\, T_1 \mid T^{-1}\, T_2] = I_n$

Block $\mathbf{T}^{-1}\mathbf{T}_1$ can thus be written:

$$\mathbf{T}^{-1}\,\mathbf{T}_1 = \begin{bmatrix} \mathbf{I}_q \\ 0 \end{bmatrix} \begin{matrix} \downarrow q \\ \downarrow n-q \end{matrix} \Rightarrow \widetilde{\mathbf{A}} = \left[\begin{array}{c|c} \mathbf{K}_1 & \\ -- & \mathbf{T}^{-1}\mathbf{A}\mathbf{T}_2 \\ 0 & \end{array} \right]$$
$$\xrightarrow{\quad q \quad}$$

In canonical form:

$$\widetilde{\mathbf{A}} = \begin{bmatrix} \widetilde{\mathbf{A}}_{11} & \widetilde{\mathbf{A}}_{12} \\ 0 & \widetilde{\mathbf{A}}_{22} \end{bmatrix} \begin{matrix} \downarrow q \\ \downarrow n-q \end{matrix}$$
$$\xrightarrow{\quad q \quad} \xrightarrow{\quad n-q \quad}$$

Likewise, $\widetilde{\mathbf{B}} = \mathbf{T}^{-1}\,\mathbf{B}$ and $\mathbf{B}$ is extracted from $\mathbf{Q}_{\mathbf{G}}$:

$$\Rightarrow \quad \exists \mathbf{K}_2 (q \times m) \text{ so that } \mathbf{B} = \mathbf{T}_1\,\mathbf{K}_2$$
$$\Rightarrow \quad \widetilde{\mathbf{B}} = \mathbf{T}^{-1}\,\mathbf{T}_1\,\mathbf{K}_2 = \begin{bmatrix} \mathbf{K}_2 \\ 0 \end{bmatrix} \begin{matrix} \downarrow q \\ \downarrow n-q \end{matrix}$$
$$\xrightarrow{\quad m \quad}$$

Matrix $\widetilde{\mathbf{C}}$ does not have any particular characteristic.

In the new basis, the equations of state and measurement have the following canonical form:

$$\dot{\widetilde{\mathbf{x}}}(t) = \begin{bmatrix} \widetilde{\mathbf{A}}_{11} & \widetilde{\mathbf{A}}_{12} \\ 0 & \widetilde{\mathbf{A}}_{22} \end{bmatrix} \widetilde{\mathbf{x}}(t) + \begin{bmatrix} \widetilde{\mathbf{B}}_1 \\ 0 \end{bmatrix} \mathbf{u}(t)$$

$$\mathbf{y}(t) = [\,\widetilde{\mathbf{C}}_1 \quad \widetilde{\mathbf{C}}_2\,]\,\widetilde{\mathbf{x}}(t) + \mathbf{D}\,\mathbf{u}(t)$$

This form corresponds to a partitioning of the state vector into two sub-vectors $\widetilde{\mathbf{x}}_1(t)$, $\widetilde{\mathbf{x}}_2(t)$ representing the controllable part and the non-controllable part of state $\widetilde{\mathbf{x}}(t)$. The equations developed from this partitioning are:

$$\dot{\widetilde{\mathbf{x}}}_1(t) = \widetilde{\mathbf{A}}_{11}\,\widetilde{\mathbf{x}}_1(t) + \widetilde{\mathbf{A}}_{12}\,\widetilde{\mathbf{x}}_2(t) + \widetilde{\mathbf{B}}_1\,\mathbf{u}(t) \text{ controllable part (size } q)$$

$$\dot{\tilde{\mathbf{x}}}_2(t) = \tilde{\mathbf{A}}_{22}\,\tilde{\mathbf{x}}_2(t) \quad \text{non-controllable part (size } n - q)$$

EXAMPLE 2.7.– let us consider the system described by the equations:

$$\dot{\mathbf{x}}(t) = \begin{bmatrix} -1 & 0 & 0 & 0 \\ 0 & 0 & 1 & 0 \\ -1 & 0 & 0 & 1 \\ 0 & 0 & 0 & 0 \end{bmatrix} \mathbf{x}(t) + \begin{bmatrix} 1 \\ 0 \\ 1 \\ 0 \end{bmatrix} u(t) \quad y(t) = [0 \quad 0 \quad 1 \quad 0]\,\mathbf{x}(t)$$

Is this system controllable?

$$\mathbf{Q_G} = \begin{bmatrix} 1 & -1 & +1 & -1 \\ 0 & +1 & -1 & +1 \\ 1 & -1 & +1 & -1 \\ 0 & 0 & 0 & 0 \end{bmatrix} \quad \text{rank } \mathbf{Q_G} = 2 \quad \text{(columns (2), (3) and (4) are linked).}$$

$\mathbf{T} = [\mathbf{T}_1 \mid \mathbf{T}_2]\,\mathbf{T}_1$ linear combination of independent columns of $\mathbf{Q_G}$.

$$\mathbf{T}_1 = \begin{bmatrix} 1 & 0 \\ 0 & 1 \\ 1 & 0 \\ 0 & 0 \end{bmatrix} \qquad \mathbf{T} = \begin{bmatrix} 1 & 0 & 0 & 0 \\ 0 & 1 & 0 & 0 \\ 1 & 0 & 1 & 0 \\ 0 & 0 & 0 & 1 \end{bmatrix}$$

$\mathbf{T}_2$ is chosen in order to least disturb the initial state structure and in particular in order to maintain, for the new state variables, a significant physical nature.

$$\mathbf{T}^{-1} = \begin{bmatrix} 1 & 0 & 0 & 0 \\ 0 & 1 & 0 & 0 \\ -1 & 0 & 1 & 0 \\ 0 & 0 & 0 & 1 \end{bmatrix} \quad \Rightarrow \quad \tilde{\mathbf{A}} = \mathbf{T}^{-1}\mathbf{A}\,\mathbf{T} \begin{bmatrix} -1 & 0 & 0 & 0 \\ 1 & 0 & 1 & 0 \\ 0 & 0 & 0 & 1 \\ 0 & 0 & 0 & 0 \end{bmatrix}$$

$$\tilde{\mathbf{B}} = \mathbf{T}^{-1}\mathbf{B} = \begin{bmatrix} 1 \\ 0 \\ 0 \\ 0 \end{bmatrix} \qquad \text{and} \qquad \tilde{\mathbf{C}} = \mathbf{C}\,\mathbf{T} = [\,1 \quad 0 \quad 1 \quad 0\,]$$

In the new basis, we thus obtain the canonical form for the following controllability:

$$\dot{\tilde{\mathbf{x}}}(t) = \left[\begin{array}{cc:cc} -1 & 0 & 0 & 0 \\ 1 & 0 & 1 & 0 \\ \hdashline 0 & 0 & 0 & 1 \\ 0 & 0 & 0 & 0 \end{array}\right] \tilde{\mathbf{x}}(t) + \left[\begin{array}{c} 1 \\ 0 \\ \hline 0 \\ 0 \end{array}\right] u(t)$$

$$y(t) = \left[\begin{array}{cc:cc} 1 & 0 & 1 & 0 \end{array}\right] \tilde{\mathbf{x}}(t)$$

where we note the blocks of zeros characteristic for this form in $\tilde{\mathbf{A}}$ and $\tilde{\mathbf{B}}$.

2.4.4. *Scalar representation of partially controllable systems*

From the canonical form described above, we can determine the transfer matrix of the system as done in section 2.3.2:

$$H(p) = \tilde{\mathbf{C}}[p\mathbf{I} - \tilde{\mathbf{A}}]^{-1}\tilde{\mathbf{B}} + \mathbf{D} = [\tilde{\mathbf{C}}_1 \quad \tilde{\mathbf{C}}_2] \left[\begin{array}{cc} p\mathbf{I}_q - \tilde{\mathbf{A}}_{11} & -\tilde{\mathbf{A}}_{12} \\ 0 & p\mathbf{I}_{n-q} - \tilde{\mathbf{A}}_{22} \end{array}\right] \left[\begin{array}{c} \tilde{\mathbf{B}}_1 \\ 0 \end{array}\right] + \mathbf{D}$$

By using the triangular nature per blocks of the matrix to reverse, we obtain:

$$= [\tilde{\mathbf{C}}_1 \quad \tilde{\mathbf{C}}_2] \left[\begin{array}{cc} [p\mathbf{I}_q - \tilde{\mathbf{A}}_{11}]^{-1} & \mathbf{M}_{12} \\ 0 & [p\mathbf{I}_{n-q} - \tilde{\mathbf{A}}_{22}]^{-1} \end{array}\right] \left[\begin{array}{c} \tilde{\mathbf{B}}_1 \\ 0 \end{array}\right] + \mathbf{D}$$

which is a form where the calculation of matrix $\mathbf{M}_{12}$ is useless because it appears only in the final result:

$$H(p) = \tilde{\mathbf{C}}_1[p\mathbf{I}_q - \tilde{\mathbf{A}}_{11}]^{-1}\tilde{\mathbf{B}}_1 + \mathbf{D}$$

We note on this form that the degree in p of the transfer matrix is reduced *a priori* to q instead of n. Hence, the scalar response represents only the controllable part of the state.

2.5. Observability of systems

2.5.1. *General definitions*

A system is *observable* if the observation of measurement $\mathbf{y}(t)$ during a *finite period of time* (t_0, t_1) makes it possible to determine the state vector at instant t_0.

A system is *entirely* or *partially* observable depending on whether we can build all or a part of state $\mathbf{x}(t_0)$.

2.5.2. *Observability of linear and invariant systems*

Firstly, we will suppose that measurement $\mathbf{y}(t)$ performed on the system is a scalar measurement and we will deal here with the following problem.

Let us consider a free state system described by equation of state $\dot{\mathbf{x}}(t) = \mathbf{A}\,\mathbf{x}(t)$ and for which we notice the scalar output $y(t) = \mathbf{C}\,\mathbf{x}(t)$ $(\dim \mathbf{C} = 1 \times n)$.

If $\mathbf{x}_0$ is the initial state $x(0)$ $(t_0 = 0)$, how can we find state x_0 from the observation of $y(t)$ during the finite period of time $[0, t_1]$? We will answer this question by formulating output $y(t)$ according to $\mathbf{x}_0$.

We have:

$$\mathbf{x}(t) = e^{\mathbf{A}t}\mathbf{x}_0 = \sum_{i=0}^{n-1} \alpha_i(t)\mathbf{A}^i \mathbf{x}_0 \quad \text{and thus: } y(t) = e^{\mathbf{A}t}\mathbf{x}_0 = \sum_{i=0}^{n-1} \alpha_i(t)\mathbf{C}\mathbf{A}^i \mathbf{x}_0$$

By forming the scalar products:

$$< \alpha_k(t), y(t) > = \int_0^{t_1} \alpha_k(t)\,y(t)\,dt = z_k$$

$$< \alpha_k(t), y(t) > = \sum_{i=0}^{n-1} \left(\int_0^{t_1} \alpha_k(t)\,\alpha_i(t)\,dt \right) \mathbf{C}\mathbf{A}^i\,\mathbf{x}_0$$

we obtain:

$$z_k = <\alpha_k(t),\ y(t)> = \sum_{i=0}^{n-1} a_{ki}\ \mathbf{CA}^i\ \mathbf{x}_0$$

with: $a_{ki} = \displaystyle\int_0^{t_1} \alpha_k(t)\,\alpha_i(t)\,dt$

and in a matrix form:

$$
\begin{bmatrix} z_0 \\ z_1 \\ \cdot \\ \cdot \\ \cdot \\ z_{n-1} \end{bmatrix}
=
\begin{bmatrix}
a_{00} & a_{01} & \cdot & \cdot & \cdot & a_{0n-1} \\
a_{01} & \cdot & & \cdot & \cdot & \cdot \\
\cdot & & \cdot & & & \\
\cdot & & & & & \\
\cdot & & & & & \\
a_{0n-1} & \cdot & & \cdot & \cdot & a_{n-1\ n-1}
\end{bmatrix}
\begin{bmatrix} \mathbf{C} \\ \mathbf{CA} \\ \mathbf{CA}^2 \\ \vdots \\ \vdots \\ \mathbf{CA}^{n-1} \end{bmatrix}
\mathbf{x}_0
$$

size: $n \times 1$ $n \times n$ $n \times n$ $n \times 1$

We can show that, by construction, matrix $\{a_{ij}\}$ is regular. Thus, a necessary and sufficient condition for the system to be observable (thus to be able to uniquely extract $\mathbf{x}_0$ from the equation below) is that the observability matrix defined by:

$$
\mathbf{Q}_0 = \begin{bmatrix} \mathbf{C} \\ \mathbf{CA} \\ \mathbf{CA}^2 \\ \vdots \\ \vdots \\ \mathbf{CA}^{n-1} \end{bmatrix}
\quad \text{(and of size } n \times n \text{ here) is of rank } n \text{ and we obtain:}
$$

$$\mathbf{x}_0 = \begin{bmatrix} \mathbf{C} \\ \mathbf{CA} \\ \mathbf{CA}^2 \\ \vdots \\ \mathbf{CA}^{n-1} \end{bmatrix}^{-1} \begin{bmatrix} a_{00} & a_{01} & \cdot & \cdot & \cdot & a_{0n-1} \\ a_{01} & & \cdot & \cdot & \cdot & \cdot \\ \cdot & & \cdot & \cdot & \cdot & \cdot \\ \cdot & & \cdot & \cdot & \cdot & \cdot \\ \cdot & & \cdot & \cdot & \cdot & \cdot \\ a_{0n-1} & \cdot & & \cdot & \cdot & a_{n-1\,n-1} \end{bmatrix}^{-1} \begin{bmatrix} z_0 \\ z_1 \\ \cdot \\ \cdot \\ \cdot \\ z_{n-1} \end{bmatrix}$$

EXAMPLE 2.8.– let us consider a system represented by equations of state and measurement in companion form for the observation:

$$\dot{\mathbf{x}}(t) = \begin{bmatrix} 0 & 0 & \cdot & \cdot & \cdot & -b_0 \\ 1 & 0 & 0 & \cdot & \cdot & -b_1 \\ 0 & 1 & 0 & \cdot & \cdot & -b_2 \\ \cdot & \cdot & \cdot & \cdot & \cdot & \cdot \\ \cdot & \cdot & 0 & 1 & 0 & -b_{n-2} \\ \cdot & \cdot & \cdot & 0 & 1 & -b_{n-1} \end{bmatrix} \mathbf{x}(t) + \begin{bmatrix} a_0 \\ a_1 \\ \cdot \\ \cdot \\ \cdot \\ a_{n-1} \end{bmatrix} u(t)$$

and $y(t) = \begin{bmatrix} 0 & 0 & 0 & \cdots & 0 & 1 \end{bmatrix} \mathbf{x}(t)$.

The calculation of observability matrix:

$$\mathbf{Q}_0 = \begin{bmatrix} \mathbf{C} \\ \mathbf{CA} \\ \mathbf{CA}^2 \\ \vdots \\ \vdots \\ \mathbf{CA}^{n-1} \end{bmatrix} \quad \text{leads to: } \mathbf{Q}_0 = \begin{bmatrix} 0 & 0 & \cdot & \cdot & 0 & 1 \\ 0 & \cdot & \cdot & 0 & 1 & * \\ \cdot & \cdot & 0 & 1 & * & * \\ \cdot & \cdot & \cdot & \cdot & \cdot & \cdot \\ 0 & 1 & \cdot & \cdot & \cdot & * \\ 1 & * & \cdot & \cdot & * & * \end{bmatrix}$$

matrix of rank n by construction. The companion form for the observation represents entirely observable systems through the inputs. Like in the case of controllability, this result can be generalized to the multi-variable case for which observability matrix $\mathbf{Q}_0$ is of size $nl \times n$.

The observability test becomes:

$$\text{rank}\begin{bmatrix} \mathbf{C} \\ \mathbf{CA} \\ \mathbf{CA}^2 \\ \vdots \\ \vdots \\ \mathbf{CA}^{n-1} \end{bmatrix}_{nl \times n} = n$$

2.5.3. *Case of partially observable systems*

When matrix $\mathbf{Q_0}$ is not of full rank, the system is not entirely controllable and only a part of the state vector is reconstructible from the observations. If rank $\mathbf{Q_0} = r < n$, we can show (as we did for the canonical form for controllability) that there is a basis of the state space where the system admits the following canonical representation:

$$\dot{\tilde{\mathbf{x}}} = \begin{bmatrix} \tilde{\mathbf{A}}_{11} & 0 \\ \tilde{\mathbf{A}}_{21} & \tilde{\mathbf{A}}_{22} \end{bmatrix} \tilde{\mathbf{x}}(t) + \begin{bmatrix} \tilde{\mathbf{B}}_1 \\ \tilde{\mathbf{B}}_2 \end{bmatrix} \mathbf{u}(t)$$

$$\mathbf{y}(t) = \begin{bmatrix} \tilde{\mathbf{C}}_1 & 0 \end{bmatrix} \tilde{\mathbf{x}}(t) + \mathbf{D}\,\mathbf{u}(t)$$

However, the basis change matrix $\mathbf{T}$ $(\mathbf{x}(t) = \mathbf{T}\,\tilde{\mathbf{x}}(t))$ is obtained differently; we build $\mathbf{T}^{-1}$ (and not $\mathbf{T}$) into two blocks:

$$\mathbf{T}^{-1} = \begin{bmatrix} (\mathbf{T}^{-1})_1 \\ (\mathbf{T}^{-1})_2 \end{bmatrix}$$

The first block $(\mathbf{T}^{-1})_1$ is obtained by linear combinations of independent rows of matrix $\mathbf{Q_0}$ and completed by block $(\mathbf{T}^{-1})_2$ so that the ensemble is a regular matrix.

Once this matrix $\mathbf{T}^{-1}$ is calculated, we determine the matrices in the new basis like we did above, for the canonical form for controllability:

$$\widetilde{\mathbf{A}} = \mathbf{T}^{-1}\mathbf{A}\mathbf{T} \quad \widetilde{\mathbf{B}} = \mathbf{T}^{-1}\mathbf{B} \quad \widetilde{\mathbf{C}} = \mathbf{C}\mathbf{T}$$

The canonical form thus obtained entails a split of the state vector into two parties:

- $\widetilde{\mathbf{x}}_1(t)$ of size r which represents the entirely observable part of $\widetilde{\mathbf{x}}(t)$;

- $\widetilde{\mathbf{x}}_2(t)$ of size $n-r$ which represents the non-observable part of $\widetilde{\mathbf{x}}(t)$.

As shown in the equations developed below (by ignoring the term of control), $\mathbf{y}(t) = \widetilde{\mathbf{C}}_1\,\widetilde{\mathbf{x}}_1(t)$:

- $\dot{\widetilde{\mathbf{x}}}_1(t) = \widetilde{\mathbf{A}}_{11}\,\widetilde{\mathbf{x}}_1(t)$ entirely observable part (size r);

- $\dot{\widetilde{\mathbf{x}}}_2(t) = \widetilde{\mathbf{A}}_{21}\,\widetilde{\mathbf{x}}_1(t) + \widetilde{\mathbf{A}}_{22}\,\widetilde{\mathbf{x}}_2(t)$ non-observable part (size $n-r$).

2.5.4. *Case of partially controllable and partially observable systems*

If rank $\mathbf{Q_G} = q < n$ and rank $\mathbf{Q_0} = r < n$, we show that there is a basis of the state space where the system admits the following minimum canonical representation:

- $\widetilde{\mathbf{x}}_1(t)$ represents the controllable , non-observable part of $\widetilde{\mathbf{x}}(t)$;

- $\widetilde{\mathbf{x}}_2(t)$ represents the controllable, observable part of $\widetilde{\mathbf{x}}(t)$;

- $\widetilde{\mathbf{x}}_3(t)$ represents the non-controllable, non-observable part of $\widetilde{\mathbf{x}}(t)$;

- $\widetilde{\mathbf{x}}_4(t)$ represents the non-controllable, observable part of $\widetilde{\mathbf{x}}(t)$.

$$\dot{\widetilde{\mathbf{x}}} = \begin{bmatrix} \widetilde{\mathbf{A}}_{11} & \widetilde{\mathbf{A}}_{12} & \widetilde{\mathbf{A}}_{13} & \widetilde{\mathbf{A}}_{14} \\ 0 & \widetilde{\mathbf{A}}_{22} & 0 & \widetilde{\mathbf{A}}_{24} \\ 0 & 0 & \widetilde{\mathbf{A}}_{33} & \widetilde{\mathbf{A}}_{34} \\ 0 & 0 & 0 & \widetilde{\mathbf{A}}_{44} \end{bmatrix} \widetilde{\mathbf{x}}(t) + \begin{bmatrix} \widetilde{\mathbf{B}}_1 \\ \widetilde{\mathbf{B}}_2 \\ 0 \\ 0 \end{bmatrix} \mathbf{u}(t)$$

$$\mathbf{y}(t) = \begin{bmatrix} 0 & \widetilde{\mathbf{C}}_2 & 0 & \widetilde{\mathbf{C}}_4 \end{bmatrix} \widetilde{\mathbf{x}}(t) + \mathbf{D}\,\mathbf{u}(t)$$

2.6. Bibliography

[AND 90] ANDERSON B.D.O., MOORE J.B., *Linear Optimal Control*, Prentice-Hall, 1990.

[AZZ 88] D'ÁZZO J.J., HOUPIS H., *Linear Control System. Analysis and Design. Conventional and Modern*, McGraw-Hill, 1988.

[DOR 95] DORATO P., ABDALLAH C., CERONE V., *Linear-Quadratic Control. An Introduction*, Prentice-Hall, 1995.

[FAU 84] FAURRE P., ROBIN M., *Eléments d'Automatique*, Dunod, Paris, 1984.

[FRA 94] FRANKLIN G.F., POWELL J.D., EMANI-NAEINI A., *Feedback Control of Dynamic Systems*, Addison-Wesley Publishing Co., 1994.

[FRI 86] FRIEDLAND B., *Control Systems Design*, McGraw-Hill, 1986.

[KAI 80] KAILATH T., *Linear Systems*, Prentice-Hall, 1980.

[LAR 93] DE LARMINAT P., *Commande des systèmes linéaires*, Hermes, Paris, 1993.

Chapter 3

Discrete-Time Systems

3.1. Introduction

Generally, a signal is a function (or distribution) with support in the time space T, and with value in the vector space E, which is defined on R. Depending on whether we have a continuous-time signal or a discrete-time signal, the time space can be identified with the set of real numbers R or with the set of integers of Z. A discrete system is a system which transforms a discrete signal, noted by u, into a discrete signal noted by y. The class of systems studied in this chapter is the class of time-invariant and linear discrete (DLTI) systems. Such systems can be described by the recurrent equations [3.1] or [3.2][1]:

$$\begin{cases} x(k+1) = Ax(k) + Bu(k) \\ \ \ y(k) = Cx(k) + Du(k) \end{cases} \tag{3.1}$$

$$y(k) + \cdots a_n y(k-n) = b_0 u(k) + \cdots b_n u(k-n) \tag{3.2}$$

where signals u, x and y are sequences with support in Z ($k \in Z$) and with value in R^m, R^n and R^p respectively. They represent the input, the state and the output of the system (see the notations used in Chapters 2 and 3). A, B, C, D, a_i, b_i are appropriate size matrices with coefficients in R:

Chapter written by Philippe CHEVREL.

[1] We can show the equivalence of these two types of representations (see Chapter 2).

$$A \in R^{nxn}, B \in R^{nxm}, C \in R^{pxn}, D \in R^{pxm}, a_i \in R^{pxp}, b_i \in R^{pxm} \qquad [3.3]$$

If equations [3.1] and [3.2] can represent intrinsically discrete systems, such as a μ-processor or certain economic systems, they are, most often, the result of discretization of continuous processes. In fact, let us consider the block diagram of an automated process, through a computer control (see Figure 3.1). Seen from the computer, the process to control, which is supplied with its upstream digital-analog and downstream analog-digital converters (ADC), is a *discrete system* that converts the discrete signal u into a discrete signal y. This explains the importance of the discrete system theory and its development, which is parallel to the development of digital μ-computers.

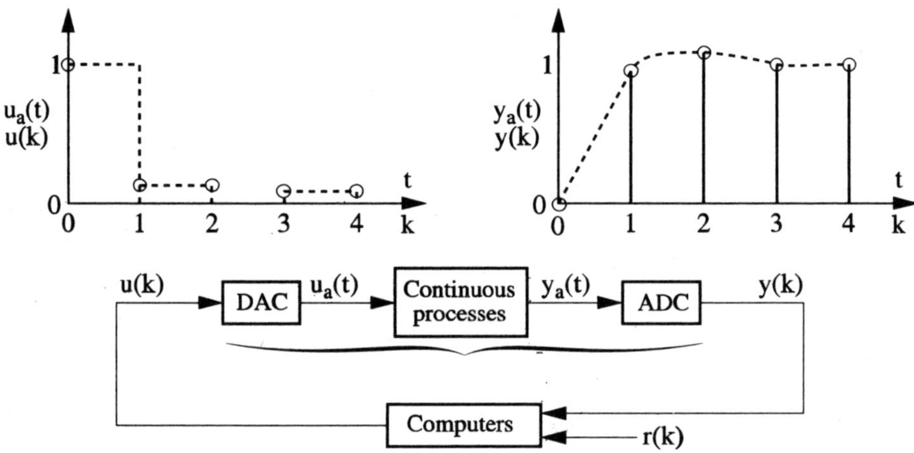

Figure 3.1. *Computer control*

This chapter consists of three distinct parts. The analysis and manipulation of signals and discrete-time systems are presented in sections 3.2 and 3.3. The discretization of continuous-time systems and certain concepts of the sampling theory are dealt with in section 3.4.

3.2. Discrete signals: analysis and manipulation

3.2.1. *Representation of a discrete signal*

A discrete-time signal[2] is a function $x(.)$ with support in $T = Z$ and with value in $E = R^n$, $n \in N$.

We will talk of a scalar signal if $n = 1$, of a vector signal in the contrary case and of a causal signal if $x(k) = 0$, $\forall k \in Z^-$. Only causal signals will be considered in what follows. There are several ways to describe them: either explicitly, through an analytic expression (or by tabulation), like in the case of elementary signals defined by equations [3.4] to [3.6], or, implicitly, as a solution of a recurrent equation (see equation [3.7]):

$$\text{Discrete impulse[3]:} \qquad \delta(k) = \begin{cases} 1 & \text{if } k = 0 \\ 0 & \text{if } k \in Z^* \end{cases} \qquad [3.4]$$

$$\text{Unit-step function:} \qquad \Gamma(k) = \begin{cases} 1 & \text{if } k \in Z^+ \\ 0 & \text{if } k \in Z^{-*} \end{cases} \qquad [3.5]$$

$$\text{Geometrical sequence:} \qquad g(k) = \begin{cases} a^k & \text{if } k \in Z^{+*} \\ 0 & \text{if } k \in Z^{-*} \end{cases} \qquad [3.6]$$

It will be easily verified that the solution of equation [3.7] is the geometrical sequence [3.6] previously defined. Hence, the geometrical sequence has, for discrete-time signals, a role similar to the role of the exponential function for continuous-time signals.

$$\text{First order recurrent equation:} \quad \begin{cases} x(k+1) = ax(k) \\ \quad x(0) = 1 \end{cases} \qquad [3.7]$$

---

2 Unlike a continuous-time signal, which is a function with real number support ($T = R$).
3 We note that if the continuous-time impulse or Dirac impulse is defined only in the distribution sense, it goes differently for the discrete impulse.

3.2.2. *Delay and lead operators*

The concept of an operator is interesting because it enables a compact formulation of the description of signals and systems. The manipulation of difference equations especially leads back to a purely algebraic problem.

We will call "operator" the formal tool that makes it possible to univocally associate with any signal $x(\cdot)$ with support in T another signal $y(\cdot)$, itself with support in T. As an example we can mention the "lead" operator, noted by q [AST 84]. Defined by equation [3.8], it has a role similar to that of the "derived" operator for continuous-time signals. The delay operator is noted by q^{-1} for obvious reasons (identity operator: $1 \stackrel{\Delta}{=} q \circ q^{-1}$).

$$x: T \to E \atop k \to x(k) \quad \Leftrightarrow \quad qx: T \to E \atop k \to x(k+1) \qquad [3.8] \qquad x: T \to E \atop k \to x(k) \quad \Leftrightarrow \quad q^{-1}x: T \to E \atop k \to x(k-1)$$

Table 3.1. *Backwards-forwards shift operators*

Any operator f is called *linear* if and only if it converts the entire sequence $x_1(k) + \lambda x_2(k)$, $\lambda \in R$ into the sequence $y_1(k) + \lambda y_2(k)$ with $y_1 \stackrel{\Delta}{=} f(x_1)$ and $y_2 \stackrel{\Delta}{=} f(x_2)$.

It is called *stationary* if it converts any entire delayed or advanced sequence $x(k-r)$, $r \in Z$ into the sequence $y(k-r)$, with $y \stackrel{\Delta}{=} f(x)$ (formally, $f(q^{-r}x) = q^{-r} f(x)$).

The *gain* of the operator is induced by the standard used in the space of the signals considered (for example, L_2 or L_∞). The gain of the lead operator is unitary.

These definitions will be useful in section 3.3. Except for the lead operator, operator $\delta_T \stackrel{\Delta}{=} \dfrac{1-q^{-1}}{T}$ and operator $w \stackrel{\Delta}{=} (q+1)^{-1} \circ (q-1)$ will be used sometimes.

3.2.3. *z-transform*

3.2.3.1. *Definition*

The *z-transform* represents one of the main tools for the analysis of signals and discrete systems. It is the discrete-time counterpart of the Laplace transform. The *z-transform* of the sequence $\{x(k)\}$, noted by $X(z)$, is the bound, when it exists, of the sequence: $X(z) \overset{\Delta}{=} \sum\limits_{-\infty}^{\infty} x(k)z^{-k}$ where z is a variable belonging to the complex plan.

For a causal signal, the *z-transform* is given by [3.9] and we can define the convergence radius R of the sequence (the sequence is assumed to be entirely convergent for $|z| > \mathrm{R}$).

$$X(z) \overset{\Delta}{=} Z\{x(k)\} \overset{\Delta}{=} \sum\limits_{0}^{\infty} x(k)z^{-1} \qquad |z| > \mathrm{R} \tag{3.9}$$

$X(z)$ is the function that generates the numeric sequence $\{x(k)\}$. We will easily prove the results of Table 3.2.

$x(k)$	$X(z)$	R	$x(k)$	$X(z)$	R
$\delta(k)$	1	∞	$a^k \Gamma(k)$	$\dfrac{z}{z-a}$	$\lvert a\rvert$
$\Gamma(k)$	$\dfrac{z}{z-1}$	1	$a^k \sin(\omega k)\Gamma(k)$	$\dfrac{z\sin\omega}{z^2 - (2a\cos\omega)z + a^2}$	$\lvert a\rvert$
$k\,\Gamma(k)$	$\dfrac{z}{(z-1)^2}$	1	$a^k \cos(\omega k)\Gamma(k)$	$\dfrac{z^2 - z\cos\omega}{z^2 - (2a\cos\omega)z + a^2}$	$\lvert a\rvert$

Table 3.2. *Table of transforms*

3.2.3.2. *Inverse transform*

The inverse transform of $X(z)$, which is a rational fraction in z, can be obtained for the simple forms by simply reading through the table. In more complicated cases, a previous decomposition into simple elements is necessary. We can also calculate the sequence development of $X(z)$ by polynomial division according to the decreasing powers of z^{-1} or apply the method of deviations, starting from the definition of the inverse transform:

$$x(k) = Z^{-1}(X(z)) = \frac{1}{2\pi j} \int_C X(z) z^k dz \qquad [3.10]$$

where C is a circle centered on 0 including the poles of $X(z)$.

3.2.3.3. *Properties of the z-transform*[4]

We will also show, with no difficulties (as an exercise), the various properties of the *z-transform* that can be found below. The convergence rays of the different sequences are mentioned. We note by R_x the convergence ray of the sequence associated with the causal sequence $x(k)$.

P1: *z-transform* is linear ($R_{ax+by} = \max(R_x, R_y)$)

$$Z(\{ax(k) + by(k)\}) = aX(z) + bY(z), \quad \forall a, b \in R$$

P2: delay theorem ($R_{q^{-r}x} = R_x$)

$$Z(\{q^{-r}x(k)\}) = z^{-r}X(z), \quad \forall r \in Z^+$$

P3: lead theorem ($R_{q^n x} = R_x$)

$$Z(\{q^n x(k)\}) = z^n \left[X(z) - \sum_{0}^{n-1} x(k)z^{-k} \right], \qquad \forall n \in Z^+$$

In particular: $Z(\{x(k+1)\}) = zX(z) - x(0)$

P4: initial value theorem

If $x(k)$ has $X(z)$ as a transform and if $\lim_{z \to \infty} X(z)$ exists, then:

$$x(0) = \lim_{z \to \infty} X(z)$$

---

4 Note: the various manipulated signals are assumed to be causal.

P5: final value theorem

$$\text{If } \lim_{k \to \infty} x(k) \text{ exists, then: } \lim_{k \to \infty} x(k) = \lim_{z \to 1} (1 - z^{-1}) X(z)$$

P6: discrete convolution theorem ($R_{x_1 * x_2} = \max(R_{x_1}, R_{x_2})$)

Let us consider two causal signals $x_1(k)$ and $x_2(k)$ and their convolution integral

$$x_1 * x_2(n) = \sum_{k=-\infty}^{+\infty} x_1(n-k) x_2(k) = \sum_{k=0}^{n} x_1(n-k) x_2(k).$$ We have:

$$Z(\{x_1 * x_2(n)\}) = X_1(z) \ X_2(z)$$

P7: multiplication by k ($R_{kx} = R_x$)

$$Z(\{kx(k)\}) = -z \frac{dX(z)}{dz}$$

P8: multiplication by a^k ($R_{a^k x} = |a| R_x$)

$$Z(\{a^k x(k)\}) = X(a^{-1} z)$$

3.2.3.4. *Relations between the Fourier-Laplace transforms and the z-transform*

The aim of this section is not to describe in detail the theory pertaining to the Fourier transform. More information on this theory can be found in [ROU 92]. Only the definitions are mentioned here, that enable us to make the comparison between the various transforms.

	Continuous signal: $x_a(t)$	**Discrete signal: $x(k)$**
Fourier transform	$X_F(\Omega) = \int_{-\infty}^{\infty} x_a(t) e^{-j\Omega t} dt$	$X_F(\omega) = \sum_{k=-\infty}^{+\infty} x(k) e^{-j\omega k}$
Laplace transform/ z-transform	$X_a(p) = \int_{-\infty}^{\infty} x_a(t) e^{-pt} dt$ $p \in C$	$X(z) = \sum_{k=-\infty}^{+\infty} x(k) z^{-k}$ $z \in C$

Table 3.3. *Synthesis of the various transforms*

Hence, if we suppose that $X(z)$ exists for $z = e^{j\omega}$, the signal discrete Fourier transform $x(k)$ is given by $X_F(\omega) = X(e^{j\omega})$, whereas in the continuous case, ω is a homogenous impulse at a time inverse, the *discrete impulse* ω_d (also called reduced impulse) is adimensional. The relations between the two transforms will become more obvious in section 3.4 where the discrete signal is obtained through the sampling of the continuous signal.

3.3. Discrete systems (DLTI)

A *discrete system* is a system that converts an incoming data sequence $u(k)$ into an outgoing sequence $y(k)$. Formally, we can assign an operator f that transforms the signal u into a signal y ($y(k) = f(u)(k)$, $\forall k \in Z$). The system is called *linear* if the operator assigned is linear. It is *stationary* or *time-invariant* if f is stationary (see section 3.2). It is *causal* if the output at instant $k = n$ depends only on the inputs at previous instants $k \le n$. It is called BIBO-*stable* if for any bound-input corresponds a bound-output and this, irrespective of the initial conditions. Formally: ($\sup_{k} u(k) < \infty \Rightarrow \sup_{k} (fu)(k) < \infty$). In this chapter we will consider only time-invariant linear discrete systems. Different types of representations can be envisaged.

3.3.1. *External representation*

The representation of a system with the help of relations between its only inputs and outputs is called *external*.

3.3.1.1. *Systems defined by a difference equation*

Discrete systems can be described by difference equations, which, for a DLTI system, have the form:

$$y(k) + \cdots a_n y(k-n) = b_0 u(k) + \cdots b_n u(k-n) \tag{3.11}$$

We will verify, without difficulty, that such a system is linear and time-invariant (see the definition below). The coefficient in $y(k)$ is chosen as unitary in order to ensure for the system the property of *causality* (only the past and present inputs affect the output at instant k). The *order* of the system is the order of the difference equation, i.e. the number of past output samples necessary for the calculation of the present output sample. From the initial conditions $y(-1), \cdots, y(-n)$, it is easy to recursively calculate the output of the system at instant k.

3.3.1.2. *Representation using the impulse response*

Any signal $u(\cdot)$ can be decomposed into a sum of impulses suitably weighted and shifted:

$$u(k) = \sum_{i=-\infty}^{\infty} u(i)\,\delta(k-i)$$

On the other hand, let $h(\cdot)$ be the signal that represents the impulse response of the system (formally: $h = f(\delta)$). The response of the system to signal $q^{-i}\delta$ is $q^{-i}h$ due to the property of stationarity. Hence, linearity leads to the following relation:

$$y(k) = \sum_{i=-\infty}^{\infty} u(i)\,h(k-i) = \sum_{i=-\infty}^{\infty} h(i)\,u(k-i) = h * u(k) \tag{3.12}$$

The output of the system is expressed thus as the convolution integral of the impulse response h and of the input signal u. We can easily show that the system is causal if and only if $h(k) = 0,\ \forall k < 0$. In addition, it is BIBO-stable if and only if

$$\sum_{i=0}^{\infty} |h(i)| < \infty.$$

3.3.2. *Internal representation*

In section 3.3.1.1 we saw that a difference equation of order n would require n initial conditions in order to be resolved. In other words, these initial conditions characterize the initial state of the system. In general, the instantaneous state $x(k) \in R^n$ sums up the past of the system and makes it possible to predict its future. From the point of view of simulation, the size of $x(k)$ is also the number of variables to memorize for each iteration. Based on the recurrent equation [3.11], the state vector can be constituted from the past input and output samples. For example, let us define the ith component of $x(k)$, $x_i(k)$, through the relation:

$$x_i(k) = \sum_{j=i}^{n} [b_j u(k-j+i-1) - a_j y(k-j+i-1)] \tag{3.13}$$

Then we verify that the state vector satisfies the recurrent relation of first order [3.14a] called equation of state and that the system output is obtained from the observation equation [3.14b]:

$$x(k+1) = \underbrace{\begin{pmatrix} -a_1 & 1 & 0 & 0 \\ \vdots & 0 & \ddots & 0 \\ \vdots & 0 & 0 & 1 \\ -a_n & 0 & \cdots & 0 \end{pmatrix}}_{A} x(k) + \underbrace{\begin{pmatrix} b_1 \\ \vdots \\ \vdots \\ b_n \end{pmatrix}}_{B} u(k) \qquad [3.14a]$$

$$y(k) = \underbrace{(1 \quad 0 \quad \cdots \quad 0)}_{C} x(k) + \underbrace{b_0}_{D} u(k) \qquad [3.14b]$$

We note that the iterative calculation of $y(k)$ requires only the initial state $x(0) \overset{\Delta}{=} x_0$ (obtained according to [3.13] from C.I. $\{y(-1), \cdots, y(-n)\}$) and the past input samples $\{u(i),\ 0 \le i < k\}$. As in the continuous case, this state representation[5] is defined only for a basis change and the choice of its parameterization is not without incidence on the number of calculations to perform. In addition, the characterization of structural properties introduced in the context of continuous-time systems (see Chapters 2 and 4), such as controllability or observability, are valid here.

The evolution of the system output according to the input applied and initial conditions is simply obtained by solving [3.14]:

$$y(k) = \underbrace{CA^k x_0}_{y_l(k)} + \underbrace{\sum_{i=0}^{k-1} CA^{k-1-i} Bu(i) + Du(k)}_{y_f(k)} \qquad [3.15]$$

$y_l(k)$ and $y_f(k)$ designate respectively the free response and the forced response of the system. Unlike the continuous case, the solution involves a sum, and not an integration, of powers of A and not a matrix exponential function. Each component $x_i(k)$ of the free response can be expressed as a linear combination of terms, such as $\rho_i(k)\ \lambda_i^k$, where $\rho_i(\cdot)$ is a polynomial of an order equal to $n_i - 1$, where n_i is the multiplicity order of λ_i and i[th] the eigenvalue of A.

---

5 A canonical form called *controllable companion*.

Based on the previous definitions, the system is necessarily BIBO-stable if the spectral ray of A, $\rho(A)$, is lower than the unit (i.e. all values of A are included in the unit disc). The other way round is true only if the triplet (A,B,C) is controllable and observable, i.e. if the realization considered is minimal. If $\rho(A)$ is strictly lower than 1, the system is called asymptotically stable, i.e. it verifies the following property:

$$u(\cdot) \equiv 0 \quad \Rightarrow \quad \lim_{k \to +\infty} \left\| x(k) \right\| = 0, \quad \forall x_0$$

Another way to verify that this property is satisfied is to use Lyapunov's theory for the discrete-time systems. The next result is close to the result for continuous-time systems in Chapter 2.

THEOREM 3.1.– *the system described by the recurrence* $x(k+1) = Ax(k)$, $x(0) = x_0$ *is asymptotically stable if and only if:*

$$\exists Q = Q^T > 0 \ \text{ and } \ \exists P = P^T > 0 \ \text{ solution of equation}^6\text{: } A^T PA - P = Q$$

3.3.3. *Representation in terms of operator*

The description and manipulation of systems as well as passing from one type of representation to another can be standardized in a compact manner by using the concept of operator introduced in section 3.2.2.

Let us suppose, in order to simplify, that signals u and y as causal. Hence, we will be interested only in the evolution of the system starting from zero initial conditions. In this case we can identify the manipulations on the systems to operations in the body of rational fractions whose variable is an operator. The lead operator q and the mutual operator q^{-1} are natural and hence very dispersed. Starting, successively, from representations [3.11], [3.12] and [3.15], we obtain expressions [3.16], [3.17] and [3.18] of operator $H(q)$ which are characteristic for system $y(k) = H(q)u(k)$):

$$H(q) = \frac{b_0 + \cdots + b_n q^{-n}}{1 + a_1 + \cdots + a_n q^{-n}} = \frac{b_0 q^n + \cdots + b_n}{q^n + a_1 q^{n-1} + \cdots + a_n} \tag{3.16}$$

---

6 Called a discrete-time Lyupanov equation. N.B.: $Q > 0 \Leftrightarrow Q$ is defined positive.

$$H(q) = \sum_{i=-\infty}^{+\infty} h(i)q^{-i} = \sum_{i=0}^{+\infty} h(i)q^{-i} \qquad [3.17]$$

$$H(q) = C(qI - A)^{-1}B + D = \sum_{i=1}^{+\infty} CA^{i-1}Bq^{-i} + D \qquad [3.18]\,[7]$$

The relation between the various representations is thus clarified:

$$h(0) = b_0 = D, \quad h(1) = -a_1 b_0 + b_1 = CB, \quad \cdots \quad h(k) = CA^{k-1}B \qquad [3.19]$$

The use of this formalism makes it possible to reduce the serialization or parallelization of two systems to an algebraic manipulation on the associated operators. This property is illustrated in Figure 3.2. In general, we can define an algebra of diagrams which makes it possible to reduce the complexity of a defined system from interconnected sub-systems.

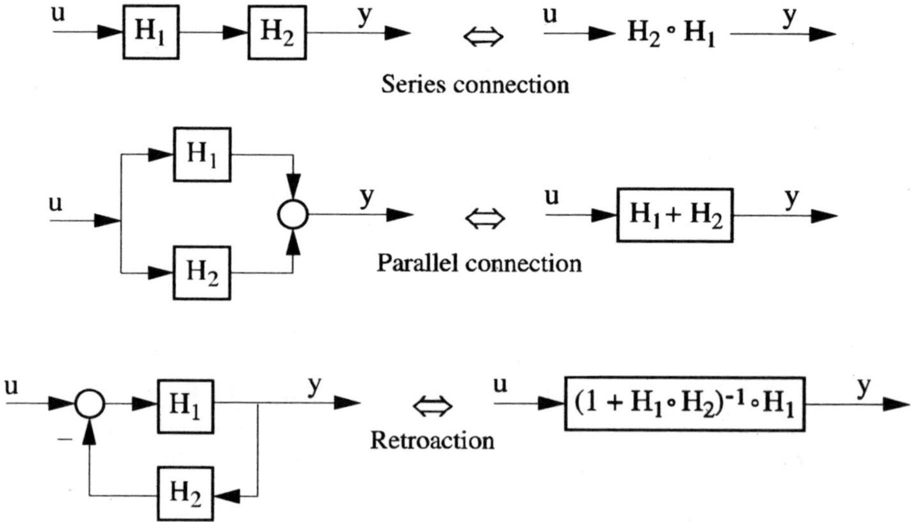

Figure 3.2. *Interconnected systems*

---

7 The system is causal.

NOTE 3.1.– acknowledging the initial conditions, which is natural in the state formalism and more suitable to the requirements of the control engineer, makes certain algebraic manipulations illicit. This point is not detailed here but for further details see [BOU 94, QUA 99].

THEOREM 3.2.– *a rational SLDI, i.e. that can be described by [3.16], is BIBO-stable if and only if one of the following propositions is verified*:

   – *the poles of the reduced form of H(q) are in modules strictly less than 1;*

   – *the sequence h(k) is completely convergent;*

   – *the state matrix A of any minimal realization of H(q) has all its values strictly less than 1 in module.*

These propositions can be extended to the case of a multi-input/multi-output DLTI system (see [CHE 99]).

NOTE 3.2.– the Jury criterion (1962) [JUR 64] makes it possible to obtain the stability of the system [3.16] without explicitly calculating the poles of $H(q)$, by simple examination of coefficients $a_1, a_2, \cdots, a_n$ with the help of Table 3.4.

$$
\begin{array}{ccccc}
a_0 & a_1 & a_2 & \cdots & a_n \\
a_n & a_{n-1} & & & a_0 \\
b_0 & b_1 & & & \\
b_{n-1} & b_{n-2} & & & \\
c_0 & c_1 & & & \\
c_{n-1} & & & &
\end{array}
\qquad
\text{with: } b_0 = \frac{a_0^2 - a_n^2}{a_0}
\qquad
b_1 = \frac{a_1 a_0 - a_{n-1} a_n}{a_0}
$$

$$
b_k = \frac{a_k a_0 - a_n a_{n-k}}{a_0}
$$

$$
c_k = \frac{b_k b_0 - b_{n-1} b_{n-1-k}}{b_0}
$$

Table 3.4. *Jury table*

The first row is the simple copy of coefficients of $a_1, a_2, \cdots, a_n$, the second row reiterates these coefficients inversely, the third row is obtained from the first two by calculating in turns the determinant formed by columns 1 and n, 2 and n, etc. (see expression of b_k), the fourth row reiterates the coefficients of the third row in inverse order, etc. The system is stable if and only if the first coefficients of the odd rows of the table $(a_0, b_0, c_0,$ etc.) are all strictly positive.

NOTE 3.3.– the class of rational systems that can be described by [3.16] or [3.18] is a sub-class of DLTI systems. To be certain of this, let us consider the system characterized by the irrational transfer: $H(q) = \ln(1+q^{-1})$. This DLTI system, whose impulse response is zero in 0 and such that $h(k) = \frac{1}{k}$ for $k \in Z^{+}$ cannot be descibed by [3.16] or [3.18].

The use of lead and delay operators is not universal. Certain motivations that will be mentioned in section 3.4 will lead to sometimes prefer other operators [GEV 93, MID 90].

Use of operator δ_{τ}

Operator $\delta_{\tau} \triangleq \dfrac{q-1}{\tau}$, $\tau \in R$ [MID 90] represents an interesting alternative to the lead operator. It is easy to pass from parameterized transfer $H(q)$ by coefficients $\{a_i, \ b_i, \ i \in \{1, \cdots n\}\}$ to parameterized transfer $H_{\delta}(\delta)$ by coefficients $\{a_{\delta_i}, \ b_{\delta_i}, \ i \in \{1, \cdots n\}\}$. Then we can work exclusively with this operator and obtain, by analogy with [3.14], a realization in the state space of the form:

$$\delta_{\tau}x(k) = \underbrace{\begin{pmatrix} -a_{\delta_1} & 1 & 0 & 0 \\ \vdots & 0 & \ddots & 0 \\ \vdots & 0 & 0 & 1 \\ -a_{\delta_n} & 0 & \cdots & 0 \end{pmatrix}}_{A_{\delta}} x(k) + \underbrace{\begin{pmatrix} b_{\delta_1} \\ \vdots \\ \vdots \\ b_{\delta_n} \end{pmatrix}}_{B_{\delta}} u(k) \qquad [3.20a]$$

$$y(k) = \underbrace{\begin{pmatrix} 1 & 0 & \cdots & 0 \end{pmatrix}}_{C_{\delta}} x(k) + \underbrace{b_{\delta_0}}_{D_{\delta}} u(k) \qquad [3.20b]$$

However, the simulation of this system requires a supplementary stage consisting of calculating at each iteration $x(k+1) = x(k) + \tau \, \delta \, x(k)$. Finally, from the point of view of simulation, the parameterization of the system according to matrices $A_{\delta}, B_{\delta}, C_{\delta}, D_{\delta}$ differs from the usual parameterization only by the addition of the intermediary variable $\delta x(k)$ in the calculations. We easily reciprocally pass to the representation in q by writing:

$$\begin{cases} qx(k) = (I + \tau A_{\delta})x(k) + \tau B_{\delta}u(k) \\ \quad y(k) = C_{\delta}x(k) + D_{\delta}u(k) \end{cases} \qquad [3.21]$$

Hence, we have the equivalences:

$$A \leftrightarrow \left(I + \tau A_\delta\right) \qquad\qquad B \leftrightarrow \tau B_\delta$$
$$C \leftrightarrow C_\delta \qquad\qquad D \leftrightarrow D_\delta \qquad\qquad [3.22]$$

$$H_q(q) \leftrightarrow H_\delta(\delta) \overset{\Delta}{=} C_\delta\left(\delta I - A_\delta\right)^{-1} B_\delta + D_\delta = \sum_{i=1}^{+\infty} C_\delta A_\delta^{\,i-1} B_\delta \delta^{-i} + D_\delta$$

Combined use of operators γ and δ_τ

We use, this time together, operator δ_τ, which was previously defined, and operator $\gamma = \dfrac{q+1}{2}$ in order to describe the recurrence:

$$\begin{cases} \delta_\tau x(k) = A_w \gamma x(k) + B_w u(k) \\ y(k) = C_w x(k) + D_w u(k) \end{cases} \qquad\qquad [3.23]$$

where matrices[8] A_w, B_w, C_w, D_w are linked to matrices A, B, C, D of the q representation based on equation [3.24]. We note that the condition of reversibility of matrix $(A + I)$ is required and that this condition is always satisfied if the discretized system is the result of the discretization of a continuous system (see section 3.4.3).

$$A_w = \frac{2}{\tau}(A - I)(A + I)^{-1} \qquad\qquad B_w = \frac{2}{\tau}(A + I)^{-1} B \qquad [3.24]$$
$$C_w = C \qquad\qquad\qquad D_w = D$$

Representations [3.20] and [3.23] have certain advantages over the q representation that will be presented in section 3.4.6. We should underline from now that the "$\gamma - \delta_\tau$" representation makes it possible to unify many results of the theory of systems traditionally obtained through different paths, depending on whether we deal with continuous or discrete-time systems [RAB 00]. In particular, the theorem of stability (see Theorem 3.1) is expressed as in continuous-time.

THEOREM 3.3.– *the system described by the recurrence $\delta_\tau x(k) = A_w \gamma\, x(k), \gamma x(0) = z_0$ is asymptotically stable if and only if $\exists Q = Q^T > 0$ and $\exists P = P^T > 0$ solution of equation[9]: $A_w^T P + P A_w = -Q$.*

---

8 Index w is used here in order to establish the relation with the W transform [FRA 92] and the operator: $w_\tau = \gamma^{-1} \circ \delta_\tau$ [RAB 00].

9 We recognize here a Lyapunov continuous-time equation.

3.3.4. *Transfer function and frequency response*

Let us consider a stable system defined through its impulse response $h(.)$ or by operator $H(q) = \sum\limits_{i=0}^{+\infty} h(i)q^{-i}$. We assume it is excited through the sinusoidal input of reduced impulse ω_d : $u(k) = e^{jk\omega_d}$.

The output is obtained by applying the theorem of discrete convolution:

$$y(k) = \sum_{l=0}^{+\infty} h(l)e^{j(k-l)\omega_d} = \underbrace{\left[\sum_{l=0}^{+\infty} h(l)e^{-jl\omega_d}\right]}_{H(e^{j\omega_d})} e^{jk\omega d}\underline{\Delta}\kappa(\omega_d)e^{j(k\omega_d+\varphi(\omega_d))}$$

Hence, the output of a DLTI system excited by a sinusoidal input of impulse ω_d is a sinusoidal signal of the same impulse. It is amplified by factor $\kappa(\omega_d) = \left|H(e^{j\omega_d})\right|$ and phase shifted of angle $\varphi(\omega_d) = \arg H(e^{j\omega_d})$. Very naturally, its static gain[10] g_{sta} is obtained for $\omega_d = 0$ or in the same way $z = 1$: $g_{sta} = \kappa(0) = H(1)$. We note that $H(e^{j\omega_d})$ is the discrete Fourier transform of the impulse response of the system and it can be obtained (see Table 3.3) from its z-transform, $H(z)$ for $z = e^{j\omega_d}$. We will often talk of *transfer function* in order to arbitrarily designate $H(z)$ or $H(q)$. However, it is important to keep in mind that $H(q)$ is an operator, whereas $H(z)$ is a complex number.

The drawing of module and of phase of $H(e^{j\omega_d})$ according to ω_d represents the Bode diagram of discrete systems. For the discrete case, there are no simple asymptotic drawings, which largely limits its use from the point of view of the design. In addition, the periodicity of function $e^{j\omega_d}$, of period 2π , induces that of frequency response $H(e^{j\omega_d})$. This property should not seem mysterious. It simply results from the fact that $\{u_l(k) = e^{j(\omega_d+2\pi l)k} , l = 1, 2, \cdots\}$ represent in reality different ways of writing the same signal. We even speak of "alias" in this case. The response of the system to each of these aliases is thus identical. In addition, it is

---

10 Ratio between input and output in static state ($u(k) = 1 = e^{jk\omega_d}\Big|_{\omega_d=0}$, $\forall k \in Z$).

easily proven that module $\kappa(\omega_d) = \left| H(e^{j\omega_d}) \right|$ and phase $\varphi(\omega_d) = \arg H(e^{j\omega_d})$ are respectively even and odd functions of ω. The place of the Bode diagram is to draw (by using the PC in practice) only on the interval $[0, \pi]$. However, an approximate drawing can be obtained by applying the rules of asymptotic drawing presented in the context of continuous-time systems (see Chapter 1), by using a rational approximation of $z = e^{j\omega}$. We will use for example the W transform, $z \leftrightarrow \dfrac{1+w}{1-w}$, and we will draw $H\left(\dfrac{1+w}{1-w}\right)_{w = j\omega_d}$ in the place and instead of $H(e^{j\omega_d})$.

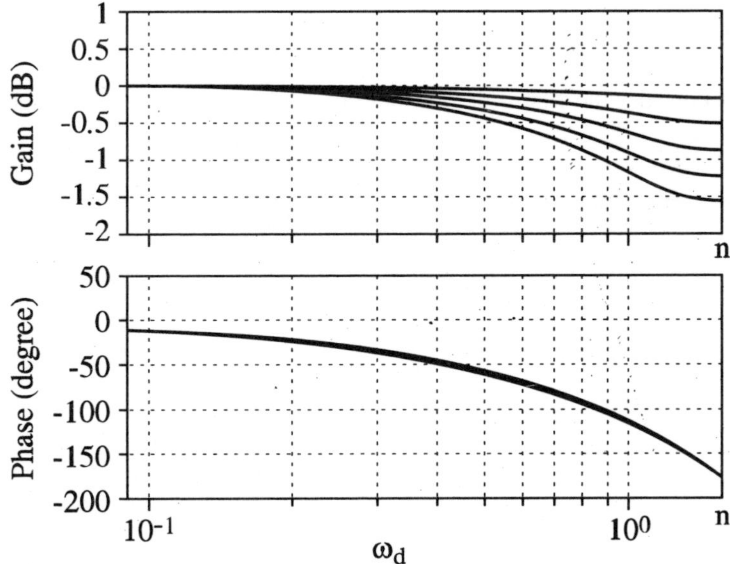

Figure 3.3. *Bode diagram*

Let us consider the case of a first order system given by its transfer function $H(z) = \dfrac{1-a}{z-a}$, $|a| < 1$. The bandwidth of the Bode diagram drawn in Figure 3.3 is more important if a is "small". This result can be linked to the time response of this same system studied in the next section.

In addition, the frequency response from the δ transfers can be written:

$$\kappa(\omega) = \left| H_\delta\left(\frac{e^{j\omega_d} - 1}{\tau}\right) \right| \quad \text{and} \quad \varphi(\omega_d) = \arg H_\delta\left(\frac{e^{j\omega_d} - 1}{\tau}\right).$$

3.3.5. *Time response of basic systems*

A DLTI system of an arbitrarily high order can be decomposed into serialization or parallelization of first and second order systems (see Chapter 1). Hence, it is interesting to outline the characteristics of these two basic systems.

3.3.5.1. *First order system*

Let us consider the first order system described by: $y(k) = \dfrac{b}{q - a} u(k)$. We can associate with it:

– the recurrent equation: $y(k + 1) = ay(k) + bu(k)$;

– the impulse response: $h(k) = ba^{k-1}$, $k \in \mathbb{N}^*$ and $h(0) = 0$;

– the unit-step response[11]: $y(k) = \dfrac{b}{1 - a}(1 - a^k)$, $k \in \mathbb{N}$.

3.3.5.2. *Second order system*

Let us consider the second order system described by: $y(k) = \underbrace{\dfrac{b_1 q + b_2}{q^2 + a_1 q + a_2}}_{H(q)} u(k)$.

We can associate with it the *recurrent equation*:

$$y(k + 2) = -a_1 y(k + 1) - a_2 y(k) + b_1 u(k + 1) + b_2 u(k)$$

---

11 It can be obtained either from the recurrent equation or by inverse z-transform of:

$$Y(z) = \frac{b}{z - a}\Gamma(z), \text{ with } \Gamma(z) = \frac{z}{z - 1} .$$

In addition, if we note by p_{d1} and by p_{d2} the poles and z_{d1} the zero of $H(q)$, we have, if the poles are *conjugated complex numbers* $p_{d1} = \rho e^{j\omega_d}$ and $p_{d1} = \rho e^{-j\omega_d}$, the *unit-step response*:

$$y(k) = H(1) + \alpha \rho^k \sin(\omega_d k + \varphi)$$

with: $\alpha = \dfrac{b_1}{\rho \sin \omega} \left(\dfrac{\rho^2 - 2 z_{d1} \rho \cos \omega + z_{d1}^2}{\rho^2 - 2\rho \cos \omega + 1} \right)^{\frac{1}{2}}$.

More generally, and according to the situation of the poles, there are various types of unit-step responses (see Figure 3.4), which are stable or unstable depending on whether the poles belong or not to the unit disc.

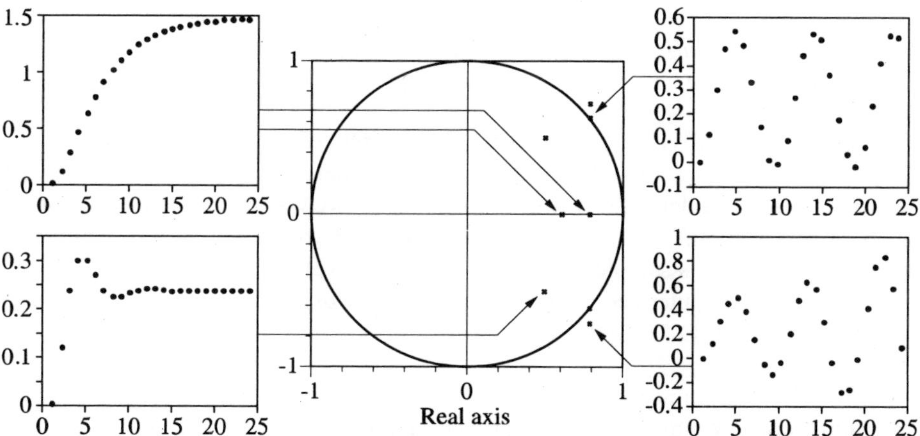

Figure 3.4. *Relation between the poles and the second order unit-step response*

3.4. Discretization of continuous-time systems

The diagram in Figure 3.1, process excluded, represents a typical chain of digital processing of the signal that traditionally proceeds in several stages.

The ADC periodically retrieves the values of the analog signal $y_a(t)$ at the instants $t_k = kT$. It returns the discrete signal $y(k)$ consisting of successive samples of $y_a(t)$. This sampling operation can be standardized by the identity: $y(k) = y_a(kT)$. In reality, the ADC also carries out the digitization of the signal (the digitized signal can have only a finite number of values). We will not discuss in what follows these quantification errors for simplicity reasons and we will note by E_T the *sampling operator*: $y = E_T(y_a)$.

The computer is the processing unit that obtains signal $u(k)$ from signals $y(k)$ and $r(k)$. The system is discrete.

The ADC converts the discrete time signal $u(k)$ into the analog signal $u_a(t)$. Several types of blockers can be considered but currently the most widespread is the 0 *order blocker*, to which we will associate *operator* B_0, that maintains constant the sample value for a sampling period (see Figure 3.1): $u_a(kT + \tau) = u(k)$, $\forall k \in Z$, $\forall \tau \in [0,T[$.

An alternative that will not be considered here consists of using a so-called "first order" blocker, operating a linear extrapolation of the input signal from the 2 last samples:

$$u_a(kT + \tau) = u(k) + \tau \frac{u(k) - u(k-1)}{T}, \quad \forall k \in Z, \quad \forall \tau \in [0,T[\qquad [3.25]$$

Another point of view is to consider the discretized process of input $u(k)$ and output $y(k)$. It is a discrete system whose model can be obtained from the continuous-time model. The way it is obtained as well as its properties are at the heart of this section. The study of conditions under which we can reconstitute an analog signal from its samples (sampling theory) as well as the analysis of problems specific to the computerized control will not be discussed here. However, for more information, see [PIC 77, ROU 82].

The ADC is supposed to retrieve samples periodically. We note by T the sampling period. We also consider that ADC and DAC are synchronized.

3.4.1. *Discretization of analog signals*

We have previously defined the sampling operation by $x = E_T(x_a)$ by making the continuous-time signal $x_a(t)$ correspond to the discrete-time signal $x(k)$. We can define the Laplace transform of $x_a(t)$ and the *z-transform* of $x(k)$ and pass directly from the first one to the second one due to Table 3.5 and operator $\mathbf{Z}$:

$$X_a(p) \overset{L^{-1}}{\rightarrow} x_a(t) \overset{E_T}{\rightarrow} x(k) \overset{Z}{\rightarrow} X(z) \qquad [3.26]$$

$$\underbrace{}_{\mathbf{Z}}$$

The inverse operation is possible only by issuing hypotheses[12] on the original signal. Otherwise, the relation is not bijective and several analog signals can lead to the same discrete signal. In this case we talk of alias ($\sin \omega_d k$ and $\sin(\omega_d + 2\pi)k$ are two aliases of a same signal).

$X_a(p)$	$x_a(t), t \geq 0$	$x(k) = x_a(kT)$ $k \geq 0$	$X(z)$
$\dfrac{1}{p}$	$\Gamma_a(t)$	$\Gamma(k)$	$\dfrac{z}{z-1}$
$\dfrac{1}{p+\alpha}$	$e^{-\alpha t}$	$e^{-\alpha kT}$	$\dfrac{z}{z-e^{-\alpha T}}$
$\dfrac{\omega}{(p+\alpha)^2 + \omega^2}$	$e^{-\alpha t}\sin \omega t$	$e^{-\alpha kT}\underbrace{\sin \omega kT}_{\omega_d k}$	$\dfrac{ze^{-\alpha T}\sin \omega_d}{z^2 - 2e^{-\alpha T}\cos \omega_d z + e^{-2\alpha T}}$
$\dfrac{p+\alpha}{(p+\alpha)^2 + \omega^2}$	$e^{-\alpha t}\cos \omega t$	$e^{-\alpha kT}\underbrace{\cos \omega kT}_{\omega_d k}$	$\dfrac{z(z - e^{-\alpha T}\cos \omega_d)}{z^2 - 2e^{-\alpha T}\cos \omega_d z + e^{-2\alpha T}}$

Table 3.5. *Table of transforms*

3.4.2. *Transfer function of the discretized system*

Let (Σ) be the continuous system defined by transfer $H(s)$ where s is the derivation operator. Let (Σ_d) be the discrete system obtained from (Σ) by adding a downstream sampler and an upstream 0 *order blocker*, like in Figure 3.1 (if (Σ) designates the continuous process and (Σ_d) the discretized process). If the two converters are synchronized and at a pace equal to the sampling period T, we obtain for (Σ_d) the transfer $H_T(q) = E_T \circ H(s) \circ B_0$. We obtain without difficulty[13], from the relation $y(k) = H_T(q)u(k)$, the following relation:

$$Y(z) = H_T(z)U(z), \quad \text{with: } H_T(z) = (1 - z^{-1})\ \mathbf{Z}\left(\frac{H(p)}{p}\right) \qquad [3.27]$$

---

12 Shannon condition.

13 At input we apply a discrete impulse which will then successively undergo the various transformations, $h(\cdot) = (E_T \circ H(s) \circ B_0 \circ \delta)(\cdot)$, then we obtain $H_T(z)$ from the z-transform of this impulse response.

Due to this relation and Table 3.5, we will be able to easily obtain the transfer function of the discretized system and, consequently, its frequency response. It is also shown (see next section) that if p_c is a pole of the continuous system, then $p_d = e^{Tp_c}$, whereas the poles of the discretized system $\{p_{di}, \ i = 1, \cdots n\}$ are obtained from the poles of the continuous system $\{p_{ci}, \ i = 1, \cdots n\}$ using the relation:

$$p_{di} = e^{Tp_{ci}} \quad \forall i = 1, \cdots n \tag{3.28}$$

Consequently, the stability of the continuous system $(\mathrm{Re}\,(p_{ci}) < 0)$ leads to the stability of the discretized system $(|p_{di}| < 1)$.

3.4.3. *State representation of the discretized system*

Let us consider this time the continuous system (Σ) which is described by the state representation:

$$\begin{cases} \dot{x}_a(t) = A\,x_a(t) + Bu_a(t) \\ y_a(t) = Cx_a(t) + Du_a(t) \end{cases} \tag{3.29}$$

Let $u(k)$ be the input sequence of the discretized model. It is transformed into the constant analog signal $u_a(t)$ fragmented by the 0 *order blocker* before "attacking" the continuous system. We try to express the relation between $u(k)$ and the sampled output and state vectors $x(k) = x_a(kT)$ and $y(k) = y_a(kT)$. We have, between the sampling instants $t_k = kT$ and $t_{k+1} = (k+1)\,T$, $u_a(t) = u(k)$ and consequently $\dot{u}_a(t) = 0$. Equation [3.29] is then rewritten between these two instants:

$$\begin{cases} \begin{pmatrix} \dot{x}_a(t) \\ \dot{u}_a(t) \end{pmatrix} = \begin{pmatrix} A & 0 \\ 0 & 0 \end{pmatrix} \begin{pmatrix} x_a(t) \\ u_a(t) \end{pmatrix} \\ y_a(t) = \begin{pmatrix} C & D \end{pmatrix} \begin{pmatrix} x_a(t) \\ u_a(t) \end{pmatrix} \end{cases} \tag{3.30}$$

Based on the solution of this differential equation, we obtain:

$$\begin{cases} \begin{pmatrix} x_a(t_{k+1}) \\ u_a(t_{k+1}) \end{pmatrix} = e^{\begin{pmatrix} A & 0 \\ 0 & 0 \end{pmatrix} T} \begin{pmatrix} x_a(t_k) \\ u_a(t_k) \end{pmatrix} = \begin{pmatrix} A_T & B_T \\ 0 & I \end{pmatrix} \begin{pmatrix} x_a(t_k) \\ u_a(t_k) \end{pmatrix} \\ y_a(t_k) = (C \quad D) \begin{pmatrix} x_a(t_k) \\ u_a(t_k) \end{pmatrix} \end{cases}$$

[3.31]

Finally, we can associate the state representation with the discretized system:

$$\begin{cases} x(k+1) = A_T\, x(k) + B_T u(k) \\ \quad\ y(k) = Cx(k) + Du(k) \end{cases} \quad \text{with:} \quad \begin{cases} A_T = e^{AT} = I + A\Psi T \\ B_T = B\Psi T \\ \Psi = I + \dfrac{AT}{2!} + \dfrac{A^2 T^2}{3!} + \cdots \end{cases}$$

[3.32][14]

We obviously have, for $H(s) = C(sI - A)^{-1}B + D$, the transfer of the discrete system given by $H_T(q) = C(qI - A_T)^{-1}B_T + D$ and we find again relation [3.28] because the poles of $H(p)$ and $H_T(z)$ are also the eigenvalues of matrices A and A_T, if we suppose the minimal realizations.

3.4.4. *Frequency responses of the continuous and discrete system*

The frequency responses of the continuous system and its discretization are given by $H(j\omega)$ and $H_T(e^{j\omega T})$. We can also mention here that if the impulse ω is expressed in rad/TU[15], the discrete impulse $\omega_d = \omega T$ is without size. The two frequency responses are very similar in low frequency, i.e. if $\omega T << \pi$. They are necessarily different in high frequency, since the frequency response of the discretized system is periodic, contrary to the one of the continuous system.

---

14 If A is non-singular, we also have $B_T = A^{-1}(e^{AT} - I)B$.

15 TU: Time Unit.

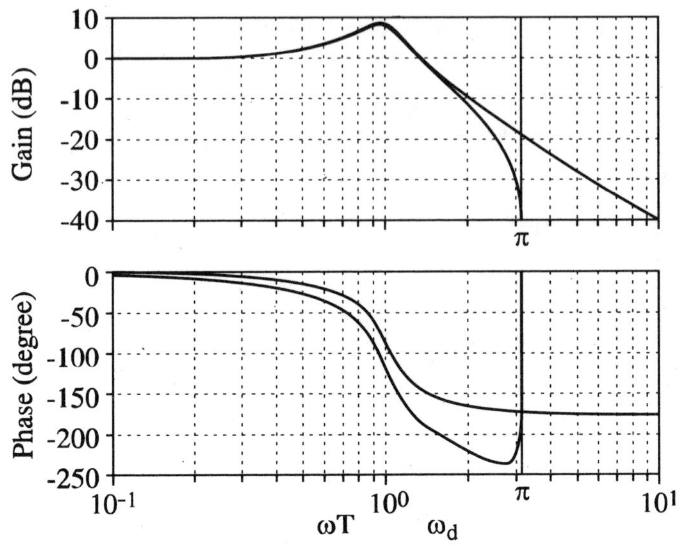

Figure 3.5. *Bode diagram of the continuous system and its discretization*

3.4.5. *The problem of sub-sampling*

Let us consider the case of a "standardized" pendulum subjected to a torque u. In the absence of friction, it will be controlled by equation $\dfrac{d^2}{dt^2}\theta + \theta = u$.

If we choose $x^T = \left[\theta,\ \dfrac{d\theta}{dt}\right]$ as state vector, we obtain the state representation:

$$\begin{cases} \dot{x} = \begin{pmatrix} 0 & 1 \\ -1 & 0 \end{pmatrix} x + \begin{pmatrix} 0 \\ 1 \end{pmatrix} u \\ y = (1 \quad 0) \end{cases}$$. We easily verify that this system is controllable and

observable. The discretized system at time T is controlled by:

$$\begin{cases} x(t_{k+1}) = \begin{pmatrix} \cos T & \sin T \\ -\sin T & \cos T \end{pmatrix} x(t_k) + \begin{pmatrix} 1 - \cos T \\ \sin T \end{pmatrix} u(t_k) \\ y = (t_k) = (1 \quad 0)\, x(t_k) \end{cases}$$

It is also controllable and observable except if $T = \pi$, which corresponds to drawing a sample every half-period oscillation of the pendulum. This pace is obviously insufficient. In this case we talk of sub-sampling.

If we change the perspective angle and if we use the formalism of transfer functions, the loss of controllability or observability illustrated above translates into a deterioration in the order of the discretized transfer. Let us illustrate this point with the help of the following example where $H_T(z)$ designates the discretized transfer obtained by transfer discretization $H(p)$ with the sampling period T.

$$H(p) = \frac{\alpha^2 + \omega^2}{(p+\alpha)^2 + \omega^2} \leftrightarrow H_T(z) = \frac{Az + B}{z^2 - 2e^{-\alpha T}\cos\omega_d z + e^{-2\alpha T}}$$

$$\text{with:} \begin{cases} A = 1 - e^{-\alpha T}\cos\omega T - \dfrac{\omega}{\alpha}e^{-\alpha T}\sin\omega T \\[2mm] B = e^{-2\alpha T} + \dfrac{\omega}{\alpha}e^{-\alpha T}\sin\omega T - e^{-\alpha T}\cos\omega T \end{cases}$$

We obtain for $\omega T = \pi$ the first order transfer $H_T(z) = \dfrac{1 + e^{-\alpha T}}{z + e^{-\alpha T}}$. We note that this transfer has a negative real pole and that its impulse response is an alternated sequence that converges towards 0 if $\alpha > 0$. Such a behavior for a first order system does not have a continuous time equivalent.

3.4.6. *The problem of over-sampling*

We talk of over-sampling if the sampling period is "very small" with respect to the dominating pole p_{c_D} of the continuous system, or even when $\left| T p_{c_D} \right| \ll 1$. The pole corresponding to the discrete system $p_{d_D} = e^{T p_{c_D}}$ is in this case very close to the unit. Hence, it is important to have a high numeric precision for the value of this pole because if not the discretized system will be considered wrong with respect to stability. For $T \ll 1$, the state matrices (see equations [3.32]) are such that $\left\| A_d - I \right\| \ll 1$ and $\left\| B_d \right\| \ll 1$. This point is illustrated by Example 3.1.

EXAMPLE 3.1.– let us consider the continuous system:

$$\left\{ \begin{array}{l} \dot{x}_a(t) = -\alpha\, x_a(t) + \alpha\, u_a(t) \\ \quad\ y_a(t) = x_a(t) \end{array} \right. \qquad (\Sigma)$$

We can obtain the state representation of the discretized system of equations [3.32]. For a sampling period $T = 10^{-6}\,s$ and $\alpha \in \{1,2\}$, we obtain a very good approximation:

$$\left\{ \begin{array}{l} x(k+1) = (1 - \alpha \times 10^{-6})\, x(k) + \alpha \times 10^{-6} u(k) \\ \quad\ y(k) = x(k) \end{array} \right. \qquad (\Sigma_d)$$

$\alpha = 2$ represents a system "twice as fast" as the case $\alpha = 1$ (the time constant of the continuous system is $\tau = \frac{1}{\alpha}$). For all that, the coefficients of the realizations obtained by discretization differ only by their sixth decimal in the two cases ($A_d \approx 1$, $B_d \approx 0$).

Such an over-sampling ($\alpha T \approx 10^{-6}$) is a source of numeric error unless there is high precision in the storing of coefficients. However, we note that this problem is not intrinsic and depends on the choice of the operator enabling the description of the discrete system. (Σ_d) can alternatively be described with the help of the operator δ_T or $\delta_T - \gamma$ (see section 3.3.3). From equations [3.32] and [3.22] we obtain here with high precision:

$$\delta_T x(k) = -\alpha\, x(k) + \alpha\, u(k) \qquad (\Sigma_d)$$

and with even higher precision (equation [3.24]):

$$\delta_T x(k) = -\alpha\, \gamma x(k) + \alpha\, u(k) \qquad (\Sigma_d)$$

Even if very simple, this example makes it possible to appreciate the interest in choosing these operators in order to prevent the risks of numeric errors pertaining to over-sampling. The coefficients of the discrete model have the same order of magnitude as the ones of the continuous model and this for an arbitrarily small sampling period. For a more detailed analysis, see [GEV 93, MID 90, SWI 98].

Finally, for the simulation algorithm to be complete, $x(k+1) = x(k) + T\,\delta_T x(k)$ or $x(k+1) = \gamma x(k) + \frac{T}{2}\,\delta_T x(k)$ should be calculated at each iteration.

3.5. Conclusion

In this chapter, we introduced the basic concepts that make it possible to understand the discrete-time signals and systems. We saw that the *z-transform* has in the case of discrete-time a role similar to the Laplace continuous-time transform and that the *lead* operator was substituted to the derivation operator. The different concepts introduced in the previous chapters in terms of representation or structural properties of the systems could then be transposed without difficulty for the case of discrete-time systems.

After briefly analyzing the behavior of basic discrete systems, we presented in short the issue of sampling passage from continuous-time signals and systems to discrete-time signals and systems. Our goal was to provide the basics that will make it possible to deal with (Chapters 8, 12 and 13) the digital simulation of continuous systems and their control by the computer. We deliberately ignored certain results, however essential, in signal theory but which were not strictly necessary in the context of this work.

3.6. Bibliography

[AST 84] ASTRÖM K.J., WITTENMARK B., *Computer Controlled Systems*, Prentice Hall, 1984.

[BOR 93] BORNE P., DAUPHIN-TANGUY G., RICHARD J.P., ROTELLA F., ZAMBETTAKIS I., *Analyse et Régulation des Processus Industriels, vol. 2, régulation numérique*, Technip, 1993.

[BOU 94] BOURLÈS H., "Semi-Cancellable Fractions in System Theory", *IEEE Trans. on Autom. Control*, vol. AC-39, no. 10, 1994.

[BRO 92] BROWN R.G., HWANG P.Y.C., *Introduction to random signal and applied Kalman Filtering*, John Wiley & Sons, 1992.

[CHE 99] CHEN C.T., *Linear system theory and design*, Oxford University Press, 1999.

[FEU 96] FEUER A., GOODWIN G.C., *Sampling in Digital Signal Processing and Control*, Birkhäuser, 1996.

[FRA 92] FRANKLIN G.F., POWELL J.D., WORKMAN M.L., *Digital Control of Dynamic Systems*, Addison Wesley, 1992.

[GEV 93] GEVERS M., LI G., *Parametrizations in Control, Estimation and Filtering Problems*, Springer Verlag, 1993.

[JUR 64] JURY E.I., *Theory and application of the z-transform*, John Wiley, 1964.

[LAR 02] DE LARMINAT P. (ed.), *Commande des systèmes linéaires*, Hermès, IC2 series, Paris, 2002.

[LJU 99] LJUNG L., *System Identification – Theory for the User*, Prentice Hall, 1999.

[MID 90] MIDDLETON R.H., GOODWIN G.C., *Digital Control and Estimation: A Unified Approach*, Prentice Hall, New York, 1990.

[PIC 77] PICINBONO B., *Eléments de théorie du signal*, Dunod, Paris, 1977.

[QUA 99] QUADRAT A., Analyse algébrique des systèmes de contrôle linéaires multidimensionnels, Thesis, Ecole nationale des ponts et chaussées, 1999.

[RAB 00] RABAH R., BERGEON B., DUSSER X., "On state-space representation for linear discrete-time systems in Hilbert spaces", *Proc. MTNS* (*Mathematical Theory of Networks and Systems*), Perpignan, 2000.

[ROU 92] ROUBINE E., *Distributions signal*, Eyrolles, Paris, 1992.

[SHA 84] SHANNON C.E., "Communication in the presence of noise", *Proc. IEEE*, 72, no. 9, p. 1192-1201, USA, 1984.

[SWI 98] SWIBER Z., "Realization using the γ-operator", *Automatica*, vol. 34, no. 11, p. 1455-1457, 1998.

Chapter 4

Structural Properties of Linear Systems

4.1. Introduction: basic tools for a structural analysis of systems

Any physical system has limitations in spite of the various possible control actions meant to improve its dynamic behavior. Some structural constraints may appear very early during the analysis phases. The following example illustrates the importance of the location of zeros with respect to the solution of a traditional control problem which is the pursuit of model, by dynamic pre-compensation. Being given a transfer procedure equal to:

$$t(p) = \frac{p-1}{(p+1)^3}$$

is it possible to find a compensator $c(p)$, so that the compensated procedure has a transfer equal to the one of the model previously fixed, $t_m(p)$? It is well known that the model to pursue cannot be chosen entirely freely. Indeed, the pursuit equation $t(p)c(p) = t_m(p)$ imposes that the model must have the same unstable zero as the procedure, otherwise the compensator will have to simplify it and hence an internal instability will occur. In addition, the relative degree of the model (the degree of difference between denominator and numerator; we will refer to it later on as the infinite zero order) cannot be lower than 2, otherwise the compensator will not be appropriate.

Chapter written by Michel MALABRE.

The object of this chapter is to describe certain structural properties of linear systems that condition the resolution of numerous control problems. The plan is the following.

After a brief description of certain main geometric and polynomial tools, useful for a structural analysis of the systems (section 4.1), we will describe the Kronecker canonical form of a matrix pencil, which, when we particularize it to different pencils (input-state, state-output and input-state-output) gives us directly, but with a common perspective, the controllable and observable canonical forms (of Brunovsky) and the canonical form of Morse (section 4.2). The following section (section 4.3) illustrates the invariance properties of the various structures of these canonical forms (indices of controllability, of observability, finite and infinite zeros) and of the associated transformation groups (basis changes, state returns, output injections). Two "traditional" control problems are considered (disturbance rejection and diagonal decoupling) and the fundamental role played by certain structures (invariant infinite and finite zeros, especially the unstable ones) is illustrated with respect to the existence of solutions, the existence of stabilizing solutions and flexibilities offered in terms of poles positions (concept of fixed poles). This is illustrated in section 4.4. Section 4.5 enumerates a few conclusions and lists the main references.

4.1.1. *Vector spaces, linear applications*

Let $\mathcal{X}$ and $\mathcal{Y}$ be real vector spaces of finite dimension and $\mathcal{V} \subset \mathcal{X}$ and $\mathcal{W} \subset \mathcal{Y}$, two sub-spaces. Let $\mathbf{L}: \mathcal{X} \to \mathcal{Y}$ be a linear application. $\mathbf{L}\mathcal{V}$ designates the image of $\mathcal{V}$ by $\mathbf{L}$ and $\mathbf{L}^{-1}\mathcal{W}$ designates the reverse image of $\mathcal{W}$ by $\mathbf{L}$:

$$\mathbf{L}\mathcal{V} := \{y \in \mathcal{Y} \text{ such that } \exists x \in \mathcal{V} \text{ and } \mathbf{L}x = y\} \tag{4.1}$$

$$\mathbf{L}^{-1}\mathcal{W} := \{x \in \mathcal{X} \text{ such that } \mathbf{L}x \in \mathcal{W}\} \tag{4.2}$$

With this notation, image $\mathrm{Im}\mathbf{L}$ and core $\mathrm{Ker}\mathbf{L}$ of $\mathbf{L}$ can also be written: $\mathrm{Im}\mathbf{L} = \mathbf{L}\mathrm{X}$ and $\mathrm{Ker}\mathbf{L} = \mathbf{L}^{-1}\{0\}$. Naturally, the notation chosen for the reverse image should not lead to the impression that $\mathbf{L}$ would be necessarily reversible.

EXAMPLE 4.1.– let us suppose that $\mathbf{L} = \begin{bmatrix} 1 & 0 \\ 2 & 0 \end{bmatrix}$, and $\mathcal{W}$ is the main straight line

$\begin{bmatrix} 1 \\ 0 \end{bmatrix}$: $\mathbf{L}^{-1}\mathcal{W} = \mathbf{L}^{-1}\{0\} = Ker\mathbf{L} = \begin{bmatrix} 0 \\ 1 \end{bmatrix}$.

Let $\mathbf{V}$ be a basis matrix of $\mathcal{V}$ and $\mathbf{W}^t$ a basis of the canceller at the left of $\mathcal{W}$ (i.e. a maximal solution of equation $\mathbf{W}^t\mathcal{W} = \{0\}$), a basis of $\mathbf{L}\mathcal{V}$ is obtained by directly preserving only the independent columns of $\mathbf{LV}$. A basis of $\mathbf{L}^{-1}\mathcal{W}$ is obtained by calculating a basis of core $Ker(\mathbf{W}^t\mathbf{L})$.

4.1.2. *Invariant sub-spaces*

Let $\mathbf{A}: \mathcal{X} \to \mathcal{X}$ be an endomorphism (linear application of a space within itself). Let n be the size of $\mathcal{X}$. A sub-space $\mathcal{V} \subset \mathcal{X}$ is called $\mathbf{A}$-invariant if and only if $\mathbf{A}\mathcal{V} \subset \mathcal{V}$. This concept is adapted to the study of trajectories of an autonomous dynamic system, which is described in continuous-time or discrete-time by:

$$\dot{\mathbf{x}}(t) = \mathbf{A}\,\mathbf{x}(t) \qquad \text{or} \qquad \mathbf{x}(k+1) = \mathbf{A}\mathbf{x}(k) \qquad\qquad [4.3]$$

Indeed, any state trajectory initiated in an $\dot{\mathbf{A}}$-invariant $\mathcal{V}$ sub-space remains indefinitely in $\mathcal{V}$. $\mathbf{A}$-invariant sub-spaces form a closed family for the addition and intersection of sub-spaces (the sum and intersection of two $\mathbf{A}$-invariant sub-spaces are $\mathbf{A}$-invariant). Consequently, for any $\mathcal{L} \subset \mathcal{X}$ sub-space, there is a bigger $\mathbf{A}$-invariant (unique) sub-space included in $\mathcal{L}$, noted by $\mathcal{L}^*$, and a smaller $\mathbf{A}$-invariant (unique) sub-space containing $\mathcal{L}$, noted by $\mathcal{L}_*$, obtained as the bound of algorithms [4.4] and [4.5]:

$$\mathcal{L}^0 = \mathcal{X},\ \mathcal{L}^1 = \mathcal{L},\ \mathcal{L}^2 = \mathcal{L} \cap \mathbf{A}^{-1}\mathcal{L}, \cdots,\ \mathcal{L}^{i+1} = \mathcal{L} \cap \mathbf{A}^{-1}\mathcal{L}^i \Rightarrow \mathcal{L}^n = \mathcal{L}^* \qquad [4.4]$$

$$\mathcal{L}_0 = \{0\},\ \mathcal{L}_1 = \mathcal{L},\ \mathcal{L}_2 = \mathcal{L} + \mathbf{A}\mathcal{L}, \cdots,\ \mathcal{L}_{i+1} = \mathcal{L} + \mathbf{A}\mathcal{L}_i, \Rightarrow \mathcal{L}_n = \mathcal{L}_* \qquad [4.5]$$

The concept of $\mathbf{A}$-invariant sub-space also makes it possible to decompose the dynamics of an autonomous system of the type [4.3] into two parts, and to describe what happens inside and "outside" sub-space $\mathcal{V}$. If we choose as first vectors of a basis of $\mathcal{X}$ the vectors obtained from a basis of $\mathcal{V}$ and if we complete this partial basis, the property of $\mathbf{A}$-invariance of $\mathcal{V}$ is translated through a zero block in the matrix representing $\mathbf{A}$ in this basis:

$$\mathbf{A} = \begin{bmatrix} \mathbf{A}_{\mathcal{V}} & \mathbf{A}_{12} \\ 0 & \mathbf{A}_{\mathcal{X}/\mathcal{V}} \end{bmatrix} \qquad\qquad [4.6]$$

where $\mathbf{A}_{\mathcal{V}}$ represents the restriction of $\mathbf{A}$ to $\mathcal{V}$ and $\mathbf{A}_{\mathcal{X}/\mathcal{V}}$ represents the complementary dynamics (more rigorously this is a representative matrix for the application in quotient $\mathcal{X}/\mathcal{V}^1$).

For controlled dynamic systems, where $\mathcal{X}$ and $\mathcal{U}$ designate, respectively, the state space and the control space described by:

$$\dot{\mathbf{x}}(t) = \mathbf{A}\,\mathbf{x}(t) + \mathbf{B}\,\mathbf{u}(t) \quad \text{or} \quad \mathbf{x}(k+1) = \mathbf{A}x(k) + \mathbf{B}\,\mathbf{u}(k) \qquad [4.7]$$

the $(\mathbf{A},\mathbf{B})$-invariance characterizes the property of having the capability to force trajectories to remain in a given sub-space, due to a suitable choice of the control law. A sub-space $\mathcal{V}$ of $\mathcal{X}$ is $(\mathbf{A},\mathbf{B})$-invariant if and only if $\mathbf{A}\mathcal{V} \subset \mathcal{V} + \text{Im}\mathbf{B}$. Similarly, $\mathcal{V}$ is $(\mathbf{A},\mathbf{B})$-invariant if and only if there is a state return (non-unique): $\mathbf{F}: \mathcal{X} \to \mathcal{U}$ such that $(\mathbf{A}+\mathbf{BF})\mathcal{V} \subset \mathcal{V}$. The sum of the two $(\mathbf{A},\mathbf{B})$-invariant sub-spaces is $(\mathbf{A},\mathbf{B})$-invariant, but this is not true for the intersection. For any sub-space $\mathcal{L} \subset \mathcal{V}$ there is a bigger $(\mathbf{A},\mathbf{B})$-invariant (unique) sub-space included in $\mathcal{L}$ and noted by $\mathcal{V}^*(\mathbf{A},\mathbf{B},\mathcal{L})$. It can be calculated as the bound of the non-increasing algorithm [4.8]:

$$\mathcal{V}^0 = \mathcal{X},\ \mathcal{V}^1 = \mathcal{L},\ \mathcal{V}^{i+1} = \mathcal{L} \cap \mathbf{A}^{-1}(\mathcal{V}^i + \text{Im}\,\mathbf{B}) \Rightarrow \mathcal{V}^n = \mathcal{V}^*(\mathbf{A},\mathbf{B},\mathcal{L}) \qquad [4.8]$$

For the analyzed dynamic systems, where $\mathcal{X}$ and $\mathcal{Y}$ designate the state space and the observation space, and described by:

$$\begin{aligned}
\dot{\mathbf{x}}(t) &= \mathbf{A}\,\mathbf{x}(t) & \mathbf{x}(k+1) &= \mathbf{A}x(k) \\
\mathbf{y}(t) &= \mathbf{C}\,\mathbf{x}(t) & \text{or} \quad \mathbf{y}(k) &= \mathbf{C}x(k)
\end{aligned} \qquad [4.9]$$

The $(\mathbf{C},\mathbf{A})$-invariance is a dual property of the $(\mathbf{A},\mathbf{B})$-invariance and is linked to the use of output injection. A sub-space $\mathcal{S}$ of $\mathcal{X}$ is $(\mathbf{C},\mathbf{A})$-invariant if and only if there is an output injection (non-unique) $\mathbf{K}: \mathcal{Y} \to \mathcal{X}$ such that $(\mathbf{A}+\mathbf{KC})\mathcal{S} \subset \mathcal{S}$. Similarly, $\mathcal{S}$ is $(\mathbf{C},\mathbf{A})$-invariant if and only if $\mathbf{A}(\mathcal{S} \cap \text{Ker}\mathbf{C}) \subset \mathcal{S}$. The intersection of two $(\mathbf{C},\mathbf{A})$-invariant sub-spaces is $(\mathbf{C},\mathbf{A})$-invariant, but this is not true for the sum. For any $\mathcal{L} \subset \mathcal{X}$ sub-space, there is a smaller $(\mathbf{C},\mathbf{A})$-invariant (unique) sub-space

---

1 Given $\mathcal{V} \subset \mathcal{X}$, the quotient $\mathcal{X}/\mathcal{V}$ represents the set of equivalence classes for the relation of equivalence $\mathcal{R}$ defined on $\mathcal{X}$ by $\forall x \in \mathcal{X},\ \forall y \in \mathcal{X}: x\mathcal{R}y \Leftrightarrow x\text{-}y \in \mathcal{V}$. We can visualize (abusively) $\mathcal{X}/\mathcal{V}$ as the set of vectors of X that are outside of $\mathcal{V}$.

containing $\mathcal{L}$ and noted by $\mathcal{S}_*(\mathbf{C},\mathbf{A},\ \mathcal{L})$. It can be calculated as the bound of the following non-decreasing algorithm:

$$S_0 = \{0\}, \ \mathcal{S}_1 = \mathcal{L}, \ \mathcal{S}_{i+1} = \mathcal{L} + \mathbf{A}(\mathcal{S}_i \cap \ \mathrm{Ker}\ \mathbf{C}) \Rightarrow \mathcal{S}_n = \mathcal{S}_*(\mathbf{C},\mathbf{A},\mathcal{L})$$

$$[4.10]$$

4.1.3. *Polynomials, polynomial matrices*

A polynomial matrix is a polynomial whose coefficients are matrices, or, similarly, a matrix whose elements are polynomials, for example:

$$\begin{bmatrix} 0 & 1 \\ 0 & 0 \end{bmatrix} p^2 + \begin{bmatrix} 1 & 2 \\ 1 & 0 \end{bmatrix} p + \begin{bmatrix} 1 & -1 \\ 0 & 1 \end{bmatrix} = \begin{bmatrix} p+1 & p^2+2p-1 \\ p & 1 \end{bmatrix} \qquad [4.11]$$

A polynomial matrix is called unimodular if it is square, reversible and polynomial reverse. A square polynomial matrix is unimodular if and only if its determinant is a non-zero scalar.

For example: $\begin{bmatrix} 1 & p \\ 0 & 1 \end{bmatrix}$ is unimodular, its reverse being equal to $\begin{bmatrix} 1 & -p \\ 0 & 1 \end{bmatrix}$.

In the study of structural properties of a given dynamic system of the following type (with $n \times n$ $\mathbf{A}$ matrix):

$$\begin{aligned} \dot{x}(t) &= \mathbf{A}\ \mathbf{x}(t) + \mathbf{B}\ \mathbf{u}(t) & \qquad \mathbf{x}(k+1) &= \mathbf{A}\mathbf{x}(k) + \mathbf{B}\ \mathbf{u}(k) \\ \mathbf{y}(t) &= \mathbf{C}\ \mathbf{x}(t) \qquad \text{or} & \mathbf{y}(k) &= \mathbf{C}\mathbf{x}(k) \end{aligned} \qquad [4.12]$$

intervene several polynomial matrices with an unknown factor p. The best known is certainly the $[p\mathbf{I}\text{-}\mathbf{A}]$ characteristic matrix that makes it possible to extract information on the poles. Other polynomial matrices make it possible to characterize properties such as controllability/obtainability, observability/detectability, or concepts grouping together state, control and output, especially in relation to the zeros of the system. These are, respectively, the matrices:

$$[p\mathbf{I}-\mathbf{A} \quad -\mathbf{B}], \qquad \begin{bmatrix} p\mathbf{I}-\mathbf{A} \\ -\mathbf{C} \end{bmatrix} \quad \text{and} \quad \begin{bmatrix} p\mathbf{I}-\mathbf{A} & -\mathbf{B} \\ -\mathbf{C} & 0 \end{bmatrix} \qquad [4.13]$$

All these polynomial matrices, which only make the two monomials in p^0 and p^1 appear, are called matrix pencils. All have the form $[p\mathbf{E}\text{-}\mathbf{H}]$, with $\mathbf{E}$ and $\mathbf{H}$ not necessarily square or of full rank. Two pencils, formed by matrices of the same size, $[p\mathbf{E}\text{-}\mathbf{H}]$ and $[p\mathbf{E'}\text{-}\mathbf{H'}]$, are said to be equivalent in the Kronecker sense if and only if there are two reversible constant matrices $\mathbf{P}$ and $\mathbf{Q}$ such that $[p\mathbf{E'}\text{-}\mathbf{H'}] = \mathbf{P}\,[p\mathbf{E}\text{-}\mathbf{H}]\,\mathbf{Q}$. $\mathbf{P}$ and $\mathbf{Q}$ are the basis changes in the departure space $\mathcal{X}$ and in the arrival space $\mathcal{X}$.

We will analyze, with the help of these matrix pencils, several structural properties of systems [4.12]. This will be done progressively in our work, from the simplest (pole beams) to the most complete (system matrix).

4.1.4. *Smith form, companion form, Jordan form*

The poles of system [4.12] are given by the eigenvalues of $\mathbf{A}$ (see Chapter 2). It is well known that these eigenvalues are linked to the dynamic operator $\mathbf{A}$ and not only to certain of its matrix representations. More precisely, the eigenvalues of $\mathbf{A}$ are not changed if we replace $\mathbf{A}$ by $\mathbf{A'} = \mathbf{T}^{-1}\mathbf{A}\mathbf{T}$, where $\mathbf{T}$ designates any basis change matrix in X. When such a relation is satisfied, we say that $\mathbf{A}$ and $\mathbf{A'}$ are equivalent. This relation is also written $\mathbf{T}^{-1}[p\mathbf{I}\text{-}\mathbf{A}]\mathbf{T} = [p\mathbf{I}\text{-}\mathbf{A'}]$ and thus $\mathbf{A}$ and $\mathbf{A'}$ are equivalent matrices if and only if the beams $[p\mathbf{I}\text{-}\mathbf{A}]$ and $[p\mathbf{I}\text{-}\mathbf{A'}]$ are equivalent in the Kronecker sense. An important interest in any equivalence notion, besides the division into separate equivalence classes that it induces on the set considered, is to represent each class by a particular element, called canonical form. In the case of $[p\mathbf{I}\text{-}\mathbf{A}]$ type beams, we know well the companion form type canonical forms (see Chapter 2) or Jordan form. These forms are in fact obtained directly from the famous Smith form which is developed for the general polynomial matrices. In practice, it is quite easy to show from Binet-Cauchy formulae that, for any given size k, two equivalent beams $[p\mathbf{I}\text{-}\mathbf{A}]$ and $[p\mathbf{I}\text{-}\mathbf{A'}]$ have the same HCF (the highest common factor) of all the non-zero minors of order k. Let us note by $\alpha_1(p)$, $\alpha_2(p)\dots$, $\alpha_n(p)$ these different HCFs for k = 1 to n. Polynomials $\alpha_i(p)$ can be divided ascendantly ($\alpha_1(p)$ divides $\alpha_2(p)$ which divides $\alpha_3(p)\dots$).

Let us introduce the following quotients: $\beta_1(p) = \alpha_1(p)$, $\beta_2(p) = \alpha_2(p)\alpha_1(p)$, ..., $\beta_n(p) = \alpha_n(p)/\alpha_{n-1}(p)$. Polynomials $\beta_i(p)$ can be divided ascendantly as well. Polynomials $\beta_i(p)$ which are different from 1 are called invariant polynomials of $[p\mathbf{I}\text{-}\mathbf{A}]$ (or of $\mathbf{A}$). The last one (the highest degree one) is the minimal polynomial of $\mathbf{A}$ (it is the smallest degree polynomial which cancels $\mathbf{A}$). The product of all $\beta_i(p)$ is $\alpha_n(p)$, which is characteristic polynomial of $\mathbf{A}$. The Smith form of $[p\mathbf{I}\text{-}\mathbf{A}]$ is the diagonal of $\beta_i(p)$. The invariant polynomials can be written in an extended form, or in a factorized form where the n eigenvalues of $\mathbf{A}$ appear (certain powers l_{ij} being then equal to 0):

$$\beta_i(p) = a_{i0} + a_{i1}p + \ldots + a_{ik_i-1}p^{k_i-1} + p^{k_i} = (p-p_0)^{l_{0i}}(p-p_1)^{l_{1i}}\ldots(p-p_n)^{l_{ni}} \quad [4.14]$$

From the point of view of terminology, the p_i singularities are called eigenvalues of **A**, (internal) poles of the dynamic system [4.12] and zeros of the beam [p**I**-**A**]. The companion form of **A** contains as many diagonal blocks as $\beta_i(p)$ which are different from 1 and for each block, of size $k_i \times k_i$, all terms are zero except for the over-diagonal which is full of "1" and the last line consisting of coefficients $-a_{ij}$ of $\beta_i(p)$. The Jordan form of **A** contains, for each eigenvalue p_i, as many blocks as $\beta_j(p)$ having a factor $(p-p_i)^{l_{ij}}$. Each basic block of this type, of size $l_{ij} \times l_{ij}$ has all its terms zero except for the diagonal which is full of "p_i" and the over-diagonal which is full of "1".

Polynomials $\beta_j(p)$ are called invariant polynomials of **A**. The factors of these polynomials, i.e. $(p-p_i)^{l_{ij}}$ are the invariant factors of **A**. The set of all $\beta_j(p)$, as well as the set of all invariant factors, form complete invariants under the relation of equivalence, i.e. under the action of basis changes (meaning that two square matrices of the same size are equivalent if and only if they have exactly the same invariant polynomials).

4.1.5. *Notes and references*

The basic tools for the "geometric" approach of automatic control engineering (invariant sub-spaces) were introduced by Wonham, Morse, Basile and Marro at the beginning of the 1970s; in particular see [BAS 92, WON 85], as well as [TRE 01]. Numerous complements on the "polynomial" tools leading to Smith, Jordan or companion forms can be found in [GAN 66], as well as in [WIL 65], which is an almost incontrovertible work for everything relative to eigenvalues.

4.2. Beams, canonical forms and invariants

The pole beam associated with the dynamic system [4.12] is a [p**E**-**H**] type beam, but with the two following particularities: **E** and **H** are square and **E** is reversible. Before considering the general case, we will transitorily suppose **E** and **H** as square, but **E** as not systematically reversible. This extension should be brought closer to the more general class of implicit systems called regular, i.e. the systems described by:

$$\mathbf{J}\dot{\mathbf{x}}(t) = \mathbf{A}\,\mathbf{x}(t) + \mathbf{B}\,\mathbf{u}(t) \qquad\qquad \mathbf{J}\mathbf{x}(k+1) = \mathbf{A}\mathbf{x}(k) + \mathbf{B}\,\mathbf{u}(k)$$

$$\mathbf{y}(t) = \mathbf{C}\,\mathbf{x}(t) \qquad \text{or} \qquad \mathbf{y}(k) = \mathbf{C}\mathbf{x}(k)$$

$$[4.15]$$

with $\mathbf{J}$ not forcibly reversible, but $[p\mathbf{J}\text{-}\mathbf{A}]$ "regular[2]", i.e. with a rank equal to n. In the case of continuous-time systems, such models particularly make it possible to manipulate the differentiators. For example, the following system describes a pure differentiator:

$$\begin{bmatrix} 0 & 1 \\ 0 & 0 \end{bmatrix} \dot{\mathbf{x}}(t) = \begin{bmatrix} 1 & 0 \\ 0 & 1 \end{bmatrix} \mathbf{x}(t) + \begin{bmatrix} 0 \\ -1 \end{bmatrix} \mathbf{u}(t) \; ; \; \mathbf{y}(t) = [1 \quad 0]\mathbf{x}(t)$$

$$[4.16]$$

It has indeed for transfer $\mathbf{C}(p\mathbf{J}\text{-}\mathbf{A})^{-1}\mathbf{B} = p$. This system has a pole infinity of order 1.

A $[p\mathbf{E}\text{-}\mathbf{H}]$ type regular square beam, with E and H as linear applications of $\mathcal{X}$ toward $\mathcal{X}$ and two isomorphic spaces of size n, will also have finite and infinite zeros. Among the most compact methods to illustrate these finite and infinite zeros of $[p\mathbf{E}\text{-}\mathbf{H}]$, we can use the Weierstrass canonical form. We easily can, by using the basis changes in $\mathcal{X}$ and in $\mathcal{X}$, which are $\mathbf{P}$ and $\mathbf{Q}$ respectively, transform the departure beam into its Weierstrass canonical form. It is a diagonal form with two main blocks separating the infinite zeros from the finite zeros:

$$\mathbf{P}[p\mathbf{E} - \mathbf{H}]\,\mathbf{Q} = \begin{bmatrix} p\mathbf{N} - \mathbf{I} & \mathbf{0} \\ \mathbf{0} & p\mathbf{I} - \mathbf{M} \end{bmatrix} \text{ with } \mathbf{N} \text{ nilpotent} \qquad [4.17]$$

Hence, the structure of infinite zeros of $[p\mathbf{E}\text{-}\mathbf{H}]$ is given by the Jordan structure of $\mathbf{N}$ (in zero because $\mathbf{N}$ has only zero eigenvalues). To better understand the fact that the singularities in "0" of $\mathbf{N}$ represent infinite singularities for the beam, it is sufficient to write $p\mathbf{N}\text{-}\mathbf{I} = p(1/p\mathbf{I}\text{-}\mathbf{N})$. In addition, the structure of finite zeros of $[p\mathbf{E}\text{-}\mathbf{H}]$ is given by the structure of $[p\mathbf{I}\text{-}\mathbf{M}]$, as in section 4.1.4. For example, the Weierstrass form of a generalized pole beam for a [4.15] type system with two infinite poles, one of order 1 and the other of order 2, and two finite poles, in $p = -1$ and $p = 0$ respectively is given by:

$$[p\mathbf{E} - \mathbf{H}] = \begin{bmatrix} a & 0 & 0 & 0 \\ 0 & b & 0 & 0 \\ 0 & 0 & c & 0 \\ 0 & 0 & 0 & d \end{bmatrix} \text{ with } a = \begin{bmatrix} -1 & p & 0 \\ 0 & -1 & p \\ 0 & 0 & -1 \end{bmatrix}, \; b = \begin{bmatrix} -1 & p \\ 0 & -1 \end{bmatrix}, \; c = [p+1], \; d = [p]$$

---

2 I.e. $\det(p\mathbf{J}\text{-}\mathbf{A})$ is not identically zero.

A way to obtain the Weierstrass form described in [4.17] is to use the following algorithms, which are very similar to algorithms [4.5] and [4.4]:

$$\mathcal{A}_1^0 = \{0\}, \quad \mathcal{A}_1^{i+1} = \mathbf{E}^{-1}\mathbf{H}\mathcal{A}_1^i, \;\Rightarrow\; \mathcal{A}_1^n = \mathcal{A}_1^* = \mathbf{E}^{-1}\mathbf{H}\mathcal{A}_1^* \qquad [4.18]$$

$$\mathcal{A}_2^0 = \mathcal{X}, \quad \mathcal{A}_2^{i+1} = \mathbf{H}^{-1}\mathbf{E}\mathcal{A}_2^i, \;\Rightarrow\; \mathcal{A}_2^n = \mathcal{A}_2^* = \mathbf{H}^{-1}\mathbf{E}\mathcal{A}_2^* \qquad [4.19]$$

The regularity of the beam $[p\mathbf{E}\text{-}\mathbf{H}]$ can be translated:

$$\mathcal{A}_1^* \oplus \mathcal{A}_2^* = \mathcal{X}, \text{ i.e. } \mathcal{A}_1^* + \mathcal{A}_2^* = \mathcal{X} \text{ and } \mathcal{A}_1^* \cap \mathcal{A}_2^* = \{0\} \qquad [4.20]$$

$$\mathbf{E}\mathcal{A}_1^* \oplus \mathbf{H}\mathcal{A}_2^* = \underline{\mathcal{X}}, \text{ i.e. } \mathbf{E}\mathcal{A}_1^* + \mathbf{H}\mathcal{A}_2^* = \underline{\mathcal{X}} \text{ and } \mathbf{E}\mathcal{A}_1^* \cap \mathbf{H}\mathcal{A}_2^* = \{0\} \qquad [4.21]$$

This leads quite naturally to the following choice for $\mathbf{P}$ and $\mathbf{Q}$:

$$\mathbf{Q} = \left[\, \text{basis of } \mathcal{A}_1^* \,\middle|\, \text{basis of } \mathcal{A}_2^* \,\right], \quad \mathbf{P} = \left[\, \text{basis of } \mathbf{E}\mathcal{A}_1^* \,\middle|\, \text{basis of } \mathbf{H}\mathcal{A}_2^* \,\right],$$

$$[4.22]$$

4.2.1. *Matrix pencils and geometry*

In the general case, $[p\mathbf{E}\text{-}\mathbf{H}]$ is a rectangular beam, with no particular hypothesis of rank, either on $\mathbf{E}$ or on $\mathbf{H}$. This means that apart from the previously defined finite and infinite zeros, $[p\mathbf{E}\text{-}\mathbf{H}]$ also has a non-trivial core and co-core. Polynomial vectors and co-vectors, $\mathbf{x}(p)$ and $\underline{\mathbf{x}}^{\mathbf{T}}(p)$ then exist such that: $[p\mathbf{E}\text{-}\mathbf{H}]\,\mathbf{x}(p) = 0$ and/or $\underline{\mathbf{x}}^{\mathbf{T}}(p)\,[p\mathbf{E}\text{-}\mathbf{H}] = 0$. The various possible solutions of these equations are in fact classified and ordered in terms of degrees. If $\mathbf{x}(p)$ is in the core of $[p\mathbf{E}\text{-}\mathbf{H}]$, the vector obtained by multiplying each component of $\mathbf{x}(p)$ by a same polynomial is also in the core. Hence, we will consider the lowest degree solutions possible. For example, for a beam described by:

$$[p\mathbf{E} - \mathbf{H}] = \begin{bmatrix} p & 1 & 0 \\ 0 & p & 1 \end{bmatrix}$$

a core basis vector of minimal degree can be described by $[1 \ {-}p \ p^2]^{\mathrm{T}}$, where "T" represents the transposition. Similarly, for a beam described by:

$$[p\mathbf{E} - \mathbf{H}] = \begin{bmatrix} p \\ 1 \end{bmatrix}$$

a co-core basis vector and of minimal degree can be described by $[1 \ {-}p]$.

Then, through a reduction procedure with respect to these first solutions, we consider the following solutions of superior degree, but the lowest one possible, and so on. The result is that only the sequence of successive degrees is essential in order to properly describe the core and co-core in a canonical form.

In order to describe the complete structure of a beam in its most general form, algorithms [4.18] and [4.19] are sufficient. An important difference with respect to the previous regular case is that, in general:

$$\mathcal{A}_1^* \cap \mathcal{A}_2^* \neq \{0\} \quad \text{when the core is } \neq \{0\} \text{ and}$$

$$\mathbf{E}\mathcal{A}_1^* + \mathbf{H}\mathcal{A}_2^* \neq \underline{\mathcal{X}} \quad \text{when the co-core is } \neq \{0\}$$

This geometric description is provided in the following section.

4.2.2. *Kronecker's canonical form*

The main result for "any" beam is the following.

Two beams $[p\mathbf{E}\text{-}\mathbf{H}]$ and $[p\mathbf{E}'\text{-}\mathbf{H}']$ are equivalent in Kronecker's sense, i.e. there are basis change matrices $\mathbf{P}$ and $\mathbf{Q}$ such that $[p\mathbf{E}'\text{-}\mathbf{H}'] = \mathbf{P} \, [p\mathbf{E}\text{-}\mathbf{H}] \, \mathbf{Q}$, if and only if $[p\mathbf{E}\text{-}\mathbf{H}]$ and $[p\mathbf{E}'\text{-}\mathbf{H}']$ have the same Kronecker's canonical form.

Kronecker's canonical form of a beam $[p\mathbf{E}\text{-}\mathbf{H}]$ is a beam characterized only from $\mathbf{E}$ and $\mathbf{H}$. This form can possibly contain identically zero columns and/or rows (this happens when in the core and/or the co-core there are constant vectors) and in addition it has a block-diagonal structure with four types of blocks:

– finite elementary divisor blocks (also called finite zeros): these are (for example) Jordan blocks, of size $k_{ij} \times k_{ij}$, associated with $(p\text{-}a_i)^{k_{ij}}$ type monomials. (We can also choose companion type blocks.) For example:

$$\begin{bmatrix} p+1 & 1 \\ 0 & p+1 \end{bmatrix} \quad \text{for the monomial } (p+1)^2, \text{ etc.} \tag{4.23}$$

– minimal index blocks per non-zero columns: these are rectangular blocks, of size $\varepsilon_I \times (\varepsilon_I + 1)$, having the form:

$$[p \quad 1] \text{ for } \varepsilon = 1, \begin{bmatrix} p & 1 & 0 \\ 0 & p & 1 \end{bmatrix} \text{ for } \varepsilon = 2, \text{ etc.} \qquad [4.24]$$

– minimal index blocks per non-zero rows: these are rectangular blocks, of size $(\eta_I + 1) \times \eta^i$, which are identical to minimal index blocks per columns, but simply transposed, thus:

$$\begin{bmatrix} p \\ 1 \end{bmatrix} \text{ for } \eta = 1, \quad \text{etc.} \qquad [4.25]$$

– infinite elementary divisor blocks (also called infinite zeros): these are square blocks, of size $v_i \times v_i$, with a diagonal full of "1" and an over-diagonal full of "p", i.e. having the form:

$$[1] \text{ for } v = 1, \begin{bmatrix} 1 & p \\ 0 & 1 \end{bmatrix} \text{ for } v = 2, \text{ etc.} \qquad [4.26]$$

Kronecker's canonical form is fully characterized by the list of polynomials $(p-a_i)^{k_{ij}}$ and by the three lists of integers $\{\varepsilon_i\}$, $\{\eta_i\}$ and $\{v_i\}$. These four lists form full invariants for the beams under the action of basis changes in the departure and arrival spaces. An example of Kronecker's canonical form (the index "K" is used to indicate that the beam is in its Kronecker's canonical form) is given below, corresponding to the list of invariants: $\{(p-a_i)^{k_{ij}}\} = \{p-3\}$, $\{\varepsilon_i\} = \{2\}$, $\{\eta_i\} = \{1\}$ and $\{v_i\} = \{2\}$:

$$[p\mathbf{E}_K - \mathbf{H}_K] = \begin{bmatrix} a & 0 & 0 & 0 \\ 0 & b & 0 & 0 \\ 0 & 0 & c & 0 \\ 0 & 0 & 0 & d \end{bmatrix} \text{ with } a = [p-3], \ b = \begin{bmatrix} p & 1 & 0 \\ 0 & p & 1 \end{bmatrix}, \ c = \begin{bmatrix} p \\ 1 \end{bmatrix}, \ d = \begin{bmatrix} 1 & p \\ 0 & 1 \end{bmatrix}$$

$$[4.27]$$

Now, due to the two algorithms [4.18] and [4.19], we can provide the geometric characteristics of these invariants. For this, we will use the following notations: given a list of positive integers $\{n_i\}$, $I = 1$ to l, ordered in a non-increasing manner (i.e. $n_i \geq n_{i+1}$), we associate with it the list $\{p_j\}$ which is defined by $p_j = \text{card}\{n_i \geq j\}$, where "card" represents the cardinal number, i.e. the total number of elements in the group. We note that the correspondence between the two lists $\{n_i\}$, $i = 1$ to l and $\{p_j\}$, $j = 1$ to h is a bijection. Indeed, it is easy to verify that list $\{n_i\}$ also satisfies $n_i = \text{card}\{p_j \geq i\}$ and consequently $l = p_1$ and $h = n_1$.

The geometric characteristics of Kronecker's invariants are given below. We note at this level that these characteristics establish the invariance of the four lists under the action of $\mathbf{P}$ and $\mathbf{Q}$ basis changes in the departure and arrival spaces. Indeed, the sizes of intermediary sub-spaces are clearly invariant when we replace $\mathbf{E}$ and $\mathbf{H}$ by $\mathbf{PEQ}$ and $\mathbf{PHQ}$:

– minimal indices per columns:

$$\forall \mu \geq 1, \ \ \text{card} \ \{\varepsilon_i \geq \mu\} = \dim \ (\mathcal{A}_2^* \cap \mathcal{A}_1^{\mu+1}) - \dim \ (\mathcal{A}_2^* \cap \mathcal{A}_1^{\mu}) \qquad [4.28]$$

– minimal indices per rows:

$$\forall \mu \geq 1, \ \ \text{card} \ \{\eta_i \geq \mu\} = \dim \ (\mathcal{A}_1^* + \mathcal{A}_2^{\mu-1}) - \dim \ (\mathcal{A}_1^* + \mathcal{A}_2^{\mu}) \qquad [4.29]$$

– infinite elementary divisors:

$$\forall \mu \geq 1, \ \ \text{card} \ \{\nu_i \geq \mu\} = \dim \ (\mathcal{A}_2^* + \mathcal{A}_1^{\mu}) - \dim \ (\mathcal{A}_2^* + \mathcal{A}_1^{\mu-1}) \qquad [4.30]$$

– finite elementary divisors. From the definitions of algorithms [4.18] and [4.19] it is easy to verify that, not only:

$$\mathbf{H}\mathcal{A}_2^* \subset \mathbf{E}\mathcal{A}_2^*, \ \text{but also:} \ \mathbf{H}(\mathcal{A}_2^* \cap \mathcal{A}_1^*) \subset \mathbf{E}(\mathcal{A}_2^* \cap \mathcal{A}_1^*)$$

In addition: $\dim \ (\mathcal{A}_2^*) - \dim \ (\mathcal{A}_2^* \cap \mathcal{A}_1^*) = \dim \ (\mathbf{E}\mathcal{A}_2^*) - \dim \ (\mathbf{E}(\mathcal{A}_2^* \cap \mathcal{A}_1^*))$.

The finite elementary divisors of the beam $[p\mathbf{E}-\mathbf{H}]$ are then given by the finite elementary divisors (in the sense of Smith's form; see section 4.1.4) of the next square operator, double restriction of $\mathbf{H}$ to two quotient spaces (in the departure and arrival spaces):

$$\hat{\mathbf{H}}: \ \mathcal{A}_2^* / \mathcal{A}_2^* \cap \mathcal{A}_1^* \to \mathbf{E}\mathcal{A}_2^* / \mathbf{E}(\mathcal{A}_2^* \cap \mathcal{A}_1^*) \qquad [4.31]$$

These general results on "any" beam will be now focused on some interesting cases that will differently clarify certain structural properties of [4.12] type systems.

4.2.3. *Controllable, observable canonical form (Brunovsky)*

Let us go back a little to the controlled dynamic systems without output equation, with $\mathcal{X}$ and $\mathcal{U}$ representing the state space and the control space. In order not to have to distinguish controllability and obtainability, we will limit ourselves here to continuous-time spaces, as described in [4.7]:

$$\dot{\mathbf{x}}(t) = \mathbf{A}\,\mathbf{x}(t) + \mathbf{B}\,\mathbf{u}(t) \tag{4.32}$$

We can "naturally" associate with this system the controllability beam $[p\mathbf{I}-\mathbf{A} - \mathbf{B}]$, i.e. for which $\mathbf{E} = [\mathbf{I}\quad\mathbf{0}]$ and $\mathbf{H} = [\mathbf{A}\quad\mathbf{B}]$. Due to the subjectivity of $\mathbf{E}$, Kronecker's form of the controllability beam can have only two types of invariants, i.e. minimal indices per columns and finite elementary divisors (indeed, for the other types of blocks see [4.25] and [4.26], the block sub-matrix in $\mathbf{E}$ is not of full rank per row and hence it cannot be a part of the global subjective $\mathbf{E}$). These invariants have a tighter connection with more traditional concepts, such as the controllability indices and the non-controllable poles. More exactly, we can easily show that the minimal indices per columns of the controllability beam are exactly equal to the controllability indices of the pair $(\mathbf{A}, \mathbf{B})$. The finite elementary divisors of the controllability beam correspond exactly to the non-controllable dynamics (with multiplicities considered through the invariant factors) of the pair $(\mathbf{A}, \mathbf{B})$. This will be mentioned in section 4.3. Before, we will characterize the group of transformations acting on the dynamic system [4.32] and that is equivalent to the group of basis changes on the left and right on $[p\mathbf{I}-\mathbf{A} - \mathbf{B}]$.

"Kronecker's" group of transformations acting on the controllability beam $[p\mathbf{I}-\mathbf{A} - \mathbf{B}]$ corresponds identically to the "feedback" group acting on the pair $(\mathbf{A}, \mathbf{B})$, in other words formulated:

$$\exists \mathbf{P}\ \&\ \mathbf{Q}\ \text{reversible such that:}\quad \mathbf{P}[p\mathbf{I}-\mathbf{A}\ -\mathbf{B}]\mathbf{Q} = [p\mathbf{I}-\mathbf{A}'\ -\mathbf{B}']$$

$$\Leftrightarrow \exists \mathbf{T}\ \&\ \mathbf{G}\ \text{reversible}\ \&\ \exists \mathbf{F}\ \text{such that:}\quad \mathbf{A}' = \mathbf{T}^{-1}(\mathbf{A}+\mathbf{BF})\mathbf{T},\quad \mathbf{B}' = \mathbf{T}^{-1}\mathbf{BG}$$

(To be sure, it is sufficient to note that $\mathbf{P}=\mathbf{T}^{-1}$ and $\mathbf{Q} = \begin{bmatrix} \mathbf{T} & \mathbf{0} \\ \mathbf{FT} & \mathbf{G} \end{bmatrix}$.)

Kronecker's canonical form of a controllability beam $[p\mathbf{I}-\mathbf{A}\ -\mathbf{B}]$ thus contains only minimal index blocks per columns and, possibly, blocks of finite elementary divisors. In order to show the quasi-immediate relation that exists between this Kronecker's form and the more traditional controllability canonical forms (like Brunovsky's form) we will take an example for which the minimal indices per columns are equal to $\{\varepsilon_1\} = \{1\}$, $\{\varepsilon_2\} = \{2\}$, and a finite elementary divisor is equal to $\{p+2\}$:

$$[p\mathbf{E}_K - \mathbf{H}_K] = \begin{bmatrix} a & 0 & 0 \\ 0 & b & 0 \\ 0 & 0 & c \end{bmatrix} \text{ with } a = [p \quad 1],\ b = \begin{bmatrix} p & 1 & 0 \\ 0 & p & 1 \end{bmatrix},\ c = [p+2]$$

Since this form is associated with $[p\mathbf{I}-\mathbf{A}\ -\mathbf{B}]$, we can write it differently so that it maintains a controllability beam form, which will be noted by $[p\mathbf{I}-\mathbf{A}_c\ -\mathbf{B}_c]$. This is easily obtained by switching all the constant columns in the last positions. The pair $(\mathbf{A}_c, \mathbf{B}_c)$ thus obtained is in Brunovsky's controllable canonical form and, just by reading it, we note that the controllable space is of size 3, the pole in $\{-2\}$ is non-controllable and the controllability indices are 1 and 2 (see section 4.3):

$$\mathbf{A}_c = \begin{bmatrix} 0 & 0 & 0 & 0 \\ 0 & 0 & 1 & 0 \\ 0 & 0 & 0 & 0 \\ 0 & 0 & 0 & -2 \end{bmatrix} \quad \mathbf{B}_c = \begin{bmatrix} 1 & 0 \\ 0 & 0 \\ 0 & 1 \\ 0 & 0 \end{bmatrix}$$

The general structure of matrices $(\mathbf{A}_c, \mathbf{B}_c)$ in Brunovsky's canonical form is the following:

$$\mathbf{A}_c = \begin{bmatrix} diag\{\mathbf{A}_{ci}\} & 0 \\ 0 & \mathbf{A}_{non\ c} \end{bmatrix} \quad \mathbf{B}_c = \begin{bmatrix} diag\{\mathbf{B}_{ci}\} \\ 0 \end{bmatrix}$$

with:

$$\mathbf{A}_{ci} = \begin{bmatrix} 0 & 1 & 0 & \ldots & 0 \\ 0 & 0 & 1 & 0\ldots & 0 \\ \ldots & \ldots & \ldots & \ldots & \ldots \\ 0 & 0 & 0 & 0 & 1 \\ 0 & 0 & 0 & 0 & 0 \end{bmatrix}, \quad \mathbf{B}_{ci} = \begin{bmatrix} 0 \\ 0 \\ \ldots \\ 0 \\ 1 \end{bmatrix} \qquad [4.33]$$

Blocks $\mathbf{A}_{ci}$ are of size $\varepsilon_i \times \varepsilon_i$; blocks $\mathbf{B}_{ci}$ are of size $\varepsilon_I \times 1$; the remaining matrix $\mathbf{A}_{non\ c}$ (that can be described, for example, in Jordan's form; see section 4.1.4) is of size $\left(n - \sum_i \varepsilon_i\right) \times \left(n - \sum_i \varepsilon_i\right)$. It does not exist if the system is controllable: it describes the non-controllable dynamics; integers ε_i are the controllability indices of the pair $(\mathbf{A}, \mathbf{B})$.

What has just been illustrated for controllability is also applicable and in a dual way to observability.

Let us go back a little to the dynamic systems without a term of control, with $\mathcal{X}$ and $\mathcal{Y}$ designating the state space and the observation space respectively. We will limit ourselves here to continuous-time systems as described in section 4.9:

$$\begin{aligned} \dot{\mathbf{x}}(t) &= \mathbf{A}\,\mathbf{x}(t) \\ \mathbf{y}(t) &= \mathbf{C}\,\mathbf{x}(t) \end{aligned} \qquad [4.34]$$

We can "naturally" associate the observability beam with this system:

$$\begin{bmatrix} p\mathbf{I} - \mathbf{A} \\ -\mathbf{C} \end{bmatrix} \qquad [4.35]$$

i.e. for which $\mathbf{E} = [\mathbf{I}\ \mathbf{0}]^{T}$ and $\mathbf{H} = [\mathbf{A}^{T}\ \mathbf{C}^{T}]^{T}$. Due to the injectivity of $\mathbf{E}$, Kronecker's form of the observability beam can have only two types of invariants, i.e. row minimal indices and finite elementary divisors (indeed, for the other types of blocks see [4.24] and [4.26], the block sub-matrix in $\mathbf{E}$ is not of column full rank, and hence it cannot be a part of the global injective $\mathbf{E}$). These invariants have a tighter connection with more traditional concepts, such as the observability indices and the non-observable poles. More exactly, we can easily show that the minimal indices per rows of the observability beam are exactly equal to the observability indices of the pair $(\mathbf{C}, \mathbf{A})$. The finite elementary divisors of the observability beam correspond exactly to the non-observable dynamics (with multiplicities considered for the invariant factors) of the pair $(\mathbf{C}, \mathbf{A})$. This will be mentioned in section 4.3. Before this, we will characterize the group of transformations acting on the dynamic system [4.34] and that is equivalent to the group of basis changes on the left and right on $[p\mathbf{I}-\mathbf{A}^{\mathrm{T}}\ -\mathbf{C}^{\mathrm{T}}]^{\mathrm{T}}$.

"Kronecker's" transformation group acting on the observability beam $[p\mathbf{I}-\mathbf{A}^{\mathrm{T}}\ -\mathbf{C}^{\mathrm{T}}]^{\mathrm{T}}$ corresponds identically to the "injection" group acting on the pair $(\mathbf{C}, \mathbf{A})$, in other words formulated:

$$\exists \mathbf{P}\ \&\ \mathbf{Q}\ \text{reversible such as}:\quad \mathbf{P}\begin{bmatrix} p\mathbf{I}-\mathbf{A} \\ -\mathbf{C} \end{bmatrix}\mathbf{Q} = \begin{bmatrix} p\mathbf{I}-\mathbf{A}' \\ -\mathbf{C}' \end{bmatrix}$$

$$\Leftrightarrow \exists \mathbf{T}\ \&\ \mathbf{H}\ \text{reversible}\ \&\ \exists \mathbf{R}\ \text{such as}:\quad \mathbf{A}' = \mathbf{T}^{-1}(\mathbf{A}+\mathbf{R}\mathbf{C})\mathbf{T},\quad \mathbf{C}' = \mathbf{H}\mathbf{C}\mathbf{T}$$

$$\text{(To be sure, it is sufficient to note that: } \mathbf{P} = \begin{bmatrix} \mathbf{T}^{-1} & \mathbf{T}^{-1}\mathbf{R} \\ \mathbf{0} & \mathbf{H} \end{bmatrix}, \text{ and } \mathbf{Q} = \mathbf{T}.)$$

Kronecker's canonical form of an observability beam $[p\mathbf{I}-\mathbf{A}^{\mathrm{T}}\ -\mathbf{C}^{\mathrm{T}}]^{\mathrm{T}}$ thus contains only blocks of minimal indices per rows and, possibly, blocks of finite elementary divisors. In order to show the quasi immediate relation that exists between this Kronecker's form and the more traditional observability canonical forms (like Brunovsky's form) we will take an example for which the minimal indices per rows are equal to $\{\eta_1\} = \{1\}$, $\{\eta_2\} = \{2\}$ and a finite basic divisor is equal to $\{p+5\}$:

$$[p\mathbf{E}_K - \mathbf{H}_K] = \begin{bmatrix} a & 0 & 0 \\ 0 & b & 0 \\ 0 & 0 & c \end{bmatrix} \text{ with } \quad a = \begin{bmatrix} p \\ 1 \end{bmatrix},\ b = \begin{bmatrix} p & 0 \\ 1 & p \\ 0 & 1 \end{bmatrix},\ c = [p+5]$$

Since this form is associated with $[p\mathbf{I}-\mathbf{A}^{\mathrm{T}}\ -\mathbf{C}^{\mathrm{T}}]^{\mathrm{T}}$, we can write it differently so that it maintains an observability beam form, which will be noted by $[p\mathbf{I}-\mathbf{A}_o^{\mathrm{T}}\ -\mathbf{C}_o^{\mathrm{T}}]^{\mathrm{T}}$. This is easily obtained by switching all the constant rows in the last positions:

$$\mathbf{A_o} = \begin{bmatrix} 0 & 0 & 0 & 0 \\ 0 & 0 & 1 & 0 \\ 0 & 0 & 0 & 0 \\ 0 & 0 & 0 & -5 \end{bmatrix} \quad \mathbf{C_o} = \begin{bmatrix} 1 & 0 & 0 & 0 \\ 0 & 1 & 0 & 0 \end{bmatrix}$$

The pair $(\mathbf{A_o}, \mathbf{C_o})$ thus obtained is in Brunovsky's observable canonical form. The unnoticeable space is of size 1. The pole in {-5} is un-observable and the unobservable indices are 1 and 2 (see section 4.3).

The general structure of matrices $(\mathbf{A_o}, \mathbf{C_o})$ in Brunovsky's canonical form is the following:

$$\mathbf{A_o} = \begin{bmatrix} diag\{\mathbf{A}_{oi}\} & 0 \\ 0 & \mathbf{A}_{non\ o} \end{bmatrix} \quad \mathbf{C_o} = \begin{bmatrix} diag\{\mathbf{C}_{oi}\} \\ 0 \end{bmatrix}, \text{ with:}$$

$$\mathbf{A}_{oi} = \begin{bmatrix} 0 & 1 & 0 & ... & 0 \\ 0 & 0 & 1 & 0... & 0 \\ ... & ... & ... & ... & ... \\ 0 & 0 & 0 & 0 & 1 \\ 0 & 0 & 0 & 0 & 0 \end{bmatrix}, \quad \mathbf{C}_{oi} = \begin{bmatrix} 1 & 0 & ... & 0 & 0 \end{bmatrix} \qquad [4.36]$$

Blocks $\mathbf{A}_{oi}$ are of size $\eta_I \times \eta_i$, blocks $\mathbf{C}_{oi}$, are of size $1 \times \eta_i$ and the remaining matrix A_{nono} (that can be described, for example, in Jordan's form, see section 4.1.4) is of size $\left(n - \sum_i \eta_i \right) \times \left(n - \sum_i \eta_i \right)$. It does not exist if the system is observable; the integers η_i are the observability indices of the pair $(\mathbf{A}, \mathbf{C})$.

Let us consider now more general dynamic systems, with $\mathbf{u}$ inputs and $\mathbf{y}$ outputs.

4.2.4. *Morse's canonical form*

The systems described by equation [4.12], i.e.:

$$\dot{\mathbf{x}}(t) = \mathbf{A}\,\mathbf{x}(t) + \mathbf{B}\,\mathbf{u}(t) \qquad\qquad \mathbf{x}(k+1) = \mathbf{A}\mathbf{x}(k) + \mathbf{B}\,\mathbf{u}(k)$$

$$\mathbf{y}(t) = \mathbf{C}\,\mathbf{x}(t) \qquad \text{or} \qquad \mathbf{y}(k) = \mathbf{C}\mathbf{x}(k)$$

have as "naturally" associated beam the following matrix, known as Rosenbrock's "system matrix":

$$[p\mathbf{E} - \mathbf{H}] = \begin{bmatrix} p\mathbf{I} - \mathbf{A} & -\mathbf{B} \\ -\mathbf{C} & 0 \end{bmatrix} \qquad [4.37]$$

For this beam:

$$\mathbf{E} = \begin{bmatrix} \mathbf{I} & 0 \\ 0 & 0 \end{bmatrix} \text{ and } \mathbf{H} = \begin{bmatrix} \mathbf{A} & \mathbf{B} \\ \mathbf{C} & 0 \end{bmatrix}$$

"Kronecker's" group of transformation acting on the system matrix [4.37] corresponds identically to the "feedback and injection" group acting on the system [4.12], in other words formulated:

$$\exists \mathbf{P} \& \mathbf{Q} \text{ reversible such as :} \quad \mathbf{P} \begin{bmatrix} p\mathbf{I} - \mathbf{A} & -\mathbf{B} \\ -\mathbf{C} & 0 \end{bmatrix} \mathbf{Q} = \begin{bmatrix} p\mathbf{I} - \mathbf{A'} & -\mathbf{B'} \\ -\mathbf{C'} & 0 \end{bmatrix}$$

$$\Leftrightarrow \exists \mathbf{T}, \mathbf{G} \& \mathbf{H} \text{ reversible } \& \exists \mathbf{F} \& \mathbf{R} \text{ such as :}$$

$$\mathbf{A'} = \mathbf{T}^{-1}(\mathbf{A} + \mathbf{BF} + \mathbf{RC})\mathbf{T}, \quad \mathbf{B'} = \mathbf{T}^{-1}\mathbf{BG}, \quad \mathbf{C'} = \mathbf{HCT}$$

(To be sure, it is sufficient to note that: $\mathbf{P} = \begin{bmatrix} \mathbf{T}^{-1} & \mathbf{T}^{-1}\mathbf{R} \\ 0 & \mathbf{H} \end{bmatrix}$ and $\mathbf{Q} = \begin{bmatrix} \mathbf{T} & 0 \\ \mathbf{FT} & \mathbf{G} \end{bmatrix}$.)

Kronecker's canonical form of a system matrix contains in general all the possible types of blocks. To visualize in terms of matrices **A**, **B** and **C** the form of the canonical representation obtained, it is sufficient, like in the previous case, to switch the rows and columns in order to move to the right all the constant columns (representative of the input matrix) and to the bottom the constant rows (representative of the output matrix). Let us take again the example [4.27] of section 4.2.2, in which there is a block of each type:

$$[p\mathbf{E}_K - \mathbf{H}_K] = \begin{bmatrix} a & 0 & 0 & 0 \\ 0 & b & 0 & 0 \\ 0 & 0 & c & 0 \\ 0 & 0 & 0 & d \end{bmatrix} \text{ with } a = [p-3], b = \begin{bmatrix} p & 1 & 0 \\ 0 & p & 1 \end{bmatrix}, c = \begin{bmatrix} p \\ 1 \end{bmatrix}, d = \begin{bmatrix} 1 & p \\ 0 & 1 \end{bmatrix}$$

The corresponding matrices have then the following form (written here by preserving the order of blocks), which is called Morse's canonical form, and noted by $(\mathbf{A}_M, \mathbf{B}_M, \mathbf{C}_M)$:

$$\mathbf{A}_M = \begin{bmatrix} 3 & 0 & 0 & 0 & 0 \\ 0 & 0 & 1 & 0 & 0 \\ 0 & 0 & 0 & 0 & 0 \\ 0 & 0 & 0 & 0 & 0 \\ 0 & 0 & 0 & 0 & 0 \end{bmatrix}, \quad \mathbf{B}_M = \begin{bmatrix} 0 & 0 \\ 0 & 0 \\ 1 & 0 \\ 0 & 0 \\ 0 & 1 \end{bmatrix}$$

$$\mathbf{C}_M = \begin{bmatrix} 0 & 0 & 0 & 1 & 0 \\ 0 & 0 & 0 & 0 & 1 \end{bmatrix}$$

The general structure of triplets $(\mathbf{A}_M, \mathbf{B}_M, \mathbf{C}_M)$ in Morse's canonical form is the following:

$$\mathbf{A}_M = \begin{bmatrix} \mathbf{A}_1 & 0 & 0 & 0 \\ 0 & \mathbf{A}_2 & 0 & 0 \\ 0 & 0 & \mathbf{A}_3 & 0 \\ 0 & 0 & 0 & \mathbf{A}_4 \end{bmatrix}, \mathbf{B}_M = \begin{bmatrix} 0 & 0 \\ \mathbf{B}_2 & 0 \\ 0 & 0 \\ 0 & \mathbf{B}_4 \end{bmatrix}, \mathbf{C}_M = \begin{bmatrix} 0 & 0 & \mathbf{C}_3 & 0 \\ 0 & 0 & 0 & \mathbf{C}_4 \end{bmatrix} \qquad [4.38]$$

where $\mathbf{A}_1$ is in Jordan's form $(\mathbf{A}_2, \mathbf{B}_2)$ in controllable canonical form [4.33], $(\mathbf{A}_3, \mathbf{C}_3)$ in observable canonical form [4.36] and $(\mathbf{A}_4, \mathbf{B}_4, \mathbf{C}_4)$ in simultaneously controllable [4.33] and observable [4.36] form.

The parts having the indices "2" and "3", which characterize certain core structures (on the right and left), have an important but very particular role in certain control or observation problems, called non-regular. We will not discuss in detail this aspect here. However, the parts having the indices "1" and "4" that are the result of finite and infinite elementary divisors of the system matrix are directly linked to invariant finite zero and infinite zero type structures, which we will deal with in section 4.3.

4.2.5. *Notes and references*

The general context of matrix pencils, and particularly Kronecker's canonical form, is detailed in [GAN 66]. The "geometric" presentation done here is mainly based on the works of [LOI 86]. A main reference work for the study of various beams associated with the analysis of linear systems, such as the system matrix, is [ROS 70]; for everything that is more particularly linked to the canonical forms presented here as derived from Kronecker's form, the reader can refer to [BRU 70, MOR 73, THO 73].

4.3. Invariant structures under transformation groups

It is exactly because they are invariant under the action of various transformation groups that the structures previously introduced have a fundamental role in the analysis and synthesis of observation and/or control systems. For example, the poles of a given system (in open loop) are invariant by basis changes but they are not so by state returns: it is well known in fact that a property equivalent to state controllability is the capability to freely modify the poles by state return. However, the invariant zeros, finite and infinite, are not at all modifiable by such actions. That is why their location conditions the resolving of traditional control problems. In the following sections, we will recall a few invariance properties of the main structures connected to linear systems.

4.3.1. *Controllability indices*

The controllability indices and the invariant factors of the non-controllable part (if it exists) of the pair $(\mathbf{A}, \mathbf{B})$ (see section 4.2.3) form a set of full invariants under the action of the transformation group $(\mathbf{T}, \mathbf{F}, \mathbf{G})$ where $\mathbf{T}$ and $\mathbf{G}$ designate the basis changes on the state and on the control, and $\mathbf{F}$ is a state return. This "feedback" group is defined by:

$$(\mathbf{A}, \mathbf{B}) \xrightarrow{\;(\mathbf{T},\mathbf{F},\mathbf{G})\;} (\mathbf{T}^{-1}(\mathbf{A}+\mathbf{BF})\mathbf{T}, \mathbf{T}^{-1}\mathbf{BG})$$

This basically means that any control law in the form of a regular state return, i.e. $\mathbf{u}(t) = \mathbf{Fx}(t) + \mathbf{Gv}(t)$, with $\mathbf{G}$ reversible, maintains these structures. Through a connection with a more "traditional" definition of controllability indices, noted by $\{c_1, c_2, ..., c_m\}$ where m is the size of the control space, we recall that the general characterization of minimal indices per columns as described in [4.28], when particularized to the controllability beam $[p\mathbf{I}-\mathbf{A}\ -\mathbf{B}]$, with $\mathbf{B}$ of full rank (injective), gives very directly:

$$card\,\{c_j\} = card\,\{c_j \geq 1\} = m := rank(\mathbf{B})$$

$$card\,\{c_j \geq i\} = rank([\mathbf{B}\ \mathbf{AB} \ldots \mathbf{A}^{i-1}\mathbf{B}]) - rank([\mathbf{B}\ \mathbf{AB} \ldots \mathbf{A}^{i-2}\mathbf{B}]), \text{ for } i \geq 2.$$

4.3.2. Observability indices

The observability indices and the invariant factors of the non-observable part (if it exists) of the pair $(\mathbf{C}, \mathbf{A})$ (see section 4.2.3) form a set of full invariants under the action of the transformation group $(\mathbf{T}, \mathbf{R}, \mathbf{H})$ where $\mathbf{T}$ and $\mathbf{H}$ designate basis changes on the state and on the output respectively and $\mathbf{R}$ is an output injection. This "injection" group is defined by:

$$(\mathbf{C}, \mathbf{A}) \xrightarrow{\ (\mathbf{T},\mathbf{R},\mathbf{H})\ } (\mathbf{T}^{-1}(\mathbf{A} + \mathbf{RC})\mathbf{T}, \mathbf{HCT})$$

A more "traditional" definition of the observability indices, noted by $\{o_1, o_2, \ldots, o_l\}$ where l is the size of the output space, can be found in connection to the general characterization of minimal indices per rows such as described in [4.29], particularized to the observability beam $[p\mathbf{I}-\mathbf{A}^T\ -\mathbf{C}^T]^T$, with $\mathbf{C}$ of full rank (subjective):

$$card\,\{o_j\} = card\,\{o_j \geq 1\} = l := rank(\mathbf{C})$$

$$card\,\{o_j \geq i\} = rank\begin{bmatrix} \mathbf{C} \\ \mathbf{CA} \\ \ldots \\ \mathbf{CA}^{i-1} \end{bmatrix} - rank\begin{bmatrix} \mathbf{C} \\ \mathbf{CA} \\ \ldots \\ \mathbf{CA}^{i-2} \end{bmatrix}, \text{ for } i \geq 2.$$

4.3.3. Infinite zeros

As introduced in section 4.2.4, Morse's canonical form, $(\mathbf{A}_M, \mathbf{B}_M, \mathbf{C}_M)$ described in [4.38], is obtained from the initial system, let us say $(\mathbf{A}, \mathbf{B}, \mathbf{C})$, by the action of an element of the "feedback and injection" transformation group, let us say $(\mathbf{T}_M, \mathbf{F}_M, \mathbf{G}_M, \mathbf{R}_M, \mathbf{H}_M)$. This form is in fact maximally non-controllable and non-observable. It is in fact important, based on its particular structure, to verify that the system transfer matrix written in Morse's canonical form will use only the part having the index "4" linked to the infinite elementary divisors and has a diagonal form:

$$C_M[p\mathbf{I}-\mathbf{A}_M]^{-1}\mathbf{B}_M = C_4[p\mathbf{I}-\mathbf{A}_4]^{-1}\mathbf{B}_4 = diag\{p^{-n_i}\}$$

where n_i, $i = 1$ to r is the size of each block in part "4" which is in controllable and observable canonical form. For example, the following form $(\mathbf{A}_4, \mathbf{B}_4, \mathbf{C}_4)$ where, to simplify writing, all the non-specified terms are zero:

$$\mathbf{A}_4 = \begin{bmatrix} 0 & & & & & \\ & 0 & 1 & & & \\ & 0 & 0 & & & \\ & & & 0 & 1 & 0 \\ & & & 0 & 0 & 1 \\ & & & 0 & 0 & 0 \end{bmatrix}, \quad \mathbf{B}_4 = \begin{bmatrix} 1 & \\ & 0 \\ & 1 \\ & 0 \\ & 0 \\ & 1 \end{bmatrix}, \quad \mathbf{C}_4 = \begin{bmatrix} 1 & & \\ & 1 & 0 \\ & & 1 & 0 & 0 \end{bmatrix}$$

corresponds to the list $\{n_i\}= \{1,2,3\}$. The corresponding system has 3 infinite zeros, of orders 1, 2 and 3. This is the result of the transfer diagonal structure of Morse's canonical form and because the transformations that lead to the canonical form of the system maintain the structure of zeros infinity.

Indeed, based on the relations:

$$\mathbf{A}_M = \mathbf{T}_M^{-1}(\mathbf{A}+\mathbf{B}\mathbf{F}_M +\mathbf{R}_M\mathbf{C})\mathbf{T}_M; \quad \mathbf{B}_M = \mathbf{T}_M^{-1}\mathbf{B}\mathbf{G}_M; \quad \mathbf{C}_M = \mathbf{H}_M\mathbf{C}\mathbf{T}_M$$

the passage from $(\mathbf{A},\mathbf{B},\mathbf{C})$ to $(\mathbf{A}_M, \mathbf{B}_M, \mathbf{C}_M)$ is reflected in the following relation:

$$\mathbf{C}_M(p\mathbf{I}-\mathbf{A}_M)^{-1}\mathbf{B}_M = \mathbf{B}_1(p)[\mathbf{C}(p\mathbf{I}-\mathbf{A})^{-1}\mathbf{B}]\mathbf{B}_2(p)$$

with:

$$\mathbf{B}_1(p) := [\mathbf{I}-\mathbf{F}_M(p\mathbf{I}-\mathbf{A})^{-1}\mathbf{B}]^{-1}\mathbf{G}_M \quad \text{and}$$

$$\mathbf{B}_2(p) := \mathbf{H}_M[\mathbf{I}-\mathbf{C}(p\mathbf{I}-\mathbf{A}-\mathbf{B}\mathbf{F}_M)^{-1}\mathbf{R}_M]^{-1}$$

Transfers $\mathbf{B}_1(p)$ and $\mathbf{B}_2(p)$ have the particular property of being biproper matrices: a biproper matrix is a proper matrix (i.e. whose bound is finite when p tends toward infinite), reversible and when reversed, also proper. A biproper matrix is no more than a unimodular matrix (see section 4.1.3), but on the ring of eigenfunctions. A scalar biproper is any transfer function in which the numerator and denominator have the same degree. A unimodular (polynomial) has neither pole nor finite zero (its Smith's form is reduced to the identity; see section 4.1.4), a biproper on the other hand has only poles and finite zeros and it cannot simplify (by product) any singularity to infinity. The behaviors at infinite of $(\mathbf{A},\mathbf{B},\mathbf{C})$ and $(\mathbf{A}_M, \mathbf{B}_M, \mathbf{C}_M)$ are thus identical. The behavior of $(\mathbf{A}_M, \mathbf{B}_M, \mathbf{C}_M)$ is roughly described by the list of $p^{-n}i$. The integers n_i, which are equal in number to the rank of the system, are called the orders of infinite zeros of the system considered.

In a purely "transfer matrix" context, we thus define Smith's canonical form to infinity, which is the canonical representation under the action of the transformation group by multiplications, on the left and right, through bipropers.

The general relations of the [4.30] type also make it possible to geometrically characterize the orders of infinite zeros.

4.3.4. *Invariants, transmission finite zeros*

As previously recalled, any multiplication of a given transfer by a unimodular preserves the finite singularities of this transfer (a unimodular has only poles and infinite zeros). The group of transformations obtained by multiplications on the right and left by unimodulars makes it possible to associate with each transfer matrix its canonical form, called Smith McMillan's form, from which the so-called transmission poles and zeros can be calculated (linked to the transfer, i.e. to the controllable and observable part of the system considered). Synthetically, we can obtain it as follows:

– write the departure transfer, let us say $\mathbf{T}(p)$, as $\mathbf{T}(p) = [1/\mathbf{d}(p)]\,\mathbf{N}(p)$, where $\mathbf{d}(p)$ is the LMCD (the lowest multiple common denominator) of all the denominators present in $\mathbf{T}(p)$;

– write $\mathbf{N}(p)$ in Smith's canonical form (by unimodular actions on the right and left);

– divide each term of the diagonal thus obtained by $\mathbf{d}(p)$ and perform all the numerators/denominators possible simplifications.

Hence, we reach a diagonal formula (always with r elements, r being the rank of the system), of type $\varepsilon_i(p)\,/\psi_i(p)$, where $\varepsilon_1(p)$ divides $\varepsilon_2(p)$, ..., divides $\varepsilon_r(p)$ and $\psi_r(p)$ divides $\psi_{r-1}(p)$,..., divides $\psi_1(p)$. The transmission poles and zeros of $\mathbf{T}(p)$ correspond to the roots, respectively, of the denominators $\psi_i(p)$ and the numerators $\varepsilon_i(p)$.

These transmission structures are related to the "open loop" transfer. They are invariant under basis changes but do not remain invariant under the action of transformations such as state return or output injection.

If we consider a transfer state realization $\mathbf{T}(p)$, let us say $(\mathbf{A},\mathbf{B},\mathbf{C})$, the invariant zeros defined from the finite elementary divisors of the associated system matrix (see section 4.2.4) are invariant under Morse's group (basis changes, state returns and output injections). If the state realization is minimal, the invariant zeros coincide with the transmission zeros. Otherwise, the transmission zeros form only a sub-group of all invariant zeros.

4.3.5. *Notes and references*

The various structures presented in this section, such as controllability/observability indices and finite/infinite zeros are described in detail in [KAI 80, ROS 70] and many other works.

4.4. **An introduction to a structural approach of the control**

The objective of this section is to illustrate, based on relatively traditional control problems, the fundamental role played by certain structures (and we will dedicate our attention to infinite and finite zeros) in the existence of solutions. We will consider in particular the disturbance rejection and the diagonal decoupling.

Let us consider a stationary linear system in which $\mathbf{u}(t)$ represents a control input with m components, $\mathbf{d}(t)$ a disturbance input with q components and $\mathbf{y}(t)$ an output to control with l components and described by the state model:

$$\begin{cases} \dot{\mathbf{x}}(t) = \mathbf{A}\mathbf{x}(t) + \mathbf{B}\mathbf{u}(t) + \mathbf{E}\mathbf{d}(t) \\ \quad\ \mathbf{y}(t) = \mathbf{C}\mathbf{x}(t) \end{cases} \qquad [4.39]$$

to which the following transfer matrices are also associated:

$$\mathbf{T_u}(p) := \mathbf{C}(p\mathbf{I} - \mathbf{A})^{-1}\mathbf{B} \quad \text{and} \quad \mathbf{T_d}(p) := \mathbf{C}(p\mathbf{I} - \mathbf{A})^{-1}\mathbf{E} \qquad [4.40]$$

The problem of disturbance rejection by state return is formulated as follows: finding, if it exists, a state return having the form $\mathbf{u}(t) = \mathbf{F}\mathbf{x}(t) + \mathbf{L}\mathbf{d}(t)$ so that, for the system thus looped, the transfer matrix between $\mathbf{d}(p)$ and $\mathbf{y}(p)$ is identically zero. When disturbance $\mathbf{d}(t)$ is not measured, we impose $\mathbf{L} = \mathbf{0}$. The problem of disturbance rejection with internal stability consists of researching, if they exist, $\mathbf{F}$ solutions so that, in addition, $(\mathbf{A} + \mathbf{B}\mathbf{F})$ is stable.

The problem of diagonal decoupling by regular state return is formulated as follows: finding, if it exists, a regular state return having the form $\mathbf{u}(t) = \mathbf{F}\mathbf{x}(t) + \mathbf{G}\mathbf{v}(t)$, with reversible square $\mathbf{G}$ so that, for the system thus looped, the transfer matrix between $\mathbf{v}(p)$ and $\mathbf{y}(p)$ is diagonal (with principal diagonal), i.e. in the form:

$$\mathbf{T_{F,G}}(p) := \mathbf{C}(p\mathbf{I} - \mathbf{A} - \mathbf{BF})^{-1}\mathbf{BG} = \left[diag\{h_1(p), \cdots, h_l(p)\} \vdots 0 \right]$$

The decoupling problem with internal stability consists of researching, if they exist, $\mathbf{F}$ solutions so that, in addition, $(\mathbf{A} + \mathbf{BF})$ is stable.

4.4.1. *Disturbance rejection and decoupling: existence of solutions*

The action of a state return type control law, as described in the previous section, as well as for rejection and for decoupling is translated in terms of transfer matrices by the multiplication on the right by a particular biproper matrix. Since such a transformation maintains the structure of infinite zeros, it is very natural to see conditions of existence of solutions for this type of structure appear. To illustrate this, we use the pre-compensator, which is equivalent to the control law selected.

For disturbance rejection, the transfer between $\mathbf{d}(p)$ and $\mathbf{y}(p)$ for the compensated system by the control law $\mathbf{u}(t) = \mathbf{F}\mathbf{x}(t) + \mathbf{L}\mathbf{d}(t)$ is. equal to $\mathbf{T_u}(p)\mathbf{C}(p) + \mathbf{T_d}(p)$, that we want to cancel, with:

$$\mathbf{C}(p) := [\mathbf{I} - \mathbf{F}(p\mathbf{I} - \mathbf{A})^{-1}\mathbf{B}]^{-1}[\mathbf{F}(p\mathbf{I} - \mathbf{A})^{-1}\mathbf{E} + \mathbf{L}]$$

It is easy to realize that $\mathbf{C}(p)$ is always proper, even strictly proper (i.e. the bound of $\mathbf{C}(p)$ is equal to zero when p tends toward infinity) when $\mathbf{L} = \mathbf{0}$, i.e. when the disturbance is not available for the control law.

The equation reflecting the objective of this rejection, i.e. $\mathbf{T_u}(p)\mathbf{C}(p) + \mathbf{T_d}(p) = \mathbf{0}$, can be rewritten as:

$$[\mathbf{T_u}(p) \vdots \mathbf{T_d}(p)] \begin{bmatrix} \mathbf{I} & \mathbf{C}(p) \\ 0 & \mathbf{I} \end{bmatrix} = [\mathbf{T_u}(p) \vdots \quad \mathbf{0}] \tag{4.41}$$

In this equation, the matrix where $\mathbf{C}(p)$ intervenes is biproper (since $\mathbf{C}(p)$ is proper). A necessary condition for [4.41] to have at least one proper solution is for $[\mathbf{T_u}(p) \vdots \mathbf{T_d}(p)]$ and $\mathbf{T_u}(p)$ to have exactly the same orders of infinite zeros (because this structure is invariant under multiplication by a biproper). It turns out that this condition is also sufficient. We can also show quite simply that this necessary and sufficient condition can be reduced to the comparison of two integers. We will designate "infinite rollout" the sum of orders of infinite zeros for a given system.

The disturbance rejection is solvable by $\mathbf{u}(t) = \mathbf{F}\mathbf{x}(t) + \mathbf{L}\mathbf{d}(t)$ type state return if and only if $(\mathbf{A},\mathbf{B},\mathbf{C})$ and $(\mathbf{A}, [\mathbf{B} \mid \mathbf{E}],\mathbf{C})$ have the same rank and the same infinite rollout.

Variants of this type of result exist when the disturbance is not measured, as well as when the state is not measured. In the second case, the existence of control laws is dealt with very similar measurement dynamic returns.

For the decoupling problem, the action of a regular state return $\mathbf{u}(t) = \mathbf{F}\mathbf{x}(t) + \mathbf{G}\mathbf{v}(t)$, with reversible square $\mathbf{G}$, is equivalent to the transfer multiplication $\mathbf{T_u}(p)$ by the equivalent biproper pre-compressor:

$$\mathbf{C}(p) := [\mathbf{I} - \mathbf{F}(p\mathbf{I} - \mathbf{A})^{-1}\mathbf{B}]^{-1}\mathbf{G}$$

Let us consider, in order to simplify the explanation, the case of square systems (having as many input components as control components) and reversible systems. The objective of decoupling is then expressed by the equation:

$$\mathbf{T_u}(p)\mathbf{C}(p) = diag\{h_1(p), \cdots, h_l(p)\} \tag{4.42}$$

Based on the diagonal form desired, a necessary condition for this equation to admit a biproper solution is that, on the one hand, the system is seen in its entirety, and on the other hand the reunion of all sub-systems row by row have exactly the same orders of infinite zeros (because this structure is invariant under a biproper multiplication). It turns out that this condition is equally sufficient. In addition, we can also show that this necessary and sufficient condition can be expressed with a single integer via the infinite rollout.

For a system supposed reversible on the right (i.e. whose transfer is of full rank per rows), decoupling is solvable by $\mathbf{u}(t) = \mathbf{F}\mathbf{x}(t) + \mathbf{G}\mathbf{v}(t)$ type regular state return if and only if the infinite rollout of $(\mathbf{A}, \mathbf{B}, \mathbf{C})$ is equal to the sum of infinite rollouts calculated for each row sub-system $(\mathbf{A}, \mathbf{B}, \mathbf{c}_i)$, where $\mathbf{c}_i$ designates the ith row of $\mathbf{C}$.

4.4.2. *Disturbance rejection and decoupling: existence of stable solutions*

When the (natural) constraint of internal stability is added, the unstable zeros (if they exist) will have a similar role to the one of the infinite zeros with respect to the existence of solutions. The simplest way to be sure of this is to be able to formulate the control problem from the "transfer" equation. Before that, we must of course assume that the system considered can be stabilized. We can, however, assume it is already stable in open loop (if not a first stabilizing loop is to be used). The internal stability of the compensated system is hence translated simply by the necessary stability of the compensator researched. We must then solve a [4.41] or [4.42] type equation, on the ring of eigen and stable functions, and not only of eigenfunctions. The infinite zeros or the zeros with unstable values will then intervene as fundamental ingredients for the existence of solutions. For a system given under its state description, we will designate by "infinite and unstable rollout" the integer obtained by calculating the sum of the infinite rollout with the total number of unstable invariant zeros (sum of orders of multiplicity, irrespective of the corresponding particular (unstable) location). Thus, we obtain fairly simply the following results:

– the disturbance rejection is solvable with internal stability by $\mathbf{u}(t) = \mathbf{Fx}(t) + \mathbf{Ld}(t)$ state return if and only if $(\mathbf{A}, \mathbf{B}, \mathbf{C})$ and $(\mathbf{A}, [\mathbf{B} \mid \mathbf{E}], \mathbf{C})$ have the same rank and the same infinite and unstable rollout;

– for a system assumed to be reversible on the right, decoupling is solvable with internal stability by $\mathbf{u}(t) = \mathbf{Fx}(t) + \mathbf{Gv}(t)$ type regular state return if and only if the unstable and infinite rollout of $(\mathbf{A}, \mathbf{B}, \mathbf{C})$ is equal to the sum of infinite and unstable rollouts calculated for each row sub-system $(\mathbf{A}, \mathbf{B}, \mathbf{c_i})$, where $\mathbf{c_i}$ designates the ith row of $\mathbf{C}$.

4.4.3. *Disturbance rejection and decoupling: flexibility in the location of poles/ fixed poles*

The results presented in the two previous sections are basically multi-variable in nature. They are obviously all the more valid in particular cases like, for example, in mono-variable cases.

In this broad context of multi-variable systems, when the control problem considered is solvable, "the" solution is generally non-unique. Apart from researching, among all possible solutions, at least one stabilizing solution, we are often tempted to take advantage of the remaining degrees of freedom in order to fulfill supplementary objectives and especially to target certain poles (not only stable but, for example, sufficiently damped) for the solution looped system. The question of the possible flexibility in terms of the poles' position then arises.

For the various control problems mentioned in this chapter (i.e. model pursuit, disturbance rejection or decoupling, etc.), it turns out that the simple fact of wanting to solve this "exact" problem leads to the inevitable appearance of an entire set of poles which are present in any solution. These poles are the "fixed poles" of the problem considered. Knowing them makes it possible to delimit the constraints imposed by the problem in terms of modifications of dynamics. Reaching a few "controllability" type (minimal) hypotheses, we can then find solutions that make it possible to position all the other poles, except for, obviously, these fixed poles, which, again, find their origin in the non-coincidence of certain structures of invariant finite zeros. We can in fact set the following results, which make it possible to look at the previous section as a particular case. We assume, for all the mentioned, cases that the control problem mentioned is solvable, in the sense of section 4.4.1:

– the fixed poles of disturbance rejection by state return coincide with the invariant zeros of $(\mathbf{A}, \mathbf{B}, \mathbf{C})$ which are not invariant zeros of $(\mathbf{A}, [\mathbf{B} \mid \mathbf{E}],\mathbf{C})$. When the extended pair $(\mathbf{A}, [\mathbf{B} \mid \mathbf{E}])$ is "globally" controllable (a totally natural hypothesis), all the other poles (other than these fixed poles) can be positioned by a proper choice of state return;

– the fixed poles of the decoupling by regular state return coincide with the invariant zeros of $(\mathbf{A},\mathbf{B},\mathbf{C})$ which are not in the group obtained by bringing together the invariant zeros of each row sub-system $(\mathbf{A}, \mathbf{B}, \mathbf{c}_i)$, where $\mathbf{c}_i$ designates the ith row of $\mathbf{C}$. When the pair $(\mathbf{A},\mathbf{B})$ is controllable, all the other poles (other than these fixed poles) can be positioned by a proper choice of the state return;

– the existence of stable solutions is simply equivalent to the juxtaposition of two conditions: existence of solutions (section 4.4.1) and stability of all (possible) fixed poles.

4.4.4. *Notes and references*

The disturbance rejection and the diagonal decoupling were the object of numerous contributions, i.e. [BAS 92, TRE 01, WOH 85] for geometric treatments. Additional information on the treatment of this kind of control problem based on rational equations and especially the use of various rings in order to find solutions, can be found in [VID 85]. The results pertaining to the existence conditions expressed in terms of infinite structures were the object of several theses, such as [DIO 83, MAL 85]. Among the most recent contributions that complete the presentation of the existence of stable solutions, the use of rollouts and in particular the fixed poles and the remaining degrees of freedom, we can mention [MAL 93, MAL 97] and [MAR 94, MAR 99].

4.5. Conclusion

This chapter is to be considered as an introduction to a structural approach of the control. Its main objective was to introduce, by a simultaneous use of geometric and algebraic approaches, an entire set of structures closely related to the system considered and to illustrate the fundamental role they play in solving the control problems.

The presentation was limited to the linear, stationary and finite size case. Extensions of certain results are available for the more general classes of systems, for example, non-linear [MOO 87] or with delays [RAB 99]. The object of this final section is to mention another extension in the field of optimization.

4.5.1. *Optimal attenuation of disturbance*

When the "exact" disturbance rejection, as formulated at the beginning of section 4.4, is not solvable with stability due to the presence of at least one unstable fixed pole, the designer has the alternative to tone down the objective of the control. Instead of targeting an exact rejection, we can limit ourselves to an attenuation (optimal if possible) of this disturbance, as per a certain standard.

This point of view was largely developed by several authors such as Saberi, Sannuti and Stoorvogel (see, for example, [SAB 96]). Due to a reformulation of the optimization problem into an exact problem where the matrices of the corresponding state model are slightly modified [STO 92], the solutions of the H_2-optimal attenuation problems can be obtained from the analysis of exact rejection problems. The same applies for the H_2-optimal fixed poles, which are present in any optimal solution (see [CAM 00]).

4.6. Bibliography

[BAS 92] BASILE G., MARRO G., *Controlled and conditioned invariants in linear system theory,* Prentice-Hall, Englewood Cliffs, 1992.

[BRU 0] BRUNOVSKY P., "A classification of linear controllable systems", *Kybernetika,* vol. 6, p. 173-188, 1970.

[CAM 00] CAMART J.F., Contribution à l'étude des contraintes structurelles du rejet de perturbation et du découplage: résolutions exactes et atténuations optimales, PhD Thesis, Nantes, 2000.

[DIO 83] DION J.M., Sur la structure à l'infini des systèmes linéaires, Thesis, Grenoble, 1983.

[GAN 66] GANTMACHER F.R., *Théorie des matrices*, Dunod, Paris, 1966.

[KAI 80] KAILATH T., *Linear systems*, Prentice-Hall, Englewood Cliffs, 1980.

[LAR 02] DE LARMINAT P. (ed.), *Commande des systèmes linéaires*, Hermès, IC2 series, Paris, 2002.

[LOI 86] LOISEAU J.J., Contribution à l'étude des sous-espaces presque invariants, PhD Thesis, Nantes, 1986.

[MAL 85] MALABRE M., Sur le rôle de la structure à l'infini et des sous-espaces presque invariants dans la résolution de problèmes de commande, Thesis, Nantes, 1985.

[MAL 93] MALABRE M., MARTINEZ-GARCIA J.C., "The modified disturbance rejection problem with stability: a structural approach", *Proceedings of the 2nd European Control Conference, ECC'93*, p. 1119-1124, 1993.

[MAL 97] MALABRE M., MARTINEZ-GARCIA J.C., DEL MURO CUELLAR B., "On the fixed poles for disturbance rejection", *Automatica*, vol. 33, no. 6, p. 1209-1211, 1997.

[MAR 94] MARTINEZ-GARCIA J.C., MALABRE M., "The row by row decoupling problem with stability: a structural approach", *IEEE Transactions on Automatic Control*, AC-39, no. 12, p. 2457-2460, 1994.

[MAR 99] MARTINEZ-GARCIA J.C., MALABRE M., DION J.M., COMMAULT C., "Condensed structural solutions to the disturbance rejection and decoupling problems with stability", *International Journal of Control*, vol. 72, no. 15, p. 1392-1401, 1999.

[MOR 73] MORSE A.S., "Structural invariants of linear multivariable systems", *SIAM Journal of Control & Optimization*, vol. 11, p. 446-465, 1973.

[MOO 87] MOOG C.H., Inversion, découplage et poursuite de modèle des systèmes non linéaires, Thesis, Nantes, 1987.

[RAB 99] RABAH R., MALABRE M., "On the structure at infinity of linear delay systems with application to the disturbance decoupling problem", *Kybernetika*, vol. 35, p. 668-680, 1999.

[ROS 70] ROSENBROCK H.H., *State space and multivariable theory*, John Wiley, New York, 1970.

[SAB 96] SABERI A., SANNUTI P., STOORVOGEL A.A., "H_2 optimal controllers with measurement feedback for continuous-time systems – flexibility in closed-loop pole placement", *Automatica*, vol. 32, no. 8, p. 120-1209, 1996.

[STO 92] STOORVOGEL A.A., *The H_∞ control problem: a state space approach*, Prentice-Hall, Englewood Cliffs, 1992.

[THO 73] THORP J.S., "The singular pencil of a linear dynamical system", *International Journal of Control*, vol. 18, p. 577-596, 1973.

[TRE 01] TRENTELMAN H.L., STOORVOGEL A.A., HAUTUS M., *Control theory for linear systems*, Springer Verlag, London, 2001.

[VID 85] VIDYASAGAR M., *Control system synthesis: a factorization approach*, MIT Press, Cambridge, Massachusetts, 1985.

[WIL 65] WILKINSON J.H., *The algebraic eigenvalue problem*, Clarendon Press, Oxford, 1965.

[WON 85] WONHAM W.M., *Linear multivariable control: a geometric approach*, Springer Verlag, New York, 3rd edition, 1985.

Chapter 5

Signals: Deterministic and Statistical Models

5.1. Introduction

This chapter is dedicated to signal modeling procedures and in particular to stationary random signals. After having discussed the spectral characterization of deterministic signals, with the help of the Fourier transform and energy spectral density, we will now define the power spectral density of stationary random signals. We will show that a simple modeling by linear shaper filter excited by a white noise makes it possible to approach a spectral density with the help of a reduced number of parameters and we will present a few standard structures of shaper filters. Next, we will extend this modeling to the case of linear processes with deterministic input, in which the noises and disturbances can be considered as additional stationary noises. Further on, we will present the representation in the state space of such a modeling and the relation with the Markovian processes.

5.2. Signals and spectral analysis

A continuous-time deterministic signal $y(t), t \in \Re$ is, by definition, a *function* of $\Re$ in $\mathcal{C}$:

$$y : \Re \longrightarrow \mathcal{C}$$

$$t \longmapsto y(t)$$

where variable t designates time. In short, we speak of a *continuous signal* even if the signal considered is not continuous in the usual mathematical sense.

Chapter written by Eric LE CARPENTIER.

A discrete-time deterministic signal $y[k], k \in \mathcal{Z}$ is, by definition, a *sequence* of complex numbers:

$$y = \big(y[k]\big)_{k \in \mathcal{Z}}$$

In short, we often speak of a *discrete signal*. In general, the signals considered, be they continuous-time or discrete-time, have real values, but the generalization to complex signals done here does not entail any theoretical problem.

The spectral analysis of deterministic signals consists of decomposing them into simpler signals (for example, sine curves), in the same way as a point in space is located by its three coordinates. The most famous technique is the Fourier transform, from the French mathematician J.B. Fourier (1768–1830), which consists of using cisoid functions as basic vectors.

The Fourier transform $\widehat{y}(f)$ of a continuous-time signal $y(t)$ is a function of the form $\widehat{y} : f \longmapsto \widehat{y}(f)$ of a real variable with complex number value, which is defined for any f by:

$$\widehat{y}(f) = \int_{-\infty}^{+\infty} y(t)\, e^{-j\,2\pi\,ft}\, \mathrm{d}t \tag{5.1}$$

We note from now on that if variable t is homogenous to a certain time, then variable f is homogenous to a certain frequency. We will admit that the Fourier transform is defined (i.e. the integral above converges) if the signal has finite energy. The Fourier transform does not entail any loss of information. Indeed, knowing $\widehat{y}(f), y(t)$ can be rebuilt by the following reverse formula; for any t:

$$y(t) = \int_{-\infty}^{+\infty} \widehat{y}(f)\, e^{j\,2\pi\,ft}\, \mathrm{d}f \tag{5.2}$$

The Fourier transform is in fact the restriction of the two-sided Laplace transform $\breve{y}(s)$ to the axis of complex operators: $\widehat{y}(f) = \breve{y}(j\,2\pi\,f)$ with, for any $s \in \mathcal{C}$:

$$\breve{y}(s) = \int_{-\infty}^{+\infty} y(t)\, e^{-st}\, \mathrm{d}t \tag{5.3}$$

Likewise, the Fourier transform (or normalized frequency transform) $\widehat{y}(\nu)$ of a discrete-time signal $y[k]$ is a function of the form:

$$\widehat{y} : \Re \longrightarrow \mathcal{C}$$

$$\nu \longmapsto \widehat{y}(\nu)$$

defined for any ν by:

$$\widehat{y}(\nu) = \sum_{k=-\infty}^{+\infty} y[k]\, e^{-\jmath\, 2\pi\, \nu k} \tag{5.4}$$

We will accept that the Fourier transform of a discrete-time signal is defined (i.e. the above sequence converges) if the signal has finite energy. It is periodic of period 1. It is in fact the restriction of two-sided z transform $\breve{y}(z)$ to unit circle: $\widehat{y}(\nu) = \breve{y}(e^{\jmath\, 2\pi\, \nu})$ with, for any $z \in \mathcal{C}$:

$$\breve{y}(z) = \sum_{k=-\infty}^{+\infty} y[k]\, z^{-k} \tag{5.5}$$

The Fourier transform does not entail any loss of information. Indeed, knowing $\widehat{y}(\nu)$, we can rebuild $y[k]$ by the following reverse formula; for any k:

$$y[k] = \int_{-\frac{1}{2}}^{+\frac{1}{2}} \widehat{y}(\nu)\, e^{\jmath\, 2\pi\, \nu k}\, \mathrm{d}\nu \tag{5.6}$$

The Fourier transform (continuous-time or discrete-time) verifies the following fundamental problem: it transforms the convolution integral into a simple product. Let $y_1(t)$ and $y_2(t)$ be two real variable functions; the convolution integral $(y_1 \otimes y_2)(t)$ is defined for any t by:

$$(y_1 \otimes y_2)(t) = \int_{-\infty}^{+\infty} y_1(\tau)\, y_2(t - \tau)\, \mathrm{d}\tau \tag{5.7}$$

Likewise, let $y_1[k]$ and $y_2[k]$ be two sequences; their convolution integral $(y_1 \otimes y_2)[k]$ is defined for any k by:

$$(y_1 \otimes y_2)[k] = \sum_{m=-\infty}^{+\infty} y_1[m]\, y_2[k - m] \tag{5.8}$$

The convolution integral verifies the commutative and associative properties, and the neutral element is:

- $\delta(t)$ Dirac impulse for functions ($\delta(t) = 0$ if we have $t \neq 0$, $\int_{-\infty}^{+\infty} \delta(t)\, \mathrm{d}t = 1$);
- $\delta[k]$ Kronecker sequence for sequences ($\delta[0] = 1, \delta[k] = 0$ if we have $k \neq 0$).

In addition, the convolution of a function or sequence with delayed neutral element delays it with the same quantity. It is easily verified that the Fourier transform of the convolution integral is the product of transforms:

$$(y_1 \otimes y_2)^{\frown} = \widehat{y}_1\, \widehat{y}_2 \tag{5.9}$$

On the other hand, the Fourier transform preserves the energy (Parseval theorem). Indeed, the energy of a continuous-time signal $y(t)$ or of a discrete-time signal $y[k]$ can be calculated by the square integration of the Fourier transform module $\widehat{y}(f)$ or its normalized frequency transform $\widehat{y}(\nu)$:

– continuous-time signals: $\int_{-\infty}^{+\infty} |y(t)|^2 \, \mathrm{d}t = \int_{-\infty}^{+\infty} |\widehat{y}(f)|^2 \, \mathrm{d}f$;

– discrete-time signals: $\sum_{k=-\infty}^{+\infty} |x[k]|^2 = \int_{-\frac{1}{2}}^{+\frac{1}{2}} |\widehat{x}(\nu)|^2 \, \mathrm{d}\nu$.

The function or sequence $|\,\widehat{y}\,|^2$ is called *a power spectrum*, or *energy spectral density of signal* y because its integral (or its summation) returns the energy of signal y.

The Fourier transform is defined only for finite energy signals and can be extended to periodic or impulse signals (with the help of the mathematical theory of distributions). We will give a few examples below.

EXAMPLE 5.1 (DIRAC IMPULSE). The transform of Dirac impulse is the unit function:

$$\widehat{\delta}(f) = 1_\Re(f) \qquad\qquad [5.10]$$

EXAMPLE 5.2 (UNIT CONSTANT). It is not of finite energy, but admits a Fourier transform in the sense of distribution theory, which is a Dirac impulse:

$$\widehat{1}_\Re(f) = \delta(f) \qquad\qquad [5.11]$$

EXAMPLE 5.3 (CONTINUOUS-TIME CISOID). We have the following transformation:

$$y(t) = e^{j\,2\pi\,f_0 t} \qquad \widehat{y}(f) = \delta(f - f_0) \qquad\qquad [5.12]$$

Therefore, this means that the Fourier transform of the frequency cisoid f_0 is an impulse centered in f_0. By using the linearity of the Fourier transform, we easily obtain the Fourier transform of a real sine curve, irrespective of its initial phase; in particular:

$$y(t) = \cos(2\pi\,f_0 t) \qquad \widehat{y}(f) = \frac{1}{2}[\delta(f - f_0) + \delta(f + f_0)] \qquad\qquad [5.13]$$

$$y(t) = \sin(2\pi\,f_0 t) \qquad \widehat{y}(f) = \frac{-j}{2}[\delta(f - f_0) - \delta(f + f_0)] \qquad\qquad [5.14]$$

EXAMPLE 5.4 (KRONECKER SEQUENCE). We immediately obtain:

$$\widehat{\delta}(\nu) = 1_\Re(\nu) \qquad\qquad [5.15]$$

EXAMPLE 5.5 (UNIT SEQUENCE). The Fourier transform of the constant sequence $1_{\mathcal{Z}}[k]$ is the impulse frequency comb Ξ_1:

$$\widehat{1}_{\mathcal{Z}}(\nu) = \Xi_1(\nu) = \sum_{k=-\infty}^{+\infty} \delta(\nu - k) \qquad [5.16]$$

EXAMPLE 5.6 (DISCRETE-TIME CISOID). We have the following transform:

$$y[k] = e^{j\,2\pi\,\nu_0 k} \qquad \widehat{y}(f) = \Xi_1(\nu - \nu_0) \qquad [5.17]$$

Thus, this means that the Fourier transform of the frequency cisoid ν_0 is a frequency comb centered in ν_0.

Very often, the spectral analysis of deterministic signals is reduced to visualizing the energy spectral density, but numerous physical phenomena come along with disturbing phenomena, called "noises"; for example, mechanical systems generate vibratory or acoustic signals which are not periodic and have infinite energy.

The mathematical characterization of such signals is particularly well formalized in the case of stationary and ergodic random signals:

– *random*: this means that, in the same experimental conditions, two different experiences generate two different signals. The mathematical treatment can thus be only probabilistic, the signal observed being considered as the realization of a random variable;

– *stationary*: the statistical characteristics are then independent of the time origin;

– *ergodic*: any statistical information is included in a unique realization of infinite duration.

In any case, the complete characterization of such signals is expressed with the help of the combined probability law of the values taken by the signal in different instants, irrespective of these instants and their number. For example, for a Gaussian random signal, this combined law is Gauss' probability law. For a white random signal (or independent), this combined density is equal to the product of marginals (to clarify a current confusion, we note that these two notions are not equivalent: a Gaussian signal can be white or not, a white signal can be Gaussian or not). In practice, we have the second order statistical analysis that deals only with the first and second order moments, i.e. the mean and the autocorrelation function.

A discrete-time random signal $\mathbf{y}[k], k \in \mathcal{Z}$ is called stationary in the broad sense if its mean m_y and its autocorrelation function $r_{yy}[\kappa]$ defined by:

$$\begin{cases} m_y = E(\mathbf{y}[k]) \\ r_{yy}[\kappa] = E((\mathbf{y}[k] - m_y)^* \, (\mathbf{y}[k+\kappa] - m_y)) \qquad \forall \kappa \in \mathcal{Z} \end{cases} \qquad [5.18]$$

are independent of index k, i.e. independent of the time origin. $\sigma_y^2 = r_{yy}[0]$ is the variance of the signal considered. $\frac{r_{yy}[\kappa]}{\sigma_y^2}$ is the correlation coefficient between the signal at instant k and the signal at instant $k + \kappa$. It is traditional to remain limited only to the mean and the autocorrelation function in order to characterize a stationary random signal and this even if the characterization, referred to as *of second order*, is very incomplete (it is sufficient only for the Gaussian signals).

In practice, there is only one realization $y[k]$, $k \in \mathcal{Z}$ of a random signal $\mathbf{y}[k]$ for which we can define its time mean $\langle y[k] \rangle$:

$$\langle y[k] \rangle = \lim_{N \to \infty} \frac{1}{2N + 1} \sum_{k=-N}^{N} y[k] \tag{5.19}$$

The random signal $\mathbf{y}[k]$ is called *ergodic for the mean* if mean m_y is equal to the time mean of any realization $y[k]$ of this random signal:

$$E(\mathbf{y}[k]) = \langle y[k] \rangle \qquad \text{ergodicity for the mean} \tag{5.20}$$

In what follows, we will suppose that the random signal $\mathbf{y}[k]$ is ergodic for the mean and, to simplify, of zero mean.

The random signal $\mathbf{y}[k]$ is called *ergodic for the autocorrelation* if the autocorrelation function $r_{yy}[\kappa]$ is equal to the time mean $\langle y^*[k]\, y[k + \kappa] \rangle$ calculated from any realization $y[k]$ of this random signal:

$$E\left(\mathbf{y}^*[k]\, \mathbf{y}[k + \kappa]\right) = \langle y^*[k]\, y[k + \kappa] \rangle \qquad \forall \kappa \in \mathcal{Z} \tag{5.21}$$

$$\text{ergodicity for the autocorrelation}$$

this time mean being defined for any κ by:

$$\langle y^*[k]\, y[k + \kappa] \rangle = \lim_{N \to \infty} \frac{1}{2N + 1} \sum_{k=-N}^{N} y^*[k]\, y[k + \kappa] \tag{5.22}$$

The simplest example of ergodic stationary random signal for the autocorrelation is the cisoid $a\, e^{\jmath\,(2\pi\,\nu_0 k + \phi)}$, $k \in \mathcal{Z}$ of initial phase ϕ evenly distributed between 0 and 2π, of autocorrelation function $a^2\, e^{\jmath\, 2\pi\, \nu_0 \kappa}$, $\kappa \in \mathcal{Z}$. However, the ergodicity is lost if the amplitude is also random. In practice, the ergodicity can be rigorously verified only rarely. In general, it is a hypothesis – necessary in order to obtain the second order statistical characteristics of a random signal considered from a single realization.

Under the ergodic hypothesis, the variance σ_y^2 of the signal considered is equal to the power $\langle |y[k]|^2 \rangle$ of any realization y:

$$\sigma_y^2 = \langle |y[k]|^2 \rangle = \lim_{N \to \infty} \frac{1}{2N+1} \sum_{k=-N}^{N} |y[k]|^2 \qquad [5.23]$$

i.e. the energy of the signal y multiplied by the truncation window $1_{-N,N}$ which is equal to 1 on the interval $\{-N, \ldots, N\}$ and zero otherwise, divided by the length of this interval when $N \to +\infty$. With the help of Parseval's theorem, we obtain:

$$\sigma_y^2 = \lim_{N \to \infty} \frac{1}{2N+1} \int_{-\frac{1}{2}}^{+\frac{1}{2}} |(y\, 1_{-N,N})^\frown(\nu)|^2 \, d\nu$$

$$= \int_{-\frac{1}{2}}^{+\frac{1}{2}} \left\{ \lim_{N \to \infty} \frac{1}{2N+1} |(y\, 1_{-N,N})^\frown(\nu)|^2 \right\} d\nu \qquad [5.24]$$

Hence, through formula [5.24], we have decomposed the power of the signal on the frequency axis, with the help of function $\nu \longmapsto \lim_{N \to \infty} \frac{1}{2N+1} |(y\, 1_{-N,N})^\frown(\nu)|^2$. In numerous works, we define the power spectral density (or power spectrum, or spectrum) of a stationary random signal by this function. However, in spite of the ergodic hypothesis, we can show that this function depends on the realization considered. We will define here the power spectral density (or power spectrum) S_{yy} as the mean of this function:

$$S_{yy}(\nu) = \lim_{N \to \infty} E \left(\frac{1}{2N+1} |(\mathbf{y}\, 1_{-N,N})^\frown(\nu)|^2 \right) \qquad [5.25]$$

$$= \lim_{N \to \infty} E \left(\frac{1}{2N+1} \left| \sum_{k=-N}^{N} \mathbf{y}[k]\, e^{-j\,2\pi\,\nu k} \right|^2 \right) \qquad [5.26]$$

Hence, we have two characterizations of a stationary random signal in the broad sense, ergodic for the autocorrelation. Wiener-Khintchine's theorem makes it possible to show the equivalence of these two characterizations. Under the hypothesis that the sequence $(\kappa\, r_{yy}[\kappa])$ is entirely integrable, let:

$$\sum_{\kappa=-\infty}^{+\infty} |\kappa\, r_{yy}[\kappa]| < \infty \qquad [5.27]$$

then, the power spectral density is the Fourier transform of the autocorrelation function and the two characterizations defined above coincide:

$$S_{yy}(\nu) = \widehat{r}_{yy}(\nu) \qquad [5.28]$$

$$= \sum_{\kappa=-\infty}^{+\infty} r_{yy}[\kappa]\, e^{-j\,2\pi\,\nu\kappa} \qquad [5.29]$$

Indeed, by developing expression [5.26], we obtain:

$$S_{yy}(\nu) = \lim_{N \to \infty} \frac{1}{2N+1} E\left(\sum_{n=-N}^{N} \sum_{k=-N}^{N} \mathbf{y}[n]\, \mathbf{y}^*[k]\, e^{-\jmath 2\pi \nu(n-k)} \right)$$

$$= \lim_{N \to \infty} \frac{1}{2N+1} \sum_{n=-N}^{N} \sum_{k=-N}^{N} r_{yy}[n-k]\, e^{-\jmath 2\pi \nu(n-k)}$$

$$= \lim_{N \to \infty} \frac{1}{2N+1} \sum_{\kappa=-2N}^{2N} r_{yy}[\kappa]\, e^{-\jmath 2\pi \nu \kappa}$$

$$\times \underbrace{\operatorname{card}\left\{ (n,k) \mid \kappa = n-k \text{ and } |n| \le N \text{ and } |k| \le N \right\}}_{2N+1-|\kappa|}$$

$$= \lim_{N \to \infty} \sum_{\kappa=-2N}^{2N} \left(1 - \frac{|\kappa|}{2N+1} \right) r_{yy}[\kappa]\, e^{-\jmath 2\pi \nu \kappa}$$

$$= \widehat{r}_{yy}(\nu) - \lim_{N \to \infty} \frac{1}{2N+1} \sum_{\kappa=-2N}^{2N} |\kappa|\, r_{yy}[\kappa]\, e^{-\jmath 2\pi \nu \kappa}$$

Under hypothesis [5.27], the second term above disappears and we obtain formula [5.29].

These considerations can be reiterated briefly for continuous-time signals. A continuous-time random signal $\mathbf{y}(t), t \in \Re$ is called *stationary in the broad sense* if its mean m_y and its autocorrelation function $r_{yy}(\tau)$ defined by:

$$\begin{cases} m_y = E(\mathbf{y}(t)) \\ r_{yy}(\tau) = E((\mathbf{y}(t) - m_y)^* (\mathbf{y}(t+\tau) - m_y)) \qquad \forall \tau \in \Re \end{cases} \qquad [5.30]$$

are independent of time t.

For a realization $y(t), t \in \Re$ of a random signal $\mathbf{y}(t)$, the time mean $\langle y(t) \rangle$ is defined by:

$$\langle y(t) \rangle = \lim_{T \to \infty} \frac{1}{2T} \int_{-T}^{T} y(t)\, \mathrm{d}t \qquad [5.31]$$

The ergodicity for the mean is written:

$$E(\mathbf{y}(t)) = \langle y(t) \rangle \qquad [5.32]$$

In what follows, we will suppose that the random signal $\mathbf{y}(t)$ is ergodic for the mean and, to simplify, of zero mean.

The random signal $\mathbf{y}(t)$ is ergodic for the autocorrelation if:

$$E\left(\mathbf{y}^*(t)\,\mathbf{y}(t+\tau)\right) = \langle y^*(t)\,y(t+\tau)\rangle \qquad \forall \tau \in \Re \tag{5.33}$$

this time mean being defined for any τ by:

$$\langle y^*(t)\,y(t+\tau)\rangle = \lim_{T\to\infty} \frac{1}{2T} \int_{-T}^{T} y^*(t)\,y(t+\tau)\,\mathrm{d}t \tag{5.34}$$

The power spectral density S_{yy} is expressed by:

$$S_{yy}(f) = \lim_{T\to\infty} E\left(\frac{1}{2T}|(\mathbf{y}\,1_{-T,T})^{\frown}(f)|^2\right) \tag{5.35}$$

$$= \lim_{T\to\infty} E\left(\frac{1}{2T}\left|\int_{-T}^{T} \mathbf{y}(t)\,e^{-\jmath 2\pi\,ft}\,\mathrm{d}t\right|^2\right) \tag{5.36}$$

If function $(\tau\,r_{yy}(\tau))$ is entirely integrable, let:

$$\int_{-\infty}^{+\infty} |\tau\,r_{yy}(\tau)|\,\mathrm{d}\tau < \infty \tag{5.37}$$

then the power spectral density is the Fourier transform of the autocorrelation function:

$$S_{yy}(f) = \widehat{r}_{yy}(f) \tag{5.38}$$

$$= \int_{-\infty}^{+\infty} r_{yy}(\tau)\,e^{-\jmath 2\pi\,f\tau}\,\mathrm{d}\tau \tag{5.39}$$

Power spectral density is thus a method to characterize the spectral content of a stationary random signal. For a white signal, the autocorrelation function is expressed, with $q > 0$, by:

$$r_{yy} = q\delta \tag{5.40}$$

Through the Fourier transform, we realize immediately that such a signal has a power spectral density constant and equal to q.

Under the ergodic hypothesis, for the discrete-time signals, the power spectral density can be easily estimated with the help of the periodogram; given a recording of N points $y[0],\dots,y[N-1]$ and based on expression [5.26], the periodogram is written:

$$I_{yy}(\nu) = \frac{1}{N}|(y\,1_{0,N-1})^{\frown}(\nu)|^2 \tag{5.41}$$

$$= \frac{1}{N}\left|\sum_{k=0}^{N-1} y[k]\,e^{-\jmath 2\pi\,\nu k}\right|^2 \tag{5.42}$$

where $1_{0,N-1}$ is the rectangular window equal to 1 on the interval $\{0, \ldots, N-1\}$ and zero otherwise. With regard to the initial definition of power spectral density, we lost the mathematical expectation operator as well as the limit passage. This estimator is not consistent and several variants were proposed: Bartlett's periodograms, modified periodograms, Welsh's periodograms, correlogram, etc. The major drawback of the periodogram, and more so of its variants, is the bad resolution, i.e. the capability to separate the spectral components coming from close frequency sine curves. More recently, methods based on a signal modeling were proposed, which enable better resolution performances than those of the periodogram.

5.3. Generator processes and ARMA modeling

Let us take a stable linear process with an impulse response h, which is excited by a stationary random signal e, with output y:

$$y = h \otimes e \tag{5.43}$$

Hence, we directly obtain that signal y is stationary and its autocorrelation function is expressed by:

$$r_{yy} = h \otimes h^{*-} \otimes r_{ee} \tag{5.44}$$

where h^{*-} represents the conjugated and returned impulse response ($h^{*-}(t) = (h(-t))^{*}$). Through the Fourier transform, the power spectral density of y is expressed by:

$$S_{yy} = \left| \widehat{h} \right|^{2} S_{ee} \tag{5.45}$$

In particular, if e is a white noise of spectrum q, then:

$$S_{yy} = q \left| \widehat{h} \right|^{2} \tag{5.46}$$

Inversely, given a stationary random signal y with a power spectral density S_{yy}, if there is an impulse response h and a positive real number q so that we can write formula [5.46], we say that this system is a generating process (or a shaper filter) for y. Everything takes place as if we could consider signal y as the output of a linear process with an impulse response h excited by a white noise of spectrum q.

This modeling depends, however, on any impulse response h of the shaper filter. In order to be able to obtain a modeling with the help of a finite number of parameters, we know only one solution to date: the system of impulse response h has a rational transfer function. Consequently, we are limited to the signals whose power spectral density is a rational fraction in $j\,2\pi\,f$ for continuous-time and $e^{j\,2\pi\,\nu}$ for discrete-time. Nevertheless, the theory of rational approximation indicates that we can always get as close as we wish to a function through a rational function of sufficient degrees.

Since the module of the transfer function of an all-pass filter is constant, such a filter does not enable under any circumstance to model a certain form of power spectral density. Hence, we will suppose that the impulse response filter h is causal with minimum of phase, i.e. its poles and zeros are strictly negative real parts for continuous-time and of a module strictly inferior to 1 for discrete-time.

Finally, we note that formula [5.46] is redundant, i.e. the amplitude of power spectrum S_{yy} can be set either by the value of spectrum q or by the value of the filter gain for a given frequency. Hence, it is preferable to set the impulse response h, or its Fourier transform, in a certain sense.

For discrete-time, it is usual to choose a direct transmission shaper filter ($h[0] \neq 0$) and the impulse response is normalized with $h[0] = 1$ (in this case we say that the filter is monic). The equivalent for continuous-time consists of considering an impulse response with a Dirac impulse of unitary weight at instant 0. If this condition does not entail any constraint in the case of discrete-time (a pure delay being in this case an all-pass filter), in the case of continuous-time it implies that the power spectral density of the signal is not cancelled in high frequency.

For discrete-time, the transfer function of the filter is thus written:

$$\check{h}(z) = \frac{\check{c}(z)}{\check{a}(z)} = \frac{1 + \displaystyle\sum_{n=1}^{n_c} c[n]\, z^{-n}}{1 + \displaystyle\sum_{n=1}^{n_a} a[n]\, z^{-n}} \qquad [5.47]$$

The orders n_a and n_c characterize the structure chosen. The parameter vector $\theta = [q, a[1], \ldots, a[n_a], c[1], \ldots, c[n_c]]$ is then necessary and sufficient to correctly characterize the shaper filter.

In the case of a finite impulse response filter ($n_a = 0$), we talk of an MA (*moving average*) model because signal $y[k]$ is expressed with the help of a weighted average of the input $e[k]$ on a sliding window:

$$y[k] = e[k] + c[1]\, e[k - 1] + \cdots + c[n_c]\, e[k - n_c] \qquad [5.48]$$

The MA model is particularly capable of representing the power spectrums presenting strong attenuations in the proximity of given frequencies (see Figure 5.1). Indeed, if $\check{c}(z)$ admits a zero of a module close to 1 and of argument $2\pi\, \nu_0$, then the power spectrum is almost zero in the proximity of ν_0.

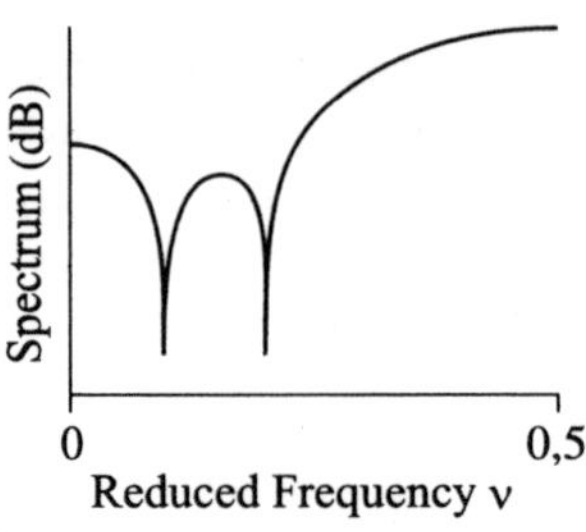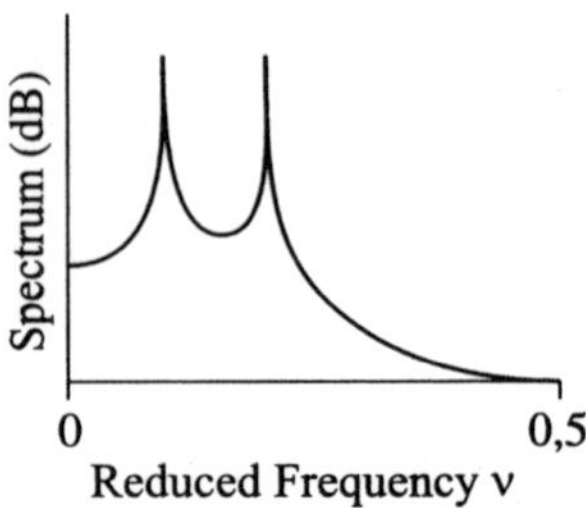

Figure 5.1. *Typical power spectrum of an MA (left)*
or AR (right) model

In the case of a single denominator ($n_c = 0$), we talk of an AR (autoregressive) model, because signal $y[k]$ at instant k is expressed with the help of a regression on the signal values at the previous instants:

$$y[k] = -a[1]\,y[k-1] - \cdots - a[n_a]\,y[k-n_a] + e[k] \qquad [5.49]$$

The AR model is particularly supported for two reasons. On the one hand, its estimation by maximum likelihood, with the help of a finite period recording of signal y, reaches an explicit solution (in the general case, we would have to call upon an optimization procedure). On the other hand, it is particularly capable of representing power spectrums presenting marked peaks in the proximity of certain frequencies, i.e. signals presenting marked periodicities (see Figure 5.1); for example, Prony's ancestral method, pertaining to the estimation of the frequency of noisy sine curves, deals with determining the argument of the poles of an AR model identified by maximum likelihood.

In the general case, we speak of an ARMA (autoregressive with moving average) model:

$$y[k] = -a[1]\,y[k-1] - \cdots - a[n_a]\,y[k-n_a]$$

$$+ e[k] + c[1]\,e[k-1] + \cdots + c[n_c]\,e[k-n_c] \qquad [5.50]$$

Finally, we note that the choice of normalization $h[0] = 1$ is not innocent. Indeed, the predictor filter of one count providing $\hat{y}[k]$, prediction of $y[k]$ on the basis of previous observations $y[k-1], y[k-2], \ldots$, obtained from the shaper filter [5.47] as:

$$\hat{y}[k] = \left(1 - \frac{\check{a}(z)}{\check{c}(z)}\right) y[k] \qquad [5.51]$$

is the optimal predictor filter, in the sense of the prediction error variance $y[k] - \hat{y}[k]$, among all linear filters without direct transmission. This prediction error is then rigorously the white sequence $e[k]$.

5.4. Modeling of LTI systems and ARMAX modeling

Let us take a linear time-invariant (LTI) system, of impulse response g. The response of this system at the known deterministic input u is $g \otimes u$, which can thus be calculated exactly. However, this is often unrealistic, because there are always signals that affect the operating mode of the system (measurement noises, non-controllable inputs). In a linear context, we will suppose here that these parasite phenomena are translated by an additional term v on the system output. The output y is then expressed by:

$$y = g \otimes u + v \qquad [5.52]$$

Hence, it is natural to propose a probabilistic context for this disturbance v and to consider it as a stationary random signal, admitting a representation by shaper filter; the output measured y is then expressed by:

$$y = g \otimes u + h \otimes e \qquad [5.53]$$

where u is the known deterministic input, e an unknown white noise of spectrum q, g the impulse response of the system and h the impulse response of the shaper filter. We suppose that h and g are the impulse responses of the systems with rational transfer function, and, to simplify, that g does not have direct transmission.

5.4.1. ARX modeling

For discrete-time, the simplest relation input-output is the following difference equation:

$$y[k] = -a[1]\, y[k-1] - \cdots - a[n_a]\, y[k - n_a]$$
$$+ b[1]\, u[k-1] + \cdots + b[n_b]\, u[k - n_b] + e[k] \qquad [5.54]$$

where the term of white noise $e[k]$ enters directly in the difference equation. This model is hence called "equation error model". Thus, the transfer functions become:

$$\breve{g}(z) = \frac{\breve{b}(z)}{\breve{a}(z)} = \frac{\displaystyle\sum_{n=1}^{n_b} b[n] z^{-n}}{1 + \displaystyle\sum_{n=1}^{n_a} a[n] z^{-n}} \qquad [5.55a]$$

$$\breve{h}(z) = \frac{1}{\breve{a}(z)} = \frac{1}{1 + \displaystyle\sum_{n=1}^{n_a} a[n] z^{-n}} \qquad [5.55b]$$

We also talk of ARX modeling, "AR" referring to the modeling of the additional noise and "X" to the exogenous input $u[k]$. Given the orders n_a and n_b, the parameter vector $\theta = [q, a[1], \ldots, a[n_a], b[1], \ldots, b[n_b]]$ fully characterizes the system. This model is not especially realistic but, as in the case of AR, we can show that the identification by maximum likelihood of an ARX model leads to an explicit solution.

5.4.2. *ARMAX modeling*

The ARX model does not give much freedom on the statistical properties of the additional noise. A solution consists of describing the equation error with the help of a running average:

$$y[k] = -a[1]\, y[k-1] - \cdots - a[n_a]\, y[k-n_a]$$
$$+ b[1]\, u[k-1] + \cdots + b[n_b]\, u[k-n_b]$$
$$+ e[k] + c[1]\, e[k-1] + \cdots + c[n_c]\, e[k-n_c] \qquad [5.56]$$

Thus, the transfer functions become:

$$\breve{g}(z) = \frac{\breve{b}(z)}{\breve{a}(z)} = \frac{\displaystyle\sum_{n=1}^{n_b} b[n] z^{-n}}{1 + \displaystyle\sum_{n=1}^{n_a} a[n] z^{-n}} \qquad [5.57a]$$

$$\breve{h}(z) = \frac{\breve{c}(z)}{\breve{a}(z)} = \frac{1 + \displaystyle\sum_{n=1}^{n_c} c[n] z^{-n}}{1 + \displaystyle\sum_{n=1}^{n_a} a[n] z^{-n}} \qquad [5.57b]$$

We talk of ARMAX modeling, "ARMA" pertaining to the modeling of the additional noise. Given the orders n_a, n_b and n_c, the parameter vector $\theta = [q, a[1], \ldots, a[n_a], b[1], \ldots, b[n_b], c[1], \ldots, c[n_c]]$ fully characterizes the system.

5.4.3. *Output error model*

In the particular case of the ARMAX model where we take $\breve{c}(z) = \breve{a}(z)$, the transfer functions become:

$$\breve{g}(z) = \frac{\breve{b}(z)}{\breve{a}(z)} = \frac{\displaystyle\sum_{n=1}^{n_b} b[n] z^{-n}}{1 + \displaystyle\sum_{n=1}^{n_a} a[n] z^{-n}} \qquad \breve{h}(z) = 1 \qquad [5.58]$$

Hence, only an additional white noise remains on the process output. We talk of an output error (OE) model. Given the orders n_a and n_b, the parameter vector $\theta = [q, a[1], \ldots, a[n_a], b[1], \ldots, b[n_b]]$ fully characterizes the system. We can show that even if this hypothesis is false (i.e. if the additional noise if colored), the identification of θ by maximum likelihood leads to an asymptotically non-biased estimation (but this estimation is not of minimal variance in this case).

5.4.4. *Representation of the ARMAX model within the state space*

We present here the reverse canonical form, in which the coefficients of transfer functions appear explicitly, which is written by assuming that $d = \max n_a, n_b, n_c$ the size of the state vector x and by possibly completing sequences a, b or c by zeros:

$$\begin{cases} x[k+1] = A\,x[k] + B\,u[k] + K\,e[k] \\ y[k] = C\,x[k] + e[k] \end{cases} \qquad [5.59]$$

$$A = \begin{bmatrix} -a[1] & 1 & 0 & \cdots & \cdots & 0 \\ \vdots & 0 & \ddots & \ddots & & \vdots \\ \vdots & \vdots & \ddots & \ddots & \ddots & \vdots \\ \vdots & \vdots & & \ddots & \ddots & 0 \\ -a[d-1] & 0 & \cdots & \cdots & 0 & 1 \\ -a[d] & 0 & \cdots & \cdots & \cdots & 0 \end{bmatrix} \qquad [5.60a]$$

$$B = \begin{bmatrix} b[1] \\ \vdots \\ \vdots \\ \vdots \\ \vdots \\ \vdots \\ b[d] \end{bmatrix} \qquad K = \begin{bmatrix} c[1] - a[1] \\ \vdots \\ \vdots \\ \vdots \\ \vdots \\ c[d] - a[d] \end{bmatrix} \qquad [5.60b]$$

$$C = \begin{bmatrix} 1 & 0 & \cdots & \cdots & \cdots & 0 \end{bmatrix} \qquad [5.60c]$$

5.4.5. *Predictor filter associated with the ARMAX model*

The one count predictor filter providing $\hat{y}[k]$, prediction of $y[k]$ on the basis of the previous observations $y[k-1], y[k-2]$, etc., and of input $u[k], u[k-1]$, etc., is obtained as:

$$\hat{y}[k] = \frac{\breve{b}(z)}{\breve{c}(z)} u[k] + \left(1 - \frac{\breve{a}(z)}{\breve{c}(z)}\right) y[k] \qquad [5.61]$$

It is the optimal predictor filter, in the sense of the second momentum of the prediction error $y[k] - \hat{y}[k]$, among all linear filters without direct transmission on $y[k]$.

This prediction error is then rigorously the white sequence $e[k]$. This is the basis of the identification methods for ARMAX models through the prediction error method. Given an input-output recording of N points $(u[k], y[k])_{0 \leq k \leq N-1}$, we will choose among all the predictor filters of the form [5.61], which is parameterized by θ and providing a prediction $y_\theta[k]$, the one that minimizes the square mean of the prediction error:

$$\dot{\theta} = \arg \min_\theta \frac{1}{N} \sum_{k=0}^{N-1} |y[k] - \hat{y}_\theta[k]|^2 \qquad [5.62]$$

This estimator is in fact the estimator of maximum likelihood in the case of a Gaussian white noise hypothesis. We note that the hypothesis of a shaper filter with minimum phase leads to a stable causal predictor.

5.5. From the Markovian system to the ARMAX model

The representation within the state space [5.59], in which the unique noise sequence $e[k]$ intervenes both on the equation of state and on the equation of measurement, is called "innovation form" or "filter form". However, by generalizing to the study of systems with m inputs ($u[k]$ is a vector with m lines) and p outputs ($y[k]$ is a vector with p lines), the random contributions are usually represented with the help of two noises $v[k]$ (the noise of the system) and $w[k]$ (the measurement noise) in a representation within the state space of size d as follows:

$$\begin{cases} x[k+1] = A\,x[k] + B\,u[k] + v[k] \\ y[k] = C\,x[k] + w[k] \end{cases} \qquad [5.63]$$

where $v[k]$ and $w[k]$ are two white noises of spectra Q and R respectively and of interspectrum S, i.e.:

$$\begin{cases} E\,(v^*[k]\,v^T[k + \kappa]) = Q\,\delta[k] \\ E\,(w^*[k]\,w^T[k + \kappa]) = R\,\delta[k] \\ E\,(v^*[k]\,w^T[k + \kappa]) = S\,\delta[k] \end{cases} \qquad [5.64]$$

Noise $v[k]$ generally represents the uncertainties on the process model or the disturbances on the exogenous input. Noise $w[k]$ generally represents the measurement noise. We talk of a Markovian system.

However, Kalman's filtering (see Chapter 7) enables us to show that it is always possible to represent such a system in the innovation form, as:

$$\begin{cases} \hat{x}[k+1] = A\,\hat{x}[k] + B\,u[k] + K\,e[k] \\ y[k] = C\,\hat{x}[k] + e[k] \end{cases} \qquad [5.65]$$

where $\hat{x}[k]$, $e[k]$ and K are the state prediction, the innovation (and we can prove it is white) and the gain of Kalman's stationary filter operating on model [5.63]. Such a form is minimal, in the sense that it entails only as many noises as measurements.

In the particular mono-input-mono-output case, we find the ARMAX model, whose canonical form is given in [5.60].

5.6. Bibliography

[KAY 81] KAY S.M., MARPLE S.L., Jr., "Spectrum analysis: a modern perspective", *Proceedings of the IEEE*, vol. 69, no. 11, p. 1380–1419, 1981.

[KWA 91] KWAKERNAAK H., SIVAN R., *Modern Signals and Systems*, Prentice-Hall, 1991.

[LAR 75] DE LARMINAT P., THOMAS Y., *Automatique des systèmes linéaires. Vol. 1. Signaux et systèmes*, Dunod, Paris, 1975.

[LAR 93] DE LARMINAT P., *Automatique. Commande des systèmes linèaires*, Hermès, Paris, 1993.

[LJU 87] LJUNG L., *System Identification: Theory for the User*, Prentice-Hall, 1987.

[MAR 87] MARPLE S.L., Jr., *Digital Spectral Analysis with Applications*, Prentice-Hall, 1987.

[MAX 96] MAX J., LACOUME J.L., *Méthodes et techniques de traitement du signal et applications aux mesures physiques. Vol. 1. Principes généraux et méthodes classiques*, Masson, Paris, 5th edition, 1996.

[OPP 75] OPPENHEIM A.V., SCHAEFER R.W., *Digital Signal Processing*, Prentice-Hall, 1975.

[PAP 71] PAPOULIS A., *Probabilités, variables aléatoires et processus stochastiques*, Dunod, Paris, 1971.

Chapter 6

Kalman's Formalism for State Stabilization and Estimation

We will show how, based on a state representation of a continuous-time or discrete-time linear system, it is possible to elaborate a negative feedback loop, by assuming initially that all state variables are measurable. Then we will explain how, if it is not the case, it is possible to build the state with the help of an observer. These two operations bring about similar developments, which use either a pole placement or an optimization technique. These two approaches are presented successively.

6.1. The academic problem of stabilization through state feedback

Let us consider a time-invariant linear system described by the following continuous-time equations of state:

$$\dot{x}(t) = A\,x(t) + B\,u(t) \quad ; \quad x(0) \neq 0 \tag{6.1}$$

where $x \in \mathbf{R}^n$ is the state vector and $u \in \mathbf{R}^m$ the control vector. The problem is how to determine a control that brings $x(t)$ back to 0, irrespective of the initial condition $x(0)$. In this chapter, our interest is mainly in the *state feedback controls*, which depend on the state vector x. A linear state feedback is written as follows:

Chapter written by Gilles DUC.

$$u(t) = -K\, x(t) + e(t) \qquad\qquad [6.2]$$

where K is an $m \times n$ matrix (Figure 6.1) and signal $e(t)$ represents the input of the looped system..

The equations of the looped system are written as follows:

$$\dot{x}(t) = (A - BK)\, x(t) + B e(t) \qquad\qquad [6.3]$$

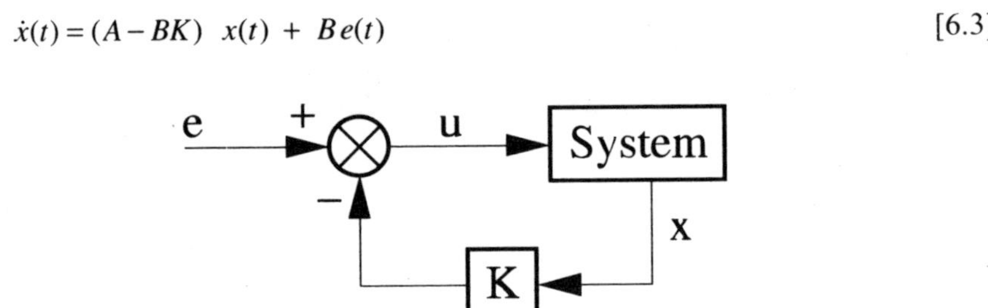

Figure 6.1. *State feedback linear control*

Hence, the state feedback control affects the dynamics of the system which depends on the eigenvalues of $A - BK$ (let us recall that the poles of the open loop system are eigenvalues of A; similarly, the poles of the closed loop system are eigenvalues of $A - BK$).

In the case of a discrete-time system described by the equations:

$$x_{k+1} = F\, x_k + G u_k \quad ; \quad x_0 \neq 0 \qquad\qquad [6.4]$$

the state feedback and the equations of the looped system can be written:

$$u_k = -K\, x_k + e_k \qquad\qquad [6.5]$$

$$x_{k+1} = (F - GK)\, x_k + G e_k \quad ; \quad x_0 \neq 0 \qquad\qquad [6.6]$$

so that the dynamics of the system depends on the eigenvalues of $F - GK$.

The research for matrix K can be done in various ways. In the following section, we will show that under certain conditions, it makes it possible to choose the poles of the looped system. In section 6.4, we will present the quadratic optimization approach, which consist of minimizing a criterion based on state and control vectors.

6.2. Stabilization by pole placement

6.2.1. *Results*

The principle of stabilization by pole placement consists of *a priori* choosing the poles preferred for the looped system, i.e. the eigenvalues of $A - BK$ in continuous-time (or of $F - GK$ in discrete-time) and then to obtain matrix K ensuring this choice. The following theorem, belonging to Wonham, specifies on which condition this approach is possible.

THEOREM 6.1.– *a real matrix K exists irrespective of the set of eigenvalues $\{\lambda_1, \cdots, \lambda_n\}$, real or conjugated complex numbers chosen for $A - BK$ (for $F - GK$ respectively) if and only if (A, B) $((F, G)$ respectively) is controllable.*

Demonstration. It is provided for continuous-time but it is similar for discrete-time as well. Firstly, let us show that the condition is sufficient: if the system is not controllable, it is possible, through passage to the controllable canonical form (see Chapter 2), to express the state equations as follows:

$$\begin{pmatrix} \dot{x}_1(t) \\ \dot{x}_2(t) \end{pmatrix} = \begin{pmatrix} A_{11} & A_{12} \\ 0 & A_{22} \end{pmatrix} \begin{pmatrix} x_1(t) \\ x_2(t) \end{pmatrix} + \begin{pmatrix} B_1 \\ 0 \end{pmatrix} u(t) \qquad [6.7]$$

By similarly decomposing the state feedback [6.2]:

$$u(t) = -(K_1 \quad K_2) \begin{pmatrix} x_1(t) \\ x_2(t) \end{pmatrix} + e(t) \qquad [6.8]$$

the equation of the looped system is written:

$$\begin{pmatrix} \dot{x}_1(t) \\ \dot{x}_2(t) \end{pmatrix} = \begin{pmatrix} A_{11} - B_1K_1 & A_{12} - B_1K_2 \\ 0 & A_{22} \end{pmatrix} \begin{pmatrix} x_1(t) \\ x_2(t) \end{pmatrix} + \begin{pmatrix} B_1 \\ 0 \end{pmatrix} e(t) \qquad [6.9]$$

so that, the state matrix being block-triangular, the eigenvalues of the looped system are the totality of eigenvalues of sub-matrices $A_{11} - B_1K_1$ and A_{22}. The eigenvalues of the non-controllable part are thus, by all means, eigenvalues of the looped system.

Let us suppose now that the system is controllable. In this part, we will assume that the system has only one control; however, the result cannot be extended to the case of multi-control systems. As indicated in Chapter 2, the equations of state can be expressed in companion form:

$$\begin{pmatrix} \dot{x}_1(t) \\ \vdots \\ \dot{x}_n(t) \end{pmatrix} = \begin{pmatrix} 0 & 1 & 0 & \cdots & 0 \\ 0 & 0 & 1 & \ddots & \vdots \\ 0 & 0 & 0 & \ddots & 0 \\ \vdots & \vdots & & \cdots & \ddots & 1 \\ -a_n & -a_{n-1} & -a_{n-2} & \cdots & -a_1 \end{pmatrix} \begin{pmatrix} x_1(t) \\ \vdots \\ x_n(t) \end{pmatrix} + \begin{pmatrix} 0 \\ \vdots \\ 0 \\ 0 \\ 1 \end{pmatrix} u(t) \qquad [6.10]$$

By writing the state feedback [6.2] as:

$$u(t) = -\begin{pmatrix} k_n & k_{n-1} & \cdots & k_1 \end{pmatrix} \begin{pmatrix} x_1(t) \\ \vdots \\ x_n(t) \end{pmatrix} + e(t) \qquad [6.11]$$

the equation of the looped system remains in companion form:

$$\begin{pmatrix} \dot{x}_1(t) \\ \vdots \\ \dot{x}_n(t) \end{pmatrix} = \begin{pmatrix} 0 & 1 & \cdots & 0 \\ 0 & 0 & \ddots & \vdots \\ \vdots & \vdots & \ddots & 1 \\ -a_n - k_n & \cdots & \cdots & -a_1 - k_1 \end{pmatrix} \begin{pmatrix} x_1(t) \\ \vdots \\ x_n(t) \end{pmatrix} + \begin{pmatrix} 0 \\ \vdots \\ 0 \\ 1 \end{pmatrix} e(t) \qquad [6.12]$$

so that the characteristic polynomial of the looped system is written:

$$\det(\lambda I - (A - BK)) = \lambda^n + (a_1 + k_1)\,\lambda^{n-1} + \cdots + (a_n + k_n) \qquad [6.13]$$

We see that, by choosing the state feedback coefficients, it is possible to arbitrarily set each characteristic polynomial coefficient so that we can arbitrarily set its roots, which are precisely the eigenvalues of the looped system. In addition, matrix K is thus uniquely determined.

Theorem 6.1 thus shows that it is possible to stabilize a controllable system through a state feedback (it is sufficient to take all λ_i with a negative real part in continuous-time, inside the unit circle in discrete-time). More generally, it shows that the dynamics of a controllable system can be randomly set for a linear state feedback.

However, in this chapter we will not deal with the practical issue of choosing the eigenvalues. Similarly, we note that for a multi-variable system (i.e. a system with several controls), the choice of eigenvalues is not enough in order to uniquely set matrix K. Degrees of freedom are also possible for the choice of the eigenvectors of matrices $A - BK$ or $F - GK$. Chapter 14 will tackle these aspects in detail.

6.2.2. Example

Let us consider the system described by the following equations of state:

$$\begin{cases} \begin{pmatrix} \dot{x}_1(t) \\ \dot{x}_2(t) \end{pmatrix} = \begin{pmatrix} 0 & 1 \\ 0 & -1 \end{pmatrix} \begin{pmatrix} x_1(t) \\ x_2(t) \end{pmatrix} + \begin{pmatrix} 0 \\ 1 \end{pmatrix} u(t) \\ y(t) = \begin{pmatrix} 1 & 0 \end{pmatrix} \begin{pmatrix} x_1(t) \\ x_2(t) \end{pmatrix} \end{cases} \qquad [6.14]$$

We can verify that this system is controllable:

$$\text{rank}\,(B \quad AB) = \text{rank} \begin{pmatrix} 0 & 1 \\ 1 & -1 \end{pmatrix} = 2 \qquad [6.15]$$

We obtain, with $K = (k_1 \quad k_2)$:

$$\det\,(\lambda I - (A - BK)) = \begin{vmatrix} \lambda & -1 \\ k_1 & \lambda + 1 + k_2 \end{vmatrix} = \lambda^2 + (1 + k_2)\,\lambda + k_1 \qquad [6.16]$$

and by identifying with a second order polynomial written in the normalized form:

$$\lambda^2 + (1 + k_2)\,\lambda + k_1 \equiv \lambda^2 + 2\,\xi\,\omega_0\,\lambda + \omega_0^2 \quad \Leftrightarrow \quad \begin{cases} k_1 = \omega_0^2 \\ k_2 = 2\,\xi\,\omega_0 - 1 \end{cases} \qquad [6.17]$$

Figure 6.2 shows the evolution of the output and the control, in response to the initial condition $x(0) = (1 \quad 1)^T$, for different values of ω_0 and ξ: the higher ω_0 is, the faster the output returns to 0, but at the expense of a stronger control, whereas the increase of ξ leads to better dynamics.

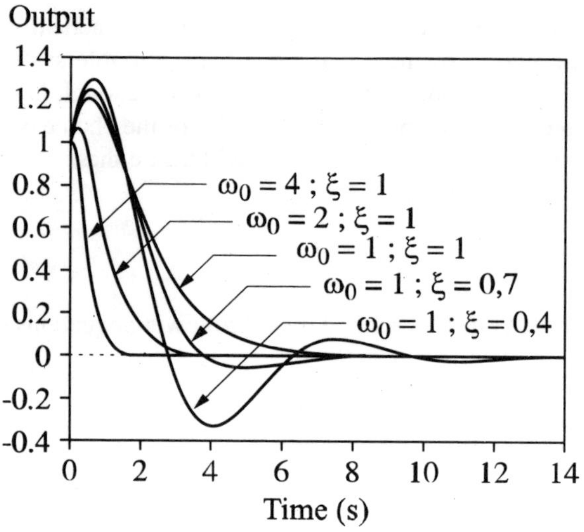

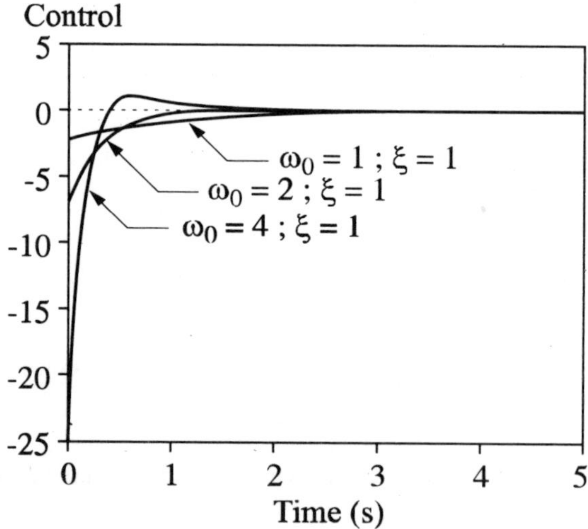

Figure 6.2. *Stabilization by pole placement*

6.3. Reconstruction of state and observers

6.3.1. *General principles*

The disadvantage of state feedback controls, like the ones mentioned in the previous chapter, is that in practice we do not always measure all the components of state vector x. In this case, we can build a dynamic system called *observer*, whose role is to rebuild the state from the information available, i.e. the controls u and all the available measures. The latter will be grouped together into a z vector (Figure 6.3).

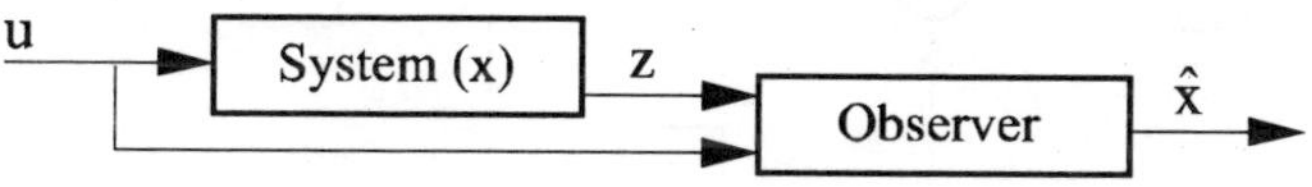

Figure 6.3. *The role of an observer*

6.3.2. *Continuous-time observer*

Let us suppose that the equations of state are written:

$$\begin{cases} \dot{x}(t) = A\,x(t) + B\,u(t) \\ z(t) = C\,x(t) \end{cases} \qquad [6.18]$$

The equations of a continuous-time observer, whose state is marked $\hat{x}(t)$, are calculated on those of the system, but with a supplementary term:

$$\begin{cases} \dot{\hat{x}}(t) = A\,\hat{x}(t) + B\,u(t) + L\,(z(t) - \hat{z}(t)) \\ \hat{z}(t) = C\,\hat{x}(t) \end{cases} \qquad [6.19]$$

The observer equation of state includes a term proportional to the difference between the real measures $z(t)$ and the reconstructions of measures obtained from the observer's state, with an L gain matrix. In the case of a system with n state variables and q measures (i.e. $\dim(x) = \dim(\hat{x}) = n$, $\dim(z) = q$), L is an $n \times q$ matrix.

Equations [6.19] correspond to the diagram in Figure 6.4: in the lower part of the figure we see equations [6.18] of the system we are dealing with. The failure term with the L gain matrix completes the diagram.

Hence, equations [6.19] can be written as follows:

$$\dot{\hat{x}}(t) = (A - L\,C)\,\hat{x}(t) + B\,u(t) + L\,z(t) \qquad [6.20]$$

which makes the observer look like a state system $\hat{x}(t)$, with the inputs $u(t)$ and $z(t)$ and with the state matrix $A - LC$. We infer that the observer is a stable system if and only if all the eigenvalues of $A - LC$ are strictly negative real parts.

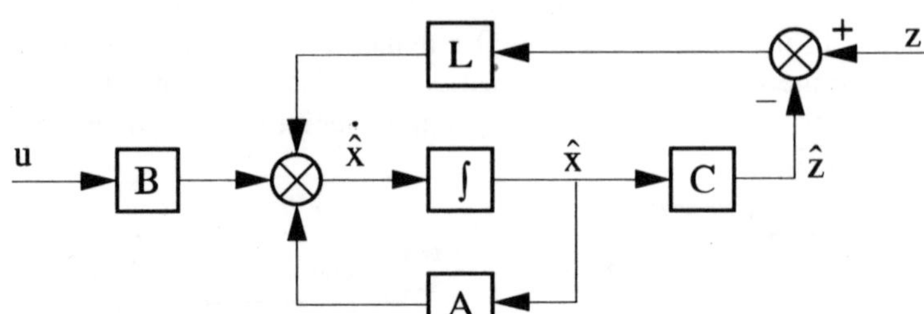

Figure 6.4. *Structure of the observer*

Let us now consider the *reconstruction error* $\varepsilon(t)$ that appears between $x(t)$ and $\hat{x}(t)$. Based on [6.18] and [6.19], we obtain:

$$\dot{\varepsilon} = \dot{x} - \dot{\hat{x}} = (Ax + Bu) - (A\hat{x} + Bu + LCx - LC\hat{x})$$

$$\dot{\varepsilon}(t) = (A - LC)\varepsilon(t) \qquad [6.21]$$

and hence the reconstruction error $\varepsilon(t)$ tends toward 0 when t tends toward infinity if and only if the observer is stable. In addition, the eigenvalues of $A - LC$ set the dynamics of $\varepsilon(t)$. Hence, the problem is to determine an L gain matrix ensuring stability with a satisfactory dynamics.

6.3.3. *Discrete-time observer*

The same principles are applied for the synthesis of a discrete-time observer; if we seek to rebuild the state of a sampled system described by:

$$\begin{cases} x_{k+1} = F x_k + G u_k \\ z_k = C x_k \end{cases} \qquad [6.22]$$

the observer's equations can be written in the two following forms:

$$\begin{cases} \hat{x}_{k+1} = F \hat{x}_k + G u_k + L(z_k - \hat{z}_k) \\ \hat{z}_k = C \hat{x}_k \end{cases} \qquad [6.23]$$

$$\hat{x}_{k+1} = (F - LC)\hat{x}_k + G u_k + L z_k \qquad [6.24]$$

From equations [6.22] and [6.23] we infer that the reconstruction error verifies:

$$\varepsilon_{k+1} = (F-L\,C)\,\varepsilon_k \qquad\qquad [6.25]$$

In order to guarantee the stability of the observer and, similarly, the convergence toward 0 of error ε_k, matrix L must be chosen so that all the eigenvalues of $F-L\,C$ have a module strictly less than 1.

According to [6.23] or [6.24], we note that the observer operates as a predictor: based on the information known at instant k, we infer an estimation of the state at instant $k+1$. Hence, this calculation does not need to be supposed infinitely fast because it is enough that its result is available during the next sampling instant.

6.3.4. *Calculation of the observer by pole placement*

We note the analogy between the calculation of an observer and the calculation of a state feedback, discussed in section 6.1: at that time, the idea was to determine a K gain matrix that would guarantee to the looped system a satisfactory dynamics, the latter being set by the eigenvalues of $A - B\,K$ (or $F - G\,K$ for discrete-time). The difference is in the fact that the matrix to determine appears on the right in product $B\,K$ (or $G\,K$), whereas it appears on the left in product $L\,C$.

However, the eigenvalues of $A-L\,C$ are the same as the ones of $A^T-C^TL^T$, expression in which the matrix to determine L^T appears on the right. Choosing the eigenvalues of $A^T-C^TL^T$ is thus exactly a problem of stabilization by pole placement: the results listed in section 6.1 can thus be applied here by replacing matrices A and B (or F and G) by A^T and C^T (or F^T and C^T) and the state feedback K by L^T.

Based on Theorem 6.1, we infer that matrix L exists for any set of eigenvalues $\{\lambda_1, \cdots, \lambda_n\}$ chosen *a priori* if and only if (A^T, C^T) is controllable. However, we can write the following equivalences:

$$(A^T, C^T) \text{ controllable} \Leftrightarrow rank\,[C^T\ A^T C^T\ \cdots (A^T)^{n-1}C^T] = n \Leftrightarrow$$

$$rank\begin{bmatrix} C \\ C\,A \\ \vdots \\ C\,A^{n-1} \end{bmatrix} = n \Leftrightarrow (C, A) \text{ observable} \qquad\qquad [6.26]$$

Hence, we can arbitrarily choose the eigenvalues of the observer if and only if the system is observable through the measures available. Naturally, the result obtained from equation [6.26] can be used for the discrete-time case by simply replacing matrix A with matrix F.

6.3.5. *Behavior of the observer outside the ideal case*

The results of sections 6.3.2 and 6.3.3, even if interesting, describe an ideal case which will never be achievable in practice. Let us suppose, for example, that a disturbance $p(t)$ is applied on the system [6.18]:

$$\begin{cases} \dot{x}(t) = A\,x(t) + B\,u(t) + E\,p(t) \\ z(t) = C\,x(t) \end{cases} \qquad [6.27]$$

but observer [6.19] is not aware of it and then a calculation identical to the one in section 6.3.2 shows that the equation obtained for the reconstruction error can be written:

$$\dot{\varepsilon}(t) = (A - L\,C)\,\varepsilon(t) + E\,p(t) \qquad [6.28]$$

so that the error does not tend any longer toward 0. If $p(t)$ can be associated with a noise, Kalman filtering techniques can be used in order to minimize the variance of $\varepsilon(t)$. We provide a preview of this aspect in section 6.5.3.

If we suppose that modeling uncertainties affect the state matrix of system [6.18], so that a matrix $A' \neq A$ intervenes in this equation, then the reconstruction error is governed by the following equation:

$$\dot{\varepsilon}(t) = (A - L\,C)\,\varepsilon(t) + (A' - A)\,x(t) \qquad [6.29]$$

so that there again the error does not tend toward 0.

NOTE 6.1.– observers [6.19] or [6.23] rebuild all state variables, operation that may seem superfluous if the measures available are of very good quality (especially if the measurement noises are negligible): from the moment the observation equation already provides q linear combinations (that we will suppose independent) of state variables, it is sufficient to reconstitute $n - q$, independent from the previous ones. Therefore, we can synthesize a reduced observer, following an approach similar to the one presented in these sections (see [FAU 84, LAR 96]). However, the physical interpretation underlined in section 6.3.2, where the observer appears naturally as a physical model of the system completed by a retiming term, is lost.

6.3.6. *Example*

Let us consider again the system described by equations [6.14]:

$$\begin{cases} \begin{pmatrix} \dot{x}_1(t) \\ \dot{x}_2(t) \end{pmatrix} = \begin{pmatrix} 0 & 1 \\ 0 & -1 \end{pmatrix} \begin{pmatrix} x_1(t) \\ x_2(t) \end{pmatrix} + \begin{pmatrix} 0 \\ 1 \end{pmatrix} u(t) \\ y(t) = \begin{pmatrix} 1 & 0 \end{pmatrix} \begin{pmatrix} x_1(t) \\ x_2(t) \end{pmatrix} \end{cases}$$

We can verify that this system is observable:

$$\text{rank} \begin{pmatrix} C \\ C\,A \end{pmatrix} = \text{rank} \begin{pmatrix} 1 & 0 \\ 0 & 1 \end{pmatrix} = 2 \qquad [6.30]$$

The observer's equations can be written, by noting $L = (l_1 \quad l_2)^T$:

$$\begin{aligned} \begin{pmatrix} \dot{\hat{x}}_1(t) \\ \dot{\hat{x}}_2(t) \end{pmatrix} &= \begin{pmatrix} 0 & 1 \\ 0 & -1 \end{pmatrix} \begin{pmatrix} \hat{x}_1(t) \\ \hat{x}_2(t) \end{pmatrix} + \begin{pmatrix} 0 \\ 1 \end{pmatrix} u(t) + \begin{pmatrix} l_1 \\ l_2 \end{pmatrix} \left(y(t) - \hat{x}_1(t) \right) \\ &= \begin{pmatrix} -l_1 & 1 \\ -l_2 & -1 \end{pmatrix} \begin{pmatrix} \hat{x}_1(t) \\ \hat{x}_2(t) \end{pmatrix} + \begin{pmatrix} 0 \\ 1 \end{pmatrix} u(t) + \begin{pmatrix} l_1 \\ l_2 \end{pmatrix} y(t) \end{aligned} \qquad [6.31]$$

The characteristic polynomial of the observer is written:

$$\det \left(\lambda I - (A - L\,C) \right) = \begin{vmatrix} \lambda + l_1 & -1 \\ l_2 & \lambda + 1 \end{vmatrix} = \lambda^2 + (1 + l_1)\,\lambda + (l_1 + l_2) \qquad [6.32]$$

and by identifying with a second order polynomial written in normalized form:

$$\begin{aligned} \lambda^2 + (1 + l_1)\lambda + (l_1 + l_2) &\equiv \lambda^2 + 2\,\xi\,\omega_0\,\lambda + \omega_0{}^2 \quad \Leftrightarrow \\ \begin{cases} l_1 = 2\,\xi\,\omega_0 - 1 \\ l_2 = \omega_0{}^2 - 2\,\xi\,\omega_0 + 1 \end{cases} & \end{aligned} \qquad [6.33]$$

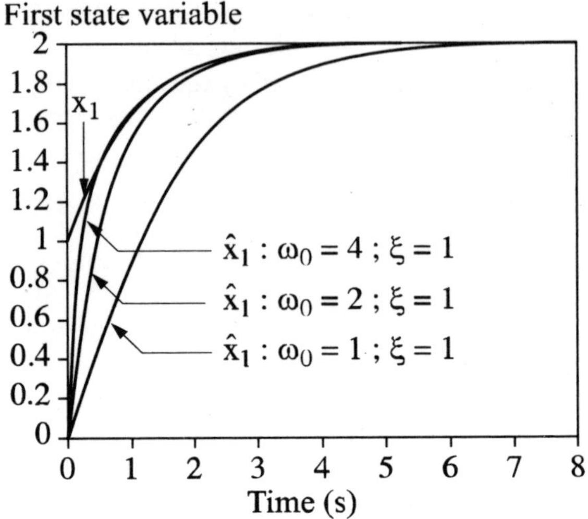

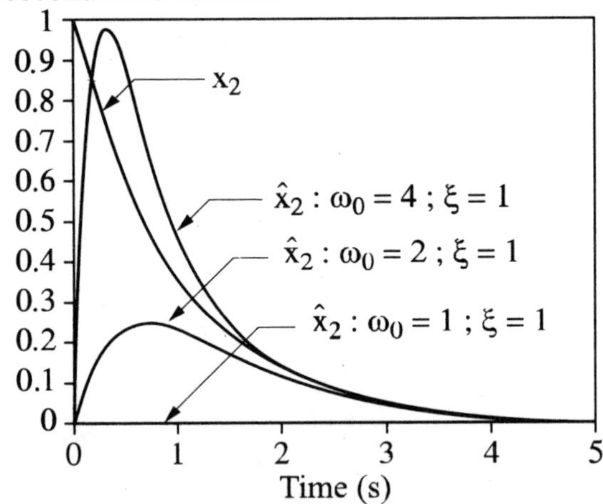

Figure 6.5. *Observer by pole placement*

Figure 6.5 shows the evolution of the two state variables in response to the initial condition $x(0) = (1 \quad 1)^T$ and the evolutions of the state variables of the observer initialized by $\hat{x}(0) = (0 \quad 0)^T$ for different values of ω_0 : the higher ω_0 is, the faster the observer's state joins the systems' state.

6.4. Stabilization through quadratic optimization

6.4.1. *General results for continuous-time*

Let us consider again system [6.1], with an initial condition $x(0) \neq 0$. The question now is to determine the control that enables to bring back $x(t)$ state to 0, while minimizing the criterion:

$$J = \int_0^\infty \left(x(t)^T Q\, x(t) + u(t)^T R\, u(t) \right) dt \qquad [6.34]$$

where Q and R are two symmetric matrices, one positive semi-defined and the other one positive defined:

$$Q = Q^T \geq 0 \; , \; R = R^T > 0 \qquad [6.35]$$

(hence, we have $x^T Q x \geq 0 \;\; \forall x$ and $u^T R u > 0 \;\; \forall u \neq 0$). Since matrix Q is symmetric, we will write it in the form $Q = H^T H$, where H is a full rank rectangular matrix.

The solution of the problem is provided by Theorem 6.2.

THEOREM 6.2.– *if conditions [6.35] are verified, and also if:*

$$\begin{cases} (A, B) \;\; \text{is stabilizable} \\ (H, A) \;\; \text{is detectable} \end{cases} \qquad [6.36]$$

there is a unique, symmetric and positive semi-defined matrix P, which is the solution of the following equation (called Riccati's equation):

$$P A + A^T P - P B R^{-1} B^T P + Q = 0 \qquad [6.37]$$

The control that minimizes criterion [6.34] is given by:

$$\begin{cases} u(t) = - K\, x(t) \\ K = R^{-1} B^T P \end{cases} \qquad [6.38]$$

It guarantees the asymptotic stability of the looped system:

$$\dot{x}(t) = (A - BK)\,x(t) \text{ is such that: } \forall\, x(0)\quad x(t) \underset{t\to\infty}{\to} 0 \qquad [6.39]$$

The value obtained for the criterion is then $J^{*} = x(0)^{T} P\,x(0)$. ∎

Elements of demonstration

The condition of stabilizability of (A, B) is clearly a necessary condition for the existence of a control that stabilizes the system. We will admit that it is also a sufficient condition for the existence of a matrix P, symmetric and positive semi-defined, solution of Riccati's equation [MOL 77]. If, moreover, (H, A) is detectable, we show that this matrix is unique [ZHO 96].

If (A, B) is stabilizable, we are sure that there is a control for which J (whose upper bound is infinite) acquires a finite value: since the non-controllable part is stable, any state feedback placing all the poles of the controllable part in the left half-plane ensures that $x(t)$ and $u(t)$ are expressed as the sum of the exponential functions that tend toward 0.

Mutually, any control $u(t)$ leading to a finite value of J ensures that $x(t)^{T} Q\,x(t)$ tends toward 0, and hence that $H\,x(t)$ tends toward 0. Since (H, A) is detectable, this condition ensures that $x(t)$ tends toward 0.

Hence, let us define the function $V\big(x(t)\big) = x(t)^{T} P\,x(t)$ where P is the positive semi-defined solution of [6.37]. We obtain:

$$\frac{dV}{dt} = (Ax + Bu)^{T} Px + x^{T} P(A\,x + Bu) =$$

$$= x^{T}(A^{T}P + PA)x + u^{T}B^{T}Px + x^{T}PBu =$$

$$= x^{T}(PBR^{-1}B^{T}P - Q)x + u^{T}B^{T}Px + x^{T}PBu =$$

$$= (u^{T} + x^{T}PBR^{-1})\,R\,(u + R^{-1}B^{T}Px) - u^{T}Ru - x^{T}Qx =$$

$$= (u - u^{*})^{T}\,R(u - u^{*}) - u^{T}Ru - x^{T}Qx$$

by noting u^* the control given by [6.38]. For any stabilizable control $u(t)$ we have:

$$J = \int_0^\infty (x^T Q x + u^T R u)\, dt =$$

$$= \int_0^\infty (-\frac{dV}{dt} + (u - u^*)^T R (u - u^*))\, dt =$$

$$= x(0)^T P x(0) + \int_0^\infty ((u - u^*)^T R (u - u^*))\, dt$$

Since R is positive defined, J is minimal for $u(t) \equiv u^*(t)$ and thus has the announced value. As indicated in the third section, the detectability of (H, A) ensures the asymptotic stability of the looped system.

NOTE 6.2.– when $P > 0$, function $V(x(t))$, which is then positive defined and whose derivative is negative defined, is a Lyapunov function (condition $P > 0$ is verified if and only if (H, A) is observable [MOL 77]).

6.4.2. General results for discrete-time

The results enabling the discrete-time quadratic optimization are the same, with a few changes in the equations describing the solution. Let us consider the system [6.4] and the criterion to minimize:

$$J = \sum_{k=0}^{\infty} (x_{k+1}^T Q x_{k+1} + u_k^T R u_k) \tag{6.40}$$

matrices Q and R having the same properties as in the previous section (particularly with the conditions [6.35]). The solution of the problem is provided by Theorem 6.3 [KWA 72].

THEOREM 6.3.– *if conditions [6.35] are verified and also if:*

$$\begin{cases} (F, G) \ \textit{is stabilizable} \\ (H, F) \ \textit{is detectable} \end{cases} \tag{6.41}$$

there is a unique matrix P, symmetric and positive semi-defined, solution of the following equation (called discrete Riccati's equation):

$$F^T P F - P - F^T P G (R + G^T P G)^{-1} G^T P F + Q = 0 \qquad [6.42]$$

The control that minimizes criterion [6.40] is given by:

$$\begin{cases} u_k = -K \; x_k \\ K = (R + G^T P G)^{-1} G^T P F \end{cases} \qquad [6.43]$$

It guarantees the asymptotic stability of the looped system:

$$x_{k+1} = (F - G K) x_k \text{ is such that}: \forall x_0 \quad x_k \underset{k \to \infty}{\to} 0 \qquad [6.44]$$

The value obtained for the criterion is then $J^* = x(0)^T P x(0)$. ∎

6.4.3. *Interpretation of the results*

The results presented above require the following notes:

– the optimization of a criterion of the form [6.34] or [6.40] does not have to be considered as a goal in itself but as a particular means to calculate a control, which has the advantage of leading to a linear state feedback;

– however, we can attempt to give a physical significance to this criterion: it creates a balance between the objective (we want to make x return to 0, the evolution of x penalizes the criterion through matrix Q) and the necessary expense (the controls u applied penalize the criterion due to matrix R);

– the choice of weighting matrices Q and R depends on the user, as long as conditions [6.35] and [6.36] or [6.41] are satisfied. Without getting into details, it should be noted that if all coefficients of Q increase, the evolution of x is even more penalized, at the expense of the evolution of u controls; thus the optimization of the criterion leads to a solution ensuring a faster dynamic behavior for the looped system, but at the expense of stronger controls. Inversely, the increase of all coefficients of R will lead to softer controls and to a slower dynamic behavior;

– the two conditions in [6.36] or [6.41] are not of the same type: in fact we can always fulfill the condition of detectability by a careful choice of matrix Q.

However, the available controls impose matrix B (or G), so that there is no way of acting on the condition of stabilizability;

– the criterion optimization provides only matrix K of expression [6.38] or [6.43]. In the absence of input ($e \equiv 0$), the control ensures the convergence towards the state of equilibrium $x = 0$. Input e makes the system evolve (in particular a constant input makes it possible to orient the system toward another point of equilibrium, different from $x = 0$).

6.4.4. *Example*

Let us consider system [6.14] and the criterion:

$$J = \int_0^\infty \left(q\, y(t)^2 + r\, u(t)^2\right) dt \quad \text{and} \quad Q = \begin{pmatrix} q & 0 \\ 0 & 0 \end{pmatrix} \quad \text{and} \quad R = r \qquad [6.45]$$

where q and r are positive coefficients. Hence, we have $H = (\sqrt{q} \quad 0)$ and we can verify that (H, A) is observable :

$$\mathrm{rank}\begin{pmatrix} H \\ H\,A \end{pmatrix} = \mathrm{rank}\begin{pmatrix} \sqrt{q} & 0 \\ 0 & \sqrt{q} \end{pmatrix} = 2 \qquad [6.46]$$

In section 6.2.2 we saw that (A, B) is controllable, so that hypotheses [6.36] are verified. The positive semi-defined solution of Riccati's equation and the state feedback matrix are written by noting $\alpha = \sqrt{qr}$ and $\beta = \sqrt{q/r}$:

$$P = \begin{pmatrix} \alpha\sqrt{1+2\beta} & \alpha \\ \alpha & r(-1+\sqrt{1+2\beta}) \end{pmatrix} \qquad [6.47]$$

$$K = (\beta \quad -1+\sqrt{1+2\beta})$$

We note that the latter depends only on the ratio q/r and not on q and r separately. Figure 6.6 shows the evolution of the control and the output, in response to the initial condition $x(0) = (1 \quad 1)^T$, for different values of q/r: the higher q/r is, the faster the output returns to 0, but at the expense of a stronger control.

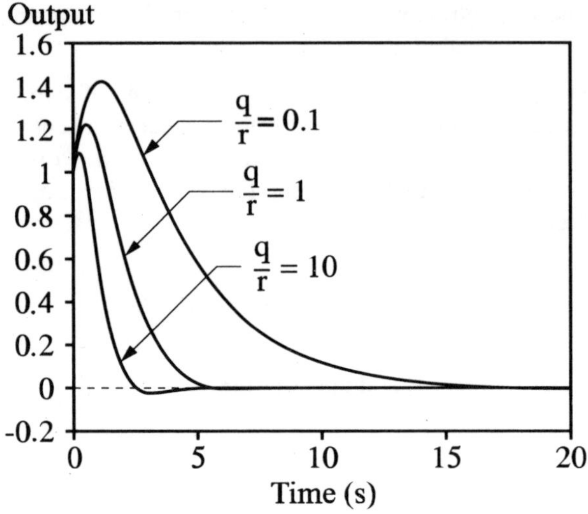

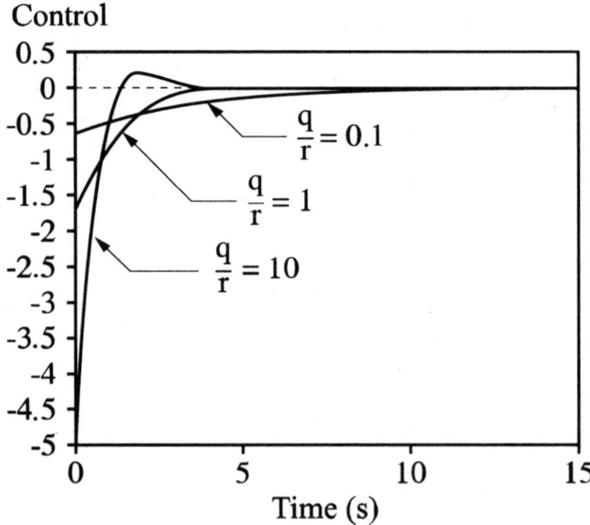

Figure 6.6. *Stabilization by quadratic optimization*

6.5. Resolution of the state reconstruction problem by duality of the quadratic optimization

6.5.1. *Calculation of a continuous-time observer*

The calculation of an observer (section 6.3) must ensure the convergence toward 0 of the reconstruction error $\varepsilon(t)$, described by:

$$\dot{\varepsilon}(t) = (A - LC)\ \varepsilon(t) \tag{6.48}$$

or, similarly, that the eigenvalues of $A - LC$ are all of negative real parts. However, in section 6.3.4 we saw that the calculation of an observer implies the calculation of a state feedback when we transpose matrices A and C of the system. In other words, if we define the following imaginary system:

$$\dot{\eta}(t) = A^T \eta(t) + C^T v(t) \tag{6.49}$$

with a state feedback, noted by $v(t) = -L^T \eta(t)$, we obtain the looped system:

$$\dot{\eta}(t) = (A^T - C^T L^T)\eta(t) \tag{6.50}$$

whose state matrix has the same eigenvalues as the observer [6.20]. Hence, there is equivalence between the stability of the looped system [6.50] and the stability of the observer.

In order to calculate matrix L^T, we can use the quadratic optimization approach presented in section 6.4. Let us define for system [6.49] a quadratic criterion:

$$J = \int_0^\infty \left(\eta(t)^T V \eta(t) + v(t)^T W v(t)\right) dt \tag{6.51}$$

where V and W are two symmetric matrices – the first is semi-positive and the second is positive definite:

$$V = V^T = J J^T \geq 0 \quad, \quad W = W^T > 0 \tag{6.52}$$

By applying Theorem 6.2, and by using the duality between stabilizability and detectability, we immediately obtain the following result.

THEOREM 6.4.– *if conditions [6.52] are verified and also if:*

$$\begin{cases} (A^T, C^T) & stabilizable \\ (J^T, A^T) & detectable \end{cases} \Leftrightarrow \begin{cases} (C, A) & detectable \\ (A, J) & stabilizable \end{cases} \qquad [6.53]$$

there is a unique matrix M, which is symmetric and positive semi-defined, solution of Riccati's equation:

$$M A^T + A M - M C^T W^{-1} C M + V = 0 \qquad [6.54]$$

The gain matrix:

$$L^T = W^{-1} C M \Leftrightarrow L = M C^T W^{-1} \qquad [6.55]$$

guarantees the asymptotic stability of the observer. ∎

We should also remember that the eigenvalues of $A - LC$ set the dynamics of the observer and of the reconstruction error $\varepsilon(t)$.

Hence, determining the observer depends only on the choice of the two new weighting matrices V and W. Like matrices Q and R of the problem of stabilization by quadratic optimization (section 6.4), their choice makes it possible to adjust the dynamics of the observer. It should particularly be noted that the increase of coefficients of V (W respectively) leads to a faster dynamics (slower, respectively).

6.5.2. Calculation of a discrete-time observer

The same approach is applicable for a discrete-time observer, for the synthesis of an observer described by equation [6.23] or [6.24]: by using the results in section 6.4.2, we obtain the results below.

THEOREM 6.5.– *if conditions [6.52] are verified and also if:*

$$\begin{cases} (C, F) & detectable \\ (F, J) & stabilizable \end{cases} \qquad [6.56]$$

there is a unique matrix M, symmetric and positive semi-defined, solution of the Riccati's discrete equation):

$$F\,M\,F^T - M - F\,M\,C^T(W + C\,M\,C^T)^{-1}C\,M\,F^T + V = 0 \qquad [6.57]$$

The gain matrix:

$$L = F\,M\,C^T(W + C\,M\,C^T)^{-1} \qquad [6.58]$$

guarantees the asymptotic stability of the observer.   ■

As for continuous-time, determining the observer depends only on the two matrices V and W: they set the eigenvalues of $F-L\,C$, on which depend the dynamics of the observer and the reconstruction error ε_k.

6.5.3. *Interpretation in a stochastic context*

In this section, we will give a short preview on the techniques of state reconstruction that can be used in a stochastic context[1]. It is interesting to realize that, in a certain measure, the results are obtained by quadratic optimization.

The system whose state we seek to rebuild is supposed to be described by:

$$\begin{cases} \dot{x}(t) = A\,x(t) + B\,u(t) + v(t) \\ z(t) = C\,x(t) + w(t) \end{cases} \qquad [6.59]$$

where $v(t)$ and $w(t)$ are white noises, of zero average and of variances:

$$E\{v(t)\,v(t)^T\} = V \;\; ; \;\; E\{w(t)\,w(t)^T\} = W \qquad [6.60]$$

Noise $v(t)$ can be interpreted as a disturbance occurring at the system input and $w(t)$ as a measurement noise. The problem is to rebuild the state of the system through an observer of the form [6.19].

---

1 More complete developments are proposed, for example, in [LAR 96].

[KWA 72] shows that the observer that ensures an average zero error $E\{x(t) - \hat{x}(t)\}$ and that optimizes the variance:

$$\Sigma(t) = E\{\varepsilon(t)\,\varepsilon(t)^T\} \quad ; \quad \varepsilon(t) = x(t) - \hat{x}(t) \qquad [6.61]$$

is given by the following equations:

$$\begin{cases} \dot{\hat{x}}(t) = A\ \hat{x}(t) + B\ u(t) + L(t)\ (z(t) - C\hat{x}(t)) \\ \hat{x}(0) = E\{x(0)\} \\ L(t) = \Sigma(t)\ C^T W^{-1} \end{cases} \qquad [6.62]$$

where $\Sigma(t)$ verifies Riccati's differential equation:

$$\begin{cases} \dot{\Sigma}(t) = \Sigma(t)\,A^T + A\,\Sigma(t) - \Sigma(t)\,C^T W^{-1} C\,\Sigma(t) + V \\ \Sigma(0) = E\{(x(0) - \hat{x}(0))\,(x(0) - \hat{x}(0))^T\} \end{cases} \qquad [6.63]$$

This observer is called *Kalman's filter* and we note that its gain varies in time. However, we have the following convergence result [KWA 72]:

THEOREM 6.6.– *if conditions [6.56] are verified, $\Sigma(t)$ solution of [6.63] tends, when $t \to \infty$, toward the unique positive semi-defined symmetric solution M of Riccati's algebraic equation [6.54].* ∎

Hence, we can interpret the observer determined in section 6.5.1 as the permanent state of Kalman's filter that optimizes the state reconstruction, considering the particular hypotheses on the noises that interfere on the system.

The same results are obtained for discrete-time [KWA 72], if we assume the system described by:

$$\begin{cases} x_{k+1} = F\,x_k + G\,u_k + v_k \\ z_k = C\,x_k + w_k \end{cases} \quad ; \quad \begin{cases} E\{v_k\,v_k^T\} = V \\ E\{w_k\,w_k^T\} = W \end{cases} \qquad [6.64]$$

Under conditions [6.56], the permanent state of Kalman's filter is there also the observer determined in section 6.5.2.

6.5.4. *Example*

Let us consider again the system described by equations [6.14]:

$$\begin{cases} \begin{pmatrix} \dot{x}_1(t) \\ \dot{x}_2(t) \end{pmatrix} = \begin{pmatrix} 0 & 1 \\ 0 & -1 \end{pmatrix} \begin{pmatrix} x_1(t) \\ x_2(t) \end{pmatrix} + \begin{pmatrix} 0 \\ 1 \end{pmatrix} u(t) \\ y(t) = \begin{pmatrix} 1 & 0 \end{pmatrix} \begin{pmatrix} x_1(t) \\ x_2(t) \end{pmatrix} \end{cases}$$

and let us calculate an observer with the weighting matrices:

$$V = \begin{pmatrix} 0 & 0 \\ 0 & v \end{pmatrix} \text{ or } J = \begin{pmatrix} 0 \\ \sqrt{v} \end{pmatrix} \text{ and } W = w \qquad [6.65]$$

where v and w are positive coefficients. We can verify that (A, J) is controllable:

$$\text{rank}\begin{pmatrix} J & AJ \end{pmatrix} = \text{rank}\begin{pmatrix} 0 & \sqrt{v} \\ \sqrt{v} & -\sqrt{v} \end{pmatrix} = 2 \qquad [6.66]$$

In section 6.3.6 we saw that (C, A) is observable, so that hypotheses [6.53] are verified. The equations of the observer are the general equations [6.19], with L solution of equations [6.54] and [6.55].

Figure 6.7 shows the evolution of the two state variables in response to the initial condition $x(0) = (1 \quad 1)^T$ and those of the state variables of the observer initialized by $\hat{x}(0) = (0 \quad 0)^T$, for different values of the ratio v/w: the higher v/w is, the faster the observer's state returns to the state of the system.

First state variable

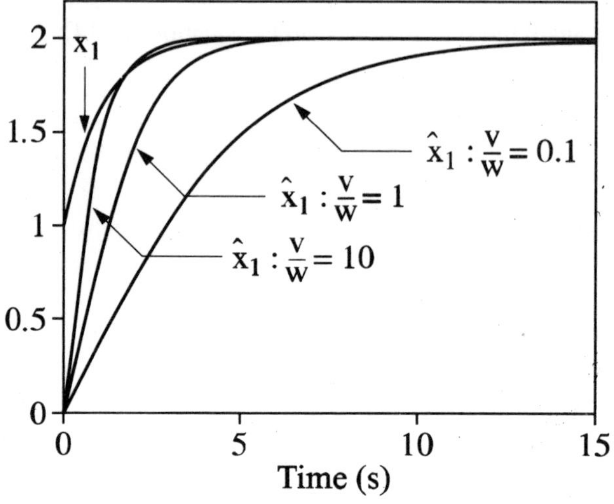

Second state variable

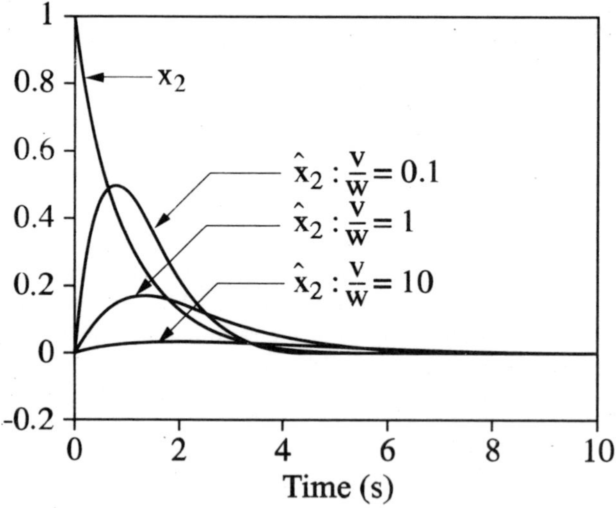

Figure 6.7. *Observer by quadratic optimization*

6.6. Control through state feedback and observers

6.6.1. *Implementation of the control*

The results of the previous sections enable to determine a control law for a system whose state is not entirely measured. We suppose that its equations are:

$$\begin{cases} \dot{x}(t) = A\,x(t) + B\,u(t) \\ z(t) = C\,x(t) \end{cases} \tag{6.67}$$

The modal control, or the optimization of a quadratic criterion, provides a state feedback control, whose general form is the following:

$$u(t) = -K\,x(t) + e(t) \tag{6.68}$$

Similarly, the modal approach, or the choice of two weighting matrices V and W provides an L gain observer.

The system control [6.67] is obtained by implementing the state feedback not from the state $x(t)$ of the system, which is not accessible, but from its reconstruction $\hat{x}(t)$ provided by the observer. Hence, it is given by the following equations, which correspond to the diagram in Figure 6.8[2]:

$$\begin{cases} \dot{\hat{x}}(t) = A\,\hat{x}(t) + B\,u(t) + L\,(z(t) - C\,\hat{x}(t)) \\ u(t) = -K\,\hat{x}(t) + e(t) \end{cases} \tag{6.69}$$

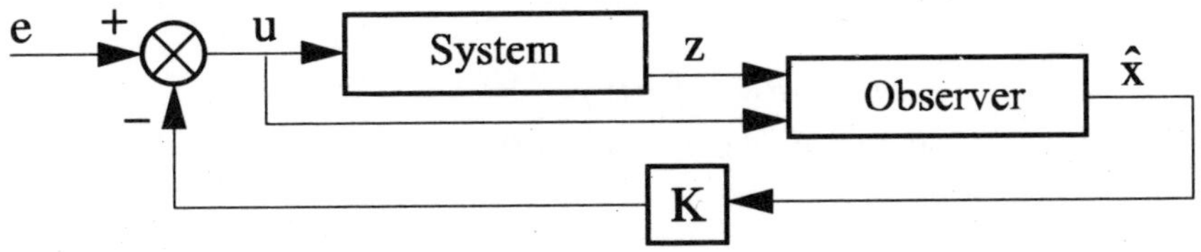

Figure 6.8. *Control by state feedback and observer*

2 The name LQG (that stands for *Linear-Quadratic-Gaussian*) control is sometimes used to designate this type of control. It obviously refers to one of the methods used for calculating the state feedback, and to the stochastic interpretation of the reconstruction carried out by the observer (section 6.5.3).

In the case of a sampled system, described by the equations:

$$\begin{cases} x_{k+1} = F\,x_k + G\,u_k \\ z_k = C\,x_k \end{cases}$$

[6.70]

the control law obtained by implementing the state feedback from its reconstruction $\hat{x}_k$ provided by the observer (following here too the principle in Figure 6.8) is given by:

$$\begin{cases} \hat{x}_{k+1} = F\,\hat{x}_k + G\,u_k + L\,(z_k - C\,\hat{x}_k) \\ u_k = -K\,\hat{x}_k + e_k \end{cases}$$

[6.71]

6.6.2. *Dynamics of the looped system*

Let us briefly discuss now the equations of state of system [6.67] looped by the control law [6.69]. For that we need $2n$ state variables because the system, as the observer, is of order n. Let us choose as state vector of the group the set of vectors x and ε together. We already know equation [6.21] describing $\varepsilon(t)$. For $x(t)$ we obtain equation [6.72]:

$$\dot{x}(t) = A\,x(t) + B\left(-K\,\hat{x}(t) + e(t)\right) = A\,x(t) + B(-K\,x(t) + K\,\varepsilon(t) + e(t))$$

$$\dot{x}(t) = (A - BK)\,x(t) + BK\varepsilon(t) + Be(t)$$

[6.72]

By considering both equations [6.72] and [6.21], we thus obtain:

$$\begin{pmatrix} \dot{x}(t) \\ \dot{\varepsilon}(t) \end{pmatrix} = \begin{pmatrix} A - BK & BK \\ 0 & A - LC \end{pmatrix}\begin{pmatrix} x(t) \\ \varepsilon(t) \end{pmatrix} + \begin{pmatrix} B \\ 0 \end{pmatrix}e(t)$$

[6.73]

In the case of a sampled system, we similarly obtain:

$$\begin{pmatrix} x_{k+1} \\ \varepsilon_{k+1} \end{pmatrix} = \begin{pmatrix} F - GK & GK \\ 0 & F - LC \end{pmatrix}\begin{pmatrix} x_k \\ \varepsilon_k \end{pmatrix} + \begin{pmatrix} G \\ 0 \end{pmatrix}e_k$$

[6.74]

The block-diagonal structure of state matrices obtained makes it possible to state that the looped system has as eigenvalues the reunion of the eigenvalues of $A - BK$

and of $A-LC$ (or $F-GK$ and $F-LC$ in discrete-time). The former have been chosen or set during the calculation of the state feedback (and in this case we have control dynamics) whereas the latter have been chosen or set during the calculation of the observer (and in this case we talk of reconstruction dynamics).

6.6.3. *Interest and limitations of this result*

The previous example justifies the approach adopted which consists of calculating the state feedback and the observer independently.

In addition, if the observer operates in perfect conditions, it theoretically maintains $\varepsilon(t)$ at 0, so that equations [6.73] and [6.74] become identical to equations [6.3] and [6.6] that we obtain if we can directly apply the state feedback!

This result must be analyzed critically. First of all it implies that the observer and the system operate exactly in the same conditions. However, we saw in section 6.3.5 that in the presence of an interference applied to the system, error $\varepsilon(t)$ does not tend toward 0 anymore. The separation of dynamics remains, however, verified in this case.

On the other hand, let us suppose that modeling uncertainties affect the system state matrix. In section 6.3.5 we established equation [6.29] that governs the reconstruction error. The equation of the looped system becomes:

$$\begin{pmatrix} \dot{x}(t) \\ \dot{\varepsilon}(t) \end{pmatrix} = \begin{pmatrix} A' - BK & BK \\ A' - A & A - LC \end{pmatrix} \begin{pmatrix} x(t) \\ \varepsilon(t) \end{pmatrix} + \begin{pmatrix} B \\ 0 \end{pmatrix} e(t) \qquad [6.75]$$

This time we see that all the system's dynamics are affected and that we cannot state anything with respect to the position of its eigenvalues. We must also add that, in certain cases, even a very small error between the two matrices can be enough to change the eigenvalues significantly.

This last point shows that it is indispensable to verify that the control is sufficiently robust to face the model errors. A detailed analysis of the robustness properties is outside the context of this work; for more details on this aspect, see [DUC 99]. A first approach consists of ensuring that the *stability margins* of the looped system are sufficiently high. In what follows we will present this aspect.

6.6.4. *Interpretation in the form of equivalent corrector*

The state feedback return and the observer are illustrated in Figure 6.8 and given in continuous-time by equations [6.69]. If, in order to simplify, we note $e(t) = 0$, equations [6.69] are written in the following form, which is obtained by carrying the expression of $u(t)$ in the first equation:

$$\begin{cases} \dot{\hat{x}}(t) = \left(A - BK - LC\right)\hat{x}(t) + L\, z(t) \\ u(t) = -K\,\hat{x}(t) \end{cases} \qquad [6.76]$$

Equations [6.76] correspond to a system with input $z(t)$ and output $u(t)$, i.e. what we usually call a corrector. By using the Laplace transform, let us calculate the transfer functions of the system and corrector. From equations [6.67] describing the system, we obtain:

$$Z(p) = G(p)\, U(p) \quad \text{with} \quad G(p) = C\,(pI - A)^{-1} B \qquad [6.77]$$

and from equations [6.76] describing the corrector:

$$U(p) = -K(p)\, Z(p) \quad \text{with} \quad K(p) = K\left(pI - A + BK + LC\right)^{-1} L \qquad [6.78]$$

The looped system can thus be represented by the negative feedback loop in Figure 6.9. The corresponding open loop transfer function is written:

$$T_{BO}(p) = K(p)\, G(p) = K\left(pI - A + BK + LC\right)^{-1} L \; C\,(pI - A)^{-1} B \qquad [6.79]$$

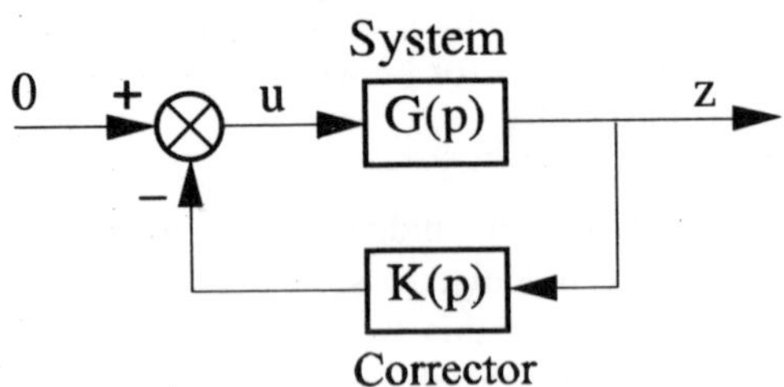

Figure 6.9. *Analysis of stability margins*

If the system has only one control[3], $T_{BO}(p)$ is a scalar transfer function, and its stability margins can be determined from its frequency response $T_{BO}(j\omega)$, traced in Bode, Black or Nyquist's planes. This analysis enables to verify that the control, which was determined by adopting a purely algebraic approach, leads to satisfactory stability margins.

The same approach is applied for a discrete-time system. The equations obtained for the state feedback control and the observer are written, by considering $e_k = 0$:

$$\begin{cases} \hat{x}_{k+1} = (F - GK - LC)\ \hat{x}_k + L\ z_k \\ u_k = -K\ \hat{x}_k \end{cases} \qquad [6.80]$$

The application of the z-transform to equations [6.70] and [6.80] makes it possible to obtain transfer functions $G(z)$ and $K(z)$ of the system and of the corrector, then the open loop transfer function:

$$T_{BO}(z) = K(z)\ G(z) = K\ (zI - F + GL + KC)^{-1}K\ C\ (zI - F)^{-1}G \qquad [6.81]$$

Hence, determining the stability margins can be done from the frequency response $T_{BO}(e^{j\omega T})$ in Bode, Black or Nyquist planes.

6.6.5. *Example*

Let us consider again system [6.14] which was our example throughout this chapter. We will use it again to illustrate what follows:

– the state feedback calculated in section 6.2.2 by the modal approach, with $\omega_0 = 2$ and $\xi = 0.7$;

– the observer calculated in section 6.3.6, with $\omega_0 = 4$ and $\xi = 0.7$.

Figure 6.10 shows the responses to an initial condition $x(0) = (1\ \ 1)^T$ with state feedback and observer: the initial condition of the observer is $\hat{x}(0) = (1\ \ 1)^T$ (top of the figure) and then $\hat{x}(0) = (0\ \ 0)^T$ (bottom of the figure). In the first case, the responses are not clear because, as the observer has been set on the system since its origin, only the control dynamic appeared; in the second case, the initial error is non-zero at $t = 0$, and the two dynamics of control and reconstruction intervene when the observer is used.

---

3 The approach is extended to the case of multi-control systems at the expense of additional developments (see [DOR 95, FRI 86]).

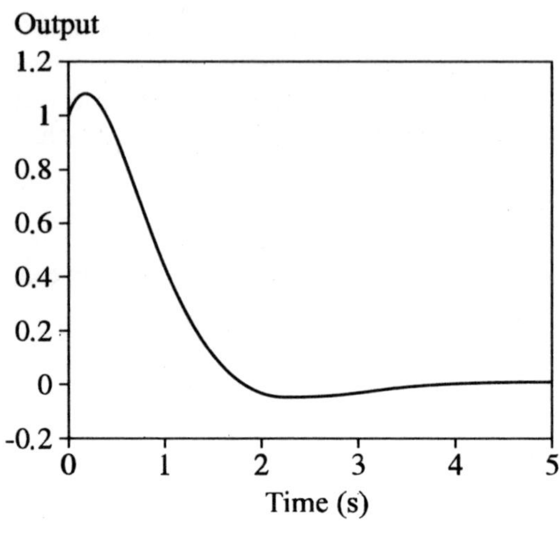

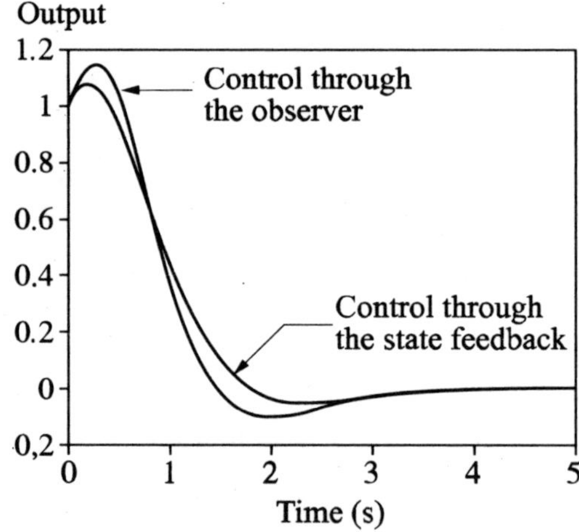

Figure 6.10. *Control by the observer, compared to control by state return*

Finally, Figure 6.11 shows Bode's diagram of $T_{BO}(j\omega)$, from where we can obtain gain and phase margins of 12.9 dB and 50.7°. These values can be considered as satisfactory.

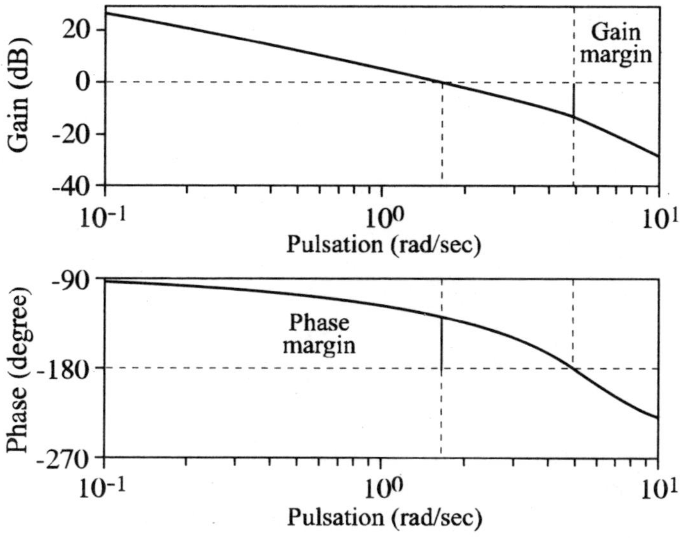

Figure 6.11. *Determination of stability margins*

6.7. A few words on the resolution of Riccati's equations

In this chapter Riccati's equations occurred several times, corresponding to continuous-time and discrete-time problems. These equations appear in numerous control engineering problems and we will finish this chapter by giving a preview on a few explicit methods of resolution.

Let us consider firstly Riccati's continuous equation:

$$X A + A^T X - X S X + Q = 0 \qquad\qquad [6.82]$$

where A, $Q = Q^T$, $S = S^T$ and $X = X^T$ are $n \times n$ real square matrices and X is the unknown factor. Neither the existence nor the uniqueness of the solution is guaranteed in the general case. In automatic control engineering, the main interest lies in the so-called "stabilizing" solution (we will see that it is unique), for which all the eigenvalues of $A - S X$ are of strictly negative real part.

Firstly, we will consider the $2n \times 2n$ Hamiltonian matrix, associated with equation [6.82]:

$$H = \begin{pmatrix} A & -S \\ -Q & -A^T \end{pmatrix} \qquad [6.83]$$

We easily verify that:

$$\begin{pmatrix} 0 & I \\ -I & 0 \end{pmatrix} \begin{pmatrix} A & -S \\ -Q & -A^T \end{pmatrix} = -\begin{pmatrix} A^T & -Q \\ -S & -A \end{pmatrix} \begin{pmatrix} 0 & I \\ -I & 0 \end{pmatrix} \qquad [6.84]$$

and thus that H is similar to $-H^T$: matrix H therefore has a symmetric spectrum with respect to the origin. As this is a real matrix, its spectrum is symmetric with respect to the real axis and hence it is also symmetric with respect to the imaginary axis. Furthermore, it is necessary to make the following hypothesis: matrix H does not have a complex eigenvalues.

Hence, we can calculate Jordan's form of H, which makes it possible to write:

$$H \begin{pmatrix} T_{11} & T_{12} \\ T_{21} & T_{22} \end{pmatrix} = \begin{pmatrix} T_{11} & T_{12} \\ T_{21} & T_{22} \end{pmatrix} \begin{pmatrix} -\Lambda & 0 \\ 0 & \Lambda \end{pmatrix} \qquad [6.85]$$

where Λ and $-\Lambda$ contain eigenvalues with strictly positive and strictly negative real parts respectively. Let us introduce the second hypothesis: matrix T_{11} is reversible.

These two hypotheses are especially verified in the problems discussed in sections 6.4.1 and 6.5.1 [ZHO 96]. Then, the stabilizing solution of equation [6.82] is:

$$X = T_{21} T_{11}^{-1} \qquad [6.86]$$

Demonstration. It is sufficient to formulate the terms of equation [6.85] where T_{11} and T_{21} intervene:

$$\begin{cases} A\,T_{11} - S\,T_{21} = -T_{11}\,\Lambda \\ -Q\,T_{11} - A^T T_{21} = -T_{21}\,\Lambda \end{cases} \Leftrightarrow$$

$$\begin{cases} A - S\,X = -T_{11}\,\Lambda\,T_{11}^{-1} \\ -Q - A^T X = -T_{21}\,\Lambda\,T_{11}^{-1} \end{cases} \Leftrightarrow$$

$$X\,A - X\,S\,X = -T_{21}\,\Lambda\,T_{11}^{-1} = -Q - A^T X \qquad [6.87]$$

In addition, the first equation of the second brace shows that $A - S\,X$ has all its eigenvalues as strictly negative real parts.

Numeric difficulties may occur when matrix H has multiple or very close eigenvalues, or even close to the imaginary axis. To improve this aspect, we can use Schur's form [BIT 91], by replacing equation [6.85] by factoring:

$$H = \begin{pmatrix} U_{11} & U_{12} \\ U_{21} & U_{22} \end{pmatrix} \begin{pmatrix} S_{11} & S_{12} \\ 0 & S_{22} \end{pmatrix} \begin{pmatrix} U_{11} & U_{12} \\ U_{21} & U_{22} \end{pmatrix}^T \qquad [6.88]$$

where matrix U is orthogonal and matrices S_{11} and S_{22} are in Schur's form (i.e. superior quasi-triangular, the real eigenvalues of H appear on the diagonal and the complex eigenvalues as 2×2 blocks), with block S_{11} corresponding to the eigenvalues with negative real part. Hence, we have:

$$X = U_{21}\,U_{11}^{-1} \qquad [6.89]$$

In the case of Riccati's discrete equation:

$$F^T X F - X - F^T X G\,(R + G^T X G)^{-1} G^T X F + Q = 0 \qquad [6.90]$$

where all matrices are real, F, $Q = Q^T$ and $X = X^T$ are of size $n \times n$, $R = R^T$ is of size $m \times m$, G is of size $n \times m$ and X is the unknown factor. The stabilizing solution, if it exists, is the one for which matrix $F - G(R + G^T X G)^{-1} G^T X F$ has all its eigenvalues of the module strictly less than 1.

The approach is a simple transposition of the continuous-time case [BIT 91, VAU 70]. We must assume that matrix F is reversible, in which case equations [6.86] and [6.89] can be applied, by using instead of [6.83] the matrix:

$$H = \begin{pmatrix} F + G\,R^{-1}G^T F^{-T} Q & -\,G\,R^{-1}G^T F^{-T} \\ -F^{-T}Q & F^{-T} \end{pmatrix} \qquad [6.91]$$

where F^{-T} is the transpose of F^{-1}. This matrix has eigenvalues that are opposite to each other, so that Jordan's form which makes it possible to write:

$$H \begin{pmatrix} T_{11} & T_{12} \\ T_{21} & T_{22} \end{pmatrix} = \begin{pmatrix} T_{11} & T_{12} \\ T_{21} & T_{22} \end{pmatrix} \begin{pmatrix} \Lambda^{-1} & 0 \\ 0 & \Lambda \end{pmatrix} \qquad [6.92]$$

where Λ and Λ^{-1} contain the eigenvalues of the module more than 1 and less than 1 respectively. Schur's form, on the other hand, is always provided by [6.88], where block S_{11} corresponds to the eigenvalues of the module less than 1.

6.8. Conclusion

In this chapter, we have presented the necessary prerequisites for the creation of controls using the state representation. We saw that by using either an approach by pole placement, or an approach by quadratic optimization, it was possible to calculate a state feedback, then to rebuild the state and finally to obtain a control from the available measures.

However, it should be clear that this chapter is left incomplete on purpose: the results presented are for the moment unusable for the majority of practical problems because they do, not deal either with the pursuit of indications or with the disturbance rejection, and that the robustness aspects are only briefly touched upon. The resolution of such problems and the resulting procedures will be in fact developed later on.

6.9. Bibliography

[BIT 91] BITTANTI S., LAUB A.J., WILLEMS J.C., *The Riccati Equation*, Springer-Verlag, 1991.

[DOR 95] DORATO P., ABDALLAH C., CERONE V., *Linear-Quadratic Control. An Introduction*, Prentice-Hall, 1995.

[DUC 99] DUC G., FONT S., *Commande H_∞ et μ-analyse*, Hermès, 1999.

[FAU 84] FAURRE P., ROBIN M., *Eléments d'Automatique*, Dunod, 1984.

[FRI 86] FRIEDLAND B., *Control System Design*, Mc Graw-Hill, 1986.

[KWA 72] KWAKERNAAK H., SIVAN R., *Linear Optimal Control Systems*, Wiley-Interscience, 1972.

[LAR 96] DE LARMINAT P., *Automatique*, 2nd edition, Hermès, 1996.

[LAR 02] DE LARMINAT P. (ed.), *Commande des systèmes linéaires*, Hermès, IC2 series, 2002.

[MOL 77] MOLINARI B.P., "The Time-Invariant Linear-Quadratic Optimal Control Problem", *Automatica*, vol. 13, p. 347-357, 1977.

[VAU 70] VAUGHAN D.R., "A non Recursive Algebraic Solution for the Discrete Riccati Equation", *IEEE Trans. Autom. Control*, AC-15, p. 597-599, 1970.

[ZHO 96] ZHOU K., DOYLE J.C., GLOVER K., *Robust and Optimal Control*, Prentice Hall, 1996.

Chapter 7

Process Modeling

7.1. Introduction

Obtaining a model of the industrial system to automate is the first task of a control engineer – and not the smallest one, as the quality of his work significantly depends on the adequacy between the model and the procedure.

To begin, we note that there are several points of view on a complex physical procedure. By *complex*, we mean a procedure with many variables and/or a procedure in which the phenomena involved and the interactions between the variables, are complicated. There is no universal model; its design depends entirely on the task for which it will be used.

Certain models pertain to the representation of physical components of installation and to their connections (*structural models*). They can be described by diagrams called PI (piping-instrumentation) which define the complete diagram of installation. Universal graphic symbols are used in order to facilitate the interpretation of this representation. Nowadays, a computer representation is also adopted: these models are generally a database and/or an object oriented representation and they are used, for example, for the maintenance of the procedure, for safety analysis, for the implementation of block diagrams, etc.

Other models are used for the design and dimensioning of the installation and for managing the various operation modes; they often refer to different functions that the installation must fulfill (*functional models*). They describe the role of each subsystem

Chapter written by Alain BARRAUD, Suzanne LESECQ and Sylviane GENTIL.

in performing the roles of the procedure, in connection with a structure and behavior of components. They are used for the design of procedure monitoring, i.e. its very high level of control: start-up, stop, failure management, manual reboot procedures, structure changes, etc.

Let us consider the example of heating a room. The goal of the system is to heat up the room. To do this, we specify several functions: generation of energy, water supply, water circulation. These functions are based on several components. For example, for water circulation, we use a heater, a pump and a control valve; for the function of generating energy, we use a boiler, a pump and a fuel tank. To all the functions enumerated above, corresponding to a normal operation, we can add a "draining" function which would correspond to taking the installation out of service.

The goal of other models is to describe the behavior of the installation (*behavioral models*); this refers to describing the evolution of physical units during all the operation phases, be it from a static or dynamic point of view.

The complete description of an installation requires the representation of continuous phenomena (main process) and of discrete aspects (discontinuous actions during changes in the operation mode, security actions, etc.). To date these two representation modes have been separated; for example, under a purely continuous angle, the synthesis of regulation loops is described by supposing that any state space is accessible and the production planning is represented in a purely discrete manner. However, nowadays, there are attempts to characterize the set in a hybrid model (combination of two aspects, continuous and discrete), but this path still has difficulties and is still the subject of research.

The behavioral model may have different objectives. The two main objectives are: the simulation of the installation in order to test its behavior in different situations offline (different control laws that the engineer seeks to compare, research into its limits, training of control operators, etc.) and the design of controls to implement. It is not necessarily the same model that is used in these two cases: the first one often requires more precision than the second one. In fact, for the majority of time, the control is calculated on a linear approximation of the system around the nominal working point because the majority of industrial systems work (in normal operating mode) in a limited range, corresponding to an optimal zone for the production. We can also use, in order to calculate the control, a highly simplified non-linear model, the intermediary between these two situations being the calculation of a set of linear models for different working points or operating modes. However, in order to design the automation of an installation in order to optimize the working points or train the operators, the model must be the most robust possible.

The complex model is based on a precise knowledge of physical, chemical, biological or other laws, describing the material phenomena governing the processes implemented in the procedure. We often speak, in this case, of a *knowledge model* or a *model based on the first principles*. It is thus quite naturally described in the form of non-linear differential equations in the dynamic case and/or in the form of algebraic equations in the static case. These equations describe the main laws of the physical world, which are in general material or energy balances. When we can reduce the differential equations to first degree equations (by possibly introducing intermediary variables), we obtain an algebraic differential state model.

The complex model can be simplified under the hypothesis of linearity, in order to obtain linear differential equations from which we can move either to a state representation or to an input-output representation by transfer function. Then, if we want, we can also use the traditional methods in order to discretize these linear models, in order to directly calculate a discrete control. The first section provides a few examples, which are trivial in comparison with the exhaustive task of the engineer for an industrial procedure, but which illustrate the methodology.

When the objective is the development of a control on a linear model, it may be simpler to directly research this model. This research can be done from specific experimentations. In that case we speak of *identification*, rather than modeling. We obtain a *model of representation*. We know well the link between the transfer function and the frequency response and it is thus easy to translate the latter into a mathematical model. However, it is basically impossible to perform a harmonic analysis on an industrial procedure – because it is incompatible with the production constraints – or with the response time of the procedure. Hence, faster means have been investigated in order to obtain these models from time characteristic responses; the most widely used is of course the unit-step response because it corresponds to a change of the working point of the installation, in other words to a current industrial practice. Therefore, a few fast graphic constructions make it possible to obtain, for a minimal cost, a transfer function close to the system. The second section deals with this aspect.

It was soon clear that, in order to make the model robust for the entire range of operation where linearization is valid, we should use input signals with a much larger spectrum than the step function, in order to excite all the modes of the system. As such we use the identification on any input-output data (but that are full of information regarding the behavior of the system); in this case, only the strong numerical methods make it possible to extract the information contained in these data sets. These methods are explained in the third section. The method that will be the most developed can in fact be used on non-linear representations and that is why we also use it in order to parameterize the knowledge methods mentioned above.

7.2. Modeling

The behavioral modeling of a continuous procedure described in what follows is based on a mathematical formalism: we search for a set of equations representing the system in the largest possible operating range. This is a task that may take several months and pertains to the multi-disciplinary teams. In fact, it requires a knowledge of physics, chemistry, biology, etc. in order to be able to understand the phenomena that the model will describe, and knowing numerical analysis in order to write the model equations in a form that is adapted to the numerical calculus. It is also necessary to have computer knowledge in order to be able to implement this calculations.

A procedure is sufficiently complex in order to be able to describe straightaway its behavior by a system of equations. In order to realize a global model, we need to decompose the general system into simpler subsystems, through a descending approach, then recombine the various models into an ascending approach. This decomposition can be found in the methodology of software development: that is why we can use the same tools in order to manage these approaches (SADT, for example). At the level of a basic subsystem, there is no optimal methodology: is it necessary to start by writing the most complicated model possible – by calling upon the description of detailed mechanisms – and later simplify it, either because we have no knowledge regarding the coefficients present at this elementary level and no possibility of estimating them in practice, or because this model is too complicated to be used? Or is it necessary to start by writing a very rough model and not complicate it unless the simulation results obtained are too inaccurate? It is obvious that the model must be the result of a compromise between precision and simplicity. When it is established, we have to *verify* it: this means that we test it to make sure there is no physical inconsistency between its behavior and the behavior of the system, due, for example, to numerical problems or to wrong initial hypotheses. Then we have to *validate* it; this means testing its adequacy with the set of tasks for which it was designed.

In order to initiate the modeling of a reasonably complex subsystem, we generally write material and/or energy balances. Therefore, it is convenient to locate the energy or material sources at the system's input, those at the output (in general connected to another subsystem), the elements that can store or lose energy or matter and those that transport them.

The *bond-graphs* are a graphic representation tool for energy transfers in a physical system, sometimes used as intermediaries between the physical description of a procedure and the writing of equations. Through a formalism reuniting fields as various as mechanics, they describe electricity and hydraulics – simply because they are based on the description of power exchange between subsystems. The graph consists of arcs connecting the stress variables e or the stream variables f whose product represents the power. Forces, torques, tension and pressure are stress variables. Speed, flow and current are stream variables. There are several main elements. The resistances dissipate energy (electric resistances, viscous friction). The capacities store energy (electric condenser, spring), as well as inertial elements (inductance, masses, moments of inertia). The transforming elements preserve power $e_1 f_1 = e_2 f_2$ while imposing a fixed ratio between streams and input and output stresses ($e_1 = ne_2$, $f_1 = f_2/n$). Finally, the junctions are of two types (called 0 and 1) depending on whether they connect elements that preserve the stress and distribute the stream or the other way round. We will not go into further detail on this method, which is dealt with in specific works (see [DAU 00], for example).

By admitting that we have conveniently traced the balance equations to write, they are general in the form of non-linear differential equations. They can be used as such in the simulation fine model, but they will not be generally linearized in order to obtain the control calculation model. The linearization is operated as follows. Let us assume that the differential equation is:

$$y^{(n)}(t) = g\big(y^{(n-1)}(t), y^{(n-2)}(t), \ldots, y(t), e(t), t\big) \qquad [7.1]$$

We represent [7.1] by a set of first order differential equations; this is in reality a possible state representation of [7.1], which is obtained by noting:

$$y_1 = y \qquad [7.2]$$

$$y_2 = \frac{dy_1}{dt} \qquad [7.3]$$

$$y_3 = \frac{dy_2}{dt} \qquad [7.4]$$

$$\vdots \qquad [7.5]$$

$$y_n = \frac{dy_{n-1}}{dt} \qquad [7.6]$$

$$\frac{dy_n}{dt} = g\big(y_n(t), y_{n-1}(t), \ldots, y_1(t), e(t), t\big) \qquad [7.7]$$

If the model is represented by several differential equations whose variables are coupled, we will generally have:

$$\begin{cases} \frac{dy_1}{dt} = f_1(y_1, y_2, \ldots, y_m, e, t) \\ \vdots \\ \frac{dy_m}{dt} = f_m(y_1, y_2, \ldots, y_m, e, t) \end{cases} \qquad [7.8]$$

We suppose that system [7.8] has a balance point Y_0, E_0 for which the derivatives are zero, i.e. which is defined by:

$$\begin{cases} y_i(t) = Y_{i0} + x_i(t) \\ e(t) = E_0 + u(t) \\ f_i(Y_{10}, Y_{20}, \ldots, Y_{m0}, E_0, t) = 0 \end{cases} \qquad [7.9]$$

Now, we try to represent the trajectory of small variations $x(t)$ and $u(t)$ by carrying [7.9] over [7.8] and by using a first order Taylor serial development, which leads to:

$$\frac{\mathrm{d}Y_{i0}(t)}{\mathrm{d}t} + \frac{\mathrm{d}x_i(t)}{\mathrm{d}t} = \frac{\mathrm{d}y_i(t)}{\mathrm{d}t} \qquad [7.10]$$

$$= f_i(Y_{10}, \ldots, Y_{m0}, E_0, t) + \frac{\mathrm{d}f_i}{\mathrm{d}y_1}(Y_{10}, \ldots, Y_{m0}, E_0, t)x_1(t)$$

$$+ \cdots + \frac{\mathrm{d}f_i}{\mathrm{d}e}(Y_{10}, \ldots, Y_{m0}, E_0, t)u(t) \qquad [7.11]$$

Hence, we find the following linear approximation:

$$\begin{bmatrix} \frac{\mathrm{d}x_1}{\mathrm{d}t} \\ \vdots \\ \frac{\mathrm{d}x_n}{\mathrm{d}t} \end{bmatrix} = \begin{bmatrix} \frac{\mathrm{d}f_1}{\mathrm{d}y_1} & \cdots & \frac{\mathrm{d}f_1}{\mathrm{d}y_n} \\ \vdots & \vdots & \vdots \\ \frac{\mathrm{d}f_n}{\mathrm{d}y_1} & \cdots & \frac{\mathrm{d}f_n}{\mathrm{d}y_n} \end{bmatrix} \begin{bmatrix} x_1 \\ \vdots \\ x_n \end{bmatrix} + \begin{bmatrix} \frac{\mathrm{d}f_1}{\mathrm{d}e} \\ \vdots \\ \frac{\mathrm{d}f_n}{\mathrm{d}e} \end{bmatrix} u(t) \qquad [7.12]$$

where the state matrix is the Jacobian of the non-linear relation vector $\mathbf{f}(\mathbf{y}, e, t)$. Therefore, we obtain a linear state representation of the non-linear system.

Simple examples

We will take the simple example of two cascade tanks supplied by a liquid volume flow rate. The tanks are the two storage elements of the matter; the incoming and

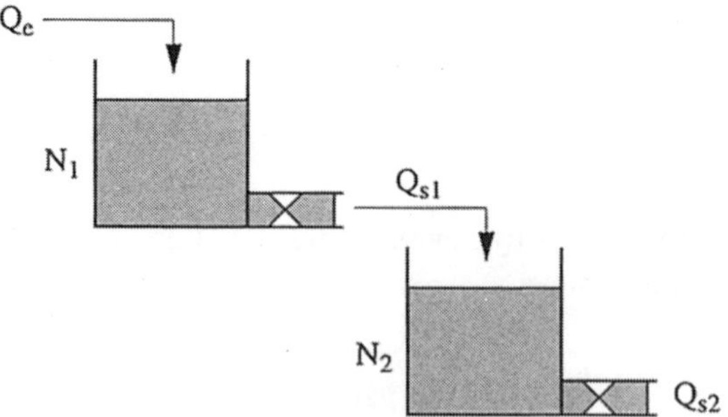

Figure 7.1. *Cascade tanks*

outgoing flows of the second tank link this subsystem to its environment. There are no losses or intermediary transport element, hence we will write two matter balance equations, one for each of the storage elements.

Let Q_e be the volume flow rate entering the first tank, Q_{s1} the volume flow rate leaving the tank, A_1 its section, N_1 the water level in the tank and K the restriction coefficient of the output tank. The outgoing flow is proportional to the square root of the pressure difference Δp at the edges of the tank, which is itself linked to the level (law of turbulent flows). Hence, we have – if the atmospheric pressure is the reference pressure:

$$Q_{s1} = K_1 \sqrt{\Delta p} = K_1 \sqrt{N_1} \qquad [7.13]$$

The same law describes the second tank, where, in order to simplify the notations, we suppose that the tanks have the same coefficient K:

$$Q_{s2} = K_2 \sqrt{\Delta p} = K_2 \sqrt{N_2} \qquad [7.14]$$

The mass balance of the tanks gives:

$$Q_e - Q_{s1} = A_1 \frac{\mathrm{d}N_1}{\mathrm{d}t} \qquad [7.15]$$

$$Q_{s1} - Q_{s2} = A_2 \frac{\mathrm{d}N_2}{\mathrm{d}t} \qquad [7.16]$$

In general, we can assume that the levels are subjected to small variations with respect to the balance given by the working points Q_{e0}, N_{10} and N_{20}. The balance is defined by:

$$Q_{e0} = K_1 \sqrt{N_{10}} = K_2 \sqrt{N_{20}} \qquad [7.17]$$

It should be noted, however, that this equation would be sufficient if we intended to size up the system, i.e. to choose the tanks (K_1, K_2 coefficients) according to the average levels and flows wanted. We write:

$$\begin{cases} N_1 = N_{10} + n_1 \\ N_2 = N_{20} + n_2 \end{cases} \qquad [7.18]$$

$$Q_e = Q_{e0} + Q_0 \qquad [7.19]$$

$$Q_{s1} = K_1 \sqrt{N_{10} + n_1} \qquad [7.20]$$

The limited development of the square root leads to:

$$Q_{s1} = K_1 \sqrt{N_{10}} \left(1 + \frac{1}{2} \frac{n_1}{N_{10}} \right) \qquad [7.21]$$

and similarly:

$$Q_{s2} = K_2\sqrt{N_{20}}\left(1 + \frac{1}{2}\frac{n_2}{N_{20}}\right) \tag{7.22}$$

Equation [7.15] thus becomes:

$$Q_{e0} + Q_0 - K_1\sqrt{N_{10}}\left(1 + \frac{1}{2}\frac{n_1}{N_{10}}\right) = A_1\frac{dN_1}{dt} = A_1\frac{dn_1}{dt} \tag{7.23}$$

If:

$$S_1 = \frac{2\sqrt{N_{10}}}{K_1} \tag{7.24}$$

then:

$$Q_0 - \frac{n_1}{S_1} = A_1\frac{dn_1}{dt} \tag{7.25}$$

Based on [7.16] and [7.22], the evolution of the level of the second tank is described by:

$$\frac{n_1}{S_1} - \frac{n_2}{S_2} = A_2\frac{dn_2}{dt} \tag{7.26}$$

where:

$$S_2 = \frac{2\sqrt{N_{20}}}{K_2} \tag{7.27}$$

The state representation of this system follows immediately:

$$X = \begin{bmatrix} n_1 \\ n_2 \end{bmatrix} \tag{7.28}$$

$$\dot{X} = \begin{bmatrix} -\frac{1}{A_1 S_1} & 0 \\ \frac{1}{A_2 S_1} & -\frac{1}{A_2 S_2} \end{bmatrix} X + \begin{bmatrix} \frac{1}{A_1} \\ 0 \end{bmatrix} Q_0 \tag{7.29}$$

$$y = \begin{bmatrix} 1 & 1 \end{bmatrix} X \tag{7.30}$$

where we will measure the two levels. The transfer function of the second level is:

$$H(s) = \frac{n_2(s)}{Q_0(s)} = \frac{S_2}{(1 + A_1 S_1 s)(1 + A_2 S_2 s)} \tag{7.31}$$

Let us take a second example: a direct current engine operated by an armature. Let R and L be the resistance and the inductance of the armature, $u(t)$ the supply voltage, $i(t)$ the armature current, $e(t)$ the back electromotive force, $\gamma(t)$ the engine torque, J

and f the inertia and frictions of the tree rotating at a speed $\omega(t)$. The electric equation of the armature is:

$$u(t) = Ri(t) + L\frac{di(t)}{dt} + e(t) \qquad [7.32]$$

The back electromotive force is proportional to speed (linear state):

$$e(t) = k_1\omega(t) \qquad [7.33]$$

The engine torque is proportional to the current:

$$\gamma(t) = k_2 i(t) \qquad [7.34]$$

Newton's law applied to the tree engine gives us the balance of the engine and working torques:

$$J\frac{d\omega(t)}{dt} = \gamma(t) - f\omega(t) \qquad [7.35]$$

The Laplace transform applied to this group of equations gives:

$$(Js + f)\Omega(s) = k_2\frac{U(s) - k_1\Omega(s)}{R + Ls} \qquad [7.36]$$

The transfer function of the of the engine system is:

$$H(s) = \frac{\Omega(s)}{U(s)} = \frac{k_2}{(Js + f)(R + Ls) + k_1 k_2} \qquad [7.37]$$

If we choose as state vector:

$$X = \begin{bmatrix} \omega \\ \frac{d\omega}{dt} \end{bmatrix} \qquad [7.38]$$

we find the state representation:

$$\dot{X} = \begin{bmatrix} 0 & 1 \\ -\frac{fR}{JL} - \frac{k_1 k_2}{JL} & -\frac{R}{L} - \frac{f}{J} \end{bmatrix} X + \begin{bmatrix} 0 \\ \frac{k_2}{LJ} \end{bmatrix} u \qquad [7.39]$$

and if we measure the speed:

$$y = \begin{bmatrix} 1 & 0 \end{bmatrix} X \qquad [7.40]$$

We know that this representation is not unique. We could have chosen as state vector:

$$X = \begin{bmatrix} \omega \\ i \end{bmatrix} \qquad [7.41]$$

which gives the state representation:

$$\dot{X} = \begin{bmatrix} -\frac{f}{J} & \frac{k_2}{J} \\ -\frac{k_1}{L} & -\frac{R}{L} \end{bmatrix} X + \begin{bmatrix} 0 \\ \frac{1}{L} \end{bmatrix} u \qquad [7.42]$$

$$y = \begin{bmatrix} 1 & 0 \end{bmatrix} X \qquad [7.43]$$

Obviously, these two representations have the same transfer function.

In order to have the control the question that arises is: can the value of all these physical parameters intervening in these knowledge models be obtained? We can use the manufacturers' documentation for small systems as the ones developed below. For more complex systems (like chemical or biotechnological systems) this task may be very complicated. That is why we determine, often directly from experimental recordings, the parameters of transfer functions (the two time constants of the tanks, for example). The following two sections describe this approach.

7.3. Graphic identification approached

When the objective of modeling is the research of a (simple) linear model in view of the control, we can use direct methods based on the use of experimental recordings. Two methods are available: the use of the harmonic response of the system or the analysis of time responses with specific excitations. The goal researched is, for a minimal cost, to obtain an input-output representation of the procedure in the form of a continual transfer function $F(p)$. Let us recall that $F(p)$ models only the dynamic part of the procedure. The time expressions will entail the initial conditions.

The first approach (harmonic response of the system) is rarely conceivable because its implementation is often incompatible with manufacturing requirements or, more so, because the response time of the procedure makes recording it particularly long and tedious.

The second approach is based on the recording of the system's response to the given excitations. In particular, we use the recording of the unit-step response, which corresponds, from a practical point of view, to a change in the operating point. Hence, from unique data, we identify the system by determining the coefficients of a standardized transfer function with a predefined structure. It is important to point out that these graphic methods do not make it possible to estimate the precision of the parameters obtained. In addition, the quality of the model depends on the operating mode (noise level, instrumentation, etc.) and on the operator (in particular during the use of graphs). It is understood that the data must be collected in the absence of saturation and that it is essential to verify the non-saturation at the level of internal regulation loops (when they exist). Finally, the graphic techniques based on the use of the unit-step response and presented below suppose that the system to identify is asymptotically stable.

The first criterion to consider for the choice of the method is whether or not the final value, noted by $s(\infty)$, which corresponds to a pseudo-periodical or a periodical response is exceeded. In Table 7.1, there is a classification of various methods presented in the remaining part of this section, as well as the standardized transfer function and the parameters to identify.

7.3.1. *Pseudo-periodic unit-step response*

From a balance position $e(t_0^-)$, $s(t_0^-)$, we apply at instant t_0 a step function ΔE. Then we obtain a unit-step response in the form of the one in Figure 7.2. This response presents a first exceedance A_1, a final value $s(\infty)$. Therefore, we use as a model the transfer function:

$$F(p) = \frac{S(p)}{E(p)} = \frac{K}{1 + 2\zeta \frac{p}{\omega_n} + \frac{p^2}{\omega_n^2}}$$

which has three parameters to identify K, ζ, ω_n (see Table 7.1). The unit-step response of the *model selected* has the form:

$$s_m(t) = s(t_0^-) + K\Delta E \left[1 - \frac{1}{\sqrt{1 - \zeta^2}} e^{-\zeta \omega_n (t - t_0)} \sin\left(\sqrt{1 - \zeta^2} \omega_n (t - t_0) + \theta \right) \right]$$

with $\tan \theta = \sqrt{1 - \zeta^2}/\zeta$, for $t \geqslant t_0$. The (relative) amplitude of the first exceedance, noted by $A_{1\%}$, depends only on damping ζ. By using the graph given in Figure 7.3, we obtain the numeric value of ζ. The angular frequency ω_n is linked to the period of oscillations. Its numeric value is obtained by using Figure 7.4.

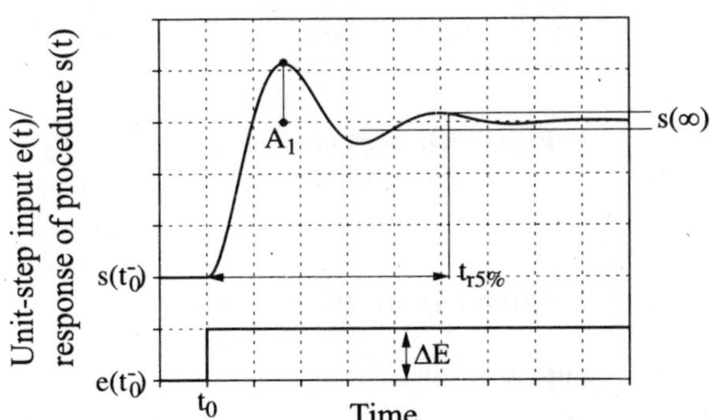

Figure 7.2. *Pseudo-periodic unit-step response*

Graphs to use

The two graphs to use are parameterized by damping ζ. Their use is described below, based on the unit-step response in Figure 7.2.

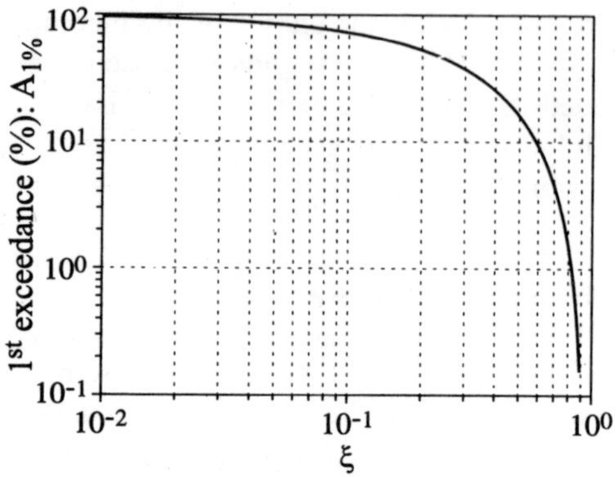

Figure 7.3. *First exceedance (percentage) according to ζ*

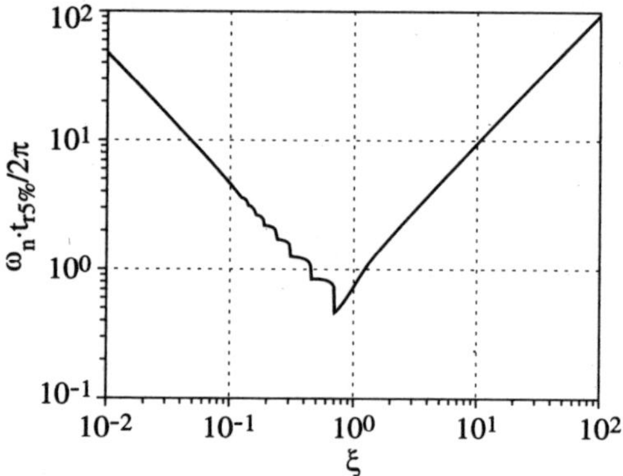

Figure 7.4. *Response time at 5% according to damping ζ*

Determining the transfer function $F(p)$: method

The method is divided up as follows:

1) determining the static gain of procedure $K = \frac{s(\infty) - s(0^-)}{\Delta E}$;

2) determining the first relative exceedance (percentage) $A_{1\%} = 100 \times \frac{A_1}{s(\infty) - s(0^-)}$;

3) with the help of Figure 7.3, determining the numeric value of ζ;

4) obtaining the response time at 5%, noted by $t_{r5\%}$;

5) by using Figure 7.4, determining the numeric value of $\frac{\omega_n \cdot t_{r5\%}}{2\pi}$, then obtaining the numeric value of the angular frequency ω_n.

Transfer function to identify	Method	Necessary tools Coefficients to identify
Pseudo-periodic unit-step response		
2nd order: $\dfrac{K}{1 + 2\zeta\frac{p}{\omega_n} + \frac{p^2}{\omega_n^2}}$		Graphs K, ζ, ω_n
Aperiodic unit-step response		
1st order: $\dfrac{K}{1 + \tau p}$		K, τ
Delayed 1st order: $\dfrac{K\,e^{-T_r p}}{1 + \tau p}$	Broïda	Graph K, T_r, τ
2nd order: $\dfrac{K}{(1 + \tau_1 p)(1 + \tau_2 p)}$	Cadwell	Graph K, τ_1, τ_2
High order: $\dfrac{K\,e^{-T_r p}}{(1 + \tau p)^n}$	Strejc	Graph K, T_r, τ, n

Table 7.1. *Graphic methods*

7.3.2. *Aperiodic unit-step response*

The method used for the identification of transfer function depends on the general features of the obtained step response. If this response does not have a horizontal tangent at instant $t = t_0$, the tester will obviously choose a 1st order model. Otherwise, the choice will have to be between three models: a delayed 1st order model (Broïda method), a 2nd order model (Cadwell's method), and a model (strictly) superior to 1 delayed or not (Strejc method). Here again, the know-how and *a priori* knowledge are very important with respect to the "quality" of the model selected. In order to simplify the expressions, we will choose from now on $t_0 = 0$. If this is not the case, a simple variable change makes it possible to have this situation.

7.3.2.1. *First order model*

When the unit-step response appears similar to that in Figure 7.5, the model used is:

$$F(p) = \frac{S(p)}{E(p)} = \frac{K}{1 + \tau p}$$

The unit-step response of the *model selected* has the form:

$$s_m(t) = s(0^-) + K\,\Delta E\left(1 - e^{-\left(\frac{t}{\tau}\right)}\right) \quad \text{for} \quad t \geqslant 0 \qquad [7.44]$$

Two parameters must be identified: the static gain of system K and the time constant τ.

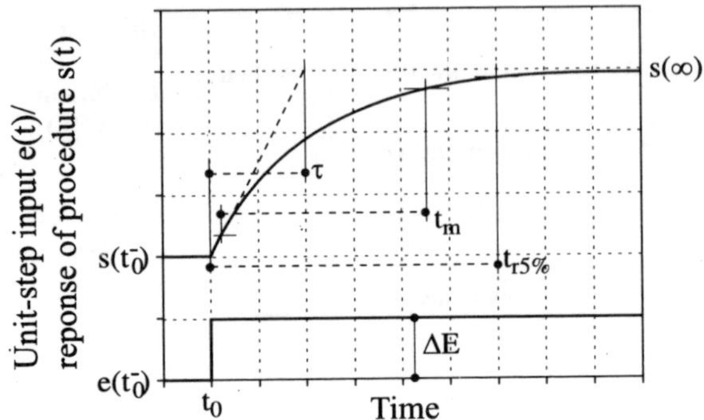

Figure 7.5. *Unit-step response for a* 1st *order system*

Determining the transfer function $F(p)$: method

The method is divided up as follows:

1) determining the static gain of procedure $K = \frac{s(\infty) - s(0^-)}{\Delta E}$;

2) obtaining the time constant τ. Two approaches are conceivable depending on whether or not we know the output final value $s(\infty)$. In both cases, we use the properties which resulted from the mathematical expression [7.44]:

a) determining τ using $s(\infty)$ (see Figure 7.5): the tangent at the unit-step response $t = 0$ intersects the horizontal line $s(\infty)$ for $t = \tau$, the response time at 5% verifies $t_{r5\%} = 3\tau$ and the rise time[1] of the system verifies $t_m = 2.2\tau$,

b) determining τ without using $s(\infty)$. We choose two instants t_1 and $t_2 = 2t_1$. We have the numeric values of $s(t_0^-)$, $s(t_1)$ and $s(t_2)$. By using [7.44], we obtain:

$$\begin{cases} s(t_1) = s(t_0^-) + K\,\Delta E\left(1 - e^{-\left(\frac{t_1}{\tau}\right)}\right) \\ s(t_2) = s(2t_1) = s(t_0^-) + K\,\Delta E\left(1 - e^{-\left(\frac{2t_1}{\tau}\right)}\right) \end{cases} \qquad [7.45]$$

We suppose that $x = e^{-\left(\frac{t_1}{\tau}\right)}$. Then $\frac{s(t_1) - s(t_0^-)}{s(t_1) - s(t_0^-)} = \frac{1-x}{1-x^2} = \frac{1}{1+x}$, which makes it possible to determine x. The numeric value of the time constant is given by $\tau = -\frac{t_1}{\ln(x)}$. Expression [7.45] then enables us to calculate K.

Other graphic methods can be used, in particular that of the semi-logarithmic plane [LAR 77]. However, since its use is not immediate, we will not present it here.

7.3.2.2. *Second order model*

We will present a unit-step response given in Figure 7.6.

---

1. Rise time is defined between instants t_1 and t_2 such as $s(t_1) = s(t_0^-) + 0.1\,K\,\Delta E$, $s(t_2) = s(t_0^-) + 0.9\,K\,\Delta E$.

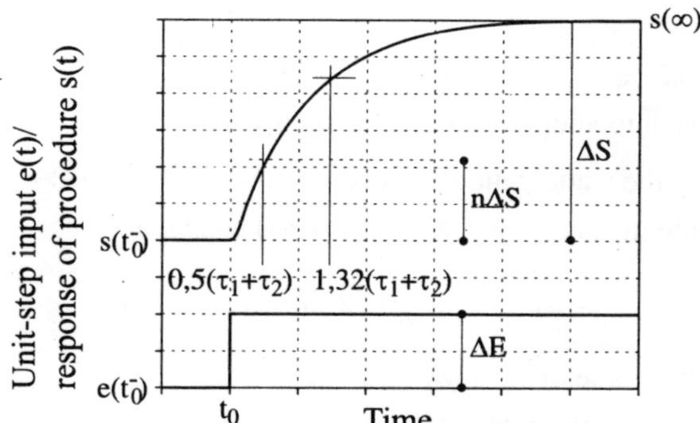

Figure 7.6. *Aperiodic unit-step response:
choice of second order model*

When the selected model is:

$$F(p) = \frac{S(p)}{E(p)} = \frac{K}{(1 + \tau_1 p)(1 + \tau_2 p)}$$

with $\tau_1 < \tau_2$, the unit-step response has the time expression:

$$s_m(t) = s(0^-) + K\,\Delta E\left(1 - \frac{\tau_1}{\tau_1 - \tau_2}e^{-\left(\frac{t}{\tau_1}\right)} + \frac{\tau_2}{\tau_1 - \tau_2}e^{-\left(\frac{t}{\tau_2}\right)}\right) \quad \text{for } t \geqslant 0$$

Determining the transfer function $F(p)$: Cadwell method

The *Cadwell method* is based on the fact that the unit-step response accepts a particular point $P_1(t_1, s_1)$ which does not depend on the ratio $x = \tau_2/\tau_1$ and another point P_2 whose coordinates (t_2, s_2) strongly depend on this ratio. The approach to adopt is the following:

1) determining the static gain of procedure $K = \frac{s(\infty) - s(0^-)}{\Delta E}$;

2) obtaining the instant $t_1 = 1.32(\tau_1 + \tau_2)$ such that:

$$s_1 = s(0^-) + 0.74\{s(\infty) - s(0^-)\}$$

3) inferring the numeric value of $\tau_{sum} = \tau_1 + \tau_2$;

4) obtaining on the curve the point (t_2, s_2) such that $t_2 = 0.5\tau_{sum}$. Inferring the value of $n_\% = \frac{s(t_2) - s(0^-)}{s(\infty) - s(0^-)}$;

5) with the help of the graph in Figure 7.7, determining the value of $\frac{1}{1+x}$. Inferring $x = \tau_2/\tau_1$. Calculating the numeric values of τ_1 and τ_2 with:

$$\begin{cases} \tau_1 = \frac{\tau_{sum}}{1+x} \\ \tau_2 = x\frac{\tau_{sum}}{1+x} \end{cases} \qquad\qquad [7.46]$$

Similar method

This method lies on an approach similar to the previous one; it was developed by Strejc. The approach to adopt is summed up below:

1) determining the static gain of procedure $K = \frac{s(\infty)-s(0^-)}{\Delta E}$;

2) obtaining the instant $t_1 = 1.2564\,(\tau_1 + \tau_2)$ such that:

$$s_1 = s(0^-) + 0.72\big\{s(\infty) - s(0^-)\big\}$$

3) obtaining the value of $\tau_{sum} = \tau_1 + \tau_2$;

4) obtaining on the curve the point (t_2, s_2) such that $t_2 = 0.3574\,\tau_{sum}$. Inferring the value of $n_\% = \frac{s(t_2)-s(0^-)}{s(\infty)-s(0^-)}$;

5) with the help of the graph in Figure 7.8, determining the value of $x = \tau_2/\tau_1$. Calculating the numeric values of τ_1 and τ_2 via the formulae of [7.46].

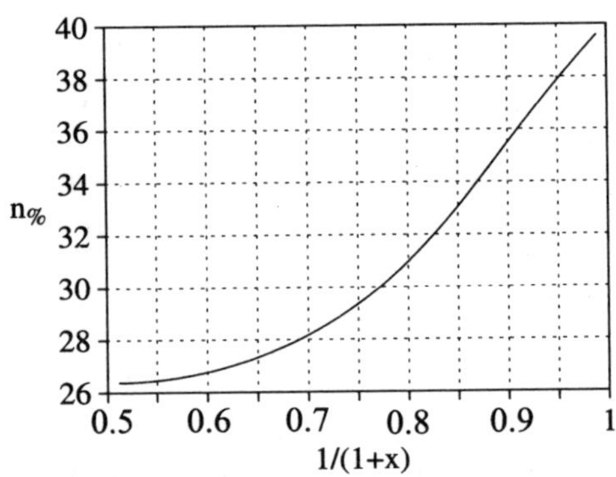

Figure 7.7. *Cadwell method,* $n_\% = f(x)$

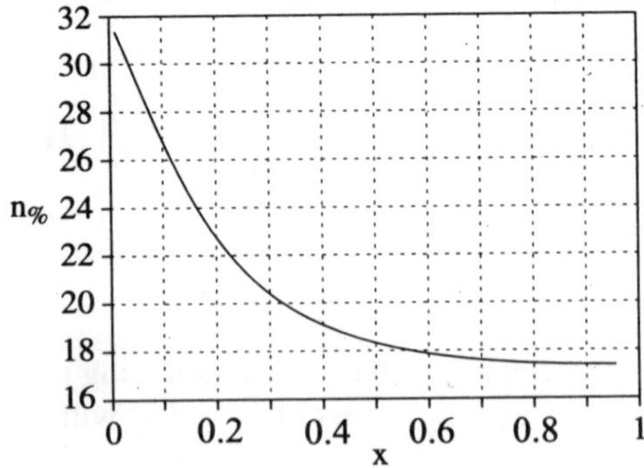

Figure 7.8. *Strejc method,* $n_\% = f(x)$

7.3.2.3. *Model of an order superior to 1*

When the unit-step response has a form similar to that in Figure 7.9, we can use the *Strejc method*. We suppose there is a low dispersion of time constants of the system – which is consistent with the Strejc hypothesis $T_u/T_n > 0.1$, where these parameters are defined below. The transfer function sought is:

$$F(p) = \frac{K\,e^{-T_r p}}{(1 + \tau p)^n}$$

There are four parameters to determine: K, T_r, τ, n. We set $t_0 = 0$.

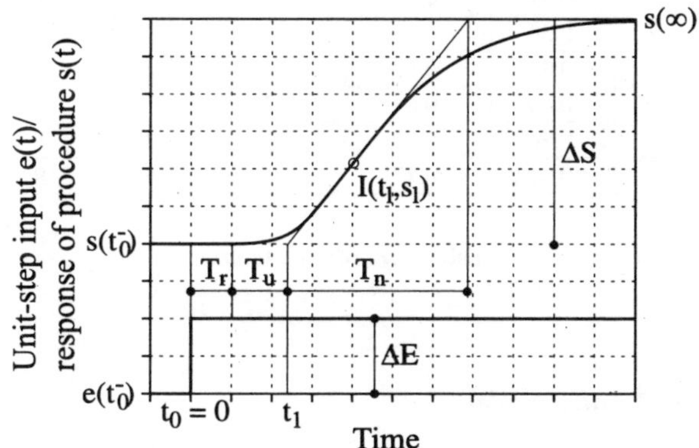

Figure 7.9. *Aperiodic unit-step response: choice of Strejc model*

Determining the transfer function $F(p)$: Strejc method

The difficulty of this method lies in the *course of the tangent to the point of inflexion* $I(t_i, s_i)$. The delay T_r can be adjusted in order to compensate the error due to the imprecision on the position of I. We do not present here the use of the graph making it possible to determine a fraction order n [DAV 65].

The approach to adopt here is summed up below:

1) determining the static gain of the procedure $K = \frac{s(\infty)-s(0^-)}{\Delta E}$;

2) tracing the tangent to the point of inflexion I. Calculating $s_i\% = \frac{s_i-s(0^-)}{s(\infty)-s(0^-)}$;

3) rounding $s_i\%$ to one of the values given in Table 7.2. Hence, we set the order n;

4) on the curve, obtaining value T_n. Inferring T_u by using the column T_u/T_n;

5) calculating $T_r = t_1 - T_u$. If T_r is negative (or very small in front of T_u, T_n), we can readjust the tangent at the inflexion point in order to obtain $T_r = 0$;

6) calculating the time constant τ by using one of the last three columns in Table 7.2.

NOTE 7.1. The calculation of τ can be done by using more than the last three columns. If dispersion is too significant, the position of the inflexion point must be modified, as well as the tangent. In practice, we can see that a slight modification of the tangent can lead to significant variations of parameters T_r, τ, n. This is explained by the fact that different triplets (T_r, τ, n) can lead to similar forms of the unit-step response.

n	$s_i\,\%$	T_u/T_n	T_n/τ	T_u/τ	t_i/τ
1	0	0	1	0	0
2	0.26	0.104	2.7	0.28	1
3	0.32	0.22	3.7	0.8	2
4	0.35	0.32	4.46	1.42	3
5	0.37	0.41	5.12	2.1	4
6	0.38	0.49	5.7	2.81	5
7	0.39	0.57	6.2	3.55	6
8	0.40	0.64	6.7	4.31	7
9	0.407	0.71	7.2	5.08	8
10	0.413	0.77	7.7	5.87	9

Table 7.2. *Strejc method*

7.3.2.4. *Delayed 1ˢᵗ order model*

We suppose that the unit-step response of the system considered can be approached by that of the delayed 1ˢᵗ order system as illustrated in Figure 7.10. The transfer function selected is:

$$F(p) = \frac{K\,e^{-T_r p}}{1 + \tau p}$$

The delay T_r can be graphically adjusted and the time constant τ can be obtained by one of the methods described in section 7.3.2.1. In order to determine the numeric values of T_r and τ, the Broïda method uses two particular points $s(t_1)$ and $s(t_2)$ obtained on the experiment curve.

Determining the transfer function $F(p)$: Broïda method

The method is divided up as follows:

1) determining the static gain of procedure $K = \frac{s(\infty) - s(0^-)}{\Delta E}$;

2) obtaining t_1, t_2 such that:

$$\begin{cases} s(t_1) = s(0^-) + 0.28\{s(\infty) - s(0^-)\} \\ s(t_2) = s(0^-) + 0.4\{s(\infty) - s(0^-)\} \end{cases}$$

[7.47]

3) calculating the time constants of the model:

$$\begin{cases} \tau = 5.5\,(t_2 - t_1) \\ T_r = 2.8\,t_1 - 1.8\,t_2 \end{cases}$$

[7.48]

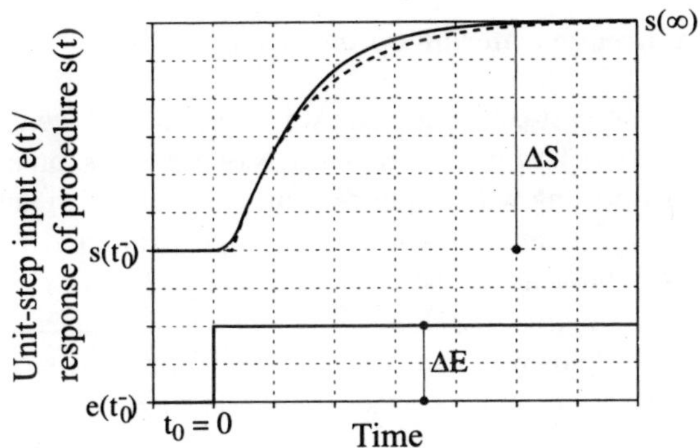

Figure 7.10. *Unit-step response, use of Broïda model*

NOTE 7.2. The unit-step response of the model selected is:

$$s_m(t) = s(0^-) + K\,\Delta E \left(1 - e^{-(\frac{t + T_r}{\tau})}\right) \quad \text{for} \quad t \geqslant 0$$

By using [7.47], we have:

$$\begin{cases} S_1 = \dfrac{s(t_1) - s(0^-)}{s(\infty) - s(0^-)} = 0.28 = 1 - e^{-(\frac{t_1 - T_r}{\tau})} \\[2ex] S_1 = \dfrac{s(t_1) - s(0^-)}{s(\infty) - s(0^-)} = 0.4 = 1 - e^{-(\frac{t_2 - T_r}{\tau})} \end{cases}$$

or:

$$\begin{cases} -\dfrac{t_1 - T_r}{\tau} = \ln(1 - 0.28) = \ln(0.72) \\[2ex] -\dfrac{t_2 - T_r}{\tau} = \ln(0.6) \end{cases}$$

[7.49]

from which we extract $\frac{t_1 - t_2}{\tau} = \frac{\ln(0.72)}{\ln(0.6)}$. Expressions [7.48] previously given are then easily established.

7.3.3. *Partial conclusion*

A graphic approach enables us to easily obtain a continuous transfer estimation. However, this facility is limited to predefined structured models associated to the recording of a specific response. We need to be aware of the fact that it is essential to have a little, even very little, noise on the responses and, irrespective of everything, the precision of the result can be uncertain. To go beyond the constraints, only a purely numeric approach is likely to bring a positive response. This topic will be now dealt with.

7.4. Identification through criterion optimization

A first revolution, at least psychological, was to admit that we can more success-fully apprehend the research of behavioral models of a process, on a continuous-time scale by definition, by dealing with z transfer functions for which the concept of time unity disappeared, after a carefully chosen input-output sampling. There were at least two reasons for this approach: on the one hand, the problem of identification (and also the problem of simulation and control) is made much easier by the computing tool and is already well known in data analysis (linear or non-linear regression) and on the other hand, since the object of the behavioral model was, *a priori*, the syn-thesis of a computer control, the procedure was subjected to a "per piece constants" type inputs, for which the discrete model was ideally adapted (no loss of information during sampling and thus no approximation during the passage from continuous-time scale to discrete-time scale). However, this does not mean that an approach aiming to deal directly with the identification of a continuous model must be mocked, but we just need to know where not to go. To return to the initial objective of modeling, in our opinion it is necessary to have good reasons to give up the "model method" approach which consists of finding a set of parameters, for example those of a transfer, so that the output of the model "fits best" to the output of the procedure for the same input excitation. This obviously means, in the majority of cases, a non-linear optimization. However, the majority of control engineers dealing with identification have often kept ignoring this approach in order to avoid having to deal with this numeric problem, even if it has been well controlled for more than 20 years (see the IMSL, TOMS, HARWELL, NAG, etc., databases).

7.4.1. *Algorithms*

Generally speaking, the quasi-totality of algorithms is built on basic principles of the linear parametric estimation whose dynamic version – that constitutes Kalman's filter [RAD 70] – represents a superior contribution to this field. The fundamental alternative is to use the techniques of mathematical programming – in particular at the level of the non-linear optimization [GIL 81]. Because of the delicate status of these problems, their implementation required many studies and research before robust solu-tions were obtained. Hundreds of publications dealt with the algorithm developments, the majority of which focused on the conceptual aspect of the problem. This ten-dency has decreased significantly since the beginning of the 1980s, and gave way to an increase in studies on digital and computer implementation. Before approaching this question under a more pragmatic angle, we will see what we can use from the multitude of approaches proposed in other works [LAR 77, LJU 87]. Hence, we will distinguish several categories of problems and classes of methods; however, for reasons of simplicity, we will limit ourselves to the stationary context (unknown parameters constant in time) which, in reality, covers many industrial applications.

7.4.2. *Models*

7.4.2.1. *Single output system*

For each input, the model is a transfer $H(z)$, or its original h_t, the weighting sequence, i.e. the discrete impulse response. Let us recall that in this case the relation between an input u_t and an output y_t is given by:

$$y_t = -(a_1 y_{t-1} + \cdots + a_n y_{t-n}) + (b_0 u_{t-k} + \cdots + b_m u_{t-k-m}) \qquad [7.50]$$

for the transfer:

$$H(z) = \frac{z^{-k}(b_0 + \cdots + b_m z^{-m})}{1 + a_1 + \cdots + a_n z^{-n}} = \frac{B(z)}{A(z)} \qquad [7.51]$$

a_i and b_i are the parameters of the model. The integers k, m and n represent the delay (expressed in sampling intervals), the number of zeros and the number of poles respectively. In such a context, we can also impose that all transfers related to the various inputs have the same denominator. With certain methods, the user does not have a choice (all algorithms derived from least squares!).

7.4.2.2. *Multivariable system*

Here, the system fundamentally has several outputs and one or several inputs. In this case, two attitudes are possible: we either call upon a series of single output systems, and this is generally not optimal at the level of the model's complexity (number of states), or we deal with the problems directly as they stand, by using only the state representation:

$$\begin{cases} x_{t+1} = Ax_{t+1} + Bu_t \\ y = Cx_t \end{cases} \qquad [7.52]$$

where u_t is the vector input, y_t the vector output and x_t the state vector. The parameters to estimate are matrices A, B and C and possibly the initial state x_0. Here, the over-parameterization represented by this approach imposes the research of so-called specific *canonical* forms.

7.4.3. *Methods*

7.4.3.1. *Operation in a stochastic context*

This means that we try to identify at the same time, a model of the procedure and a model of the ambient noises or interferences. This second model is reduced to a so-called "shaper" filter whose input, imaginary signal, is labeled as an independent sequence. Therefore, it is traditional to use what we call an **ARMAX** structure which consists of writing an input-output relation including the imaginary signal, called here *innovation* and noted by ν_t:

$$y_t = -(a_1 y_{t-1} + \cdots + a_n y_{t-n}) + (b_0 u_{t-k} + \cdots + b_m u_{t-k-m})$$

$$+ \nu_t + c_1 \nu_{t-1} + \cdots + c_r \nu_{t-r} \qquad [7.53]$$

Parameter r is the degree of the "shaper filter".

This approach usually uses "maximum likelihood" type methods, corresponding to the hypothesis according to which ν_t is an independent Gaussian sequence, and "extended least squares", which represents a version which can be operated in real-time. The "instrumental variable" type algorithms or "generalized least squares" are alternatives in which the shaper filter becomes, for example, $C(z)/D(z)$, instead of $C(z)/A(z)$ previously. We note $C(z) = 1 + c_1 z^{-1} + \cdots + c_r z^{-r}$ and $D(z)$ another polynomial not formulated here. In all cases, the zeros and the poles of the shaper filter must be asymptotically stable.

7.4.3.2. *Operation in a deterministic context*

This time, the only objective is determining a model of the procedure. Without getting into details, we can consider that there are two basic structures, depending on whether we deal with an output error ε_t (method of the model whose output is η_t):

$$\eta_t = -(a_1 \eta_{t-1} + \cdots + a_n \eta_{t-n}) + (b_0 u_{t-k} + \cdots + b_m u_{t-k-m}) \qquad [7.54a]$$

$$\varepsilon_t = \eta_t - y_t \qquad [7.54b]$$

or an equation or prediction error e_t ("least squares" method):

$$y_t = -(a_1 y_{t-1} + \cdots + a_n y_{t-n}) + (b_0 u_{t-k} + \cdots + b_m u_{t-k-m}) + e_t \qquad [7.55]$$

We note that formalizing an output error [7.54] or a prediction error [7.55] is only a particular case of what we call *innovation* in [7.53]. In the first case, it is enough to assume that $r = n$ and $a_i = c_i, i = 1, \ldots, n$; in the other case, we simply have $r = 0$. The fact that we speak of deterministic context does not mean that we cannot assimilate the interferences and/or measurement noises to random variables, contrary to certain interpretations.

7.4.4. *Optimization criteria*

7.4.4.1. *Quadratic optimization criterion*

The formulation of the identification algorithm is expressed, from the point of view of calculation, in terms of linear regression (or linearized), i.e. it leads to a quadratic minimization problem (with respect to parameters), more often without constraints. In this case, we can carry out either a general processing of all data at the same time (offline operation), or a sequential processing (with respect to time) of the inputs-outputs enabling a real-time implementation (online). We often find in works on this subject the name "recursive" for designating this type of approach. This term seems totally inadequate if no recursion appears, in the mathematical sense of the word! By supposing that $\theta = [a_1, \ldots, a_n; b_0, \ldots, b_m]$, such a criterion is, for example, that of simple least squares:

$$J(\theta) = \sum_t e_t^2 \qquad [7.56]$$

NOTE 7.3. With this type of criterion, the estimation of *parameters* is often *biased*! In fact, for the estimation *not to be biased*, the sequence e_t must be available to an *independent* random sequence, at the optimum which is rarely the case.

7.4.4.2. *Non-linear optimization criterion*

Unlike the previous case, we consider here the hypothesis where the identification procedure leads to a non-linear regression. This time, the numeric procedure is then necessarily iterative (non-linear programming (NLP) algorithm) and thus implies an offline processing. The model method [RIC 71] consists of minimizing the criterion:

$$J(\theta) = \sum_t \epsilon_t^2 \qquad [7.57]$$

whereas the maximum likelihood method operates with:

$$J(\tilde{\theta}) = \sum_t v_t^2 \qquad [7.58]$$

where $\tilde{\theta} = [\theta, c_1, \ldots, c_r]$. Only the solutions approached for this problem can have a sequential aspect. This is the case of the extended least square method with the criterion [7.58]. Generally, these approaches are fundamentally *unbiased*.

7.4.5. *The problem of precision*

Let us note by $\hat{\theta}$ the set of parameters obtained by minimizing one of the previous criteria [7.56] to [7.58]. A basic question is to know what degree of *confidence* we can give to this result. It is necessary to make statistic hypotheses on, for example, the type of interferences focused by e_t, ϵ_t or v_t for talking of standard deviation for the parameters contained in $\hat{\theta}$. However, in a context of industrial application, we can perfectly evaluate the precision of $\hat{\theta}$ without resorting to any stochastic reasoning. The following expression, obtained from Fisher's information theory and in relation to Rao Cramer's inequalities, can be used without resorting to sophisticated theories:

$$\sigma^2(\hat{\theta}) = var(\hat{\theta}) = diag\left\{\frac{J(\hat{\theta})}{N - \dim(\hat{\theta})}\left[\frac{\partial^2 J}{\partial\hat{\theta}_i\partial\hat{\theta}_j}\right]^{-1}\right\} \qquad [7.59]$$

Here, N represents the horizon length of the summation with respect to t in the criteria. The uncertainty on each component $\hat{\theta}_i$ of $\hat{\theta}$ is then quantified by $\sqrt{(var(\hat{\theta}_i))}$. The interpretation of this value then depends of course on reasonable hypotheses that can be associated to the context. For example, if e_t, ϵ_t or v_t (depending on the method used) can be assimilated to a Gaussian independent random variable, then $\hat{\theta}_i$ is an unbiased estimator of standard deviation $\sqrt{(var(\hat{\theta}_i))}$. If, on the other hand, no such hypothesis is made, $\sqrt{(var(\hat{\theta}_i))}$ does not necessarily represent a standard deviation any longer, but remains an indirect indication of the uncertainty on $\hat{\theta}$. To understand why this argument is justified, it is sufficient to reason in a parametric space of size 1 instead of $n + m + 1(+r)$. In [7.59], let us ignore for a second the reverse of the second derivatives matrix of the criterion, which are optimally evaluated. The term $J(\hat{\theta})/(N - \dim(\hat{\theta}))$ represents the root mean square of the prediction error, of the output error or of the innovation. If we divided it by the root mean square of the measured output y_t, we would have an image of the signal-to-noise ratio. In other words, the "precision" of $\hat{\theta}$, expressed via [7.59], will improve proportionally to the S/N ratio, which is perfectly logical. The interpretation of the second term is more subtle. The altitude of J minimum is in relation with the S/N ratio. If the function to minimize J presents a minimum with a flat base, the exact position of this minimum could considerably vary, for the same level of noise or interferences, from a data recording to another. In other words, the value of $\hat{\theta}$ could considerably change from one attempt to another. Hence, for a calculation done with a given recording, the precision of $\hat{\theta}$ will be very low. Let us go back to formula [7.59]. A flat base corresponds to a second small derivative (in absolute value); its reverse is thus high, and so $var(\hat{\theta})$ is high. Conversely, when the criterion presents a sharp base, the position of the minimum will vary very little from one recording to another and the calculation obtained with any recording will provide an $\hat{\theta}$ which we can trust. In this case, the second derivative is high, its reverse is small, and $var(\hat{\theta})$ is thus small, translating a low uncertainty. In a multidimensional context, we can add to the previous theory the impact of the Hessian conditioning that increases along with an over-parameterized model and/or an input which is dynamically too poor. This is translated into an increase of the standard of the reverse – from which again a big uncertainty is obtained as expected!

7.4.6. *How to optimize*

The minimization of criteria is the milestone of identification algorithms. The absence of good numeric methods means that there are very few chances to reach a result that is certain. There are two fundamentally distinct situations to consider. If the criterion is quadratic, we have to use orthogonal factorization methods, be it a general, offline or sequential processing [LAW 77]. It is the only way to guarantee that the matrices that must be symmetric and positive definite in theory are so numerically – irrespective of the precision of calculation (simple or double). For a non-linear criterion, only iterative methods can solve the problem. Here the situation is less clear, due to the difficulty of the problem and to the multitude of possible approaches. The criteria to minimize are characterized by a reduced number of parameters to estimate, from a few units to a few dozen, their "square sum" type structure and a very high conditioning factor which can reach 10^{10}. Geometrically, this is translated into contour lines extremely extended in certain directions and very dense in others: mathematically, the conditioning is defined by:

$$\left\| \frac{\partial^2 J}{\partial \hat{\theta}_i \partial \hat{\theta}_j} \right\| \left\| \left[\frac{\partial^2 J}{\partial \hat{\theta}_i \partial \hat{\theta}_j} \right]^{-1} \right\|$$

These characteristics exclude any heuristic approach. For example, using Nelder-Mead simplex method is absurd in this situation. Likewise, it is equally inappropriate to use a method of conjugate gradients (reserved for big problems of modest conditioning) and even less appropriate to use, the method of the so-called "optimal" gradient. This last choice is the worst because it fails almost systematically. Algorithms like those of Levenberg and Marquart, usually related to "square sums" type criteria, are efficient in the 1963 initial description only if the minimum is close to zero! Moreover, they are not always robust in the case of high conditioning, even if they have been highly improved [MOR 77]. So we are left with "quasi-Newton" type algorithms. Only the factorized Hessian implementations are capable of being numerically stable in the case of high conditioning [POW 75].

7.4.7. *Partial conclusion*

The transfer approach is by far the most practical. The identification of a state model is of a complexity order of magnitude superior to the previous approach, due to the *a priori* over-parameterization that is, triplet (A, B, C). It is the problem of "canonical forms" that we have here, with the immediate incidence on the numerically undecidable character of algebraic conditions (all or nothing) of matrix rank (transforms of approximately projective type) [BAR 77]. Hence, no totally automatic procedure is surely satisfactory. Algorithms must remain guided (interactively) by the user at the level of the canonical structure. This is one of the arguments that encourages us to orient ourselves towards other approaches involved by questions of numeric conditioning (stability and robustness of calculations) without any reference to any canonical form. Passing through an impulse response is not of any practical interest nowadays, if this type of model does not lead to anything (so to speak) in terms of control. To conclude, we need to underline that the identification of a procedure is related to *open loop* and that, fundamentally, everything that has been said above supposes that the input does not depend on the output. Similarly, the models we referred to express a dynamic input-output relation, *in variation around zero*.

7.4.8. *Practical application*

Implementing a behavioral (identification) model requires a knowledge which works on three levels:

– conducting an *ad hoc* test campaign;

– determining the proper structure of the model and estimating its parameters;

– validating this modeling in relation to the final objective of the design of such a model.

7.4.8.1. *Identification protocols*

When a test campaign must be defined, a first task consists of sufficiently understanding the physical operating mode of the procedure to be modeled, in order to correctly define the magnitudes that will have the role of inputs – real or non-real inputs, but manipulable (regulator set points), non-manipulable inputs (measurable interferences) – and of outputs. Then, since the procedures are *a priori* non-linear, several test campaigns could be implemented around various characteristic working points. The actions can be of the "slot" or "symmetric step functions" type. Being able to record parts of unit-step responses is always a very interesting information with respect to *a priori* knowledge (sign and magnitude order of static gain, delay evolution, presence of a non-minimum phase, etc.). As soon as the sinusoidal input is eliminated (to be kept in mind in order to verify the linearity of the procedure within a given variation range, for example), and as soon as a certain independence is guaranteed between the various inputs, we underline that it is not necessary to introduce other constraints than that which ensures a functioning sufficiently rich in information on the dynamics of tested procedure. In particular, it is not necessary to impose a test campaign controlled by a computer – for example, in order to obtain a pseudo-random binary (or ternary) sequence type of input excitation. The argument that it is easy in these conditions to obtain by correlation an impulse response model does not have a practical value for an industrial procedure. It is also true that these inputs are excellent, even the best, and if the identification by correlation is not advisable, "carrying out" an autocorrelation on the inputs and inter-correlation for inputs and outputs is an efficient way to perform a concentration of initial data by replacing the real input-output pair by the input autocorrelation and input-output inter-correlation pair. An indispensable approach is to operate in closed loop. In this case, we replace the real input-output pairs by the input-set point and output-set point inter-correlation pairs. The set point, the external input of the looped system, can very well be again a binary sequence for its dynamic qualities, but this is still not indispensable, since the particularity of its autocorrelation function is no longer used.

7.4.8.2. *Identification tools*

A few years back, due to the multiplicity of solutions proposed in works on the subject, we could think that the difficulty is to know which algorithm to choose. The problem was to encode the method selected, because of the shortage of "handy" available solutions. Today the situation is no longer the same. There are ergonomic environments offering a reduced but realistic choice of possible alternatives: see [MAT 95], which is only one of the universally known environments. Except for a specialist in the identification field and digital techniques, we strongly advise everyone not to initiate the design of an identification tool whose first vocation would be the modeling of industrial procedures. At the level of methods, a main choice concerns the alternative between "prediction error" type approaches and "output error" type approaches, the latter being preferred by us. A second criterion refers to the available support for determining structural parameters such as k, m and n (or their equivalent including multivariable systems). Finally, it is very important to have validation means – for example, a very important one would be the accuracy obtained for the parameters. Poles, zeros, typical responses and residuals are a valuable amount of information that can be used by an identification software. Even if identification was presented here based on time information generated by the procedure, analyzing the scalar properties of a model can be a significant piece of information as well. Finally, the validation of other input-output data is easily done today by comparing, for example, the new residuals e_t, ϵ_t or v_t with respect to those obtained by an optimization with the initial data, both in terms of root mean square and autocorrelation.

7.4.8.3. *"Customizing"*

As for the unconditional aspects of the "home" software, below there are a few recommendations on what should be done and what should not be done:

– *never encode* the following type of formulae:

$$P_+ = P - Pv(1 + v^T Pv)^{-1}v^T P \qquad [7.60]$$

where P is assumed to be symmetric defined positive and where v represents a column vector. From a mathematical point of view, P_+ is defined as positive if P is. Numerically, it is not necessarily the case. Thus, for example, negative variances in Kalman's filter, the sequential least squares of any type, etc.;

– for gradient calculations, *never perform lateral finite differences, but central ones*, in the absence of the basically incontrovertible analytical form. In fact, the lateral finite difference cannot "tend" towards zero when the current point tends towards a minimum;

– never use "gradient" type non-linear optimization methods, including the conjugated gradient;

– use factorized "quasi-Newton" type methods which are incomparably more robust than Levenberg-Marquant's *original* algorithm, even if this is the reference in the "identification" environment;

– in the calculation of a criterion whose theoretical value for the current set of parameters diverges (for example, a transfer output error whose poles are, during an iteration, unstable), *interrupt the simulation* and the calculation of sensitivity functions, hence of the gradient, on a preset upper bound. In the contrary case, there is a fatal error risk in the optimization solver (irrespective of which one it is!);

– use the factorized forms for all the formulae for the update of symmetric positive defined matrix. For example, the equivalent of [7.60] is:

$$
QM = Q \begin{bmatrix} 1 & 0\cdots 0 \\ Yv & Y \end{bmatrix} = \begin{bmatrix} z & g^T \\ 0 & Z \end{bmatrix} \tag{7.61}
$$

where Y is Cholesky's factorization of P; hence, by construction, Z is that of P_+. In this approach, Q is an orthogonal matrix (Householder transformations or Givens rotations) calculated to perform the superior triangular shaping of M. Quantity g/z corresponds to the gain to be applied to the prediction error in the least squares formula.

7.5. Conclusion around an example

To conclude this brief overview of parametric identification, here is a test problem with which it is easy to practice while using the various methods of identification previously covered. The following results were obtained with the Matlab [LJU 95] "identification" tool box, but any other software *implementing the appropriate calculations* can be used.

7.5.1. *Simulated procedure*

It is totally artificial and easily reproducible. It is characterized by a transfer in z presenting the following characteristics:

– a pure delay of two sampling intervals;

– a non-minimum phase, with a zero of value 1.1;

– a pair of conjugated complex number poles highly oscillating $(0.9 \pm 0.3i)$;

– a faster mode (0.8);

– a static gain set at 5.

The working point of this procedure is, for the input, around level 50 and, for the output, around level 400. The test protocol is reduced to two unbalanced slots of different lengths – so that after their application, the procedure is around the same working point. A 20 standard deviation independent Gaussian noise was added on the calculated output. The simulation results appear in Figure 7.11 for an 81 second simulation period and a sampling interval $t_e = 0.25$ second, i.e. 325 pairs (u_t, y_t).

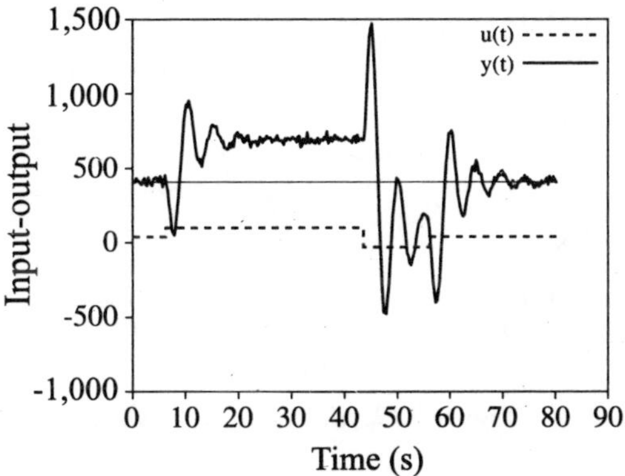

Figure 7.11. *Simulated procedure*

The reasons for these choices can be used as starting points for the "manual" construction of test campaigns. The length of the first slot enables us to "see" the unit-step response of the system. The short period of a second is there to tell us that it is not necessary to see it for a numeric approach (opposed to a graphic one) in order to benefit from the information present in the input-output data. Finally, the last one is built in such a way that the system returns to its initial working point. This makes it possible to visualize a drift and to evaluate it if necessary. In addition, the sequence of the two ascending and descending slots makes it possible to appreciate the linearity hypothesis implicitly made. Moreover, the presence of vertical fronts offers the possibility of easily evaluating an initial value of delay k. For a multi-input context, the same type of approach can be implemented, by paying attention to make these level changes at different instants for each input, so that, globally, the inputs are "independent". Obviously nothing prevents subcontracting to a program the generation of statistically independent pseudo-random binary sequences. Whoever can do a lot can also do a little, but again, this is not indispensable.

7.5.2. *In search of a model*

Due to space constraints here, we cannot present all the characteristic results of tests that progressively lead to a "good model". Briefly, upon the selection of a method, we can say in general that a noise analysis on the measurements enables us to choose between the method of the model (if the noise seems "sufficiently white") and the maximum likelihood (in the contrary case). This reasoning ensures no bias for any coherent experimental condition. The real difficulty is to rapidly estimate the structure parameters k, m and n of the transfer. Generally, it is better to start with low complexity $n = 1$ or 2 and increase n according to the results. Similarly, for m, the values 0 and 1 are rarely exceeded. The initial estimation of k is done visually – hence the interest in the inputs with at least one isolated rigid front.

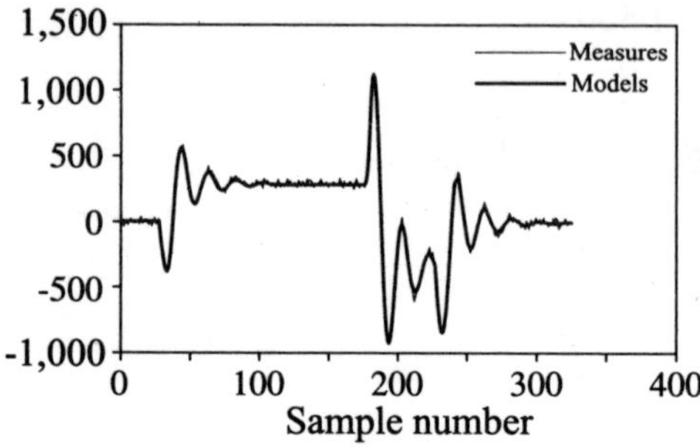

Figure 7.12. *Method of the model*

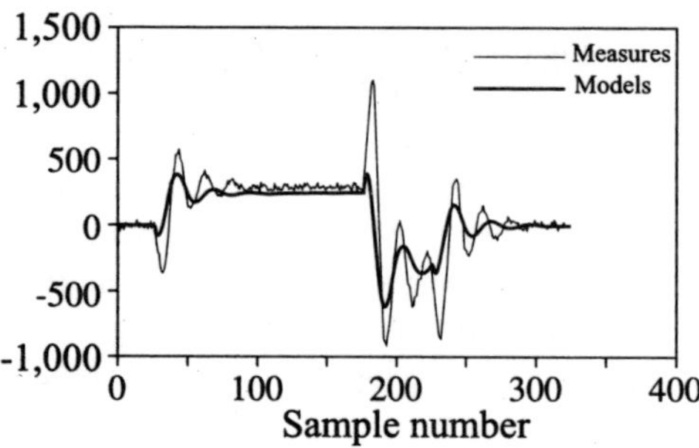

Figure 7.13. *Method of simple least squares*

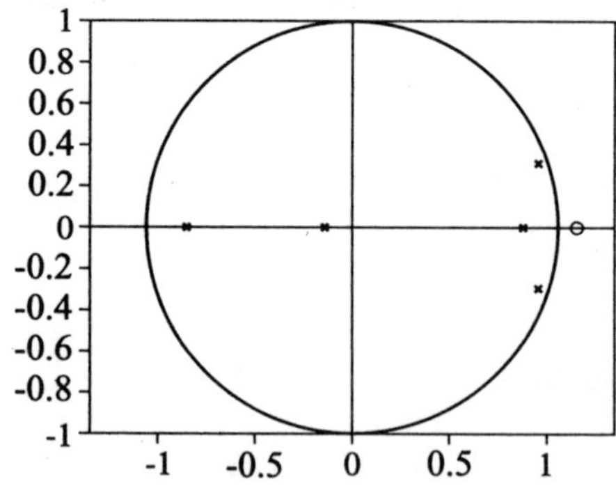

Figure 7.14. *Place of poles for a too complex model*

In order to adapt parameters k, m and n, we have several indicators:

– as long as $\sigma(a_n) > |a_n| \Rightarrow$ decrease n with 1;

– as long as $\sigma(b_m) > |b_m| \Rightarrow$, decrease m with 1;

– as long as $\sigma(b_0) > |b_0| \Rightarrow$, increase k with 1;

– if $\forall i : |\text{pole}_i| \ll 1$ and $\text{Re}(\text{pole}_i) < 0$ decrease n with 1;

– if the output error reaches a level inferior to the noise level, then there is over-parameterization.

The first three arguments express that maybe the parameters in question are zero. The following one indicates that a fast oscillating pole, which cannot be the result of a correctly sampled continuous pole, models the measurement noise rather than the procedure. In other words, we have a too complicated model which adjusts the data so well that the noise present in the measurements is no longer filtered. Finally, the fifth procedure is based on the same approach. It is equally appropriate to examine the autocorrelation (or the spectrum) of the minimized error which must be sufficiently characteristic of an independent sequence, with respect to the hypotheses justifying the choice of the method and model. Numerous other details coming from experience are to be considered and their synthesis would be difficult here. Anyone interested may refer to the works previously mentioned.

To conclude, below a few characteristic results for a good structure are listed, i.e. $k = 2, m = 1, n = 3$. Through the model method (see Figure 7.12), we have obtained the results in Table 7.3.

Parameter	Gain	Zero	$\text{Pole}_{1,2}$	Pole_3
Value	4.906	1.0967	$8.93\text{e}{-}1 + 2.96\text{e}{-}1\,i$	$7.73\text{e}{-}1$
Standard deviation	$7.3\text{e}{-}2$	$2.3\text{e}{-}3$	$1.7\text{e}{-}3$	$8.1\text{e}{-}3$

Parameter	b_0	b_1	a_1	a_2	a_3
Value	-1.141	1.251	-2.560	2.266	$-6.843\text{e}{-}1$
Standard deviation	$3.6\text{e}{-}2$	$3.8\text{e}{-}2$	$9.6\text{e}{-}3$	$1.8\text{e}{-}2$	$8.9\text{e}{-}3$

Table 7.3. *The model method*

For comparison, the least squares method leads to Figure 7.13, which clearly underlines the significant bias present in the estimation of model parameters. The parameters are those in Table 7.4.

Parameter	Gain	Zero	$\text{Pole}_{1,2}$	Pole_3
Value	4.11	1.38	$9.18\text{e}{-}1 + 2.23\text{e}{-}1\,i$	$-4.72\text{e}{-}1$
Standard deviation	$5.2\text{e}{-}1$	$5.2\text{e}{-}1$	$1.19\text{e}{-}2$	$5.0\text{e}{-}2$

Parameter	b_0	b_1	a_1	a_2	a_3
Value	$-8.93\text{e}{-}1$	1.23	-1.36	$2.50\text{e}{-}2$	$4.22\text{e}{-}1$
Standard deviation	$2.4\text{e}{-}1$	$2.4\text{e}{-}1$	$4.9\text{e}{-}2$	$8.9\text{e}{-}2$	$4.9\text{e}{-}2$

Table 7.4. *Least squares method*

Even if the standard deviations are not so good, this degradation does not make it possible to explain the non-adequacy of the model. Only the residual analysis makes it possible to explain the strongly biased character of the least squares method.

Finally, if we test a transfer with $n = 5$, two poles in the left unit semi-circle appear (see Figure 7.14), which are indicators of a not very complex model. These results were obtained with the model method.

7.6. Bibliography

[BAR 77] BARRAUD A., "Identification de systèmes multivariables par modèle d'état sous forme canonique", *Rairo Automatique*, vol.11–2, p. 161–194, 1977.

[DAU 00] DAUPHIN TANGUY G. (ed.), *Les Bond graphs*, Hermès, IC2 series, Paris, 2000.

[DAV 65] DAVOUST G., "Nomogramme pour l'analyse et le réglage des systèmes industriels", *Automatisme*, no. 4, p.3–8, 1965.

[GIL 81] GILL P.E., MURRAY W., WRIGHT M.H., *Practical Optimization*, Academic Press, 1981.

[LAR 77] DE LARMINAT P., THOMAS Y., *Automatique des systèmes linéaires. Tome 2. Identification*, Flammarion Sciences, Paris, 1977.

[LAW 77] LAWSON C.L., HANSON R.J., *Solving Least Squares Problems*, Prentice-Hall, 1977.

[LJU 87] LJUNG L., *System Identification: Theory for the User*, Prentice-Hall, 1987.

[LJU 95] LJUNG L., *System Identification Toolbox*, The Mathworks, Inc., 1995.

[MAT 95] MATLAB, The Mathworks, Inc., 1995.

[MOR 77] MORÉ J.J., "The Levenberg-Marquardt algorithm: Implementation and theory", in Watson G.A. (ed.), *Numerical Analysis*, Springer-Verlag, Lecture Notes in Mathematics 630, p. 105–116, 1977.

[POW 75] POWELL M.J.D., Subroutine VA13, Rapport CSS 15, Harwell Library, 1975.

[RAD 70] RADIX J.C., *Introduction au filtrage numérique*, Eyrolles, Paris, 1970.

[RIC 71] RICHALET J., *Identification par la méthode du modèle*, Gordon and Breach, 1971.

Chapter 8

Simulation and Implementation of Continuous Time Loops

8.1. Introduction

This chapter deals with ordinary differential equations, as opposed to partial derivative equations. Among the various possible problems, we will consider exclusively the situations with given initial conditions. In practice, the other situations – fixed final and/or intermediary conditions – can always be solved by a sequence of problems with initial conditions that we try to, by optimization, determine so that the other conditions are satisfied. Similarly, we will limit ourselves to 1st order systems (using only first order derivatives) as long as in practice we can always obtain such a system by increasing the number of equations.

We will study successively the linear and non-linear cases. Even though the linear case has by definition explicit solutions, the passage from formal expression to a virtual reality, with the objective of simulating, is not so trivial. On the other hand, in automatic control, Lyapunov or Sylvester's matrix equations, even if also linear, cannot be processed immediately, due to a prohibitive calculating time. For the non-linear case we will analyze the explicit approaches – which remain the most competitive for the systems whose dynamics remain of the same order of magnitude – and then we will finish by presenting a few explicit diagrams mainly addressing systems whose dynamics can significantly vary.

Chapter written by Alain BARRAUD and Sylviane GENTIL.

8.1.1. *About linear equations*

The specific techniques of linear differential equations are fundamentally exact integration diagrams, provided that the excitation signals are constant between two sampling instants. The only restrictions of the integration interval thus remain exclusively related to the sensitivity of underlying numerical calculations. In fact, irrespective of this integration interval, theoretically we have to obtain an exact value of the trajectory sought. In practice, this can be very different, irrespective of the precision of the machine, as soon as it is completed.

8.1.2. *About non-linear equations*

Inversely, in the non-linear case, the integration numerical diagrams can essentially generate only one approximation of the exact trajectory, as small as the integration interval, within the precision limits of the machine (mathematically, it cannot tend towards 0 here). On the other hand, we can, in theory, build integration diagrams of increasing precision, for a fixed integration interval, but whose sensitivity increases so fast that it makes their implementation almost impossible.

It is with respect to this apparent contradiction that we will try to orient the reader towards algorithms likely to best respond to the requirements of speed and accuracy accessible in simulation.

8.2. Standard linear equations

8.2.1. *Definition of the problem*

We will adopt the notations usually used to describe the state forms and linear dynamic systems. Hence, let us take the system:

$$\overset{\circ}{X}(t) = AX(t) + BU(t) \qquad [8.1]$$

Matrices A, B and C are constant and verify $A \in \mathbb{R}^{n \times n}, B \in \mathbb{R}^{n \times m}$. As for X and U, their size is given by $X \in \mathbb{R}^{n \times m}$ and $U \in \mathbb{R}^{m \times m}$. To establish the solution of these equations, we examine the free state, and then the forced state with zero initial conditions. For a free state, we have:

$$X(t) = e^{A(t-t_0)} X(t_0)$$

and for a forced state, with $X(t_0) = 0$:

$$X(t) = \int_{t_0}^{t} e^{A(t-\tau)} BU(\tau)\, \mathrm{d}\tau$$

In the end we obtain:

$$X(t) = e^{A(t-t_0)} X(t_0) + \int_{t_0}^{t} e^{A(t-\tau)} BU(\tau)\, d\tau \qquad [8.2]$$

8.2.2. *Solving principle*

Based on this well known result, the question is to simulate signal $X(t)$. This objective implies an *a priori* sampling interval, at least with respect to the storage of the calculation result of this signal. In the linear context, the integration will be done with this same sampling interval noted by h. In reference to the context of usual application of this type of question, it is quite natural to assume that the excitation signal $U(t)$ is constant between two sampling instant. More exactly, we admit that:

$$U(t) = U(kh), \quad \forall t \in [kh, (k+1)h] \qquad [8.3]$$

If this hypothesis was not verified, the next results – instead of being formally exact – would represent an approximation dependent on h, a phenomenon that is found by definition in the non-linear case. Henceforth, we will have $X_k = X(kh)$ and the same for $U(t)$. From equation [8.2], by supposing that $t_0 = kh$ and $t = (k+1)h$, we obtain:

$$X_{k+1} = e^{Ah} X_k + \left[\int_{kh}^{(k+1)h} e^{A[(k+1)h-\tau]}\, d\tau \right] BU_k \qquad [8.4]$$

This recurrence can be written as:

$$X_{k+1} = \Phi X_k + \Gamma U_k \qquad [8.5]$$

By doing the necessary changes of variables, the integral defining Γ is considerably simplified to give along with Φ the two basic relations:

$$\begin{cases} \Phi = e^{Ah} \\ \Gamma = \int_0^h e^{A\tau} B\, d\tau \end{cases} \qquad [8.6]$$

8.2.3. *Practical implementation*

It is fundamental not to try to develop Γ in any way. In particular, it is particularly inadvisable to want to formulate the integral when A is regular. In fact, in this particular case, it is easy to obtain $\Gamma = A^{-1}[\Phi - I]B = [\Phi - I]A^{-1}B$. These formulae cannot be an initial point for an algorithm, insofar as Γ could be marred by a calculation error, which is even more significant if matrix A is poorly conditioned. An elegant and robust solution consists of obtaining simultaneously Φ and Γ through the relation:

$$\begin{bmatrix} \Phi & \Gamma \\ 0 & I \end{bmatrix} = \exp\left(\begin{bmatrix} A & B \\ 0 & 0 \end{bmatrix} h \right) \qquad [8.7]$$

The sizes of blocks 0 and I are such that the partitioned matrices are of size $(m+n) \times (m+n)$. This result is obtained by considering the differential system $\overset{\circ}{W} = MW$, $W(0) = I$, with:

$$M = \begin{bmatrix} A & B \\ 0 & 0 \end{bmatrix} \tag{8.8}$$

and by calculating the explicit solution $W(h)$, via the results covered at the beginning of this section.

There are two points left to be examined: determining the sampling interval h and the calculation of Φ and Γ. The calculation of the matrix exponential function remains an open problem in the general context. What we mean is that, irrespective of the algorithm – as sophisticated as it is – we can always find a matrix whose exponential function will be marred by a randomly big error. On the other hand, in the context of simulation, the presence of sampling interval represents a degree of freedom that makes it possible to obtain a solution, almost with the precision of the machine, reaching the choice of the proper algorithm. The best approach, and at the same time the fastest if it is well coded, consists of using Padé approximants. The choice of h and the calculation of Φ and Γ are then closely linked. The optimal interval is given by:

$$h = \max_{i \in \mathbb{Z}} 2^i : \left\| \begin{bmatrix} A & B \\ 0 & 0 \end{bmatrix} 2^i \right\| < 1 \tag{8.9}$$

This approach does not suppose in practice any constraint, even if another signal storage interval were imposed. In fact, if this storage interval is bigger, we integrate with the interval given by [8.9] and we sub-sample by interpolating if necessary. If, on the contrary, it is smaller, we decrease h to return to the storage interval. The explanation of this approach lies on the fact that formula [8.9] represents an upper bound for the numerical stability of the calculation of exponential function [8.7]. Since the value of the interval is now known, we have to determine order q of the approximant which will guarantee the accuracy of the machine to the result of the exponential function. This is obtained very easily via the condition:

$$q = \min i : \|Mh\|^{2i+1} e_i \leqslant \epsilon, \quad e_{j+1} = \frac{e_j}{4(2j+1)(2j+3)}, \quad e_1 = \tfrac{2}{3} \tag{8.10}$$

where M is given by [8.8] and ϵ is the accuracy of the machine.

NOTE 8.1. For a machine of IEEE standard (all PCs, for example), we have $q \leqslant 8$ double precision. Similarly, if $\|Mh\| \leqslant \tfrac{1}{2}$, $q = 6$ guarantees 16 decimals.

Let us return to equation [8.7] and we shall write it as follows:

$$N = e^{Mh}$$

Let $\hat{N}$ be the estimated value of N; then, $\hat{N}$ is obtained by solving the following linear system, whose conditioning is always close to 1:

$$p_q(-Mh)\hat{N} = p_q(Mh) \qquad [8.11]$$

where $p_q(x)$ is the q degree polynomial defined by:

$$p_q(x) = \sum_{i=0}^{q} \alpha_i x^i, \quad a_k = \prod_{i=1}^{q} \frac{q+i-k}{k!} \qquad [8.12]$$

In short, the integration of [8.1] is done from [8.5]. The calculation of Φ and Γ is obtained via the estimation $\hat{N}$ of N. Finally, the calculation of $\hat{N}$ goes through that of the upper bound of sampling interval [8.9], the determination of the order of the Padé approximant [8.10], the evaluation of the corresponding polynomial [8.12] and finally the solving of the linear system [8.11].

NOTE 8.2. We can easily increase the value of the upper bound of sampling interval if $\|B\| > \|A\|$. It is enough to standardize controls $U(t)$ in order to have $\|B\| < \|A\|$. Once this operation is done, we can again improve the situation by changing M in $M - \mu I$, with $\mu = tr(M)/(n+m)$. We have in fact $\|M - \mu I\| < \|M\|$. The initial exponential function is obtained via $N = e^{\mu}e^{(M-\mu I)h}$.

NOTE 8.3. From a practical point of view, it is not necessary to build matrix M in order to create the set of calculation stages. This point will be explored – in a more general context – a little later (see section 8.3.3). We can finally choose the matrix standard L_1 or L_∞, which is trivial to evaluate.

8.3. Specific linear equations

8.3.1. *Definition of the problem*

We will now study Sylvester differential equations whose particular case is represented by Lyapunov differential equations. These are again linear differential equations, but whose structure imposes in practice a specific approach without which they basically remain unsolvable, except in the academic samples. These equations are written:

$$\overset{\circ}{X}(t) = A_1 X(t) + X(t)A_2 + D, \quad X(0) = C \qquad [8.13]$$

The usual procedure here is to assume $t_0 = 0$, which does not reduce in any way the generality of the statement. The size of matrices is specified by $A_1 \in \mathbb{R}^{n_1 \times n_1}$, $A_2 \in \mathbb{R}^{n_2 \times n_2}$ and $X, D, C \in \mathbb{R}^{n_1 \times n_2}$. It is clear that based on [8.13], the equation remains linear. However, the structure of the unknown factor does not enable us to directly apply the results of the previous section. From a theoretical point of view, we can, however, return by transforming [8.13] into a system directly similar to [8.1], via Kronecker's product, but of a size which is not usable for the majority of the time ($n_1 n_2 \times n_1 n_2$). To set the orders of magnitudes, we suppose that $n_1 = n_2$. The memory cost of such an approach is then in n^4 and the calculation cost in n^6. It is clear that we must approach the solution of this problem differently. A first method consists of noting that:

$$X(t) = e^{A_1 t}(C - E)e^{A_2 t} + E \qquad [8.14]$$

verifies [8.13], if E is the solution of Sylvester algebraic equation $A_1 E + E A_2 + D = 0$. Two comments should be noted here. The first is that we shifted the difficulty without actually solving it – because we must calculate E, which is not necessarily trivial. Secondly, the non-singularity of this equation imposes constraints on A_1 and A_2 which are not necessary in order to be able to solve the differential equation [8.13].

8.3.2. *Solving principle*

A second richer method consists of seeing that:

$$X(t) = \int_0^t e^{A_1 \tau}(A_1 C - C A_2 + D)e^{A_2 \tau}\, d\tau + C \qquad [8.15]$$

is also solution of [8.13], without any restriction on the problem data. Now we will examine how to calculate this integral by using the techniques specified for the linear standard case. For this, we have:

$$\begin{cases} Q = A_1 C - C A_2 + D \\ Y(t) = \int_0^t e^{A_1 \tau} Q e^{A_2 \tau}\, d\tau \end{cases} \qquad [8.16]$$

Thus, we have:

$$Y(t) = V(t)^{-1}W(t) \qquad [8.17]$$

with:

$$\exp\left(\begin{bmatrix} -A_1 & Q \\ 0 & A_2 \end{bmatrix} t\right) = \begin{bmatrix} V(t) & W(t) \\ 0 & Z(t) \end{bmatrix} = S(t) \qquad [8.18]$$

It is clear that $S(t)$ is the solution of the standard linear differential equation:

$$\frac{d}{dt}\begin{bmatrix} V(t) & W(t) \\ 0 & Z(t) \end{bmatrix} = \begin{bmatrix} -A_1 & Q \\ 0 & A_2 \end{bmatrix}\begin{bmatrix} V(t) & W(t) \\ 0 & Z(t) \end{bmatrix}, \quad S(0) = I$$

However, by formulating it, we have:

$$\begin{cases} \overset{\circ}{V} = -A_1 V, \quad V(0) = I \\ \overset{\circ}{W} = -A_1 W + QZ, \quad W(0) = 0 \\ \overset{\circ}{Z} = A_2 Z, \quad Z(0) = I \end{cases}$$

which thus gives:

$$\begin{cases} V(t) = e^{-A_1 t} \\ W(t) = \int_0^t e^{A_1(t-\tau)} Q e^{A_2 \tau}\, \mathrm{d}\tau \\ Z(t) = e^{A_2 t} \end{cases} \qquad [8.19]$$

From [8.19], we have:

$$W(t) = e^{-A_1 t} Y(t) = V(t) Y(t)$$

which leads to the announced result [8.17]. The solution $X(t)$, to the initial condition, is identified with $Y(t)$, because we have $X(t) = Y(t) + C$.

The particular case of Lyapunov equations represents a privileged situation, as long as the inversion of $V(t)$ disappears. In fact, when we have:

$$A_2 = A_1^T = \dot{A} \qquad [8.20]$$

there is:

$$Z(t) = e^{A_1^T t} \quad \Rightarrow \quad V(t)^{-1} = Z^T(t)$$

from where:

$$Y(t) = Z^T(t) W(t) \qquad [8.21]$$

8.3.3. *Practical implementation*

Again, everything lies on a calculation of matrix exponential function. Let us suppose again that:

$$M = \begin{bmatrix} -A_1 & Q \\ 0 & A_2 \end{bmatrix} \qquad [8.22]$$

The argument previously developed for the choice of integration interval is applied without change in this new context, including the techniques mentioned in Note 8.2. However, we have to note that, in the case of Lyapunov equations, we necessarily have

$\mu = 0$. Since the integration interval is fixed, the order of the Padé approximant is still given by [8.10]. In practice, it is useful to examine how we can calculate the matrix polynomials [8.11]. Hence:

$$\begin{cases} p_q(Mh) = \begin{bmatrix} N_1 & N_{12} \\ 0 & N_2 \end{bmatrix} \\ p_q(-Mh) = \begin{bmatrix} D_1 & D_{12} \\ 0 & D_2 \end{bmatrix} \end{cases}$$

We have the approximation of $S(h)$ [8.18]:

$$\widehat{S(h)} = \begin{bmatrix} D_1 & D_{12} \\ 0 & D_2 \end{bmatrix}^{-1} \begin{bmatrix} N_1 & N_{12} \\ 0 & N_2 \end{bmatrix} \sim \begin{bmatrix} V(h) & W(h) \\ 0 & Z(h) \end{bmatrix}$$

By developing:

$$\begin{cases} V(h) = e^{A_1 h} \sim D_1^{-1} N_1 = \Phi_1 \\ W(h) \sim D_1^{-1}(N_{12} - D_{12} D_2^{-1}) N_2 \\ Z(h) = e^{A_2 h} \sim D_2^{-1} N_2 = \Phi_2 \end{cases} \qquad [8.23]$$

Based on [8.17], we have:

$$Y(h) \sim N_1^{-1}(N_{12} - D_{12} D_2^{-1}) N_2 = Y_1 \qquad [8.24]$$

Considering definition $Y(t)$, we have:

$$Y_{k+1} = \Phi_1 Y_k \Phi_2 \qquad [8.25]$$

a recurrence relation which gives the sought trajectory by addition of initial condition C.

8.4. Stability, stiffness and integration horizon

The simulation context is by definition to simulate reality. The reality manages limited quantities and, consequently, the differential equations that we simulate are dynamically stable when they must be calculated on high time horizons. On the contrary, the dynamically unstable equations can only be used on very short periods of time, in direct relation to the speed with which they diverge. Let us go back to the previous situation – by far the most frequent one. Let us exclude for the time being the presence of a complex integrator (zero pole) and let us deal with the asymptotically stable case, i.e. when all the poles are of strictly negative real part. The experimental duration of a simulation is naturally guided by the slowest time constant T_M of the signal or its cover (if it is of the damped oscillator type). On the other hand, the constraint on the integration interval [8.9] will be in direct relation with the slowest time constraint T_m (of the signal and its cover). Let us recall that:

$$T = \frac{-1}{\text{Re}(\lambda)}, \quad \text{Re}(\lambda) < 0 \qquad [8.26]$$

where λ designates an eigenvalue, or pole of the system, and T the corresponding time constraint, and that, on the other hand, for a matrix A:

$$\|A\| > \max_i |\lambda_i| \qquad [8.27]$$

It is clear that we are in a situation where we want to integrate in a horizon that is as long as T_M is high, with an integration interval that is as small as T_m is low. This relation between the slow and fast dynamics is called *stiffness*.

DEFINITION 8.1. *We call stiffness of a system of asymptotically stable linear differential equations the relation:*

$$\rho = \frac{T_M}{T_m} = \frac{\text{Re}(\lambda_M)}{\text{Re}(\lambda_m)} \qquad [8.28]$$

where λ_M and λ_m are respectively the poles with the highest and smallest negative real part, of absolute value.

NOTE 8.4. For standard linear systems [8.1], the poles are directly the eigenvalues of A. For Sylvester equations [8.13], the poles are eigenvalues of $M = I_{n_2} \otimes A_1 + A_2^T \otimes I_{n_1}$, i.e. the set of pairs $\lambda_i + \mu_j$ where λ_i and μ_j are the eigenvalues of A_1 and A_2.

The stiff systems ($\rho \gg 100$) are by nature systems which are difficult to numerically integrate. The higher the stiffness, the more delicate the simulation becomes. In such a context, it is necessary to have access to dedicated methods, making it possible to get over the paradoxical necessity of advancing with very small integration intervals, which are imposed by the presence of very short temporal constants, even when these fast transients disappeared from the trajectory.

However, these dedicated techniques, which are fundamentally designed for the non-linear differential systems, remain incontrovertible in the stiff linear case. In fact, in spite of their closely related character, they represent algorithms as highly efficient as the specific exact diagrams of the linear case, previously analyzed.

8.5. Non-linear differential systems

8.5.1. *Preliminary aspects*

Before directly considering the calculation algorithms, it is useful to introduce a few general observations. Through an extension of the notations introduced at the beginning of this chapter, we will deal with equations of the form:

$$\overset{\circ}{x}(t) = f(x,t), \quad x(t_0) = x_0 \qquad [8.29]$$

Here, we have, *a priori*, $x, f \in \mathbb{R}^n$. However, in order to present the integration techniques, we will assume $n = 1$. The passage to $n > 1$ remains trivial and essentially pertains to programming. On the other hand, as we indicated in the introduction, we will continue to consider only the problems with given initial conditions. However, the question of uniqueness can remain valid. For example, the differential equation $\overset{\circ}{x} = x/t$ presents a "singular" point in $t = 0$. In order to define a unique trajectory among the set of solutions $x = at$, it is necessary to impose a condition in $t_0 \neq 0$. The statement that follows provides a sufficient condition of existence and uniqueness.

THEOREM 8.1. *If $\overset{\circ}{x}(t) = f(x,t)$ is a differential equation such that $f(x,t)$ is continuous on the interval $[t_0, t_f]$ and if there is a constant L such that $|f(x,t) - f(x^*,t)| \leqslant L|x - x^*|$, $\forall t \in [t_0, t_f]$ and $\forall x, x^*$, then there is a unique function $x(t)$ continuously differentiable such that $\overset{\circ}{x}(t) = f(x,t)$, $x(t_0) = x_0$ being fixed.*

NOTE 8.5. We note that:

- L is called a Lipschitz constant;
- $f(x,t)$ is not necessarily differentiable;
- if $\partial f/\partial x$ exists, the theorem implies that $|\partial f/\partial x| < L$;
- if $\partial f/\partial x$ exists and $|\partial f/\partial x| < L$, then the theorem is verified;
- written within a scalar notation ($n = 1$), these results are easily applicable for $n > 1$.

We will suppose in what follows that the differential equations treated verify this theorem (Lipschitz condition).

8.5.2. *Characterization of an algorithm*

From the instant when trajectory $x(t)$ remains formally unknown, only the approximants of this trajectory can be rebuilt from the differential equation. On the other hand, the calculations being done with a finite precision, we will interpret the result of each calculation interval as an error-free result of a slightly different (disturbed) problem. The question is to know whether these inevitable errors will or will not mount up in time to completely degenerate the approached trajectory. A first response is given by the following definition.

DEFINITION 8.2. *An algorithm is entirely stable for an integration interval h and for a given differential equation if an interference δ applied to estimation x_n of $x(t_n)$ generates at future instants an interference increased by δ.*

An entirely stable algorithm will not suffer interferences induced by the finite precision of calculations. On the other hand, this property is acquired only for a given problem. In other terms, such a solver will perfectly operate with the problem for which it was designed and may not operate at all for any other problem. It is clear that this property is not constructive. Here is a second one that will be the basis for the design of all "explicit" solvers, to which the so-called Runge-Kutta incontrovertible family of diagrams belong.

Initially, we introduce the reference linear problem:

$$\overset{\circ}{x} = \lambda x, \quad \lambda \in \mathbb{C} \tag{8.30}$$

DEFINITION 8.3. *We call a region of absolute stability the set of values $h > 0$ and $\lambda \in \mathbb{C}$ for which an interference δ applied to the estimate x_n of $x(t_n)$ generates at future instants an interference increased by δ.*

We substituted a predefined non-linear system for an imposed linear system. The key of the problem lies in the fact that any unknown trajectory $x(t)$ can be locally estimated by the solution of [8.30], $x(t) = a\,e^{\lambda t}$, on a time interval depending on the precision required and on the non-linearity of the problem to solve. This induces calculation intervals h and the faster the trajectory varies locally, the lower these calculation intervals are, and vice versa.

We will continue to characterize an integration algorithm by now specifying the type of approximation errors and their magnitude order according to the calculation interval. To do this, we will use the following notations, with an integration interval h, supposed constant for the time being:

$$\begin{cases} t_n = nh, \quad t_0 = 0 \\ x_n \text{ approximation of } x(t_n) \end{cases} \tag{8.31}$$

DEFINITION 8.4. *We call a local error the error made during an integration interval.*

DEFINITION 8.5. *We call a global error the error detected at instant t_n between the trajectory approached x_n and the exact trajectory $x(t_n)$.*

Let us formalize these errors, whose role is fundamental. Let t_n be the current instant. At this instant, the theoretical solution is $x(t_n)$ and we have an approached solution x_n. It is clear that the global error e_n can be evaluated by:

$$e_n = x_n - x(t_n) \tag{8.32}$$

Now, let us continue with an interval h in order to reach instant t_{n+1}. At instant t_n, x_n can be considered as the exact solution of the differential equation that we solve, but with another initial condition. Let $u_n(t)$ be this trajectory, solution of $\overset{\circ}{u}_n = f(u_n, t)$, with by definition $u_n(t_n) = x_n$. If the integration algorithm made it possible to solve the differential equation exactly, we would have, at instant t_{n+1}, $u_n(t_{n+1})$. In reality, we obtain x_{n+1}. The difference between these two values is the error made during a calculation interval; it is the local error:

$$d_n = x_{n+1} - u_n(t_{n+1}) \qquad [8.33]$$

There is no explicit relation between these two types of error. Even if we imagine that the global error is higher than the local error, the global error is not the accumulation of local error. The mechanism connecting these errors is complex and its analysis goes beyond the scope of this chapter. On the other hand, it is important to remember the next result, where expression $\mathbb{O}(h)$ must be interpreted as a function of h for which there are two positive constants k and h_0, independent from h, such that:

$$|\mathbb{O}(h)| \leqslant kh, \quad \forall |h| \leqslant h_0 \qquad [8.34]$$

THEOREM 8.2. *For a given integration algorithm, if the local error verifies* $d_n = \mathbb{O}(h^{p+1})$, *then the global error has a magnitude order given by* $e_n = \mathbb{O}(h^p)$, $p \in \mathbb{N}$.

NOTE 8.6. The operational algorithms have variable intervals; in this case, the magnitude order of the global error must be taken with an average interval on the horizon of calculation considered. In practice, the conclusion remains the same. The global error is of a higher magnitude order than the local error.

Since the integration interval is intuitively small (more exactly, the product $h\lambda$) to obtain high precision, it is legitimate to think that the higher p is [8.32], the better the approximant built by the solver will be. This reasoning leads to the following definition.

DEFINITION 8.6. *We call an order of an integration algorithm the integer p appearing in the global error.*

Therefore, we tried building the highest order algorithms, in order to obtain by definition increasing quality precisions, for a given interval. Reality is much less simple because, unfortunately, the higher the order is, the less the algorithms are numerically stable. Hence, there is a threshold beyond which we lose more – due to the finite precision of calculation – than what the theory expects to gain. It is easy to realize that the order of solvers rarely exceeds $p = 6$. There are two key words to classify the integration algorithms into four categories: the algorithms are "single-interval" or "multi-interval" on the one hand and on the other hand "implicit" or "explicit". We will limit ourselves here to "single-interval" explicit algorithms and we will finish with the implicit techniques in general.

8.5.3. *Explicit algorithms*

Explicit algorithms are the family of methods that are expressed as follows:

$$\begin{cases} v_i = hf\left(x_n + \sum_{j=1}^{i-1} b_{ij}v_j, t_n + a_i h\right), & i = 1, 2, \ldots, r \\ x_{n+1} = x_n + \sum_{i=1}^{r} c_i v_i \end{cases} \qquad [8.35]$$

The algorithm is explicit as long as a v_i depends only on with $j < i$. Finally, it is at one step because x_{n+1} only depends on x_n. Parameter r represents the cost of algorithm calculation, measured in the number of times where we evaluate the differential equation during an integration interval. The order of the method is a non-decreasing function of r. The triangular matrix $B = [b_{ij}]$ and vectors $a = [a_i]$ and $c = [c_i]$ are the algorithm parameters, chosen with the aim of creating the highest possible order diagram that is also the most numerically stable. Euler's 1st order method, and more generally all Runge-Kutta type algorithms, meet the formulation [8.35]. We still have to determine the field of absolute stability of a p order algorithm. This stability is judged by definition on the reference equation, which is parameterized by dynamics λ. From the current point (t_n, x_n), the exact value at the next instant t_{n+1} will be $e^{\lambda h}x_n$. For a disturbance applied to instant t_n to be non-increasing in the future, the condition is simply $|e^{\lambda h}| \leqslant 1$. If the algorithm is of p order, this means that $e^{\lambda h}$ is approached by its serial development in Taylor series at p order. The stability domain described in the complex plane $\mu = \lambda h$ is then defined by:

$$\left|\sum \frac{\mu^i}{i!}\right| \leqslant 1 \qquad [8.36]$$

For Euler's method, $p = 1$, we find the unit circle centered in $\mu = -1$. What is remarkable is that the stability field of an explicit algorithm does not depend on the formulae used to implement it (here B, a, c), but directly on the order that characterizes it! In a system of non-linear equations, the role of λ is kept by the eigenvalues of Jacobian $\partial f/\partial x$; on the other hand, the magnitude order for the local and global error is, by definition, guaranteed only for a value of μ belonging to the stability field of the method [8.36]. The constraint on the integration interval is thus operated by the "high" λ (absolute value of negative real value), i.e. by fast transients, even if they became negligible in the trajectory $x(t)$! *This is the fundamental reason as to why an explicit method is not applied in order to integrate a stiff system.*

Among all possible diagrams, we have chosen one corresponding to the best compromise between performance and complexity, and present in all the best libraries. It is the Runge-Kutta-Fehlberg solver, whose particularity is to simultaneously offer a 5 and 4 order estimation, for the same cost ($r = 6$) like the more traditional diagram of 5 order only. From this situation we will have a particularly competitive automatic management of the integration interval, based on the idea that when we are in the domain of absolute stability, the 5 order estimate can be, with respect to 4 order, the exact trajectory for estimating the local error. In addition, it is thus possible to verify the compatibility of its magnitude order with what the theory expects. The parameters of this solver are:

$$
a = \begin{bmatrix} 0 \\ 1/4 \\ 3/8 \\ 12/13 \\ 1 \\ 1/2 \end{bmatrix}, \quad
c = \begin{bmatrix} 16/135 \\ 0 \\ 6{,}656/12{,}825 \\ 28{,}561/56{,}430 \\ -9/50 \\ 2/55 \end{bmatrix}, \quad
c^* = \begin{bmatrix} 25/216 \\ 0 \\ 1{,}408/2{,}565 \\ 2{,}197/4{,}104 \\ -1/5 \\ 0 \end{bmatrix}
$$

$$
B = \begin{bmatrix}
0 & 0 & 0 & 0 & 0 & 0 \\
1/4 & 0 & 0 & 0 & 0 & 0 \\
3/32 & 9/32 & 0 & 0 & 0 & 0 \\
1{,}932/2{,}197 & -7{,}200/2{,}197 & 7{,}296/2{,}197 & 0 & 0 & 0 \\
439/216 & -8 & 3{,}680/513 & -845/4{,}104 & 0 & 0 \\
-8/27 & 2 & -3{,}544/2{,}565 & 1{,}859/4{,}104 & -11/40 & 0
\end{bmatrix}
$$

Here, parameter c^* is the second value of c, leading to a 4 order estimation, the 5 order estimation being provided by c.

8.5.4. *Multi-interval implicit algorithms*

The complexity of these techniques is another matter [LON 95]. Firstly, let us consider the implicit version of single-interval methods, which are directly obtained from the explicit case [8.35]:

$$
\begin{cases}
v_i = hf\left(x_n + \sum_{j=1}^{i} b_{ij} v_j, t_n + a_i h\right), & i = 1, 2, \ldots, r \\
x_{n+1} = x_n + \sum_{i=1}^{r} c_i v_i
\end{cases}
\tag{8.37}
$$

As long as v_i depends now on itself, its calculation implies the solving of a non-linear (static) system, precisely the one defining the differential equation to be solved. To simplify the future notations, we say:

$$
f_n = f(x_n, t_n)
\tag{8.38}
$$

A multi-interval method will be then written:

$$
x_{n+1} = \sum_{i=1}^{r} \alpha_i x_{n+1-i} + h \sum_{i=0}^{r} \beta_i f_{n+1-i}
\tag{8.39}
$$

The method is implicit for $\beta_0 \neq 0$ and explicit for $\beta_0 = 0$. Apart from the difficulty related to the implicit case already mentioned, a multi-interval algorithm cannot start, due to its reference to a past that does not exist at the beginning of the simulation. The first points are thus always calculated by a single-interval method.

Due to its particular context, solving non-linear systems intervening in the implicit structures is not done as a priority by a standard solver but rather by a specific approach consisting of using in parallel an explicit diagram, with the role of *predictor*, and an implicit diagram, called a *corrector*. In these two phases, we usually add a third one, called an *estimator*. By noting these stages P, C and E, each calculation interval is built on a $P(EC)^m E$ type structure that we interpret as being constituted of a prediction initial phase followed by m estimation-correction iterations and by an estimation final phase. This leads to the general diagram:

$$\begin{cases} P : x_{n+1}^{(0)} = \sum_{i=1}^{r} \alpha_i x_{n+1-i} + h \sum_{i=1}^{r} \beta_i f_{n+1-i}, \quad k = 0 \\ \text{as long as } k \leqslant m \\ \quad E : f_{n+1}^{(k)} = f\left(x_{n+1}^{(k)}, t_{n+1}\right) \\ \quad C : x_{n+1}^{(k+1)} = \sum_{i=1}^{r} \alpha_i x_{n+1-i}^{(k)} + h \sum_{i=0}^{r} \beta_i f_{n+1-i}^{(k)}, \quad k = k+1 \\ \text{as long as in the end} \\ E : x_{n+1} = x_{n+1}^{(m)}, \quad f_{n+1} = f\left(x_{n+1}^{(m)}, t_{n+1}\right) \end{cases} \qquad [8.40]$$

The number of m iterations is often imposed *a priori* or obtained from a convergence criterion on $|x_{n+1}^{(k+1)} - x_{n+1}^{(k)}|$. If we consider formula [8.39], we have $2r + 1$ degrees of freedom. Hence, we are capable, by choosing correctly the parameters of the method (the α_i and β_i), of building an exact solver for the polynomial trajectories $x(t)$ of a degree inferior than or equal to $2r$. From this we obtain a local error in $\mathbb{O}(h^{2r+1})$, i.e. a method of order $2r$ – therefore much more than what a single-interval explicit method could have expected. Unfortunately, the diagrams of $2r$ order are numerically unstable – there is phase difference $(EC)^m$. We prove that it is impossible to build numerically stable algorithms of an order greater than $r + 1$, for r odd, and of order greater than $r + 2$, for r even.

The field of absolute stability is always characterized from the reference equation $\overset{\circ}{x} = \lambda x$. Based on [8.39], by introducing an $\alpha_0 = -1$, we can rewrite this relation as:

$$\sum_{i=0}^{r} \alpha_i x_{n+1-i} + h \sum_{i=0}^{r} \beta_i f_{n+1-i} = 0 \qquad [8.41]$$

Applied to our reference linear equation, always with $\mu = h\lambda$, we get:

$$\sum_{i=0}^{r} (\alpha_i + \beta_i \mu) x_{n+1-i}$$

If we apply interference δ_n to x_n, the δ_n are governed by the same recurrence and thus evolve as z_i^n, where z_i is a root of the polynomial:

$$p(z) = \sum_{i=0}^{r} \alpha_i z^{r-i} + \mu \sum_{i=0}^{r} \beta_i z^{r-i} \qquad [8.42]$$

Consequently, the field of absolute stability of multi-interval methods (implicit or not depending on the value of β_0) is the set of $\mu \in \mathbb{C}$ so that the roots of the polynomial [8.42] verify $|z_i| \leqslant 1$. It is important to know that the field of absolute stability of implicit methods is always larger (often 10 times more) than that of explicit methods of the same order – hence their interest in spite of their high complexity. On the other hand, the more the order increases, the more the field is reduced. Hence, the designer will again have compromises to make. Many diagrams were suggested in books and it is obviously impossible to try to make a synthesis. For example, we chose the implicit Adams-Moulton method. This strategy corresponds to the following parameters, for $r = 1, \ldots, 6$ with $\alpha_i = 0$ $(i = 2, \ldots, r)$, $\alpha_1 = 1$ and β_i, according to the following table:

$$
\begin{bmatrix}
r=1 & r=2 & r=3 & r=4 & r=5 & r=6 \\
1 & 1/2 & 5/12 & 9/24 & 251/720 & 475/1{,}440 \\
 & 1/2 & 8/12 & 19/24 & 646/720 & 1{,}427/1{,}440 \\
 & & -1/12 & -5/24 & -264/720 & -798/1{,}440 \\
 & & & 1/24 & 106/720 & 482/1{,}440 \\
 & & & & -19/720 & -173/1{,}440 \\
 & & & & & 27/1{,}440
\end{bmatrix}
$$

In all cases, the order of the method is $r + 1$. The interval management is done along with a selection of the order within the possible range, here $r \leqslant 6$. This consists of comparing the magnitude order of the local error theoretically stipulated and with that estimated with the help of devised differences and then adapting the interval. The order is selected in such a way as to maintain the highest interval, while remaining in the field of absolute stability. The professional codes are accompanied in reality by heuristic methods which are often very sophisticated and the result of long experience, which give these tools the best compromise between cost, performance and robustness. As long as the stability field continues to intervene, there is nothing solved with respect to the stiff systems. Then, what do these techniques bring with respect to single-interval explicit methods, with a much less complicated design? They potentially offer better performances, in terms of the cost/order ratio, a better flexibility, the variation of the interval *and order* at the same time, and especially a chance to get away from this field of absolute stability – due to which it will finally be possible to integrate these stiff systems.

8.5.5. *Solver for stiff systems*

For non-linear systems, stiffness is always defined by [8.28], but this time the λ are the eigenvalues of the Jacobian $\partial f / \partial x$. This means that stiffness is a characteristic of the differential system, variable in time! A specific notion of stability had to be introduced in order to deal with this particular case, which is extremely frequent in the industrial applications of simulation. It is the S-stability (*"stiff" stability*). S-stability is expressed in terms of absolute stability for the area of the complex plane μ defined by $\mathrm{Re}(\mu) < d < 0$ and a precision constraint in the area $\{d < \mathrm{Re}(\mu) < a, a > 0 ; |\mathrm{Im}(\mu)| < c\}$. The subtlety of the approach lies in the fact that no particular precision is required for the area $\mathrm{Re}(\mu) < d < 0$, since it is *de facto* acquired by the choice of parameter d. In fact, always with respect to the local reference equation of dynamic λ, when a fast transient would impose a very small interval, we verify that it has become negligible at the following instant. We have $|e^{\mu}| = |e^{\lambda h}| < e^d$. Conversely, a has the potential increase of the trajectory and c is there to express that in oscillating phase, a minimum of points are necessary to follow, with a given precision, a pseudo-period. Gear [GEA 71], who is at the origin of all the developments for the integration of stiff systems, proved that we could build S-stable single-interval implicit algorithms for $2 \leqslant r \leqslant 6$. The counterpart is that solving a non-linear system by the phase $(EC)^m$ [8.40] was not convergent anymore, due to the interval increase authorized by the S-stability and precisely forbidden by the absolute stability to which this diagram referred. This time, we must solve the non-linear system through a more traditional approach, like Newton type, i.e. by calculating the Jacobian of the system of differential equations $\partial f / \partial x$ and then by solving the system that it induces in order to obtain the correction whose role was held by phase C. Gear's diagrams are summed up in the following table, which refers to the general relation [8.39], by noting that only β_0 is non-zero:

$$\begin{bmatrix} k & 2 & 3 & 4 & 5 & 6 \\ \beta_0 & 2/3 & 6/11 & 12/25 & 69/137 & 60/147 \\ \alpha_1 & 4/3 & 18/11 & 48/25 & 300/137 & 360/147 \\ \alpha_2 & -1/3 & -9/11 & -36/25 & -300/137 & -450/147 \\ \alpha_3 & & 2/11 & 16/25 & 200/137 & 400/147 \\ \alpha_4 & & & -3/25 & -75/137 & -225/147 \\ \alpha_5 & & & & 12/137 & 72/147 \\ \alpha_6 & & & & & -10/147 \end{bmatrix} \qquad [8.43]$$

8.5.6. *Partial conclusion*

The single-interval explicit methods are by far the simplest. On the other hand, the interval automatic management represents, in the majority of cases, their weak point. This difficulty is intrinsically linked to the fact that we have only *a priori* a given order diagram – without a Runge-Kutta-Fehlberg algorithm presented. On the contrary, the multi-interval methods are compatible with the interval automatic management, because they easily offer multiple order diagrams. However, they cannot operate without using single-interval methods. They are equally more delicate to implement, due to the iterative aspect that characterizes them (implicit form). When, in the course of integration, stiffness increases, the solver must call on Gear's parameters, with the necessity of using the Jacobian of the system in a Newton type iterative diagram. It is clear that, for a difficult problem, there is no viable simple solution.

If one is not a specialist in the field, it is strongly advisable to use the calculation databases dedicated to this type of problem [SHA 97]. In any case, the objective of this chapter was not to make the reader become a specialist, but to make him an adequate and critical user of tools that he may have to use.

8.6. Discretization of control laws

8.6.1. *Introduction*

A very particular case of numerical simulation consists of implementing the control algorithms on the calculator. There are actually several methods of doing the synthesis of a regulator. A simple method consists of experimentally determining the discrete transfer function of the system, as we saw in the previous chapter. It is then natural to directly calculate a discrete control and to implement it as such on the calculator in real-time. The other approach consists of starting with a continuous model experimentally obtained, or from the knowledge model. Then we can discretize this model (z transform, discretization of state representation) and we find ourselves in the previous situation. We can also choose to delay the discretization until the last moment in order to benefit from all the know-how of the continuous control. Then we calculate a continuous regulator that will have to be simulated on the control calculator in real-time by a difference equation. In this last case, we generally choose to have a low sampling period (with respect to the dynamics of the procedure) and we generally use very simple simulation algorithms that we could even call simplistic! We will mention some of them in this section.

NOTE 8.7. In order to take into account the presence of the pair zero order blocker/sampler in the sampled loop, it is advisable to approach it by a pure delay of a sampling half-period $e^{-\frac{T}{2}s}$ or by the transfer [BES 99]:

$$B_0'(s) = \left(1 - \frac{T}{2}s\right) \tag{8.44}$$

It is clear that this transfer is negligible as soon as the frequency corresponding to the sampling half-period is placed in a sufficiently high frequency band with respect to the transfer cross-over frequencies of the system. However, it makes it possible to consider the phase difference brought about by the presence of the blocker and explains that the results obtained with the numerical regulator are sometimes different from those obtained with the continuous regulator.

8.6.2. *Discretization*

The continuous regulator is often obtained in the form of a transfer function, describing a differential equation. We seek to replace, in this equation, the differentiation operator by a numerical approximation. Depending on the approximations chosen for representing the differentiation (and hence the integration), we find various difference equations to be programmed in the calculator.

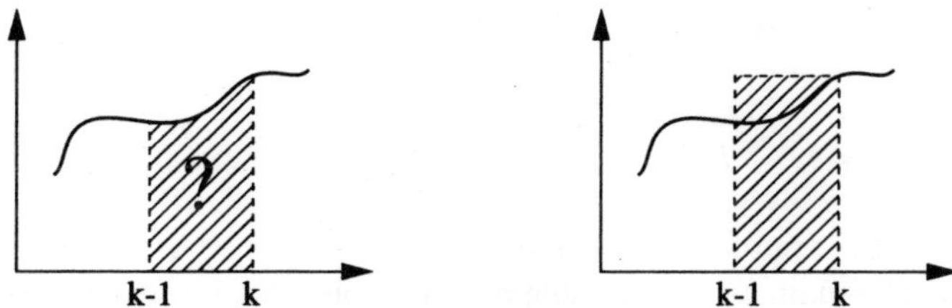

Figure 8.1. *Superior rectangle method*

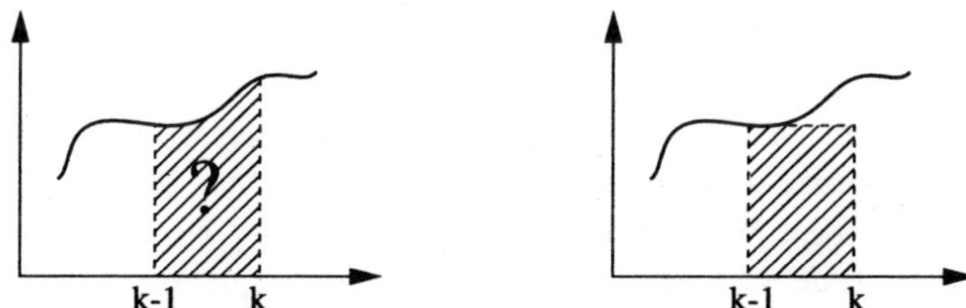

Figure 8.2. *Inferior rectangle method*

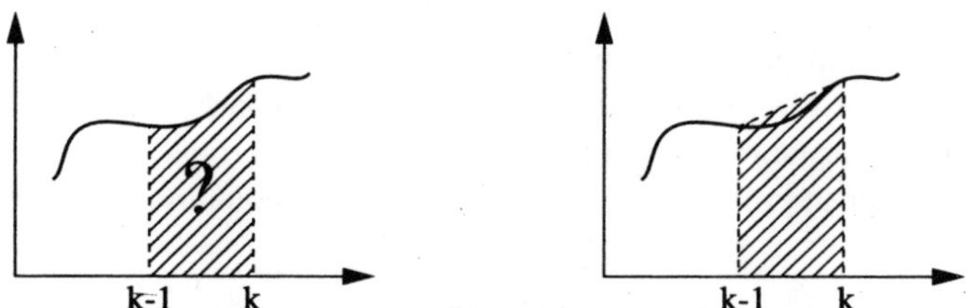

Figure 8.3. *Trapezoid method*

Let us consider first the approximation of a signal derivative $y(t)$ by a rear difference calculation:

$$\frac{\mathrm{d}y(t)}{\mathrm{d}t}\bigg|_{t=kT} \approx \frac{y_{kT} - y_{(k-1)T}}{T} \tag{8.45}$$

This derivation method corresponds to the approximation of an integral calculation by the technique known as the *superior rectangle method* or *Euler's first method* [BES 99], which is illustrated in Figure 8.1:

$$I_{kT} = I_{(k-1)T} + T y_{kT} \tag{8.46}$$

We find a second Euler's method, which is called an *inferior rectangle method* (Figure 8.2). It is based on the rear difference calculation for the derivative:

$$\left.\frac{dy(t)}{dt}\right|_{t=kT} \approx \frac{y_{(k+1)T} - y_{kT}}{T} \qquad [8.47]$$

$$I_{kT} = I_{(k-1)T} + Ty_{(k-1)T} \qquad [8.48]$$

We finally find an approximation known under various names: Tustin's approximation, trapezoid approximation or Padé approximation. This last calculation is equivalent to the integration by the trapezoid method (Figure 8.3):

$$I_{kT} = I_{(k-1)T} + \frac{T}{2}y_{(k-1)T} + \frac{T}{2}y_{kT} \qquad [8.49]$$

We will now deal with the relations that these approximations impose between the continuous and numerical transfer functions of the regulator. We know that the continuous derivation corresponds to the Laplace operator s. As for the delay of a sampling period within a difference equation, it is represented by the operator z^{-1}. If we seek the z transform of equation [8.45], we find $\frac{1-z^{-1}}{T}Y(z)$. Therefore, we can conclude that each time we find a derivation in the time equation of the regulator, i.e. the operator s in its transfer function, we will have, in the sampled transfer function, the operator $\frac{1-z^{-1}}{T}$:

$$s = \frac{z-1}{Tz} \qquad [8.50]$$

Under these conditions, it is easy to deduce, from the continuous transfer function of the regulator, the discrete transfer function, which can be then used in order to find the difference equation simulating the regulator numerically. Therefore, we will easily verify that the approximation by front difference (equation [8.47]) returns to the substitution:

$$s = \frac{z-1}{T} \qquad [8.51]$$

and the trapezoid method [8.49] to the substitution:

$$s = \frac{2}{T}\frac{z-1}{z+1} \qquad [8.52]$$

The approximation of the inferior rectangle does not maintain the stability of the continuous transfer function that is digitized. In fact, the transformation (equation [8.51]) transposes the left half-plane of plane s into an area in the poles plane in z which goes beyond the unit circle (see Figure 8.4). For this reason, this is a little used method.

With the transformation of the superior rectangle [8.50], the left half-plane in s is transposed into plane z into an area situated within the unit circle (Figure 8.5).

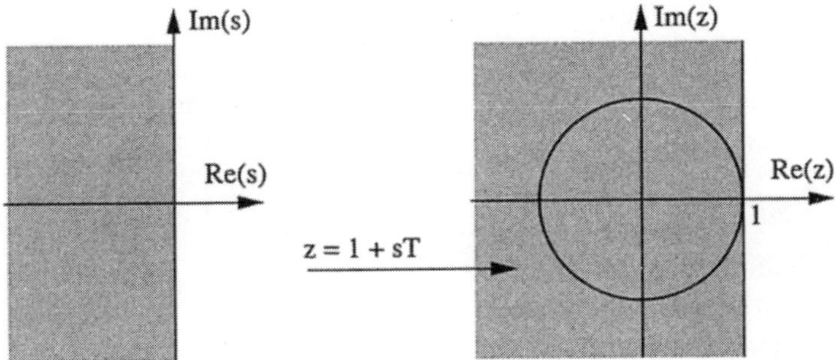

Figure 8.4. *Transformation of the inferior rectangle*

For this reason, this method is preferred to the approximation of the inferior rectangle. We can also note its advantage with respect to the latter in the calculation of the integral [8.46], which makes the value of the magnitude integrate at instant k and not at instant $(k-1)$ as in the equation [8.48].

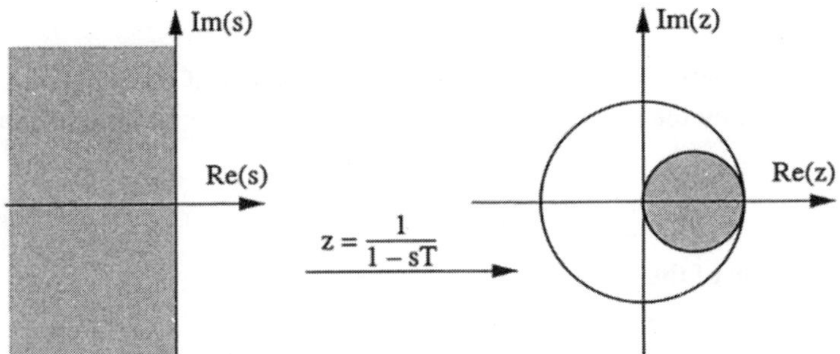

Figure 8.5. *Transformation of the superior rectangle*

Finally, we note that Tustin transformation transposes the left half-plane s within the unit circle in plane z, which guarantees the same stability properties before and after the transformation (Figure 8.6). We have the same transformation of the complex plane as with the theoretical value $z = e^{Ts}$, whose [8.52] is precisely a 1st order Padé approximant.

8.6.3. *Application to PID regulators*

With these transposition tools, if the continuous regulator is given as a transfer function, it is enough to choose the approximation desired and to perform the corresponding substitutions. Let us take the example of the classic PID regulator. Let $e(t)$

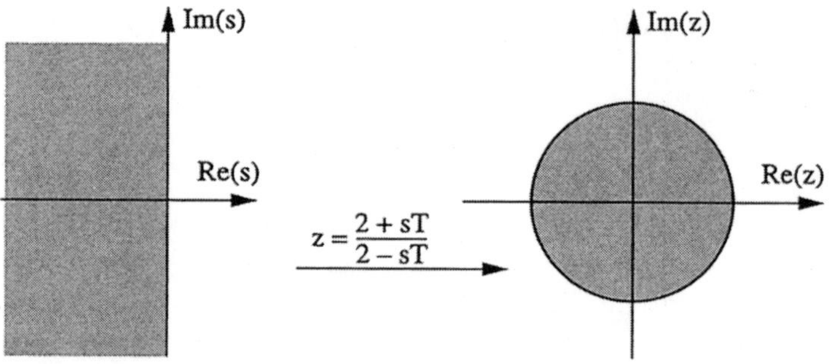

Figure 8.6. *Trapezoid transformation*

be the displacement signal at its input and $u(t)$ the forward signal at its output. Its transfer function is:

$$U(s) = K_P \left[1 + \frac{1}{sT_i} + \frac{sT_d}{1 + s\frac{T_d}{N}} \right] E(s) \qquad [8.53]$$

where K_P, T_i, T_d and N represent the setting parameters of the regulator. By using the approximation by the superior rectangle method, the integral term becomes:

$$\frac{1}{sT_i} = \frac{T}{T_i} \frac{z}{z - 1} \qquad [8.54]$$

The calculation of the filtered derivative gives:

$$sT_d = \frac{T_d}{T} \frac{z - 1}{z} \qquad [8.55]$$

$$1 + s\frac{T_d}{N} = 1 + \frac{Td}{NT} \frac{z - 1}{z} \qquad [8.56]$$

whose ratio is:

$$\frac{Td}{T} \frac{z - 1}{z + \frac{T_d}{NT}(z - 1)} \qquad [8.57]$$

and equation [8.53] becomes:

$$\frac{U(z)}{E(z)} = K_P \left[1 + \frac{T}{T_i} \frac{z}{z - 1} + \frac{T_d}{T} \cdot \frac{z - 1}{z + \frac{T_d}{NT}(z - 1)} \right] \qquad [8.58]$$

that we can write in the standard form:

$$K_P \left[1 + \frac{T}{T_i} \frac{z}{z - 1} + \frac{T_{dd}}{T} \cdot \frac{z - 1}{z - \gamma} \right] \qquad [8.59]$$

with:

$$\gamma = \frac{T_d}{NT + T_d} \qquad [8.60]$$

and:

$$\frac{T_{dd}}{T} = \frac{NT_d}{NT + T_d} \qquad [8.61]$$

We still need to write the difference equation corresponding to this transfer in order to have the programming algorithm of the numerical PID. We can, for example, set equation [8.59] at the common denominator:

$$K(z) = \frac{r_0 z^2 + r_1 z + r_2}{(z-1)(z-\gamma)} = \frac{r_0 z^2 + r_1 z + r_2}{z^2 + s_1 z + s_2} \qquad [8.62]$$

with:

$$r_0 = 1 + \frac{T}{T_i} + \frac{T_{dd}}{T} \qquad [8.63]$$

$$r_1 = 1 + \gamma + \frac{T}{T_i}\gamma + \frac{2T_{dd}}{T} \qquad [8.64]$$

$$r_2 = \gamma + \frac{T_{dd}}{T} \qquad [8.65]$$

and the difference equation is expressed as:

$$u_{kT} = -s_1 u_{(k-1)T} - s_2 u_{(k-2)T} + r_0 e_{kT} + r_1 e_{(k-1)T} + r_2 e_{(k-2)T} \qquad [8.66]$$

It is often suggested to separate the three terms – proportional, integral and derivative – in the coding, which leads to create three intermediary actions u_p, u_i and u_d, that we sum up to obtain the global action:

$$\begin{cases} U_p(z) = K_P E(z) \\ U_i(z) = K_p \frac{T}{T_i} \frac{z}{z-1} E(z) \\ U_d(z) = K_p \frac{T_{dd}}{T} \frac{z-1}{z-\gamma} E(z) \end{cases} \qquad [8.67]$$

$$\begin{cases} u_{p\,kt} = K_P e_{kT} \\ u_{i\,kt} = u_{i\,(k-1)T} + K_P \frac{T}{T_i} e_{kT} \\ u_{d\,kt} = \gamma u_{d\,(k-1)T} + K_P \frac{T_{dd}}{T}(e_{kT} - e_{(k-1)T}) \end{cases} \qquad [8.68]$$

This encoding enables us to act separately on each action; we can, for example, disconnect the derived action if we prefer a PI to a PID; and we can also limit the integral action in order to prevent it from saturating the actuators (*antireset windup*).

In expression [8.68], we saw that the action derived will depend on the error variation on a sampling period. If the latter is very low, it is possible that the variance on the error becomes of the same magnitude order as the noise, the rounding errors or the quantization errors. Thus, it is not reasonable to have too small a sampling period. Another comment should be made on the numerical realization of the integral term, according to the variance multiplied by the sampling period. When we are too close to the reference, the variance becomes low and, if the sampling period itself is low, the correction of the integral action may become zero if it is inferior to the quantization threshold. Therefore, we can notice a static error which is theoretically impossible when we have an integrator in the direct chain of control. One solution can be to increase the length of words intervening in the calculations of the integral action. A second solution consists of storing the part of e_{kT} not taken into account after the product by $\frac{T}{T_i}$ in order to add it to the value $e_{(k+1)T}$ of the next sampling period [LON 95].

8.7. Bibliography

[BES 99] BESANÇON A., GENTIL S., "Réglage de régulateurs PID analogiques et numériques", *Techniques de l'ingénieur*, Traité Mesures et contrôle, R7 416, 1999.

[GEA 71] GEAR C.W., *Numerical Initial Value Problems in Ordinary Differential Equations*, Prentice-Hall, New Jersey, 1971.

[LON 95] LONCHAMP R., *Commande numérique des systèmes dynamiques*, Presses polytechniques et universitaires romandes, 1995.

[SHA 75] SHAMPINE L.F., GORDON M.K., *Computer Solution of Ordinary Differential Equations: The Initial Value Problem*, W.H. Freeman and Company Publishers, 1975.

[SHA 97] SHAMPINE L.F., REICHELT M.W., *The SIAM Journal on Scientific Computing*, vol. 18–1, 1997.

Part 2

System Control

Chapter 9

Analysis by Classic Scalar Approach

9.1. Configuration of feedback loops

9.1.1. *Open loop – closed loops*

The block diagram of any closed loop control system (Figure 9.1) consists of an action chain and of a reaction (or feedback) chain which makes it possible to elaborate an error signal $\varepsilon(t)$, the difference between the input magnitude $e(t)$ and the measured output magnitude $r(t)$. The output of the system is $s(t)$.

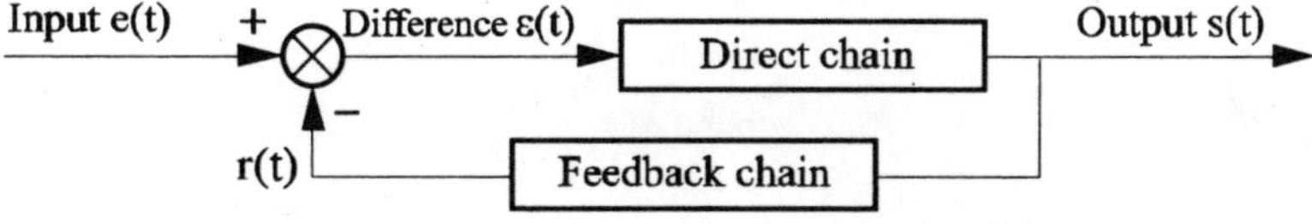

Figure 9.1. *Block diagram of a feedback control*

When the system is subjected to interferences $b(t)$, its general structure is represented by Figure 9.2 by supposing that its working point in the direct chain is known. We designate by $E(p)$, $R(p)$, $\varepsilon(p)$ and $S(p)$ the Laplace transforms of the input, the measurement, difference and the output respectively (see Figure 9.2).

Chapter written by Houria SIGUERDIDJANE and Martial DEMERLÉ.

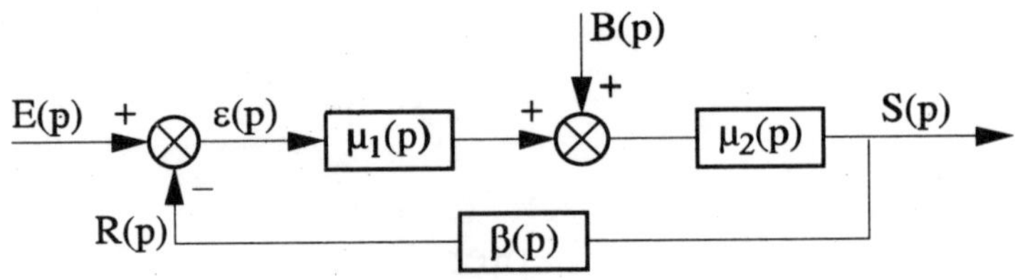

Figure 9.2. *General block diagram*

The open loop transfer function of this chain is the product of transfer functions of all its elements; it is the ratio:

$$\frac{R(p)}{\varepsilon(p)} = \mu_1(p)\mu_2(p)\beta(p) \tag{9.1}$$

The closed loop transfer function of this chain is the ratio $\dfrac{S(p)}{E(p)}$ with:

$$\varepsilon(p) = E(p) - R(p) \tag{9.2}$$

We have:

$$S(p) = \mu_2(p)(B(p) + \mu_1(p)\,\varepsilon(p)) \tag{9.3}$$

and:

$$R(p) = \beta(p)S(p) \tag{9.4}$$

by using equation [9.2] and by eliminating $R(p)$ and $\varepsilon(p)$ of equations [9.3] and [9.4], we obtain:

$$S(p) = \frac{\mu_1(p)\mu_2(p)}{1 + \mu_1(p)\mu_2(p)\beta(p)} E(p) + \frac{\mu_2(p)}{1 + \mu_1(p)\mu_2(p)\beta(p)} B(p) \tag{9.5}$$

When the interferences are zero, $B(p) = 0$, the transfer function of the looped system is then:

$$\frac{S(p)}{E(p)} = \frac{\mu_1(p)\mu_2(p)}{1 + \mu_1(p)\mu_2(p)\beta(p)}$$

[9.6]

or simply:

$$\frac{S(p)}{E(p)} = \frac{\mu(p)}{1 + \mu(p)\beta(p)}$$

[9.7]

by supposing that $\mu(p) = \mu_1(p)\mu_2(p)$.

The transfer function with respect to the interference input is obtained by having $E(p) = 0$ in equation [9.5]:

$$\frac{S(p)}{B(p)} = \frac{\mu_2(p)}{1 + \mu_1(p)\mu_2(p)\beta(p)}$$

[9.8]

9.1.2. *Closed loop harmonic analysis*

Bandwidth

The bandwidth of a system is the interval of angular frequencies for which the module of open loop harmonic gain is more than 1 in arithmetic value:

$$|\mu(j\omega)\beta(j\omega)| \gg 1$$

[9.9]

Approximate trace

In order to simplify the determination of closed loop transfer functions, we can use the approximations [9.10] and [9.11]:

$$\text{If } |\mu(j\omega)\beta(j\omega)| \gg 1 \quad \text{then} \quad \frac{S(j\omega)}{E(j\omega)} \approx \frac{1}{\beta(j\omega)}$$

[9.10]

$$\text{If } |\mu(j\omega)\beta(j\omega)| \ll 1 \quad \text{then} \quad \frac{S(j\omega)}{E(j\omega)} \approx \mu(j\omega)$$

[9.11]

Figure 9.3 shows, in Bode plane, the approximate trace (in full line), of the gain curve of the closed loop frequency response.

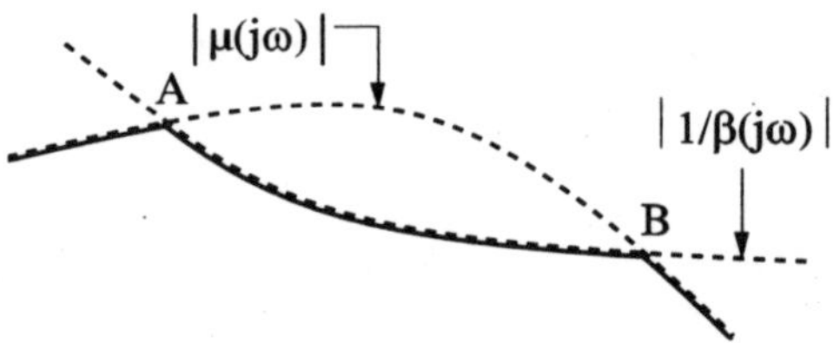

Figure 9.3. *Approximate trace of the closed loop harmonic response*

Point A, in particular for integrator systems, is often rejected at $\omega = 0$.

9.1.2.1. *Black-Nichols diagram*

The Black-Nichols diagram makes it possible to graphically pass from the open loop system to the closed loop system. This diagram corresponds to a unitary feedback. The chart in Figure 9.4 is usable for open loop gains going from –40 dB to +40 dB and a phase difference between 0° and –360°.

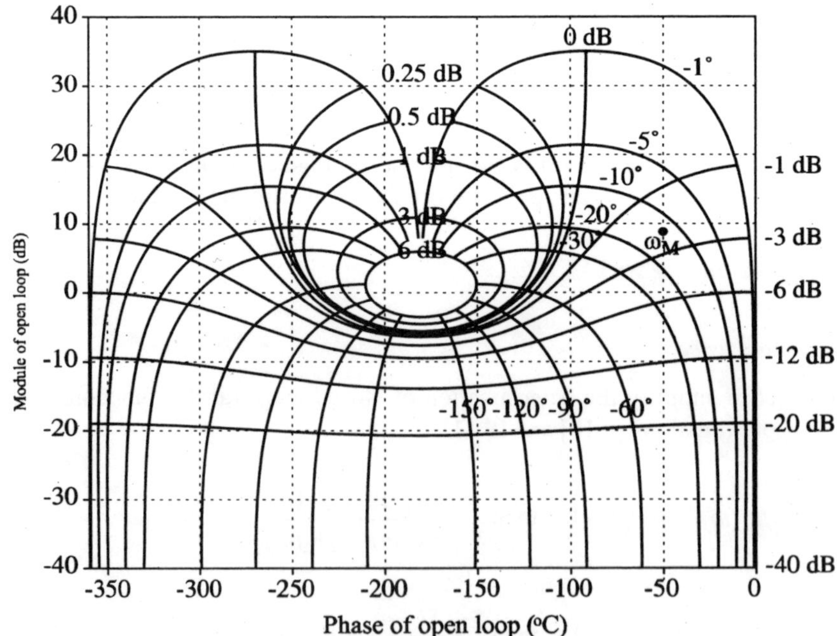

Figure 9.4. *Black-Nichols diagram*

When the feedback is not unitary, the transfer function can be re-written as a unitary feedback gain transfer function that is divided by the return gain β because:

$$F(p) = \frac{\mu}{1+\mu\beta} = \frac{\mu\beta}{1+\mu\beta}\frac{1}{\beta} \qquad [9.12]$$

9.1.2.2. *Estimation of closed loop time performances from the harmonic analysis*

The closed loop (CL) frequency response is characterized by the quality factor Q_r, also called magnification Q, i.e. the passage from the module through a maximum to an angular frequency ω_r called *resonance angular frequency*.

The time response is characterized, for a step function input, by the time of the first maximum t_m and the overflow D, as indicated in Figure 9.5a, i.e. $D = (s_{max} - s_\infty)/s_\infty$ where s_{max} represents the maximum value obtained from the output to instant t_m and s_∞ that obtained in permanent state. The overflow is expressed as a percentage.

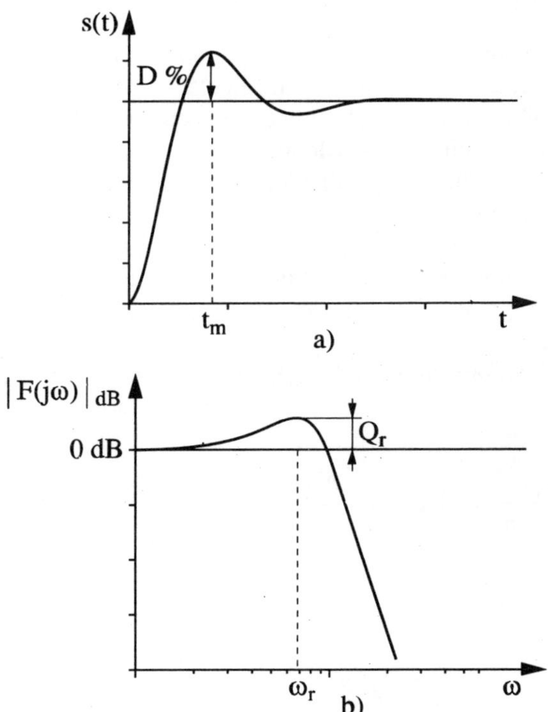

Figure 9.5. *(a) Time response in CL and (b) frequency response in CL*

When the system has good damping, let ξ be the value of the damping coefficient delimited between 0.4 and 0.7, we have the relation $\omega_c t_m \approx 3$, where ω_c represents the gap angular frequency. It is the angular frequency for which the open loop arithmetic gain is equal to the unit. For a well damped system, the quality factor Q_r has a value of less than 3 dB.

9.2. Stability

A looped system is called stable if its transfer function:

$$F(p) = \frac{\mu(p)}{1 + \mu(p)\beta(p)}$$

[9.13]

does not have poles of positive or zero real part.

In other words, the necessary and sufficient condition of stability of such a system is that $F(p)$ has all its poles with a negative real part.

When the denominator of $F(p)$ is a polynomial of order higher than 3 and does not reveal any obvious root, the analytical calculation of the roots may be fastidious. To study the stability, we then use either the geometrical criterion called Nyquist, where we reason only on the open loop in order to determine the stability of the closed loop or the so-called Routh algebraic criterion where we reason on the $F(p)$ specific equation without calculating its roots.

We can firstly show that the stability of a linear closed loop control system is connected to the diagrams of its open loop frequency response.

The transfer functions $\mu(p)$ and $\beta(p)$ are in general in the form of polynomials in p :

$$\mu(p) = \frac{N_1(p)}{D_1(p)} \text{ and } \beta(p) = \frac{N_2(p)}{D_2(p)}$$

[9.14]

then:

$$F(p) = \frac{N_1(p)D_2(p)}{D_1(p)D_2(p) + N_1(p)N_2(p)}$$

[9.15]

The system's characteristic equation is:

$$D_1(p)D_2(p) + N_1(p)N_2(p) = 0 \qquad [9.16]$$

or:

$$1 + \mu(p)\beta(p) = 0 \qquad [9.17]$$

The system is stable if equation [9.17] does not have zeros of positive or zero real part.

NOTE 9.1.– the methods presented in this chapter are valid when the open loop transfer function $\mu(p)\beta(p)$ does not result from a set of transfer functions presenting simplifications of the poles-positive real part zeros type.

9.2.1. *Nyquist criterion*

This criterion is based on the traditional property of analytical functions and it makes it possible to predict the behavior of a looped system by only knowing the open loop. To do this, we use the following Cauchy's theorem.

When a point M of affix p describes in the complex plane a closed contour C (Figure 9.6a), clockwise, surrounding P poles and Z zeros of a function $A(p)$ of the complex variable p, then the image of the point M through application A surrounds $N = Z - P$ times the origin in the same direction. We suppose that there is no singularity on C.

If we take, for example, $Z = 2$ and $P = 3$, then $N = -1$, the point M makes 1 tour around the origin, in counterclockwise direction (Figure 9.6b).

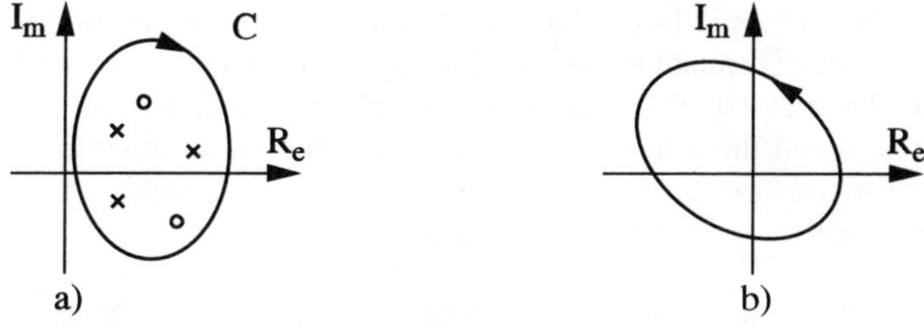

Figure 9.6. *Plane of the complex variable p and plane of $A(p)$*

The application to the Nyquist criterion leads to consider the transformation $A(p)$ as being the denominator of the transfer function of the looped system. We want this transfer function not to have poles of positive real part and hence its denominator:

$$A(p) = 1 + \mu(p)\beta(p) \qquad\qquad [9.18]$$

not to have positive real part zeros. We then choose as contour C a semicircle of infinite radius in the complex semi-plane on the right of the imaginary axis. C is called the Nyquist contour. The image of C through $A(p)$ transformation must thus surround the origin, in a counterclockwise direction, as many times as the number of unstable poles of equation [9.18] and hence of $\mu(p)\beta(p)$.

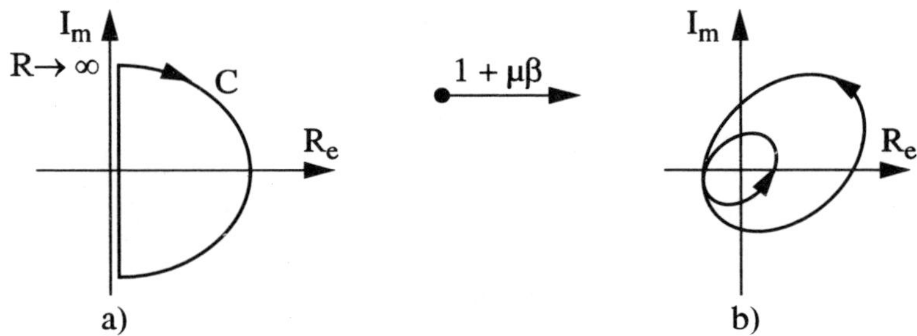

Figure 9.7. *Contour and image of the Nyquist curve*

Contour C is chosen in such as way as to surround the poles of possible zeros of $1 + \mu(p)\beta(p)$ with strictly positive real part. If C contains, for example, $Z = 1$ zero and $P = 3$ poles (Figure 9.7a), the Nyquist diagram will make $N = -2$ circuits around the origin in a clockwise direction and it will go around twice in a counterclockwise direction (Figure 9.7b).

To be stable in closed loop, it is necessary that $z = 0$, the image of C must then make $N = -P$ circuits around the origin. If the open loop transfer function $\mu(p)\beta(p)$ is stable, we have $P = 0$, the image of C through the transformation $1 + \mu(p)\beta(p)$ should not surround the origin. However, the number of circuits made around the origin in the transformation $1 + \mu(p)\beta(p)$ is equal to the number of circuits made around the critical point -1 in the transformation $\mu(p)\beta(p)$.

In what follows, we will deal only with this latter transformation and we will study the case of open loop stable systems, that of integrator systems and finally the case of open loop unstable systems.

Case 1. Open loop stable system: Nyquist diagram

Let us take the example:

$$\mu(p)\beta(p) = \frac{K}{(p+a)(p+b)(p+c)} \qquad [9.19]$$

where a, b and c are positive. Contour C and the Nyquist image corresponding curve are given by the graphs in Figure 9.8.

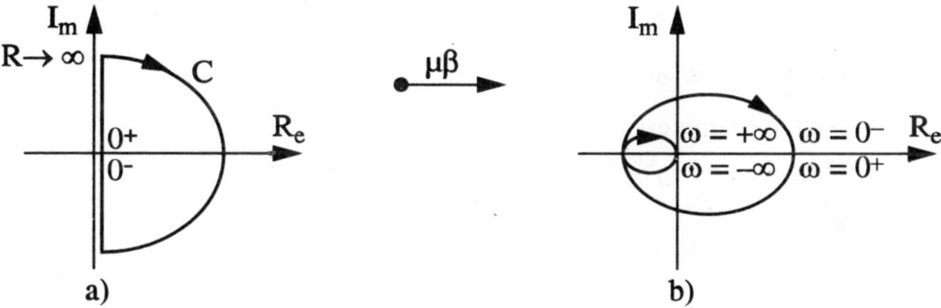

Figure 9.8. *Contour and image of Nyquist curve*

The image of the semicircle of infinitely high radius, through this transformation, is reduced to a point, which in general is the origin (the systems that can be physically created always have zero gains for infinite angular frequencies).

The image of the radius $]0^+, +\infty[$ is the curve in Figure 9.8b, corresponding to the trace of the open loop frequency response and described in the direction of increasing ω. The image of the ray $]-\infty, 0^-[$ is the symmetric curve of Nyquist place, with respect to the axis of real numbers. We can easily show this symmetry from the expression of the open loop transfer function. In fact, based on the definition:

$$\mu(p)\beta(p) = \int_0^\infty h(\theta)e^{-j\omega\theta}d\theta \qquad [9.20]$$

where $h(\theta)$ represents the impulse response of the open loop system, i.e.:

$$\mu(j\omega)\beta(j\omega) = \int_0^\infty h(\theta)\left[\cos(-\omega\theta) + j\sin(-\omega\theta)\right]d\theta \qquad [9.21]$$

we have:

$$\mu(-j\omega)\beta(-j\omega) = \int_0^\infty h(\theta)\left[\cos(\omega\theta) - j\sin(\omega\theta)\right]d\theta \qquad [9.22]$$

or:

$$\mu(-j\omega)\beta(-j\omega) = \overline{\mu(j\omega)\beta(j\omega)} \qquad [9.23]$$

Case 2. Open loop integrator system

Let us take the example:

$$\mu(p)\beta(p) = \frac{K}{p(p+a)} \qquad [9.24]$$

where a is positive.

Contour C is chosen in such a way as to exclude the origin as indicated by Figure 9.9a. This contour does not contain the unstable poles $(P = 0)$. Let C_ε be the circle of radius ε, $p = \varepsilon e^{j\theta}$ (Figure 9.9c).

The image of the ray $]0^+, +\infty[$ is the trace of the frequency response in open loop covered in the direction of increasing ω. When $\omega \to 0^+$, the gain of $\mu(p)\beta(p) \to \infty$ and the phase is of $-\pi/2$. When $\omega \to +\infty$, the gain $\to 0$ and the phase is of $-\pi$. The image of the ray $]-\infty, 0^-[$ is the symmetric curve, with respect to the axis of real numbers. The image of the semicircle of radius ε is an arc of circle of radius ∞. When $p \to 0$, $\mu(p)\beta(p) \to K/p = K/\varepsilon e^{j\theta}$. If ω increases from 0^- to 0^+, θ varies from $-\pi/2$ to $+\pi/2$ and the open loop gain is infinite.

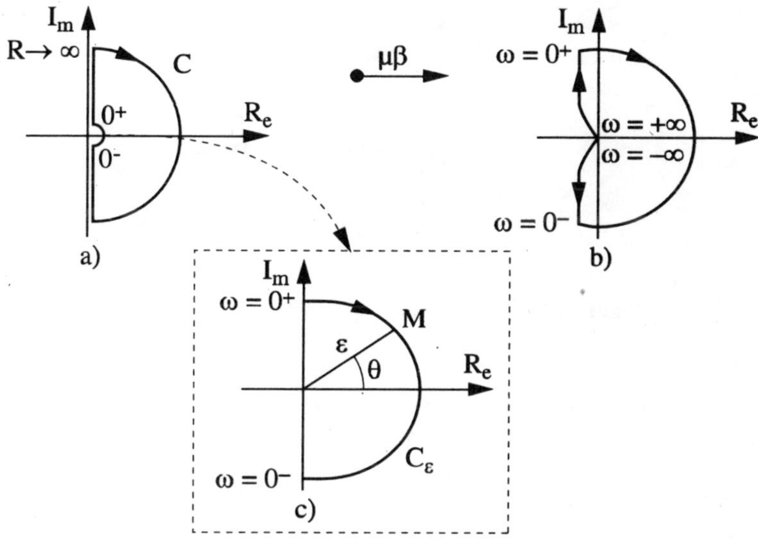

Figure 9.9. *The choice of the contour of C excludes the origin (a), the transform of the contour C (b), C_ε circle of radius ε, $p = \varepsilon e^{j\theta}$ (c)*

The image of the semicircle of an infinitely big radius is the origin of the complex plane. The transform of contour C, representing the complete Nyquist place, is represented by Figure 9.9b.

NOTE 9.2. – when the open loop transfer function $\mu(p)\beta(p)$ has terms of the form K/p^k with $k > 1$, θ varies from $-k\pi/2$ to $+k\pi/2$ and $\mu(p)\beta(p)$ describes infinite k semicircles in a clockwise direction.

Statement of Nyquist criterion

A looped system is stable (bounded input-bounded output) if and only if the Nyquist place of its open loop transfer function $\mu(p)\beta(p)$, which is described in the direction of increasing angular frequencies, does not go through the critical point − 1 and makes around it a number of circuits in the counterclockwise direction equal to the number of unstable poles of $\mu(p)\beta(p)$.

Case 3. Unstable system in open loop

EXAMPLE 9.1.– the Nyquist trace is given in Figure 9.10.

$$\mu(p)\beta(p) = \frac{K}{p(1-2p)} \tag{9.25}$$

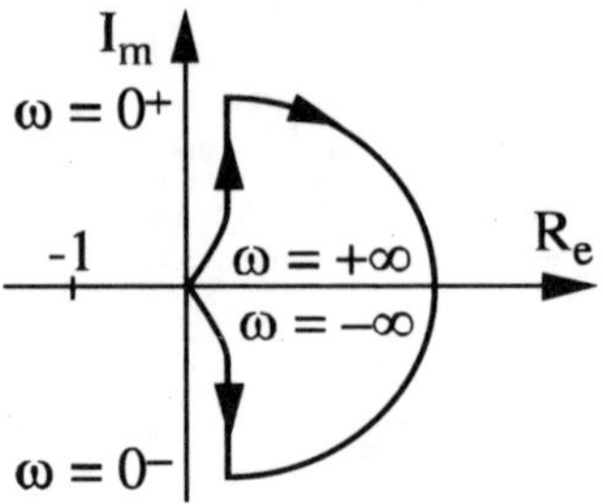

Figure 9.10. *Nyquist trace, open loop unstable system*

We have $P = 1$, this system, which is unstable in open loop, is equally unstable in closed loop because the Nyquist place does not surround once the critical point -1.

EXAMPLE 9.2.

$$\mu(p)\beta(p) = \frac{K(1+p)}{p^2(1+0.1p)}$$

[9.26]

The Nyquist trace is given in Figure 9.11a.

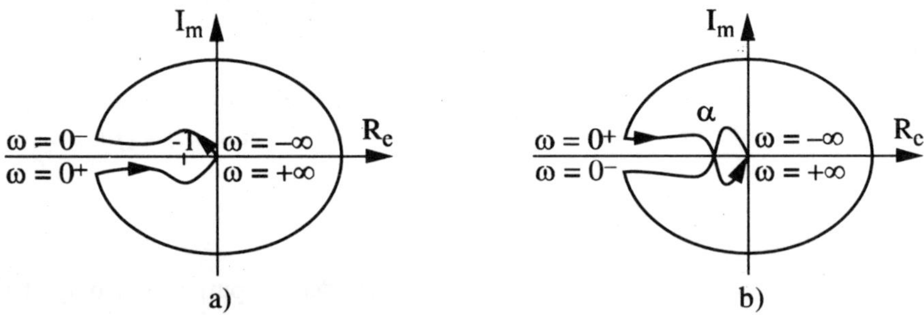

Figure 9.11. *Nyquist trace, open loop stable system (a), closed loop stable system (b)*

We have $P = 0$; this system, which is stable in open loop, is equally stable in closed loop because the Nyquist place does not surround the critical point -1.

EXAMPLE 9.3.

$$\mu(p)\beta(p) = \frac{K(1+p)(1+0.2p)}{p^2(1+5p)(1+0.1p)} \qquad [9.27]$$

The Nyquist trace is given in Figure 9.11b. We have $P = 0$, the stability of the closed loop depends on the value of K. Let α be the meeting point of the curve with the axis $-\pi$ (as indicated in the figure). If $-1 < \alpha < 0$, the looped system is stable because Nyquist place does not surround the point -1, if $\alpha < -1$, the looped system is unstable.

Simplified Nyquist criterion: reverse criterion

A simplified criterion can be deduced from the previous criterion.

A system, *stable in open loop*, is stable in closed loop if, covering the Nyquist place of the open loop in the direction of the increasing ω, leaves the critical point on the left (Figure 9.12a). If it leaves the critical point on the right, it is unstable (Figure 9.12b). If the gain curve $\mu\beta(j\omega)$ goes through the critical point, the system is oscillating (Figure 9.12c).

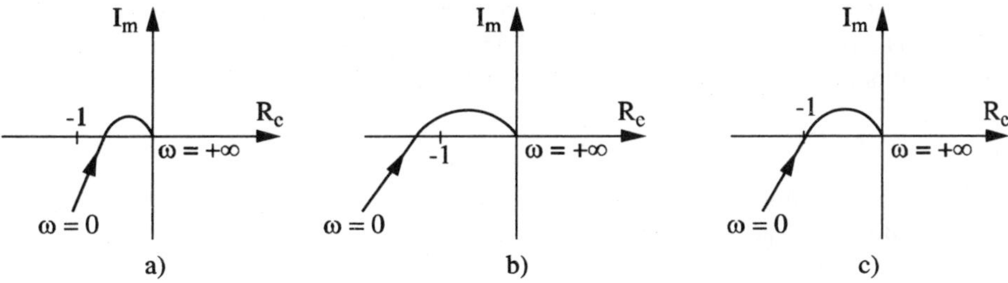

Figure 9.12. *Stable system (a), unstable system (b), oscillating system (c)*

9.2.2. *Routh's algebraic criterion*

This criterion formulates a necessary and sufficient condition so that any n degree polynomial has all its roots of strictly negative real part.

We re-write the characteristic equation [9.17] in the polynomial form:

$$1 + \mu(p)\beta(p) = a_n p^n + a_{n-1} p^{n-1} + \ldots + a_1 p + a_0 \qquad [9.28]$$

Then, we create Table 9.1 (with n + 2 rows).

1	a_n	a_{n-2}	a_{n-4}	.....
2	a_{n-1}	a_{n-3}	a_{n-5}	.....
3	$b_1=a_{n-2}-a_n\,a_{n-3}/a_{n-1}$	$b_2=a_{n-4}-a_n\,a_{n-5}/a_{n-1}$	$b_3=a_{n-6}-a_n\,a_{n-7}/a_{n-1}$	.....
4	$c_1=a_{n-3}-b_2\,a_{n-1}/b_1$	$c_2=a_{n-5}-b_3\,a_{n-1}/b_1$	$c_3=a_{n-7}-b_4\,a_{n-1}/b_1$	.....
.....	....	.....	.....	....
$n+1$	.....	0	.....	.....
$n+2$	0	0	0	0

Table 9.1. *Routh's table*

For a regular system, the number of non-zero terms decreases with 1 every 2 rows and we stop as soon as we obtain a row consisting only of zeros. The first column of Routh's table has $n + 1$ non-zero elements for a characteristic equation of n order. The roots of this equation are of strictly negative real part if and only if the terms of this first column of the table have the same sign and are not zero.

Statement of Routh's criterion

A system is stable in closed loop if and only if the elements of the first column of Routh's table have the same sign.

EXAMPLE 9.4.– let us consider again example 9.1 :

$$\mu(p)\beta(p) = \frac{K}{p(1-2p)} \tag{9.29}$$

The characteristic equation is $p(1-2p)+K = 0$ or $-2p^2 + p + K = 0$. The first two coefficients (which are the first two elements of the first column of Routh's table) are of opposite signs, the looped system is unstable.

Let us now take example 9.2:

$$\mu(p)\beta(p) = \frac{K(1+p)}{p^2(1+0.1p)}$$

[9.30]

The characteristic equation is $p^2(1+0.1p)+K(1+p)=0$ or by developing $0{,}1p^3 + p^2 + Kp + K = 0$.

Routh's table gives:

0.1	K
1	K
K-0.1K	0
K	0

The system is stable in closed loop if $K > 0$.

NOTE 9.3. – a necessary condition to have negative real part roots is that all a_i coefficients have the same sign.

9.2.3. *Stability margins*

The physical systems are represented by mathematical models which are generally not very exact. The stability of the mathematical model does not necessarily entail that of the physical system. Consequently, in order to take into account the uncertainties of the model, security margins must be defined during the theoretical study in order to ensure a satisfactory behavior to the looped system, especially when the Nyquist place of the harmonic response in open loop is near the critical point.

a) Phase margin – gain margin

The phase margin is obtained by calculating the difference between the system phase considered and $-180°$ to the gap angular frequency. The gain margin is obtained by calculating the difference between the system gain and 0 dB to the angular frequency where the phase reaches $-180°$. These margins, noted by $\Delta\phi$ and ΔG, are represented in Bode, Nyquist and Black-Nichols planes in Figure 9.13.

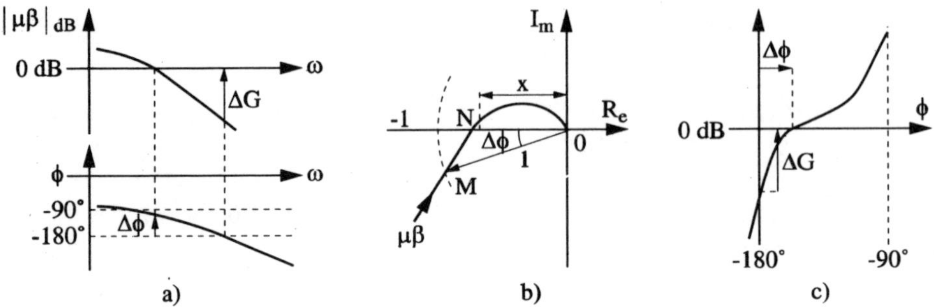

Figure 9.13. *Bode plane (a), Nyquist plane (b), Black-Nichols plane (c)*

In Bode and Black-Nichols planes:

$$\Delta\phi = \phi - (-180°) \text{ to the angular frequency } \omega_c \text{ for which } |\mu\beta| = 0 \text{ dB}$$

$$\Delta G = 0\text{dB} - |\mu\beta|_{dB} \text{ to the angular frequency } \omega = \omega_\pi \text{ for which } \phi = -180°$$

In Nyquist plane:

$$\Delta\phi \text{ is represented by the angle } (\vec{ON}, \vec{OM}) \text{ where } OM = |\mu\beta| = 1 \text{ (0dB)}$$

$$\Delta G = -20\log x \text{ where } x = |\mu\beta| \text{ to the angular frequency } \omega = \omega_\pi \text{ for which}$$
$$\phi = -180°$$

When a system is at a minimal phase difference, i.e. when all its zeros are of negative real part and of low-pass type, the only consideration of the phase margin is enough in general to ensure a convenient damping.

b) Delay margin – module margin

The delay margin, for a stable system with a phase margin $\Delta\phi$ is defined as being the ratio:

$$\tau_m = \frac{\Delta\phi}{\omega_c} \qquad\qquad [9.31]$$

where ω_c is the gap angular frequency. The delay margin represents the maximum allowed delay value leading exactly to the cancellation of the phase margin.

The module margin represents the shortest geometrical distance between curve $\mu\beta$ and point -1.

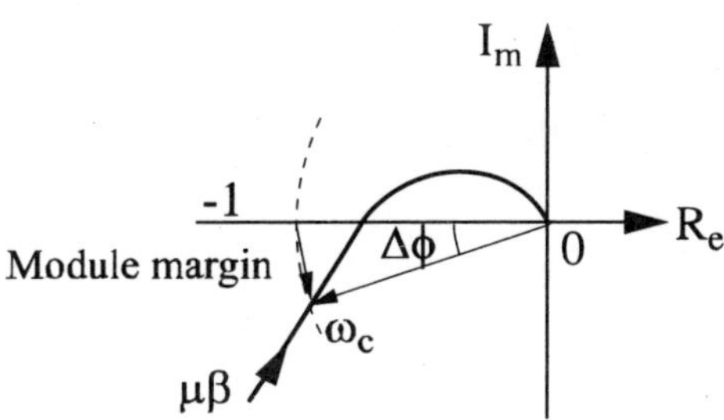

Figure 9.14. *Nyquist plane*

c) Degree of stability of a second order system

Let us consider a second degree system whose closed loop transfer function, with a unitary feedback, has the form:

$$F(p) = \frac{\omega_0^2}{p^2 + 2\xi\omega_0 p + \omega_0^2} \qquad [9.32]$$

where ξ represents the damping coefficient and ω_0 is the system's angular frequency. If ξ is low, the unit-step response is oscillating, if ξ is high, the response is strongly damped and the transient state is long.

The magnification Q_r, i.e. the maximal gain of the closed loop module curve, which can be measured directly in the Black-Nichols plane (Figure 9.15), is related to the damping coefficient by the relation:

$$Q_r = \frac{1}{2\xi\sqrt{1-\xi^2}} \qquad [9.33]$$

We note that when ξ tends towards zero, Q_r tends towards infinity. The *resonance angular frequency* is given by the relation:

$$\omega_r = \omega_0 \sqrt{1-\xi^2}$$

[9.34]

For a conveniently damped system, Q_r is less than 3 dB or a damping coefficient lower than 0.7. We can verify, with the help of a Black-Nichols diagram, that in order to have $Q_r \leq 3\text{dB}$, we need a phase margin $\Delta\phi \geq 45^0$ and a gain margin $\Delta G \geq 5\text{dB}$.

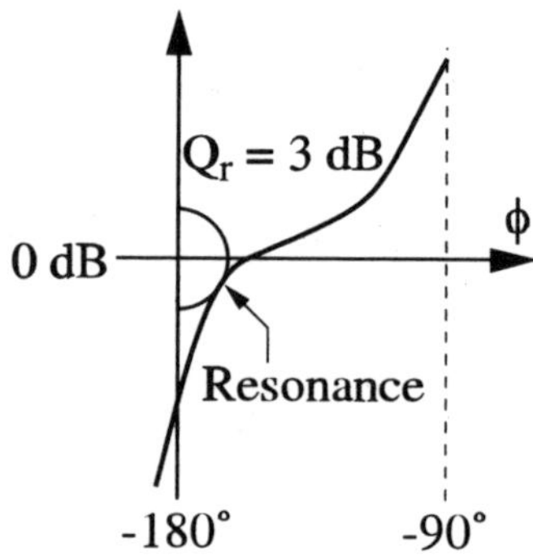

Figure 9.15. *Black-Nichols plane*

d) Degree of stability of any order system

When a system has any order transfer function, there will not be an explicit damping coefficient like in the case of second order system. However, the responses, with an equal resonance coefficient, have almost the same form. Hence, we can measure the degree of stability of any order system by its resonance factor in the Black-Nichols plane.

9.3. Precision

In section 9.1.1 we saw the general block diagram of a closed loop control system (Figure 9.2). The role of such a system is to follow the output $s(t)$ with the input $e(t)$. Signal $e(t)$ injected at the input of the system is ideally the signal that we would want to obtain at the output. In the case of a perfect feedback control, the difference $\varepsilon(t) = e(t) - r(t)$ is then zero at every instant. In the case of a real feedback control, this difference is never zero due to the time constants of physical systems, on the one hand and to the interferences acting on the system on the other hand.

The quality of the feedback control is mainly translated through its stability and the follow-up precision of the output on the input and consequently of the dynamics of this difference. The precision study makes sense only if the system is stable.

In practice, it is interesting to know the permanent error, also called static, ε_s which is the asymptotic value of the instantaneous error ε_d, called dynamic.

The specifications on the error are most often formulated in one of the following forms:

– bounded or zero static error in response to the inputs or canonical interferences (step function, ramp, harmonic signal);

– bounded dynamic error $\varepsilon_d < \varepsilon_{\max}$ for the inputs having certain given characteristics.

In the presence of an interference, due to the overlapping principle, we have:

$$\mathcal{E}(p) = \mathcal{E}_e(p) + \mathcal{E}_b(p) \tag{9.35}$$

where:

$$\mathcal{E}_e(p) = \frac{1}{1+\mu_1\mu_2\beta} E(p) \tag{9.36}$$

and:

$$\mathcal{E}_b(p) = \frac{-\mu_2\beta}{1+\mu_1\mu_2\beta} B(p) \tag{9.37}$$

These expressions are calculated by considering firstly the zero interferences $b(t) = 0$ ($e(t) \neq 0$; see Figure 9.16a) and then the zero input $e(t) = 0$ ($b(t) \neq 0$, see Figure 9.16b).

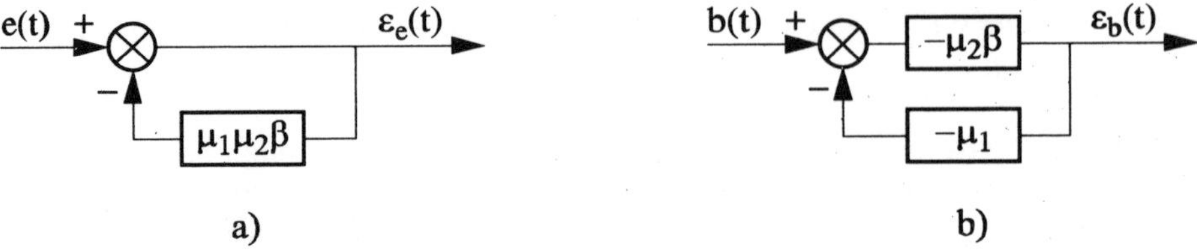

Figure 9.16. *Zero interferences b(t) = 0, e(t) ≠ 0 (a), zero input e(t) =0, b(t) ≠ 0 (b)*

9.3.1. *Permanent error*

9.3.1.1. *Step function input – zero interference*

When the input signal is a step function $E(p)$ of amplitude E_0, $E(p) = E_0 / p$. Equation [9.36] then becomes:

$$\varepsilon_e(p) = \frac{E_0}{p(1 + \mu_1\mu_2\beta)} \tag{9.38}$$

The theorem of the final value makes it possible to write:

$$\lim_{t\to\infty} \varepsilon_e(t) = \lim_{p\to 0} p\,\varepsilon_e(p) = \lim_{p\to 0} \frac{E_0}{(1 + \mu_1\mu_2\beta)} \tag{9.39}$$

This error is zero if $1 + \mu_1\mu_2\beta$ has at least one pole at the origin, i.e. if $\mu_1\mu_2\beta$ has at least one integration.

In general, $\lim_{p\to 0} \beta(p)$ is finite, hence the requirement for $\mu_1\mu_2$ to have one or more integrations.

Let us note by k the number of pure integrators; we say that the system is of class k. A system of class at least equal to 1 in open loop has a zero static error in response to an input step function.

If the system in open loop is of class 0, the static error is equal to:

$$\lim_{t\to\infty} \varepsilon_e(t) = \frac{E_0}{1 + K} \tag{9.40}$$

We notice that the error is inversely proportional to the static gain.

9.3.1.2. *Unit-step interference – zero input*

When the interference signal is a step function $B(p)$ of amplitude B_0, $B(p) = B_0 / p$. Equation [9.37] thus becomes:

$$\varepsilon_b(p) = \frac{-B_0 \mu_2 \beta}{p(1 + \mu_1 \mu_2 \beta)} \tag{9.41}$$

The theorem of the final value makes it possible to write:

$$\lim_{t \to \infty} \varepsilon_b(t) = \lim_{p \to 0} p\, \varepsilon_b(p) = B_0 \lim_{p \to 0} \frac{-\mu_2 \beta}{(1 + \mu_1 \mu_2 \beta)} \tag{9.42}$$

the following propositions will be verified:

$$\lim_{t \to \infty} \varepsilon_b(t) = 0 \Leftrightarrow \begin{cases} \lim_{p \to 0} \mu_2 \beta = 0 \\ \text{or} \\ \mu_1(p) \approx \dfrac{K}{p^k} \quad \text{when} \quad p \to 0 \quad (k \geq 1) \end{cases} \tag{9.43}$$

In general, μ_2 and β are not differentiators. The first condition is thus rarely satisfied.

A system having in open loop at least one integration *upstream from the application point of the interference* (hence in μ_1) presents a zero static error in response to an interference step function. However, if μ_1 does not have any integration, it is enough that μ_1 is an integrator system to cancel the static errors ε_e and ε_b at the same time. If μ_2 has an integration, we can introduce a supplementary integration in μ_1 in order to cancel $\varepsilon_b(\infty)$. Then we will have a second class system in open loop.

9.3.1.3. *Ramp input – zero interference*

Let us calculate the final value of the error for a ramp input $E(p) = V_0 / p^2$:

$$\varepsilon_e(p) = \frac{1}{1 + \mu_1 \mu_2 \beta} E(p) \qquad [9.44]$$

we have:

$$\lim_{t \to \infty} \varepsilon_e(t) = \lim_{p \to 0} p\ \varepsilon_e(p) = V_0 \lim_{p \to 0} \frac{1}{p\left(1 + \mu_1 \mu_2 \beta\right)} \qquad [9.45]$$

$$\lim_{t \to \infty} \varepsilon_e(t) = 0 \Leftrightarrow \mu_1 \mu_2 \beta \approx \frac{K}{p^k} \text{ when } p \to 0 \text{ with } k \geq 2 \qquad [9.46]$$

In general, $\mu_1 \mu_2$ must have at least two integrations in order to cancel the permanent error in response to an input ramp.

If the system has only one integration in direct chain, there is a finite final error:

$$\mu_1 \mu_2 \beta \approx \frac{K}{p} \text{ when } p \to 0 \ \Rightarrow\ \lim_{t \to \infty} \varepsilon_e(t) = \frac{V_0}{K} \qquad [9.47]$$

Figure 9.17 sums up the permanent errors for a step function input, (Figure 9.17a), and for a ramp input (Figure 9.17c) with a zero interference in the cases when $\mu_1 \approx K_1$ and $\mu_2 \approx K_2$ near $p = 0$. In the cases when $\mu_1 \approx K_1$ and $\mu_2 \approx K_2 / p$ near $p = 0$, Figure 9.17b shows the absence of static error but the presence of a trail error (Figure 9.17d) again with zero interference. We notice an error for a interference in step function when $\mu_1 \approx K_1$ and $\mu_2 \approx K_2$ (Figure 9.18b) whereas there is absence of error as soon as $\mu_1 \approx K_1 / p$ (Figure 9.18b).

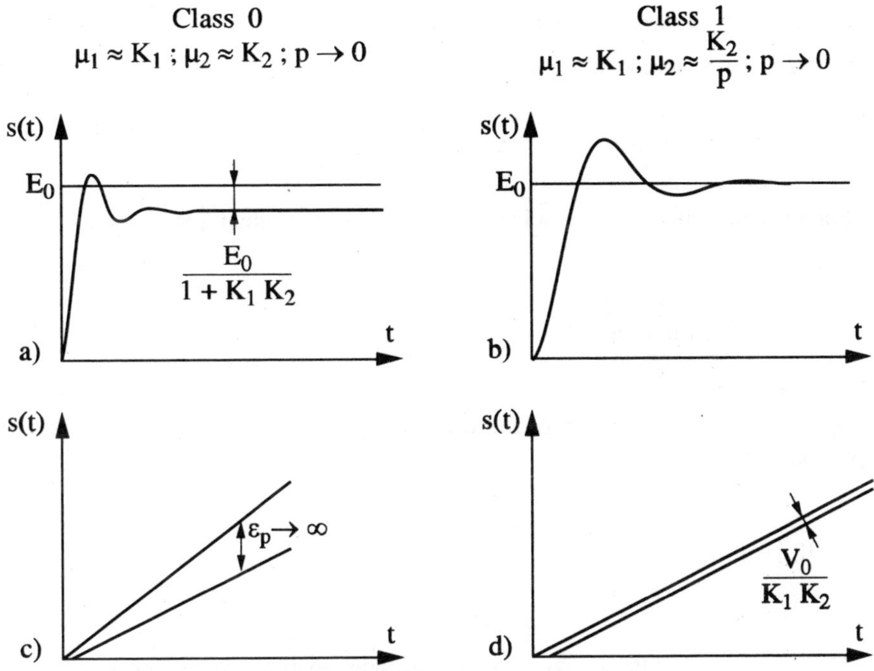

Figure 9.17. *The image of responses to a step function input and a ramp input (zero interference)*

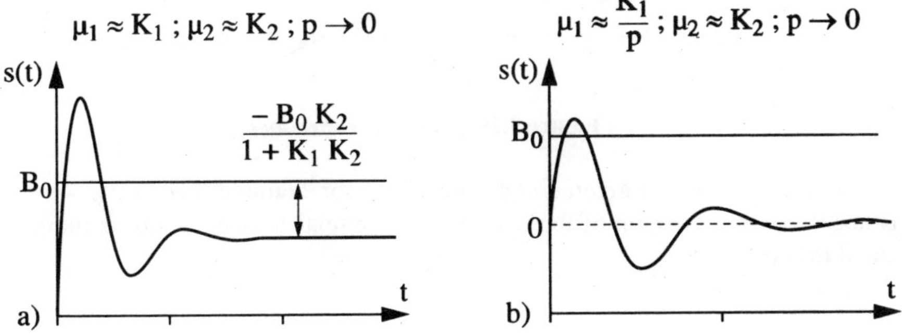

Figure 9.18. *The image of responses to an interference step function (zero input)*

9.3.1.4. *Sinusoidal input – zero interference*

Let us apply a sinusoidal input to the system:

$$e(t) = E_0 \sin \omega_0 t \qquad\qquad [9.48]$$

For a permanent state, ε_e is a harmonic signal of module $|\varepsilon_e|$ such that:

$$|\varepsilon_e| = E_0 \left| \frac{1}{1+\mu_1\mu_2\beta} \right|_{p=j\omega_0} \qquad\qquad [9.49]$$

If ω_0 is in the bandwidth of the open loop, $\left|\mu_1\mu_2\beta\right|_{p=j\omega_0} \gg 1$ we obtain the approximation:

$$|\varepsilon_e| = E_0 \left| \frac{1}{\mu_1\mu_2\beta} \right|_{p=j\omega_0} \qquad\qquad [9.50]$$

The error amplitude is inversely proportional to the gain of the open loop at angular frequency ω_0.

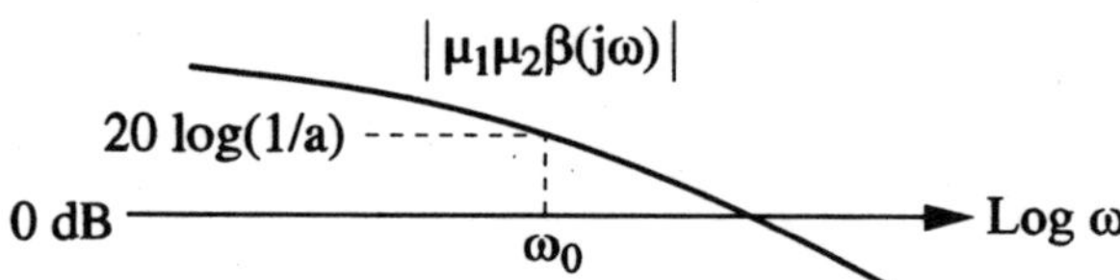

Figure 9.19. *Gain of the open loop*

In order to obtain a given a copy precision (for example, 1% to $\omega_0 = 1rd/s$), it is necessary that the gain of the open loop at angular frequency ω_0 is more than or equal to $1/a$.

9.3.2. *Transitional error*

9.3.2.1. *Case of a unitary feedback – zero interference*

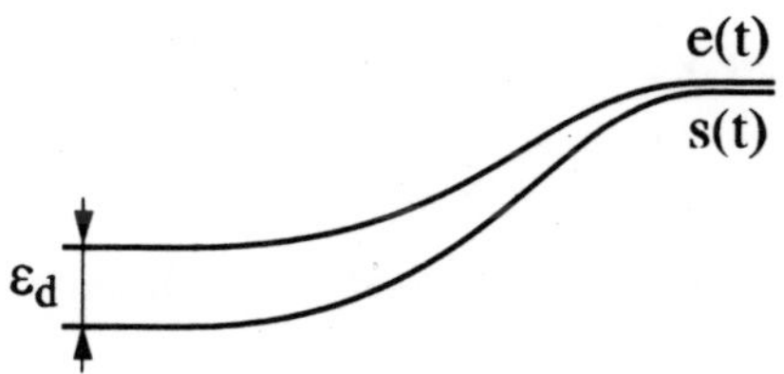

Figure 9.20. *Unitary feedback*

We seek to limit ε_d to a value $\varepsilon_{d\,\max}$.

For any input signal, the expression of the dynamic error is complex. However, if there is a limitation of the input signals at speed v and for acceleration γ, it is possible to obtain a condition characterizing the precision required.

The input signal is defined by the constraints:

$$v \leq v_M \text{ and } \gamma \leq \gamma_M \tag{9.51}$$

Among all the signals satisfying these conditions, the sine wave signal is by far the simplest. We will study this by replacing the signal by this "equivalent" sine wave signal.

Let $e(t) = E_0 \sin \omega_0 t$ be this signal. Its maximum speed and acceleration are given by:

$$v_M = E_0 \omega_0 \text{ and } \gamma_M = E_0 \omega_0^2 \tag{9.52}$$

These two equations define a sine wave signal which must be copied with an error less than $\varepsilon_{d\,\max}$. From these equations, we obtain E_0 and ω_0 :

$$E_0 = \frac{v_M^2}{\gamma_M} \text{ and } \omega_0 = \frac{\gamma_M}{v_M} \tag{9.53}$$

hence the condition:

$$\frac{1}{\left|\mu_1\mu_2\beta\right|_{p=j\frac{\gamma_M}{v_M}}} \leq \frac{\varepsilon_{d\,\max}}{\dfrac{v_M^2}{\gamma_M}}$$

[9.54]

We note that this condition is necessary and, in certain cases, it may not be sufficient.

9.4. Parametric sensitivity

Multiple factors, such as ageing or change of working points may lead to variations on the representative model parameters of the systems. It can be interesting to examine the influence of the variation of these parameters on the overall behavior of the system (stability or precision, for example).

A system is called sensitive to the variations of a parameter α if its behavior is affected by these variations. If $G(p)$ represents the transfer function of the system, the sensitivity with respect to parameter α is defined as being the ratio:

$$S_\alpha = \frac{\partial G(p)/G(p)}{\partial \alpha/\alpha}$$

[9.55]

9.4.1. *Open loop sensitivity*

Let the transfer function of an open loop system be:

$$G(p) = \frac{N(p,\alpha)}{D(p,\alpha)}$$

[9.56]

Sensitivity is thus written:

$$S_\alpha = \left(\frac{\partial N(p)}{\partial \alpha}\frac{1}{N} - \frac{\partial D(p)}{\partial \alpha}\frac{1}{D}\right)\alpha$$

[9.57]

EXAMPLE 9.5.– let us consider a first order system: $\dfrac{K}{1+Tp}$ where $K = \alpha K_0$. The sensitivity around gain K_0 is:

$$S_{\alpha_{K_0}} = \left(\frac{K_0}{\alpha K_0} - \frac{0}{1+Tp} \right) \alpha = 1 \qquad [9.58]$$

Sensitivity is 1 and any gain variation will have consequences on the output.

Let us suppose now that $T = \alpha T_0$ and therefore the sensitivity around the time constant T_0 is:

$$S_{\alpha_{T_0}} = \left(\frac{0}{K} - \frac{T_0 p}{1+\alpha T_0 p} \right) \alpha \qquad [9.59]$$

$$S_{\alpha_{T_0}} = - \frac{\alpha T_0 p}{1+\alpha T_0 p} \qquad [9.60]$$

We notice that in this case sensitivity is a function of frequency.

The responses to a step function for $K = 5$ and $K = 15$ to the nominal response corresponding to $K_0 = 10$ (Figure 9.21a) show that the system is integrally affected during the gain variations.

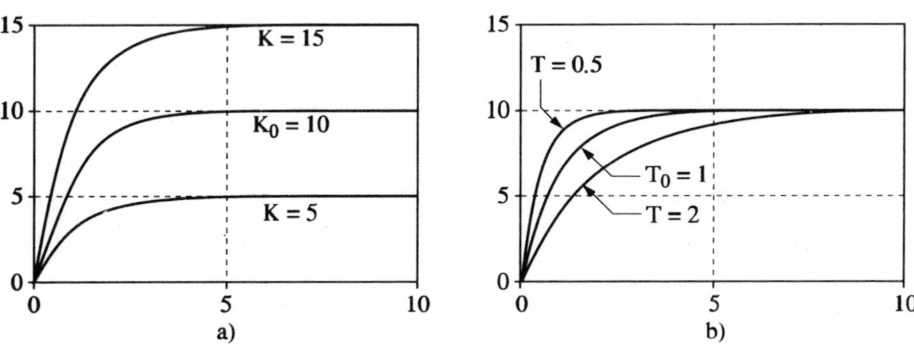

Figure 9.21. *Responses to a step function for k = 5 and k = 15 (a), for T = 0.5 and T = 2 (b)*

Similarly, we can visualize the responses to a step function for $T = 0.5$ and $T = 2$ for comparison to the nominal response corresponding to $T_0 = 1$ (Figure

9.21b). The sensitivity tends towards zero when $p \to 0$ depending on equation [9.60] and hence in permanent state the responses tend towards the value of the static gain K.

9.4.2. *Closed loop sensitivity*

Let the looped system with unitary feedback be:

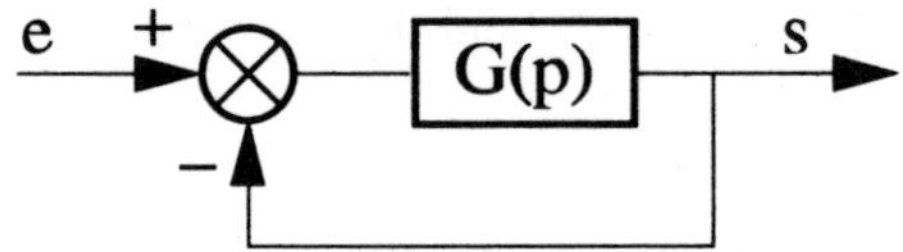

Figure 9.22. *Looped system*

The closed loop transfer function is:

$$G_{BF} = \frac{G(p)}{1+G(p)}$$

[9.61]

The sensitivity for a low variation of $G(p)$ is obtained by:

$$S_{\alpha BF} = \frac{\partial G_{BF}(p)}{\partial \alpha} \frac{\alpha}{G_{BF}}$$

[9.62]

or:

$$S_{\alpha BF} = \frac{\dfrac{\partial G(p)}{\partial \alpha}(1+G(p)) - G(p)\dfrac{\partial G(p)}{\partial \alpha}}{(1+G(p))^2} \frac{\alpha}{G(p)}(1+G(p))$$

[9.63]

$$S_{\alpha BF} = \frac{\partial G(p)}{\partial \alpha} \frac{\alpha}{G(p)(1+G(p))}$$

[9.64]

$$S_{\alpha BF} = S_\alpha \frac{1}{(1+G(p))}$$

[9.65]

This equality shows that the bigger $1 + G(p)$ is, the smaller the sensitivity of the looped system becomes.

EXAMPLE 9.6.– let us return to the system $G(p) = \dfrac{K}{1+Tp}$ previously considered and relooped with a unitary gain (see Figure 9.23).

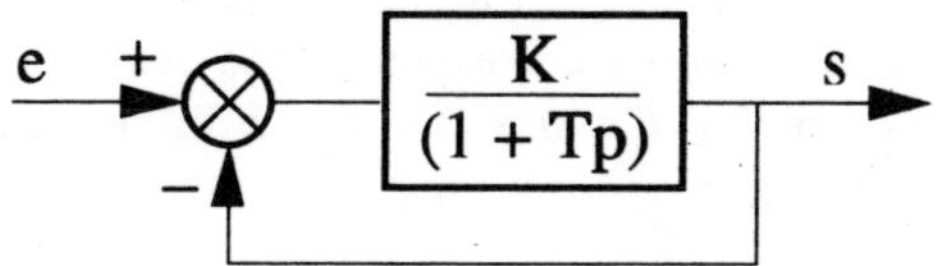

Figure 9.23. *Relooped system with unitary gain*

Let us take $K = \alpha K_0$. The closed loop transfer function is:

$$G_{BF}(p) = \frac{\alpha K_0}{1 + \alpha K_0 + Tp} \qquad [9.66]$$

which has the form:

$$G_{BF}(p) = \frac{K'}{1 + T'p} \qquad [9.67]$$

where $K' = \dfrac{\alpha K_0}{1 + \alpha K_0}$ and $T' = \dfrac{T}{1 + \alpha K_0}$.

We imagine that the sensitivities of the closed loop diagram are less than those in open loop. This is verified on the behavior of the unit-step response below. In fact, the sensitivity around gain K_0, with respect to the parameter α is:

$$S_{\alpha BF} = \frac{(1 + Tp)}{1 + \alpha K_0 + Tp} \qquad [9.68]$$

Let us suppose now that $T = \alpha T_0$ and therefore the sensitivity around the time constant T_0 is:

$$S_{\alpha BF_{T_0}} = \left(-\frac{\alpha T_0\, p}{1 + \alpha T_0\, p + K}\right)$$

[9.69]

Figure 9.24 shows the responses to a step function for $K = 5$ and $K = 15$ in comparison to the nominal response corresponding to $K_0 = 10$. Figure 9.24b shows those of the looped system for $T = 0.5$ and $T = 2$ in comparison to the nominal response corresponding to $T_0 = 1$.

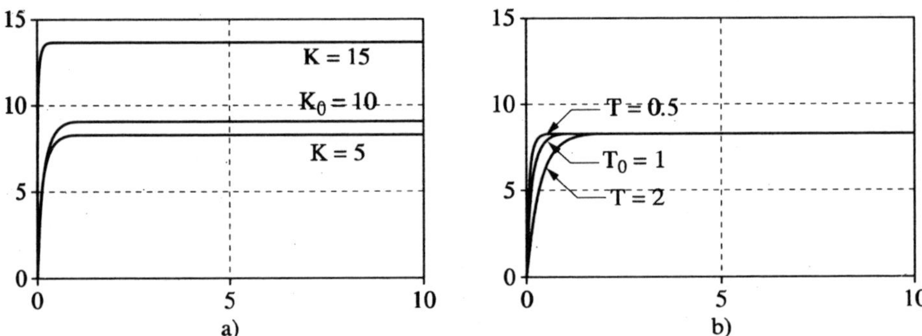

Figure 9.24. *Responses to a step function*

9.5. Bibliography

[CAR 69] DE CARFORT F., FOULARD C., *Asservissements linéaires continus*, Dunod, Paris, 1969.

[FAU 84] FAURE P., ROBIN M., *Eléments d'Automatique*, Dunod, Paris, 1984.

[GIL 70] GILLES J.C., DECAULNE P., PELEGRIN M., *Théorie des systèmes asservis*, Dunod, Paris, 1970.

[LAR 96] DE LARMINAT P., *Automatique*, 2nd edition, Hermès, Paris, 1996.

Chapter 10

Synthesis of Closed Loop Control Systems

10.1. Role of correctors: precision-stability dilemma

The correction methods covered in this chapter refer to considerations of scalar behavior. It is fundamental to understand that the specifications stipulating the *closed loop* performances will be translated by the constraints on the frequency response of the *open loop* corrected system.

The search for a compromise between stability and rapidity generally leads to imposing, on the closed loop, a behavior similar to that of a second order system having conjugated complex number poles. The choice of the damping value ξ is imposed by the required degree of stability. That is why it is indispensable to have a good knowledge of the relations between the parameters and the behavior of the second order systems in order to be able to use specifications defining the performances required from the closed loop final system. The general principle of a specification list is based on two points:

– interpretation of specifications in order to obtain the characteristics of a second order model of the closed loop needed;

– search for the constraints on the open loop introducing the behavior sought in closed loop.

For reasons of clarity, let us recall the main results of the previous chapter concerning the analysis of the behavior of systems.

Chapter written by Houria SIGUERDIDJANE and Martial DEMERLÉ.

10.1.1. *Analysis of systems' behavior*

10.1.1.1. *Static errors*

For a zero static error:

– with respect to a set point step function, there has to be at least *one* integration in the open loop;

– with respect to a set point ramp, there have to be at least *two* integrations in the open loop;

– with respect to an interference step function, there has to be *one* integration *upstream* from the input point of the interference in the open loop.

10.1.1.2. *Stability*

The analysis of stability of the looped system, in the current case of a stable system in open loop, is based on the simplified Nyquist theorem: the image of the contour by $\mu\beta$ must not surround the point -1. The distance with respect to point -1, which is expressed in terms of phase margin and gain margin provides a "measurement" of stability.

We can associate a "visual" criterion of the stability measurement with the help of the closed loop response overflow with a set point step function. For a system whose closed loop transfer function is of second order, we know how to connect the concept of unit-step response overflow to the damping coefficient ξ of the poles and consequently to the phase margin of the *open loop* (OL). We note that for a second order system, a sufficient phase margin implies a good gain margin and the only concept of phase margin is thus sufficient.

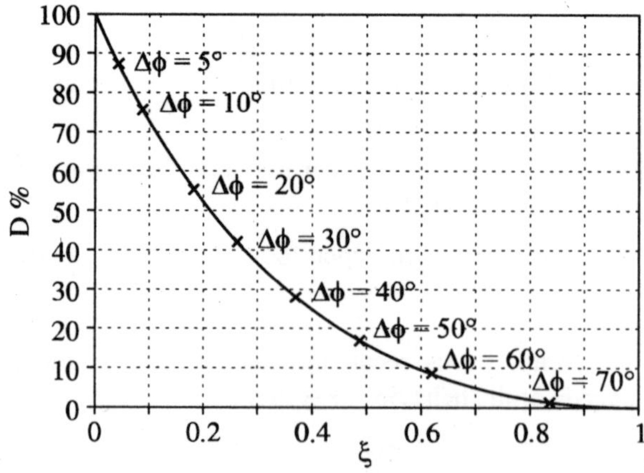

Figure 10.1. *Overflow curve based on damping*

Another way to characterize a stability measurement consists of analyzing the frequency response of the closed loop. The instability is characterized by the presence of a resonance peak.

For a system whose closed loop transfer function is of second order, we connect this concept of frequency response resonance of the *closed loop* to the damping coefficient ξ of the poles and hence to the phase margin of the *open loop*.

ξ	$\Delta\Phi$ in degrees	Resonance in dB
0.1	12	12
0.2	22	8
0.4	43	2.7
0.6	58	0.35

Table 10.1. *Damping, phase and resonance margin*

There is no more resonance from $\xi = 0.7$.

10.1.1.3. *Rapidity*

The character of rapidity can be perceived in two ways, either directly by observing a time or frequency response, or indirectly through the concept of dynamic error (a rapid system enables the pursuit of an input that rapidly varies and hence a low dynamic error; see the method of the equivalent sine curve described in the previous chapter).

Closed loop frequency behavior, bandwidth

To evaluate the rapidity of the looped system, a sinusoidal input is used whose frequency can be chosen. More often than not, the transfer function of feedback β is a constant and we can then consider two cases for the behavior in closed loop $\dfrac{\mu}{1+\mu\beta}$.

If $|\mu\beta| \gg 1$, i.e. if we are in the bandwidth in open loop, we can consider the closed loop behavior equivalent to $1/\beta$, presenting a constant gain irrespective of the frequency. For any frequency sinusoidal input in the bandwidth, we can then consider that the output is a faithful image of the input.

If $|\mu\beta| \ll 1$, i.e. we are outside the bandwidth in open loop, we can consider the behavior in closed loop equivalent to μ. In this case, the amplitude of the output is very low compared to the amplitude of the input.

A simple evaluation of rapidity consists of considering the bandwidth ω_c in closed loop. With a constant feedback β, we can then consider the closed loop gain as constant if $|\mu\beta| \gg 1$. Hence, we can associate the bandwidth in closed loop with the bandwidth in open loop. The bandwidth in open loop is a qualifier of the rapidity in closed loop.

Time behavior in unit-step response

We can also qualify the rapidity of a closed loop system by observing its response to a set point step function. For a system without overflow, we can measure the establishing time, but the dilemma precision-stability often leads to a compromise that entails a response with overflow. For a system having such a response, we use as a rapidity criterion the time necessary to reach the first maximum noted by t_m.

For a second order system we can connect the time notion of first maximum to the concept of bandwidth in open loop $t_m = \pi / \omega_c \sqrt{1 - \xi^2}$. We can use in general the relation approached: $\omega_c t_m \approx 3$ which is valid for any correct damping $(0.6 < \xi < 1)$.

Due to this relation, we can translate a first maximum time constraint of a unit-step response in closed loop into a bandwidth constraint in open loop.

Dynamic precision

In order to calculate an upper bound of the instantaneous error following the variations of any set point, we use the method of the equivalent sine curve. The result of this analysis leads to a specification of minimum gain in open loop for a certain bandwidth 0 to ω_0. Therefore, this constraint leads in general to imposing a certain bandwidth to the open loop, even if the constraint in the gain refers only to the bandwidth 0 to ω_0.

In short, we will use the following rules:

– in order to obtain a good precision, we need to:

- in static state, have one or more integrations in direct chain in order to cancel the permanent state errors,

- in dynamic state, have a high gain for a frequency band until ω_0. This makes it possible to limit the dynamic errors;

– for a good degree of stability, it is necessary that the phase margins and (or) the gain margins defined near ω_c are satisfactory;

– for good rapidity, it is necessary that the bandwidth in open loop is large and therefore generally having a high gain.

If we use the Bode diagram to represent the scalar characteristics of the *open loop* to be corrected, we could translate the above constraints depending on three frequency areas as indicated in Figure 10.2.

The previous chart underlines the contradictions between the specifications concerning precision and stability. An increase in the gain favors the precision at the expense of stability.

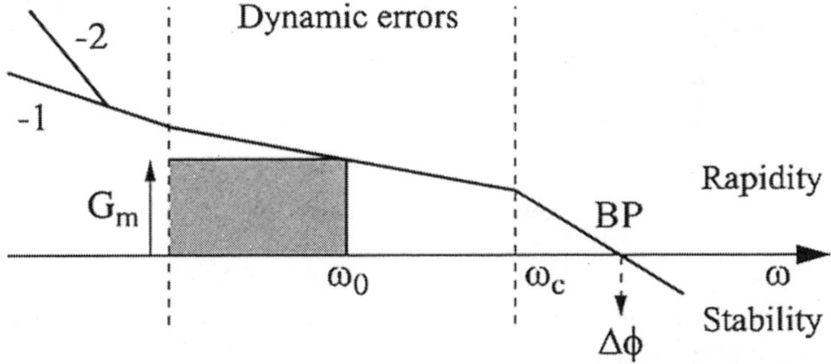

Figure 10.2. *Bode diagram of an open loop to be corrected*

The correctors or regulators have the goal to provide a control signal u to the process in order to maintain the requirements of precision and stability. They are inserted into a looped system as represented in Figure 10.3.

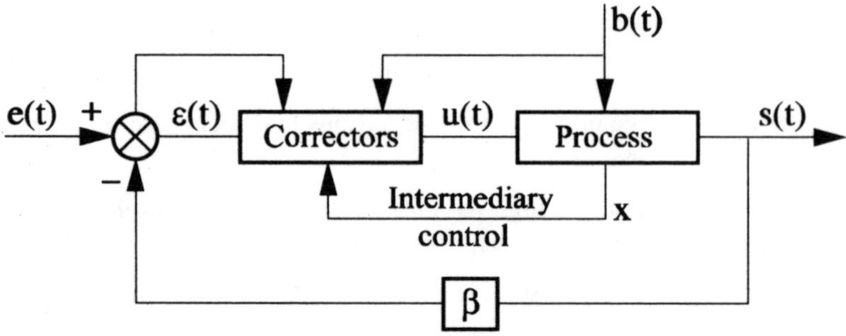

Figure 10.3. *Corrector in a looped system*

The control magnitude can be a function of ε, x, b or e. Based on the signals considered and the type of the function created, there are several types of correctors.

Firstly, we will review the topologies of the most widely used correctors.

10.1.2. *Serial correction*

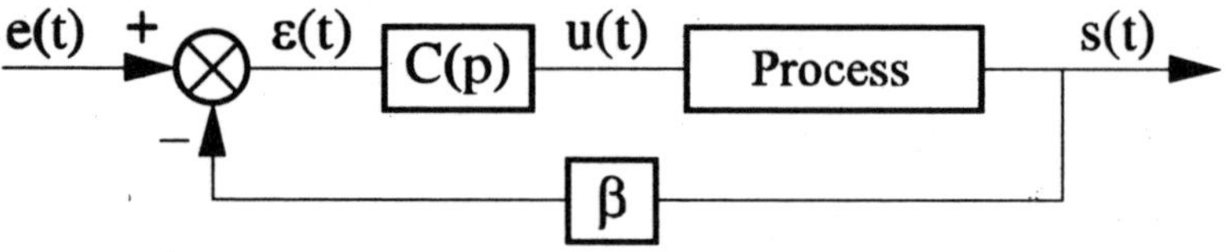

Figure 10.4. *Serial corrector*

This type of corrector is inserted in the direct chain in a serial connection with the process and provides a control signal:

$$u(t) = f(\varepsilon(t))$$

The control takes into account only the error signal.

Among the usable functions f, we can find:

– the proportional action noted by P:

$$u(t) = k\varepsilon(t) \text{ or } U(p) = k\varepsilon(p)$$

– the integral action noted by I:

$$u(t) = \frac{1}{T_i} \int_0^t \varepsilon(\theta)dt \text{ or } U(p) = \frac{\varepsilon(p)}{T_i\, p}$$

– the derived action noted by D:

$$u(t) = T_d \frac{d\varepsilon(t)}{dt} \text{ or } U(p) = T_d\, p\varepsilon(p)$$

– the phase lead action:

$$U(p) = K \frac{1+Tp}{1+aTp} \varepsilon(p) \text{ with } a<1$$

– the phase delay action:

$$U(p) = K \frac{1+Tp}{1+aTp} \varepsilon(p) \text{ with } a>1$$

A serial corrector creates the combinations of these actions more or less perfectly.

It should be noted that this corrector acts on the static precision, dynamic precision and stability.

For a first approximation:

– action I cancels the static error;

– action P increases the dynamic precision;

– action D or the phase lead tend to stabilize the system.

10.1.3. *Parallel correction*

This type of corrector is grafted in parallel on an element of the direct chain as shown in Figure 10.5.

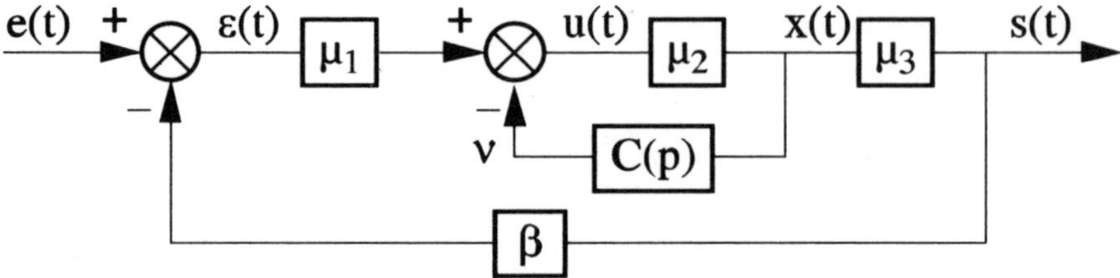

Figure 10.5. *Parallel corrector*

μ_1 can possibly be a serial corrector. x is an intermediary magnitude between the control and the output. If v contains a term of the form $\dfrac{ds(t)}{dt}$, we say that we realized a *tachymetric correction*.

It should be noted that this corrector acts essentially on the dynamic stability and precision and not on the static error (this corrector cannot introduce integration).

10.1.4. *Correction by anticipation*

These correction techniques are used only as a complement of looped correction techniques. They rely on the injection of signals in open loop in order to minimize the transitions felt by the main correction loop, following the input variation or an external interference.

10.1.4.1. *Compensation of interferences (zero input)*

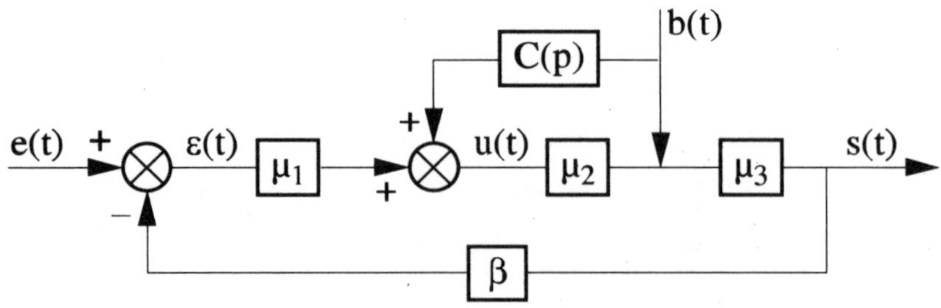

Figure 10.6. *Corrector by anticipation*

Let us suppose that the interference is measurable. It is then possible – at least theoretically – to eliminate the influence of the interference $b(t)$ with the help of a corrector by anticipation $C(p)$ described by the diagram in Figure 10.6.

In fact, it is enough to choose:

$$C(p) = -\frac{1}{\mu_2(p)}$$

to cancel the effect of $b(t)$ on the output.

We note in this case that the stability is not affected, nor is the precision with respect to the input.

Most often, $\dfrac{1}{\mu_2(p)}$ is not physically feasible, which leads to adopting approximate forms of $\dfrac{1}{\mu_2(p)}$. There is no perfect compensation of the transient state of interferences.

EXAMPLE 10.1.– if $\mu_2(p) = \dfrac{K}{p}$, we could take $C(p) = -\dfrac{1}{K}\dfrac{p}{(1+\tau p)}$ with τ less than the main time constants of μ_3.

10.1.4.2. *Compensation of the input (zero interference)*

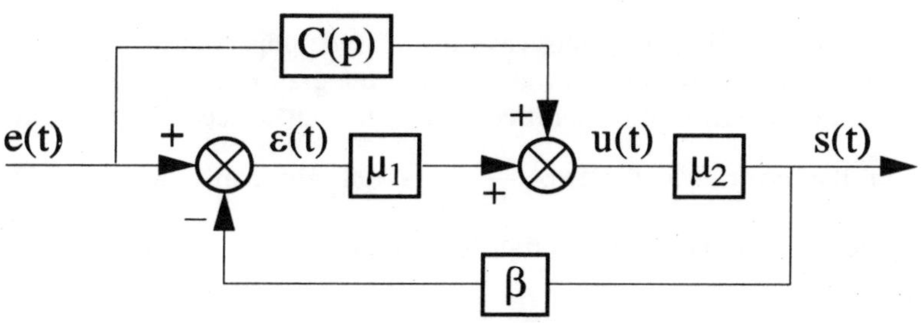

Figure 10.7. *Input compensation*

This diagram can be considered, with respect to the error, as the overlapping of the two diagrams of Figure 10.8.

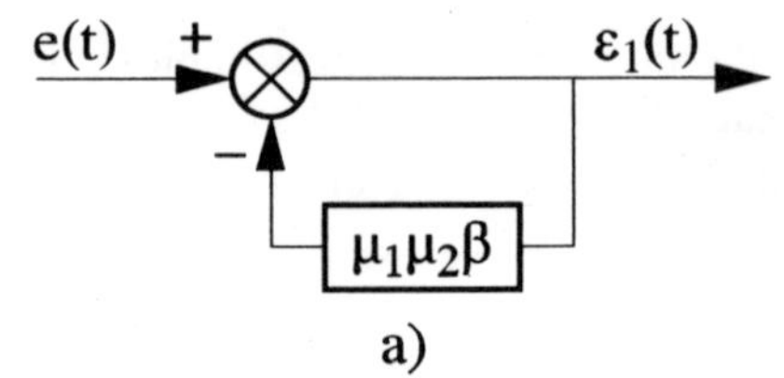

a)

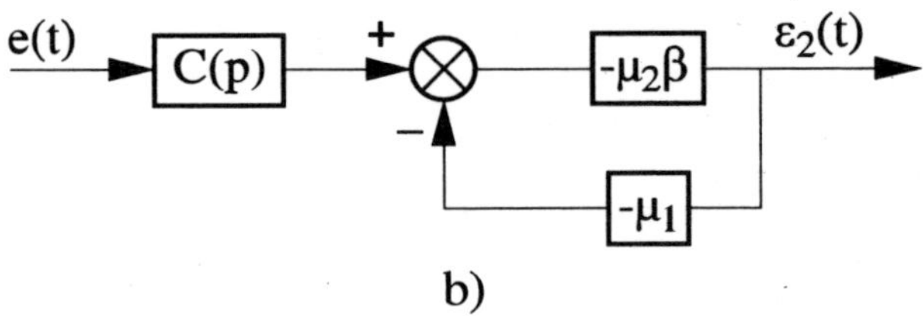

b)

Figure 10.8. *Diagram decomposition of the input compensation*

We have:

$$\varepsilon = \varepsilon_1 + \varepsilon_2 = \frac{E(p)}{1 + \mu_1 \mu_2 \beta} + \frac{E(p)C(p)(-\mu_2 \beta)}{1 + \mu_1 \mu_2 \beta}$$

In order to cancel the error with respect to the input, it is enough to choose $C(p) = \dfrac{1}{\mu_2 \beta}$.

It should be noted that in this case the closed loop control system perfectly follows the input law, without introducing the integration in the direct chain. The stability of the system is not modified, nor is the influence of interferences with respect to the error. Most often, $\dfrac{1}{\mu_2 \beta}$ is not physically feasible. Thus, the compensation will not be perfect in transient states.

10.1.5. *Conclusions*

The *feedforward correctors*, when they are feasible, do not modify the stability of the loop and compensate either the error due to the input, or the effect of interference. In a complex case (several interferences, some of which are non-measurable), the implementation of these compensators can be difficult and imperfect.

The *"parallel" correctors* generally modify an element of the direct chain, without introducing integration. They can be used in order to improve the stability.

The *"serial" correctors* modify the stability and precision with respect to the input and interferences. This mode of correction is the most widely used.

10.2. Serial correction

10.2.1. *Correction by phase lead*

10.2.1.1. *Transfer function*

Generally speaking, we call a *phase lead corrector* a corrector whose transfer function has the form:

$$C(p) = K\frac{1+Tp}{1+aTp} \text{ with } a < 1$$

This corrector can be physically created by the circuit of Figure 10.9.

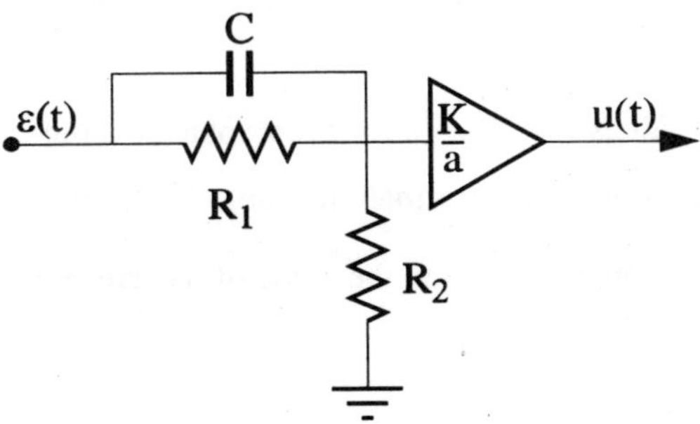

Figure 10.9. *Phase lead network*

with $T = R_1C$ and $a = \dfrac{R_2}{R_1 + R_2}$.

It should be noted that the resistive bridge alone brings about an attenuation to the low angular frequencies, which justifies the presence of an amplifier of gain $\dfrac{K}{a}$.

The Bode graph is represented in Figure 10.10.

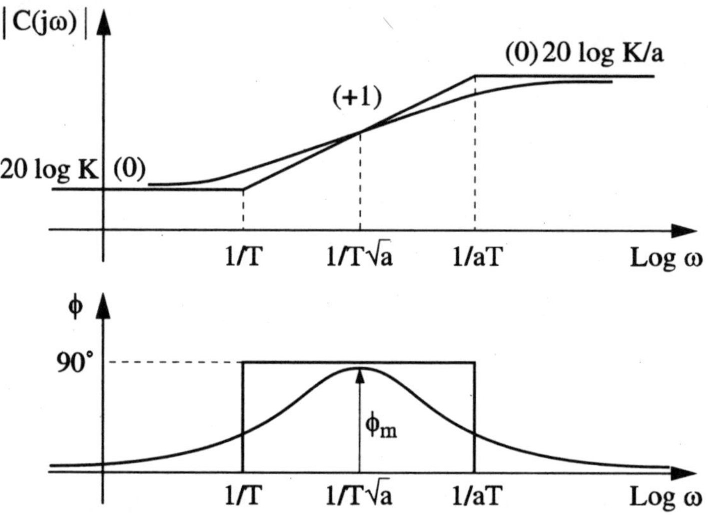

Figure 10.10. *Bode graph of a phase lead network*

The tabulated values of the maximum phase input ϕ_m (obtained for $\omega = 1/T\sqrt{a}$) are given for your information (Table 10.2), in practice we can use the formula $\sin \phi_m = \dfrac{1-a}{1+a}$ in order to find the value of a corresponding to a desired ϕ_m value.

a	1/4	1/6	1/8	1/10	1/12
ϕ_m	37°	45°	51°	55°	58°

Table 10.2. *Maximal phase lead*

10.2.1.2. *Action mechanism of a phase lead corrector*

There is no general method to configure such a corrector.

EXAMPLE 10.2.

Let us examine on a current case the problems linked to this configuration (see Figure 10.11).

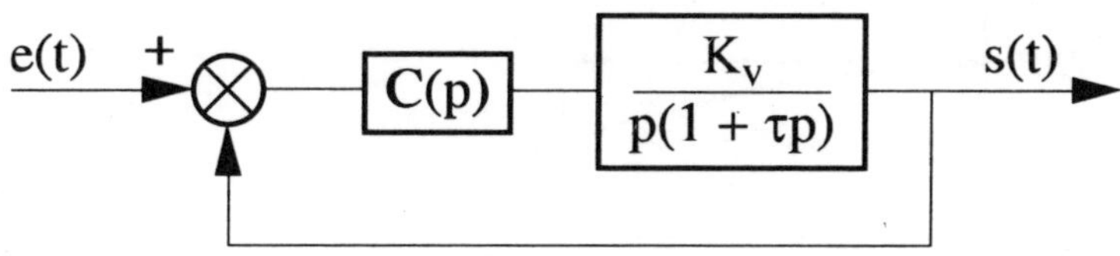

Figure 10.11. *Uncorrected system*

The OL of the uncorrected system presents an integration. *An integral action is hence useless* in order to ensure a zero static error for a step function input.

Let us suppose that we want a response time t_r imposed in closed loop, or a dynamic precision given to an angular frequency $\omega_0 << \dfrac{1}{\tau}$.

Thus, we have to introduce a minimum gain K (proportional action) in the chain that leads to an insufficient phase margin:

$$\mu\beta = \frac{KK_v}{p(1+\tau p)} \ \text{ with } \ K \ \text{ such as } \ \left|\frac{KK_v}{p(1+\tau p)}\right|_{p=j\omega_0} = G_m \, ,$$

or $\dfrac{KK_v}{\omega_0} \approx G_m$ if $\omega_0 << \dfrac{1}{\tau}$. G_m is minimum gain required at the angular frequency ω_0.

We notice on the graph in Figure 10.12 that the phase margin becomes insufficient. A correction is thus necessary, for example, a *phase lead* correction.

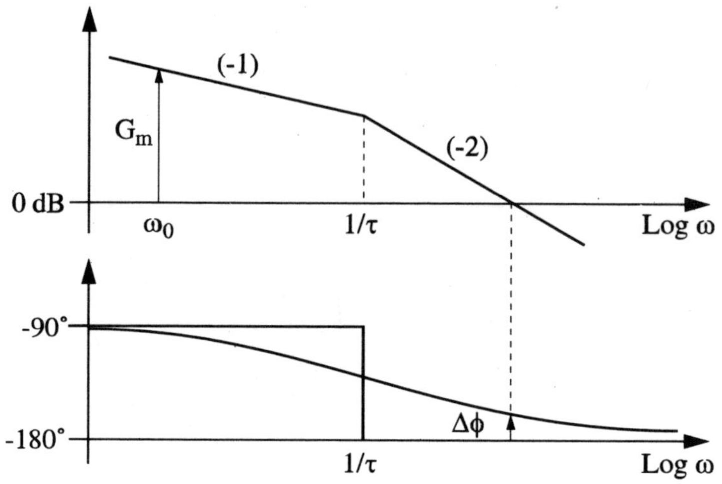

Figure 10.12. *Bode graph with insufficient phase margin*

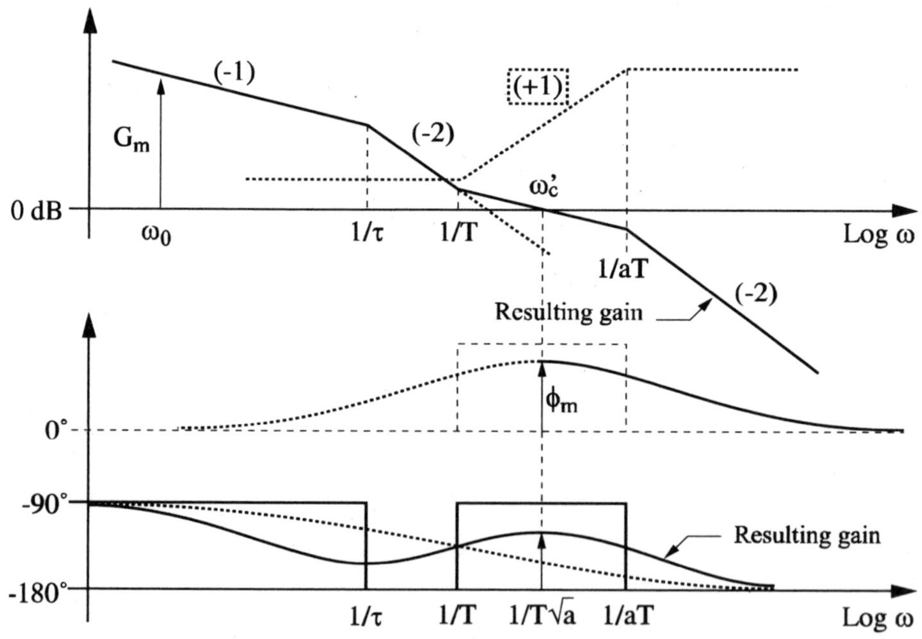

Figure 10.13. *Bode graph in OL of the corrected system*

Since this corrector is in series with $\mu\beta$, there is an addition of gain and phase curves in the Bode plane (Figures 10.10 and 10.12).

NOTE 10.1.– it should be noted that the open loop system without correction does not shift phases beyond $-180°$. Consequently, a phase lead corrector chosen for $a = 1/6$ – if it is conveniently placed – can bring a 45° phase lead, which is sufficient to ensure a convenient stability degree for the system.

For that, we cut the axis 0 dB at the angular frequency $\omega'_c = 1/T\sqrt{a}$ where the phase lead of the corrector is maximal (environment in Bode representation of the segment $[1/T, 1/aT]$). Parameter a being chosen as well as the relative position of ω'_c, we verify if the value of minimum gain requested is satisfied and if not we fail the position of ω'_c.

The resulting phase margin is calculated by adding the residual phase of the uncorrected system to ω_c with respect to $-180°$ and the ϕ_m of the corrector. It should be noted in this example that the resulting phase with respect to $-180°$ is always positive. The looped system will thus always be stable.

EXAMPLE 10.3.– it is not always like this, in particular when the OL transfer function of the uncorrected system shifts phases more than $-180°$.

Let the OL system before correction be: $\dfrac{K}{p(1+\tau_1 p)(1+\tau_2 p)}$ with $\tau_1 > \tau_2$.

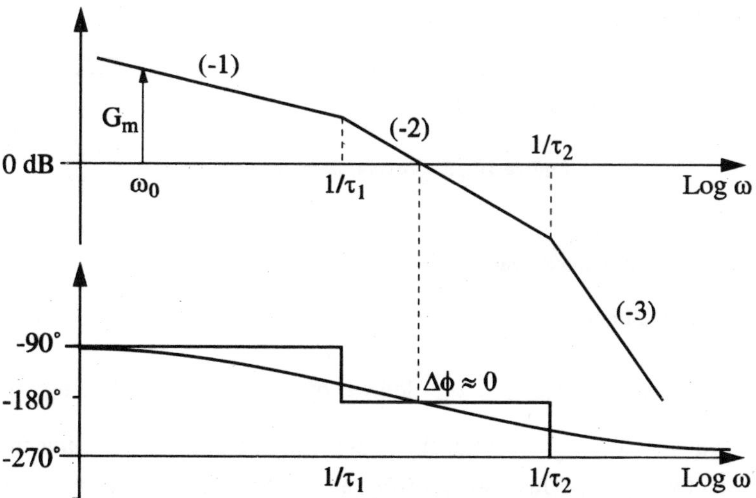

Figure 10.14. *OL system before correction*

The phase margin of the corrected system is obtained by adding to the ϕ_m of the corrector 180° plus the phase of the initial system to ω'_c:

$$\Delta\phi' = \phi(\omega'_c) + \phi_m + 180°$$

We note that this last term can be negative; hence we will need to choose a more efficient corrector (smaller a).

When the phase curve of the system before correction decreases too fast, the phase lead caused by the corrector risks being considerably contrasted at point ω'_c. The correction by phase lead can thus prove to be insufficient.

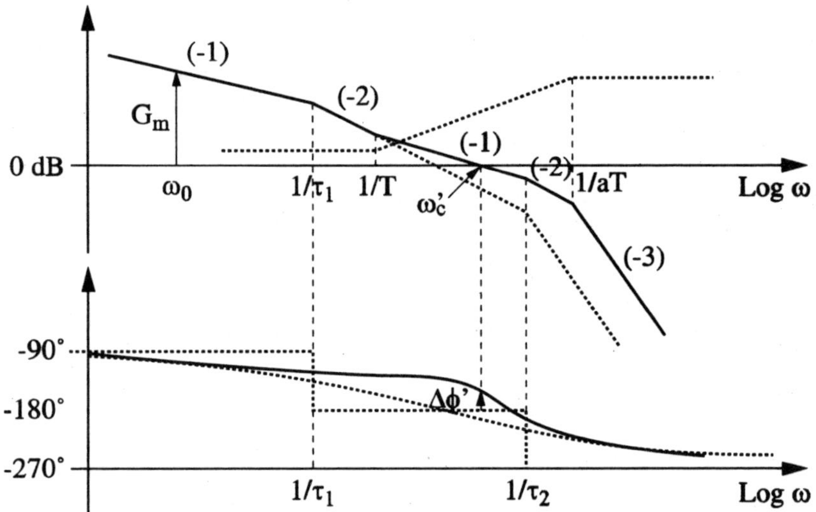

Figure 10.15. *Correction by phase lead*

In conclusion, the phase lead corrector:

– increases the bandwidth ($\omega'_c > \omega_c$);

– brings in ω'_c a phase lead ϕ_m defined by $\sin\phi_m = \dfrac{1-a}{1+a}$, if the corrector is properly centered so that its central frequency coincides with the cutting frequency of the corrected system. This choice of ω'_c avoids the complete construction of the phase in order to determine the phase margin obtained.

When the correction is perfectly set, in order to determine the final performances of the closed loop corrected system, a complete construction of the module and phase is necessary in the Bode plane. Therefore, a graph in Black-Nichols graph will make it possible to obtain the closed loop transfer function.

10.2.2. *Correction by phase delay*

10.2.2.1. *Transfer function*

Generally speaking, we call phase delay corrector a corrector whose transfer function has the form:

$$C(p) = K \frac{1 + Tp}{1 + aTp} \quad \text{with } a > 1$$

This corrector can be physically created by the circuit represented in Figure 10.16.

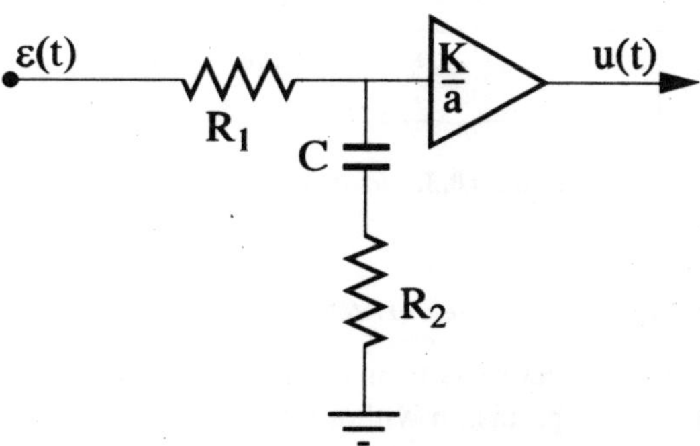

Figure 10.16. *Phase delay network*

With $T = R_2C$ and $a = \dfrac{R_1 + R_2}{R_2}$.

The phase delay corrector has the Bode diagram represented in Figure 10.17.

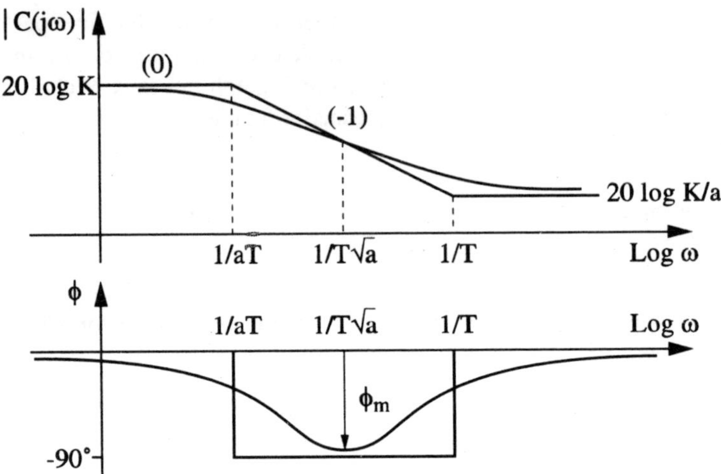

Figure 10.17. *Bode diagram of a phase delay network*

$1/a$	1/4	1/6	1/8	1/10	1/12
ϕ_m	−37°	−45°	−51°	−55°	−58°

Table 10.3. *Maximal phase delay*

10.2.2.2. *Action mechanism of these correctors*

The effect of these correctors is to increase the gain at low angular frequencies and thus improve the static precision without modifying the behavior at high angular frequencies.

In addition, we can use the phase delay corrector to make the gain fall at high angular frequencies without modifying it at low angular frequencies.

With the help of an example, let us see the effect of this corrector.

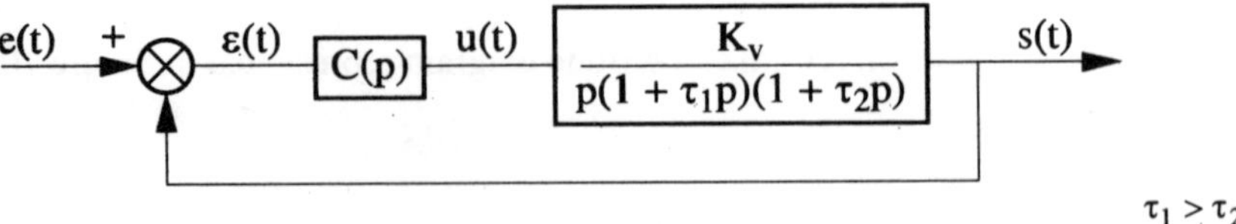

Figure 10.18. *Phase delay corrector*

The desired performances are:

– zero static error;

– dynamic precision imposed at $\omega_0 << \dfrac{1}{\tau}$;

– stability degree imposed by a phase margin $\Delta\phi$ from 45 to 50°.

The static error with respect to the input is zero (presence of one integration in the direct chain) in the absence of interferences. The graph of the OL which makes it possible to define the minimal proportional action K, ensuring the desired precision is given in Figure 10.19.

A phase lead correction is impossible for such a system because the phase of the OL decreases too fast near ω_c. Since the integral action is useless, we can then place a phase delay corrector.

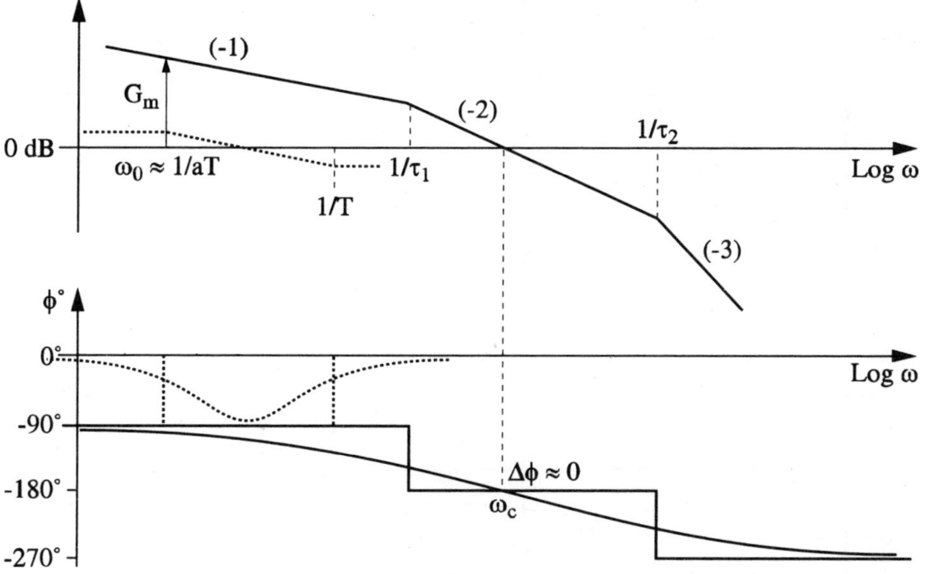

Figure 10.19. *OL graph*

This corrector must not affect the gain on the side of ω_0, hence the choice $\dfrac{1}{aT} \approx \omega_0$. The correction will be even more efficient for the stability as the gain curve will cut the axis 0 dB with the biggest slope segment possible (–1) and near its environment.

Due to these considerations, we adopt as corrector $C(p) = K\dfrac{1+Tp}{1+aTp}$ with $1/aT$ near ω_0. Parameter a will be determined by successive tests. The phase margin after correction can be estimated as follows:

– calculate the ratio τ_1/T characterizing the "length" of the slope segment (–1) defined above;

– if the new gap angular frequency ω'_c, chosen near the geometrical environment of $\left[1/T,\ 1/\tau_1\right]$, is much higher than ω_0 ($\omega'_c \geq 10\omega_0$) and much lower than $\dfrac{1}{\tau_2}$, $\left(\omega'_c \leq \dfrac{1}{10\tau_2}\right)$, then an approximation of the phase margin is given by carrying over the ratio τ_1/T in Table 10.2 for the phase lead corrector. After correction, we obtain the Bode diagram in Figure 10.20.

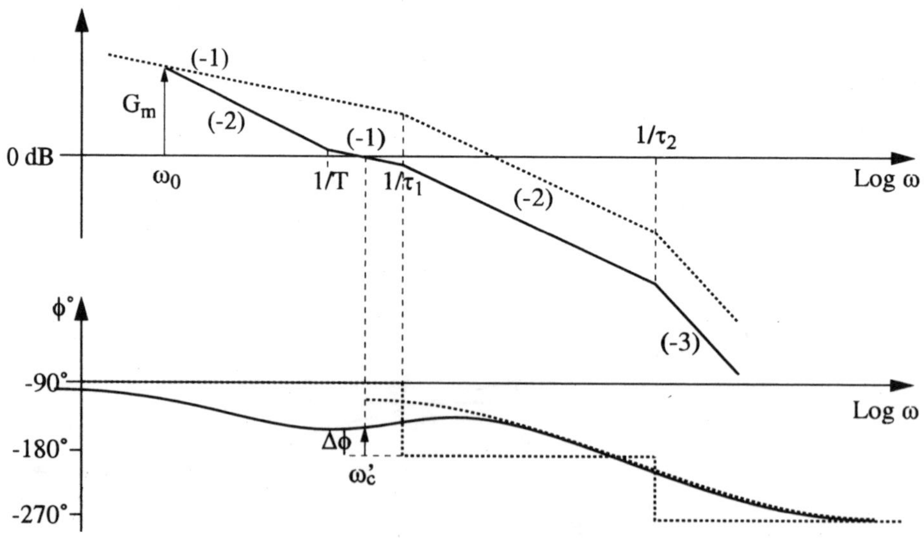

Figure 10.20. *Bode graph in OL of the corrected system*

It should be noted that near the angular frequency ω'_c, the contribution of the two breaks, (–1) to (–2) in ω_0 and (–2) to (–3) in $1/\tau_2$, is negligible. Everything takes place as if only the slopes (–2), (–1) then (–2) remained with respect to the phase. This graph is that of a double integrator in a serial connection with a phase lead corrector.

It should be noted that the phase delay correction happened in this particular example because the ratio between $1/\tau_2$ and ω_0 was high, but this is not always the case (see the combined action correction).

The phase delay corrector decreases the bandwidth ($\omega_c' < \omega_c$) and the constant stability static error (phase delay for non-integrator systems).

There again, once the correction is set, it is necessary to build with precision the total OL in order to deduce the performances in CL.

10.3. Correction by combined actions

10.3.1. *Transfer function*

Generally speaking, we call a lead-delay combined action corrector a corrector whose transfer function has the form:

$$C(p) = K \frac{(1+T_2 p)(1+T_3 p)}{(1+T_1 p)(1+T_4 p)} \text{ with } T_1 > T_2 > T_3 > T_4$$

This corrector is obtained by serially connecting a phase delay corrector $K_1 \dfrac{(1+T_2 p)}{(1+T_1 p)}$ and a phase lead corrector $K_2 \dfrac{(1+T_3 p)}{(1+T_4 p)}$ with $K_1 K_2 = K$.

The Bode diagram of such a corrector is represented in Figure 10.21.

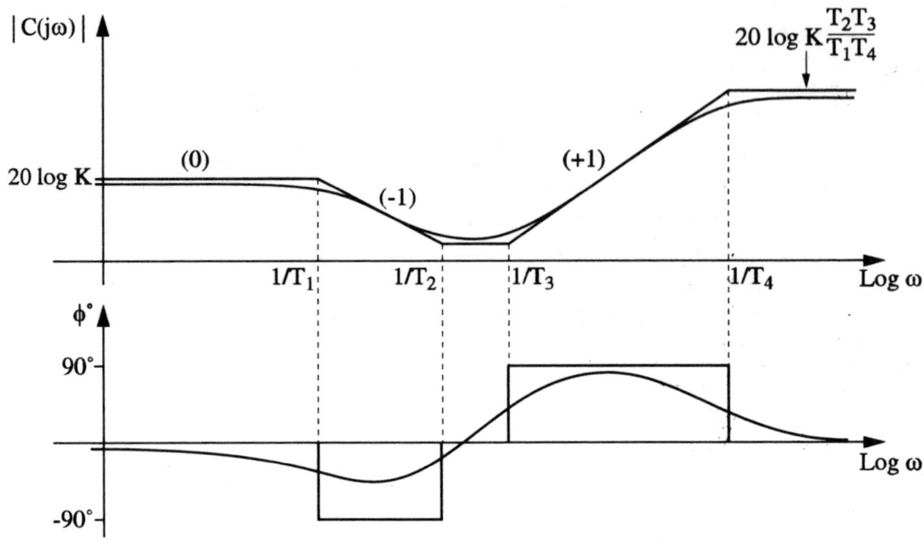

Figure 10.21. *Bode diagram of a lead-delay network*

10.3.1.1. *Action mechanism of these correctors*

These correctors combine the actions previously studied. They are used when the simple action correctors do not lead to the performances desired.

EXAMPLE 10.4.

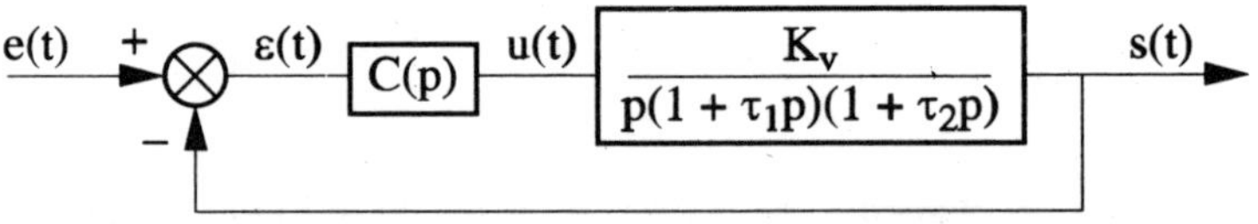

Figure 10.22. *Combined action corrector*

In Figure 10.22, we have $\tau_1 > \tau_2$ and as desired performances:

– zero static error;

– dynamic precision imposed at ω_0 (close to $\dfrac{1}{\tau_1}$);

– phase margin $\Delta\phi$ from 45 to 50°.

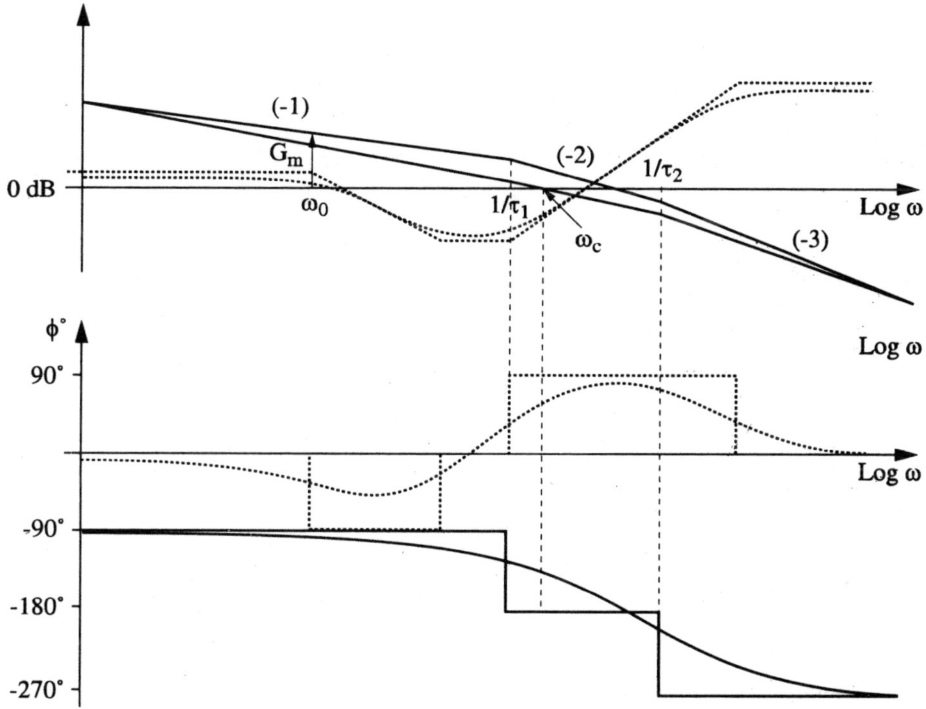

Figure 10.23. *Phase delay correction*

The phase delay correction is inefficient here (proximity of angular frequencies ω_0 and $\dfrac{1}{\tau_1}$) and therefore in order to obtain the desired stability, a phase lead term must be added.

The corrected OL is represented in Figure 10.24.

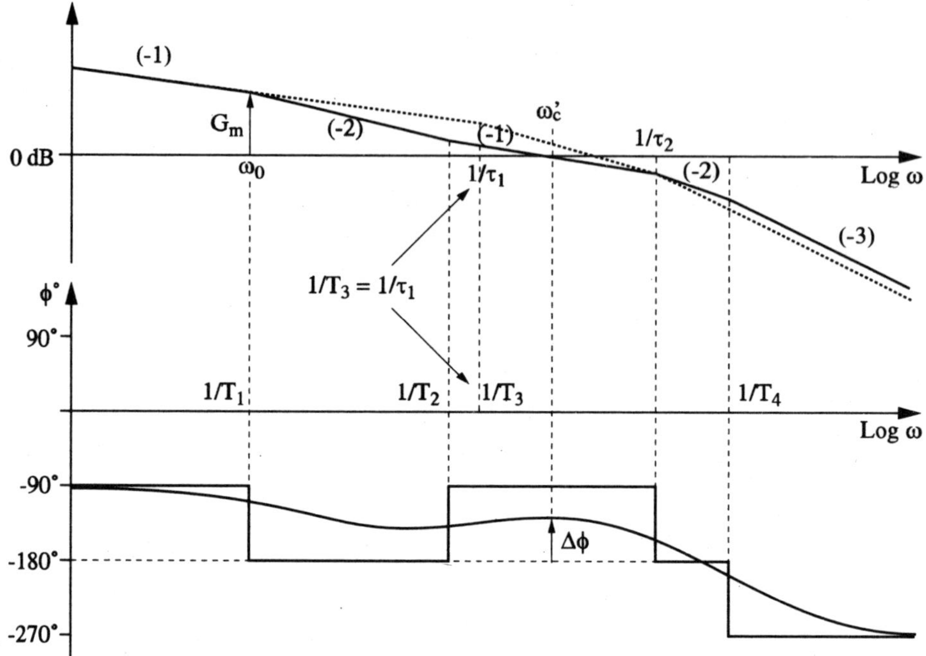

Figure 10.24. *Bode graph in OL of the corrected system*

We can estimate the phase margin of the corrected system, if ω'_c is close to the environment of slope segment (-1), by referring to Table 10.2 of the phase lead circuit. This would be even more justified in the present case since $\omega'_c > 10\omega_0$ and $\omega'_c < 1/10T_4$.

10.4. Proportional derivative (PD) correction

10.4.1. *Transfer function*

A proportional and derivative action corrector has as transfer function:

$$C(p) = K(1 + T_d p)$$

The derivative action is not physically feasible and it is approximated by a transfer function of the form: $\dfrac{T_d}{1 + \tau p}$ with $\tau \ll T_d$.

The new corrector has then as transfer function:

$$C(p) = K\left(1 + \frac{T_d p}{1 + \tau p}\right) \approx K\left(\frac{1 + T_d p}{1 + \tau p}\right)$$

Therefore, the PD corrector has a transfer function which is equivalent to that of a phase lead corrector. The same methods could be used in order to determine K, T_d and τ .

10.5. Proportional integral (PI) correction

10.5.1. *Transfer function*

A proportional and integral action corrector has as transfer function:

$$C(p) = K\left(1 + \frac{1}{T_i p}\right) = K\left(\frac{1 + T_i p}{T_i p}\right)$$

The proportional integral regulator (PI) has as its Bode diagram that in Figure 10.25.

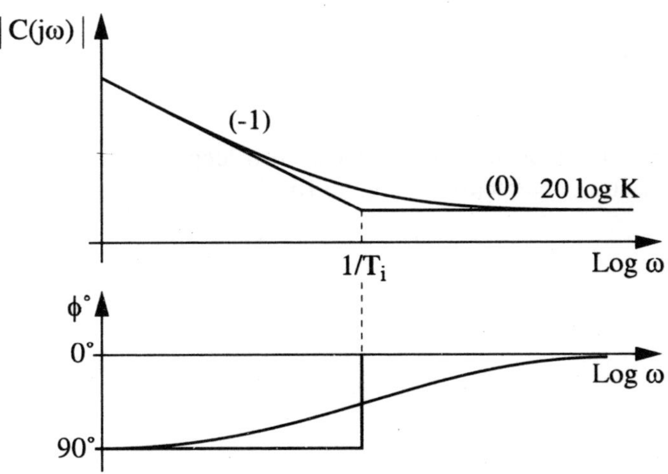

Figure 10.25. *Bode diagram of a PI corrector*

The essential function of a PI regulator is to bring one integration in the open loop. It is indispensable if the specifications stipulate that a zero static error is needed in response to a step function when the process to control does not have one. It is also indispensable if the process is an integrator but it is subjected to an interference injected upstream from the process. If we desire a zero static error for a step function interference, it is essential to add an integrator at the level of $C(p)$.

The following example (Figure 10.26) is an illustration of the correction of a non-integrator process.

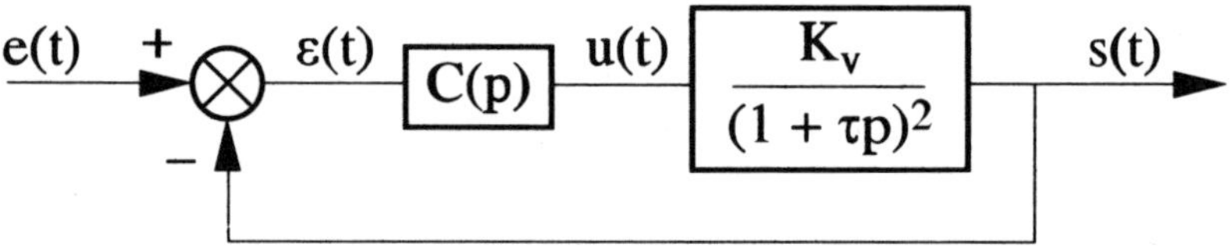

Figure 10.26. *Correction of a non-integrator process*

The desired performances are:

– zero static error;

– imposed bandwidth $\{0, \omega_c\}$;

– stability degree imposed by a phase margin $\Delta\phi$ from 45 to 50°.

An integral action is necessary in order to cancel the static error. However, it cannot be alone because then the system could not be stable.

The graph of the Bode diagram of the open loop without correction makes it possible to place a PI corrector (see Figure 10.27).

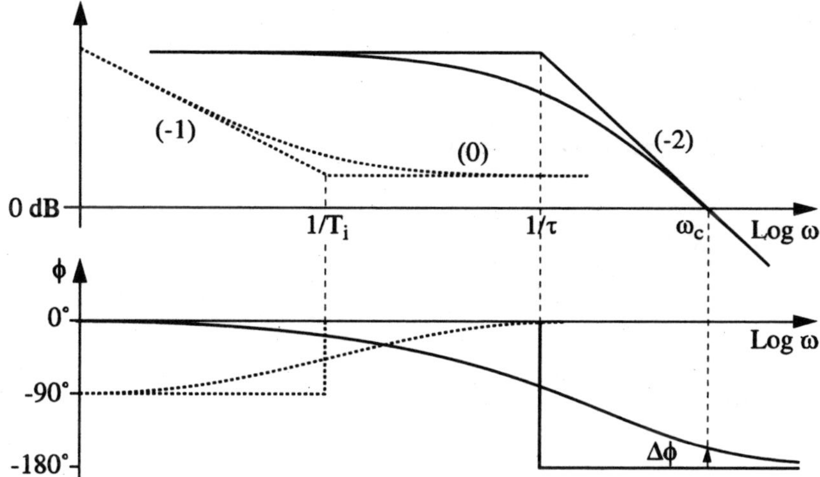

Figure 10.27. *Bode diagram of the uncorrected system in OL*

Let us introduce a PI corrector (represented by a dotted line in Figure 10.27) so that angular frequency $1/T_i'$ is near $\omega_c/10$ in order not to modify the phase margin $\Delta\phi$. The corrected curve in open loop is represented in Figure 10.28.

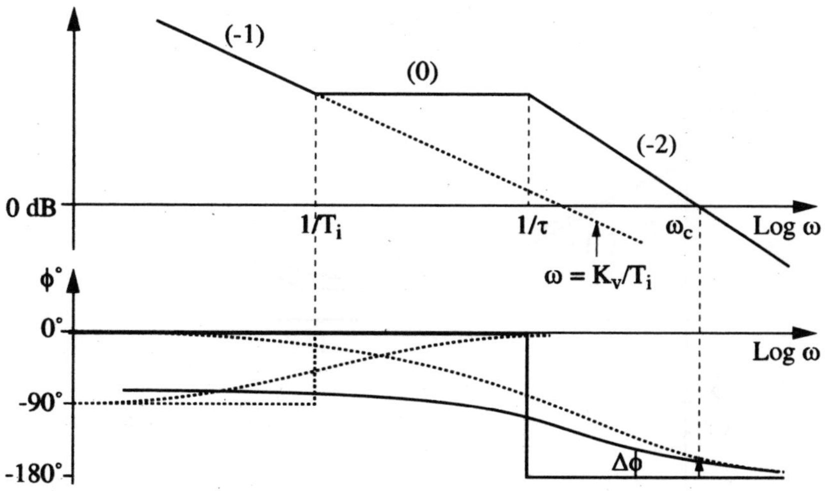

Figure 10.28. *Bode diagram of the corrected system in OL*

This corrector provides the desired performances to the closed loop.

10.6. Proportional integral proportional (PID) correction

10.6.1. *Transfer function*

An ideal PID corrector has as transfer function:

$$C(p) = K\left(1 + \frac{1}{T_i p} + T_d p\right)$$

or:

$$C(p) = K\left(\frac{1 + T_i p + T_i T_d p^2}{T_i p}\right)$$

The roots of the numerator are:

$$T_1 = \frac{T_i}{2}\left(1 + \sqrt{1 - \frac{4T_d}{T_i}}\right) \text{ and } T_2 = \frac{T_i}{2}\left(1 - \sqrt{1 - \frac{4T_d}{T_i}}\right)$$

with $T_i^2 - 4T_i T_d \geq 0$ the zeros are real and placed in $\frac{1}{T_1}$ and $\frac{1}{T_2}$.

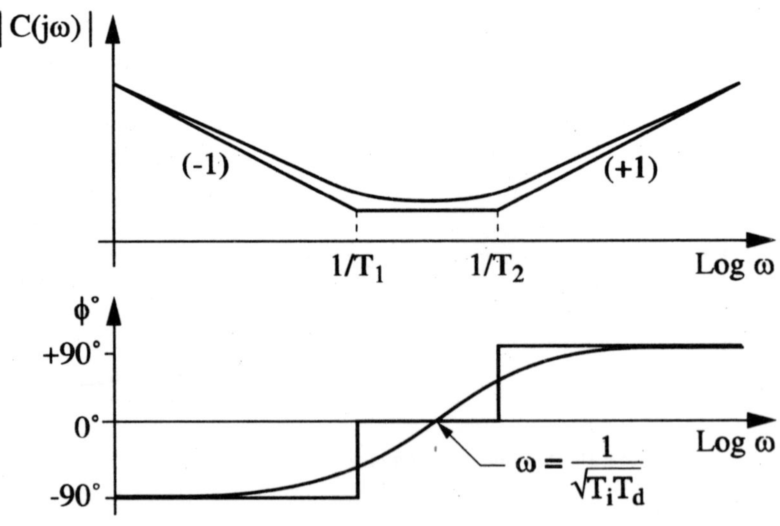

Figure 10.29. *Bode diagram of a PID corrector*

We note that $T_i = T_1 + T_2$ and $T_d = \dfrac{T_1 T_2}{T_1 + T_2}$.

If the break $\dfrac{1}{T_2}$ is much higher than $\dfrac{1}{T_1}$ we have $\dfrac{1}{T_1} \approx \dfrac{1}{T_i}$ and $\dfrac{1}{T_2} \approx \dfrac{1}{T_d}$.

In practice, the derivative action will always be approximated by $\dfrac{T_d\, p}{1 + \tau p}$.

EXAMPLE 10.5.– integrator system affected by a disturbance (for example, level regulation with flow disturbance) with $\tau_1 > \tau_2$.

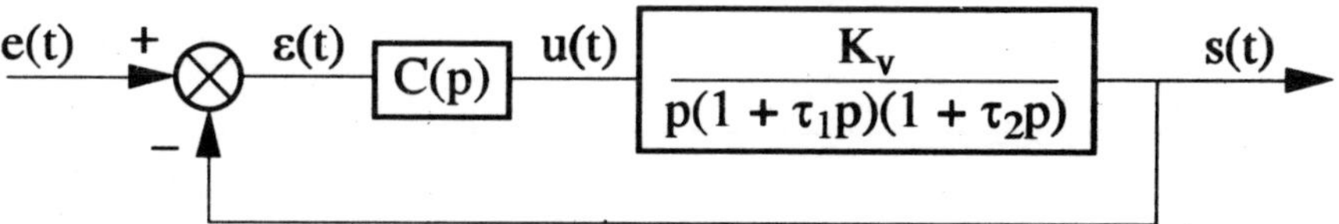

Figure 10.30. *Integrator system affected by a disturbance*

Desired performances:

– zero static error with respect to the set point and disturbance;

– phase margin $\Delta\varphi$ from 45 to 50°;

– dynamic precision imposed at angular frequency ω_0.

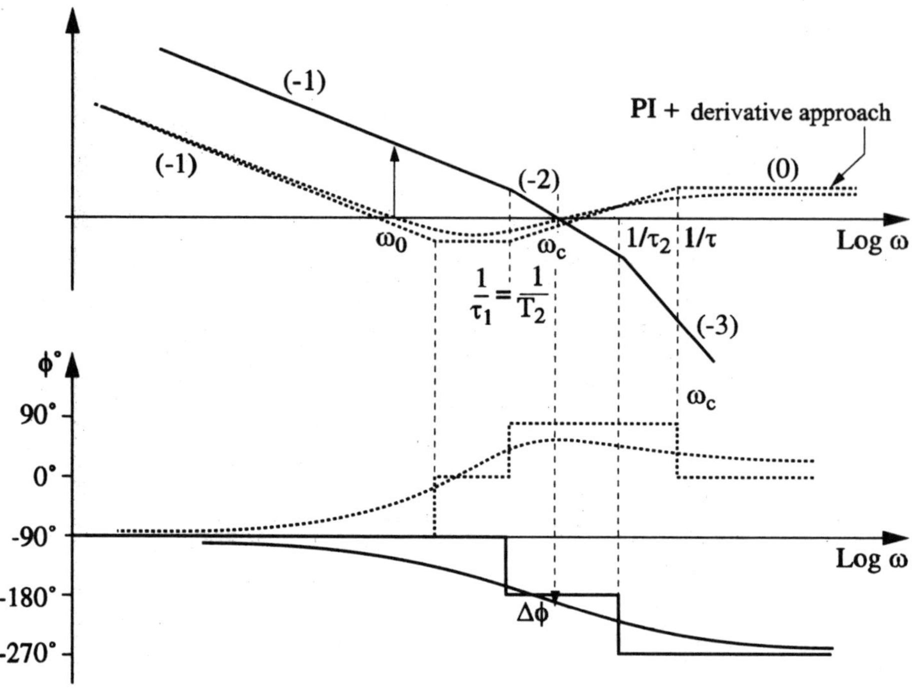

Figure 10.31. *Uncorrected system*

In order to cancel the static error with respect to the disturbance, we will have to introduce an integral action in the corrector $C(p)$. The proximity of angular frequencies ω_0, $\dfrac{1}{\tau_1}$ and $\dfrac{1}{\tau_2}$ forbids us to use a PI corrector. Let us try a PID corrector.

The double integration leads to a slope graph (-2) at low angular frequencies that has to go through point A (minimum gain at ω_0). On the other hand, the phase margin imposes to cut the axis 0 dB with a slope segment (-1) on a convenient length and follow by a slope (-2), from which we obtain the chosen corrector.

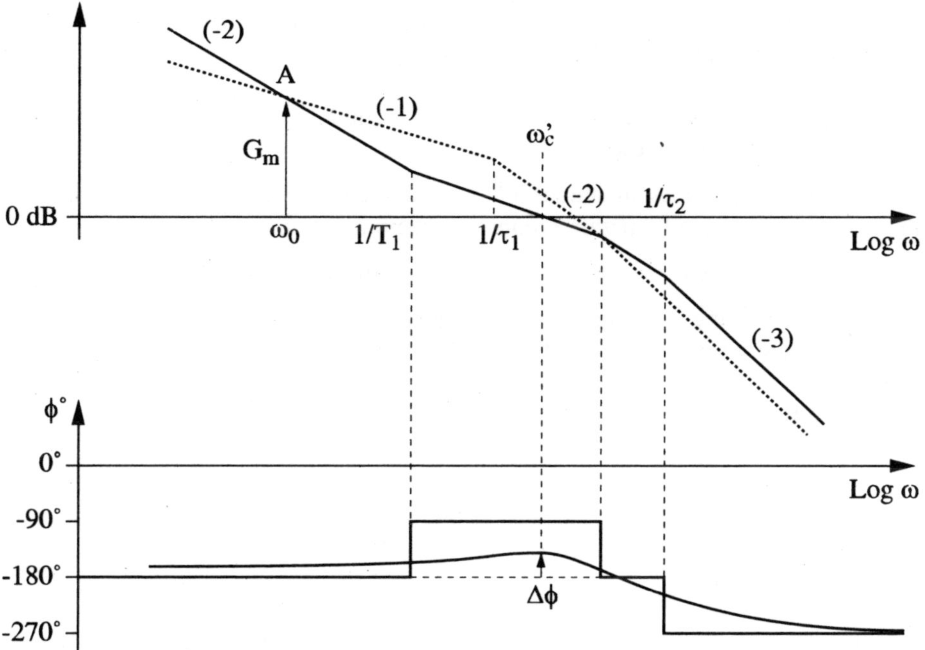

Figure 10.32. *Bode graph of the corrected system*

The same observations as above on the estimation of the phase margin apply here.

10.6.2. *Experimental adjustment method*

We will adjust a PID in order to control a system whose transfer function is unknown.

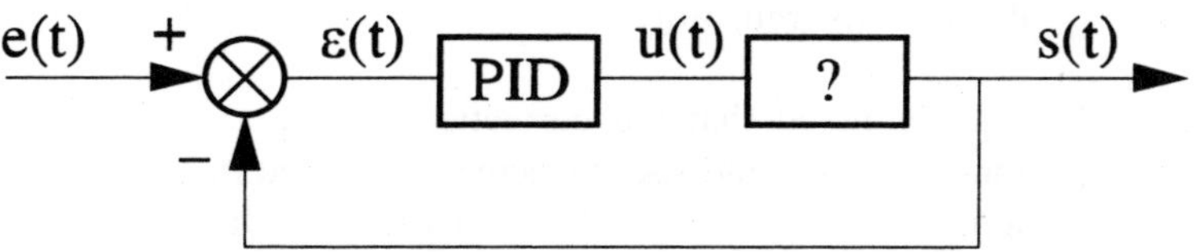

Figure 10.33. *System with unknown transfer function*

In certain cases, the transfer function of a system is unknown; hence the adjustment of a PID cannot use the methods previously presented. Thus, there are experimental adjustment methods. We shall keep in mind the Ziegler-Nichols method (sustained oscillations). This method is suitable only for a CL study. The adjustment is done as follows:

– neutralize the integral actions $(T_i = \infty)$ and derivative $(T_d = 0)$ and therefore there remains a simple loop with proportional action;

– increase gain K of the proportional action until the system becomes oscillating;

– after obtaining a balance of oscillations, increase period T_0 (expressed in seconds) of these sustained oscillations and the limit gain K_M.

The corrector can be adjusted with:

– $K = 0.6K_M$, in order to have sufficient stability margin;

– the derivation constant is chosen at $T_d = \dfrac{T_0}{8}$ whose effect is to accelerate the proportional action;

– the integration constant is defined by $T_i = \dfrac{T_0}{2}$ in order to cancel the static error.

It should be noted that here $T_i = 4T_d$, i.e.:

$$C(p) = K\left(1 + \frac{1}{T_i p} + T_d p\right) = K\frac{(1 + 2T_d p)^2}{T_i p}$$

NOTES ON THE SERIAL CORRECTION.– the methods previously described aim at quickly deciding if the given correction is adequate. On the other hand, it is obvious that there is only one corrector satisfying this problem. The choice will be made after the testing of several correctors and taking into account criteria such as: cost, technological simplicity, reliability.

It should be equally noted that the correction does not make it possible to infinitely increase the precision and speed (increase obtained in general through the increase of the gain leading to strong saturation risks). In fact, when the system saturates, the linearly defined performances are not reached (increased response time and limited precision). On the other hand, irrespective of the linearity of the system, it is not always interesting to have a too extended bandwidth, due to the amplification of noises.

10.7. Parallel correction

10.7.1. *General principle*

This type of correction is represented by the diagram in Figure 10.34.

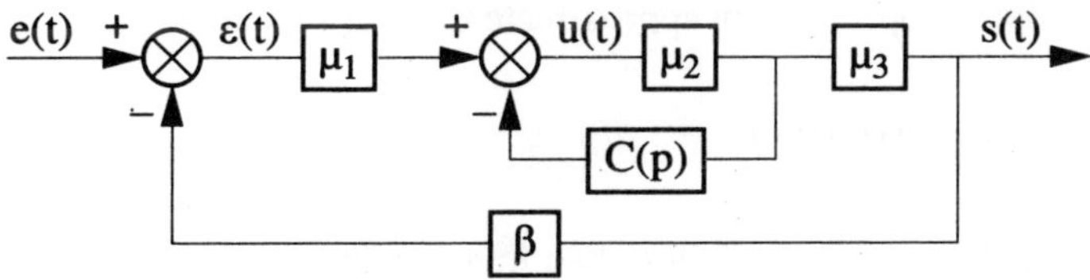

Figure 10.34. *Parallel correction*

$C(p)$ must be calculated in such a way as to obtain the performances required. These performances (precision, stability) can be translated into characteristics on the OL graph in the Bode plane. The problem is thus to determine $C(p)$ in order to obtain a corrected open loop of convenient shape.

The corrected open loop has as a transfer function:

$$\mu_1 \frac{\mu_2}{1+\mu_2 C} \mu_3 \beta$$

Let us consider the two following cases: $\left|\mu_2 C(j\omega)\right| \gg 1$ and $\left|\mu_2 C(j\omega)\right| \ll 1$.

Case 1:

$$\left|\mu_2 C(j\omega)\right| \gg 1 \;\Rightarrow\; \left|\frac{\mu_2}{1+\mu_2 C(j\omega)}\right| \approx \left|\frac{1}{C(j\omega)}\right|$$

the corrected open loop has as an approximate transfer function:

$$\frac{\mu_1(p)\mu_3(p)}{C(p)} \beta(p)$$

Case 2:

$$\left|\mu_2 C(j\omega)\right| << 1 \implies \left|\frac{\mu_2}{1+\mu_2 C(j\omega)}\right| \approx \left|\mu_2(j\omega)\right|$$

the corrected open loop has as an approximate transfer function:

$$\mu_1(p)\mu_2(p)\mu_3(p)\beta(p)$$

Let us trace the variations of these two transfer functions in the Bode plane. The curves are cut in points A and B as indicated in Figure 10.35 (one of these points can be infinitely rejected).

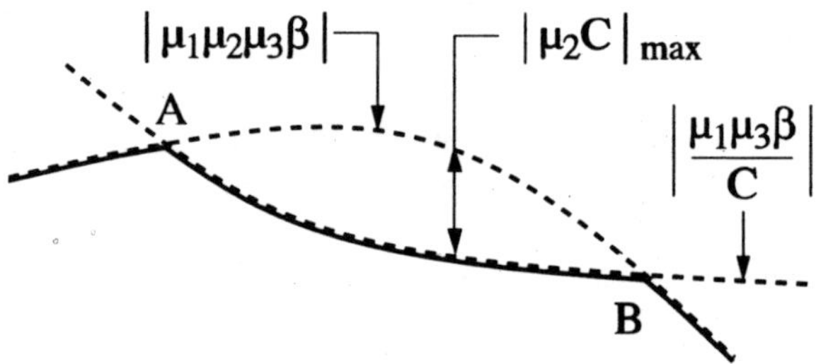

Figure 10.35. *Variation of the two transfer functions*

In these points we have:

$$\left|\mu_1\mu_2\mu_3\beta\right| = \left|\frac{\mu_1\mu_3\beta}{C}\right|$$

i.e.:

$$\left|\mu_2\right| = \left|\frac{1}{C}\right|$$

or $\left|\mu_2 C\right| = 1$.

Points A and B are thus the separations of cases 1 and 2.

Between A and B: $\left|\mu_1\mu_2\mu_3\beta\right| > \left|\dfrac{\mu_1\mu_3\beta}{C}\right|$, i.e. $\left|\mu_2 C\right| > 1$ (1st case) and outside A

and B: $\left|\mu_1\mu_2\mu_3\beta\right| < \left|\dfrac{\mu_1\mu_3\beta}{C}\right|$, i.e. $\left|\mu_2 C\right| < 1$ (2nd case).

An approximation of the corrected OL is represented with a line in Figure 10.35. This approximation may not be good near points A and B, since $\left|\mu_2 C\right|_{max}$ is low (see Figure 10.35).

10.7.2. *Simple tachymetric correction ($C(p) = \lambda p$)*

We will show in an example how this type of correction is naturally introduced.

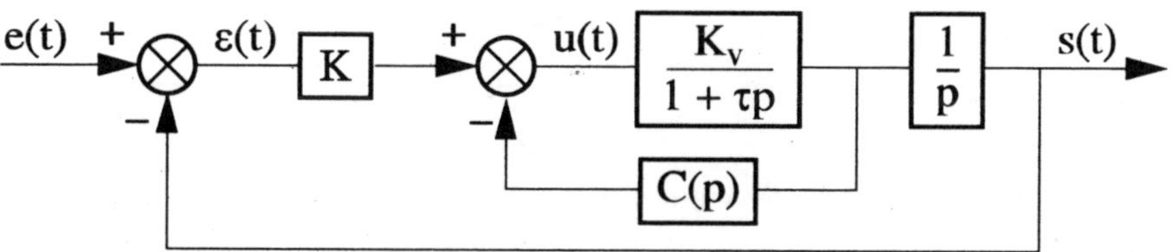

Figure 10.36. *System to correct*

The desired performances are:

– zero theoretical static error;

– minimal bandwidth $\omega_c = \dfrac{8}{\tau}$;,

– stability degree with a phase margin $\Delta\phi = 45°$.

Without $C(p)$ correction, it is necessary to choose a minimum gain K in order to maintain the constraint $\omega_c' > \omega_{c\,min}$.

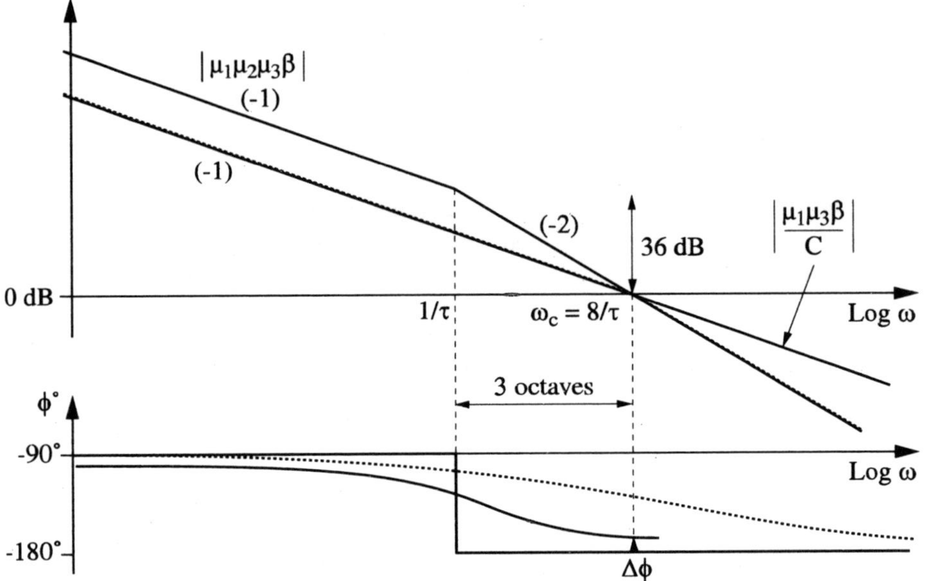

Figure 10.37. *Bode graph of the OL with 45° phase margin*

On the other hand, we have to make sure that the integration in the corrected OL is maintained (ensuring a zero static error).

Let us choose $C(p)$ so that $\dfrac{\mu_1\mu_3\beta}{C}$ has the form represented in Figure 10.37. This graph leads to a 45° phase margin (the break is set at $\omega_c = \dfrac{8}{\tau}$).

In the example of Figure 10.36, $\mu_1 = K, \mu_2 = \dfrac{K_v}{(1+\tau p)}, \mu_3 = \dfrac{1}{p}, \beta = 1$ and point A is rejected at $\omega = 0$ ($-\infty$ in the Bode plane).

Hence, we have:

$$\left|\frac{\mu_1\mu_3\beta}{C}\right| = \left|\frac{K}{Cp}\right| = \left|\frac{8}{\tau p}\right| \quad \text{(equation of the slope } (-1) \text{ cutting the axis 0 dB in}$$

$$\omega = \frac{8}{\tau})$$

By identifying:

$$C(p) = \frac{K\tau}{8}$$

Hence, we see that the output $C(p)$ is homogenous with the derivative of the output of the feedback control. We say we are dealing with a simple tachymetric correction ($C(p) = \lambda$).

In this precise case, a direct calculation by decrease of the secondary loop is preferred. Let us choose $C(p) = \lambda$.

The corrected OL has as a transfer function:

$$\frac{K}{p} \frac{\dfrac{K_v}{1+\tau p}}{1+\dfrac{\lambda K_v}{1+\tau p}} = \frac{K}{p} \frac{K_v}{(1+\lambda K_v + \tau p)} = K \frac{\dfrac{K_v}{1+\lambda K_v}}{p\left(1+\dfrac{\tau}{1+\lambda K_v} p\right)} = \frac{KK_v'}{1+\tau' p}$$

where $K_v' = \dfrac{K_v}{1+\lambda K_v}$, $\tau' = \dfrac{\tau}{1+\lambda K_v}$.

In order to have the same graph of the new OL as before, we must choose parameters K and λ in order to have 45° phase margin $\Delta\phi$, such as:

$$\omega_c = \frac{1}{\tau'} = \frac{8}{\tau}$$

$$\left| \frac{KK_v'}{p(1+\tau' p)} \right|_{p=j\omega_c} = 1 \quad \Rightarrow \quad \frac{KK_v'}{\dfrac{8}{\tau}\sqrt{2}} = 1 \quad \Rightarrow \quad KK_v' = \frac{8}{\tau}\sqrt{2}$$

$$1+\lambda K_v = 8 \quad \Rightarrow \quad \lambda K_v = 7 \quad \Rightarrow \quad K = \frac{64\sqrt{2}}{\tau K_v} \quad \text{and} \quad \lambda = \frac{7}{K_v}$$

or:

$$C(p) = \frac{7}{K_v} = \frac{7K\tau}{64\sqrt{2}} \approx \frac{K\tau}{13}$$

10.7.3. *Filtered tachymetric correction*

We will see in an example how this type of correction is naturally introduced.

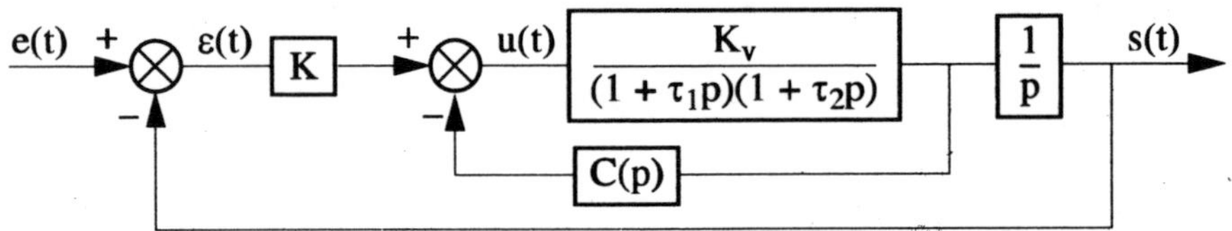

Figure 10.38. *System to correct*

The desired performances are:

– zero theoretical static error;

– minimum gain G_m at $\omega_0 = \dfrac{1}{8\tau}$;

– 45 to 50° phase margin.

Without correction, in order to have the desired gain at ω_0 we have to choose K such that:

$$\left| \frac{KK_v}{p(1+\tau_1 p)(1+\tau_2 p)} \right|_{p=j\omega_0} \geq G_m$$

where $\omega_0 << \dfrac{1}{\tau_1} << \dfrac{1}{\tau_2}$.

Hence, it is sufficient to have $\dfrac{KK_v}{\omega_0} \geq G_m$.

Let us choose $C(p)$ so that $\left|\dfrac{\mu_1 \mu_3 \beta}{C}\right|$ has the form represented in Figure 10.39.

This form was chosen in order to deform the OL according to the example in section 10.3 (correction by combined actions).

The corrected OL has the following form (we note that the quadrilateral ABXΔ is a parallelogram).

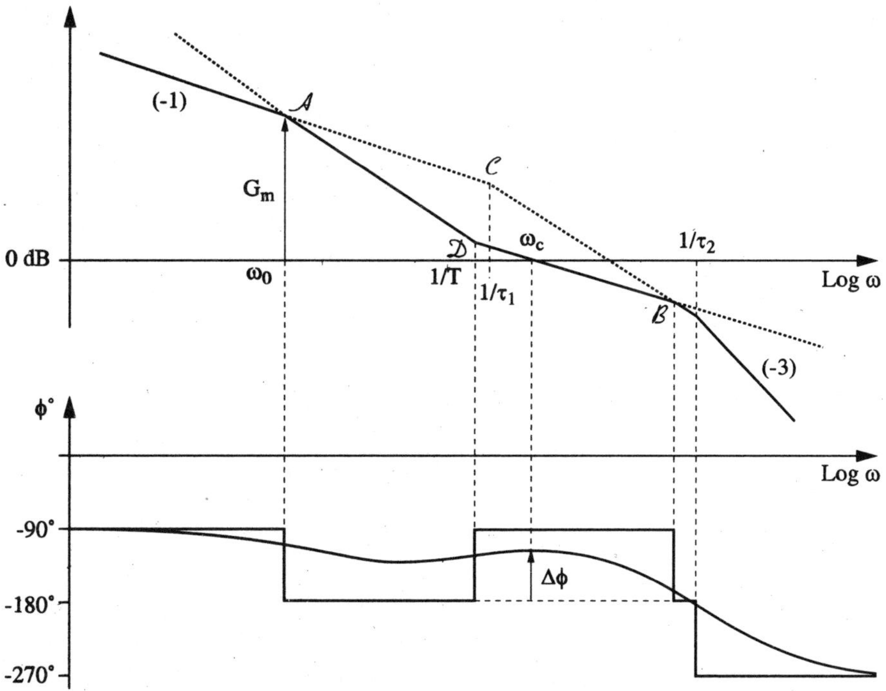

Figure 10.39. *Corrected OL*

In the example of Figure 10.38:

$$\mu_1 = K \,,\ \mu_2 = \frac{K_v}{(1+\tau_1 p)(1+\tau_2 p)} \,,\ \mu_3 = \frac{1}{p} \,,\ \beta = 1$$

and $\left|\dfrac{\mu_1\mu_3\beta}{C}\right| = \left|\dfrac{K}{Cp}\right| = \left|\dfrac{K(1+Tp)}{\lambda Tp^2}\right|$. Hence, the expression of the corrector:

$$C(p) = \frac{\lambda Tp}{1+Tp} = \lambda\frac{Tp}{1+Tp}$$

The realization goes through the reinjection of the speed through a high-pass filter created by the circuit given in Figure 10.40.

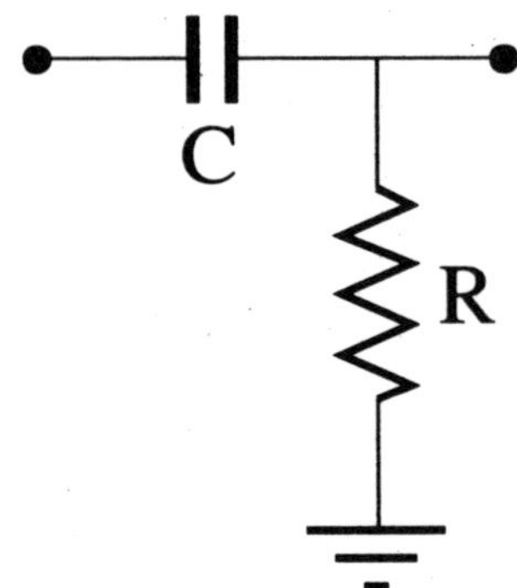

Figure 10.40. *Speed reinjection circuit*

The transfer function of this circuit is $\dfrac{RCp}{1+RCp}$.

Such a correction is called filtered tachymetric correction.

T is obtained by direct reading of the diagram and λ is determined, for example, by writing that $\left|\dfrac{\mu_1\mu_3\beta}{C}\right|$ has a gain of G_m in $\omega = \omega_0$ (point A).

$$\left|\frac{K(1+Tp)}{\lambda Tp^2}\right|_{p=j\omega_0} = G_m$$

In general, $\omega_0 << \dfrac{1}{T} \Rightarrow G_m = \dfrac{K}{\lambda T \omega_0^2}$ $\qquad\qquad \lambda = \dfrac{K}{G_m T \omega_0^2}$

It should be noted that this correction was successful because the variance between points A (ω_0) and C $\left(\dfrac{1}{\tau}\right)$ was sufficient (3 octaves). If points A and C (hence B and D) are too close (2 octaves or less), such a correction cannot provide the sufficient phase margin.

Here again there are no general criteria for the choice of the corrector. The parallel corrections prove to be particularly practical as soon as the measurement of the reinjection variables is available (speed, acceleration, etc.).

Finally, we should remember that the asymptotic graphs are approximate and that it is good to verify the corrected diagrams by using, for example, the Black-Nichols graph.

10.7.4. *Correction of delay systems: Smith predictor*

Let us assume that the system in Figure 10.41 needs to be corrected.

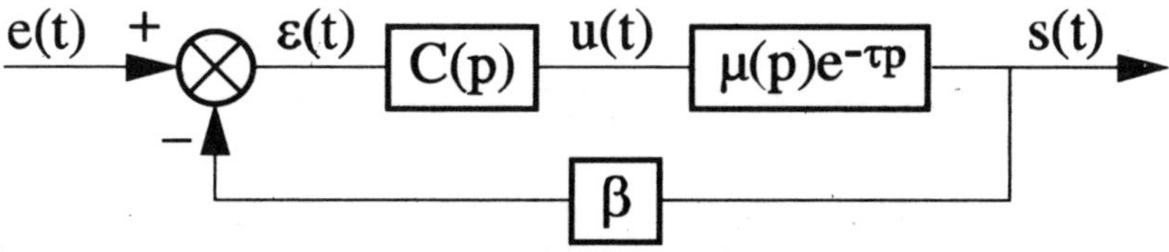

Figure 10.41. *System to be corrected*

The negative effect of the delay on stability is obvious in the fast rotation of the phase. Since the principle of causality prevents the delay compensation by a time advance, the solution proposed by Smith consists of only rejecting the delay outside the loop.

A possible structure of correction is represented in the diagram of Figure 10.42.

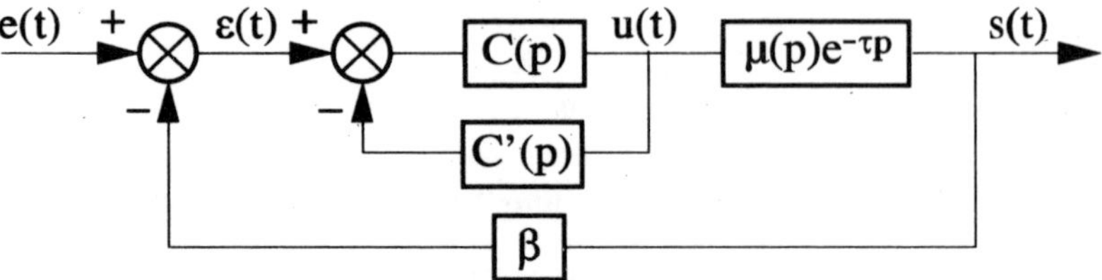

Figure 10.42. *Correction structure*

The corrected transfer function in closed loop is written:

$$\frac{\dfrac{C}{1+CC'}\,\mu e^{-\tau p}}{1+\dfrac{C}{1+CC'}\,\mu e^{-\tau p}\,\beta} = \frac{C\mu e^{-\tau p}}{1+CC'+C\mu\beta e^{-\tau p}}$$

By choosing $C' = \mu\beta(1-e^{-\tau p})$, which is physically feasible, the closed loop becomes:

$$\frac{C\mu}{1+C\mu\beta}\,e^{-\tau p}$$

An equivalent block diagram could be that in Figure 10.43.

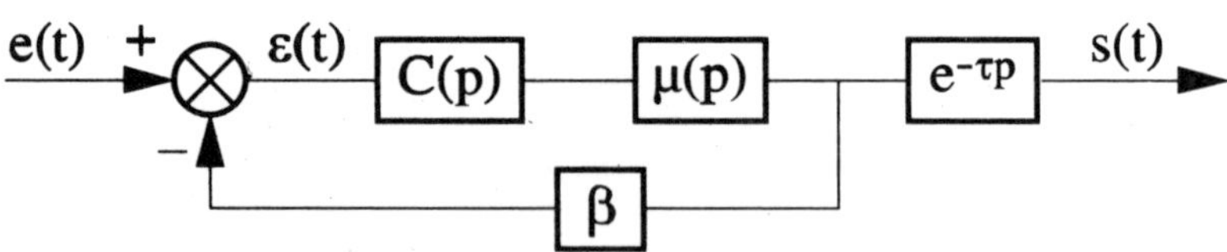

Figure 10.43. *Example of block diagram*

We notice that:

– the delay is obtained from the loop;

– its negative influence on stability is deleted;

– however, it subsists between the input and the output.

A reliability study will be necessary in order to evaluate the incidence of a variance between the pure delay of the process and the one estimated introduced in the corrector C'.

10.8. Bibliography

[CAR 69] DE CARFORT F., FOULARD C., *Asservissements linéaires continus*, Dunod, Paris, 1969.

[FAU 84] FAURE P., ROBIN M., *Eléments d'Automatique*, Dunod, Paris, 1984.

[GIL 70] GILLES J.C., DECAULNE P., PELEGRIN M., *Théorie des systèmes asservis*, Dunod, Paris, 1970.

[LAR 96] DE LARMINAT P., *Automatique,* 2nd edition, Hermès, Paris, 1996.

Chapter 11

Robust Single-Variable Control through Pole Placement

11.1. Introduction

Partially originating from the adaptive control, RST control appeared in books around 1980 [AST 90, FARG 86, LAN 93]. Curiously, this approach was systematically described for the numerical control, perhaps because of its origins mentioned above and in [KUC 79]. In fact, this polynomial approach is the traditional correction with two degrees of freedom, a combination between feedback and feedforward on the setting. The primary goal of this chapter is to replace this order in a general context and to show all the degrees of freedom available to the designer. Then, a very simple, even intuitive, methodology is proposed in order to use these degrees of freedom to achieve a certain robustness of the structure created.

11.1.1. Guiding principles and notations

Figure 11.1 shows the block diagram of the RST control. Block diagram because transfers R, S and T are polynomials and are thus not proper.

Chapter written by Gérard THOMAS.

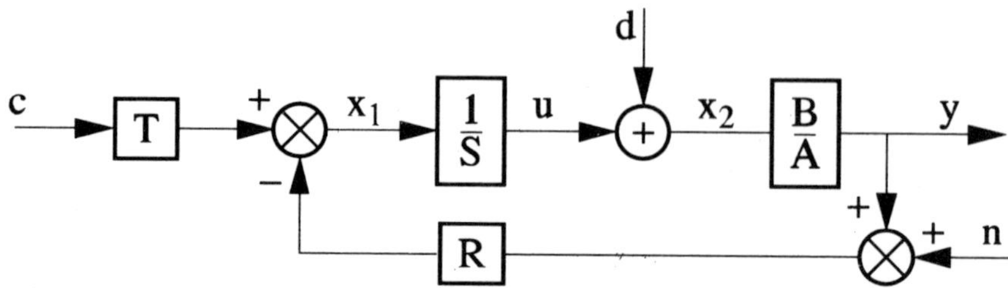

Figure 11.1. *Block diagram of RST control*

In all that follows, unless otherwise indicated, the systems studied will be discrete or continuous, i.e. will be respectively described by transfers of the z or s variable. For reasons of simplicity and coherence, the examples will be treated in the continuous case.

As for any correction structure, the designer will have to determine the correction parameters (here polynomials R, S and T) to ensure:

– internal stability [DOY 92];

– the asymptotic follow-up of a certain class of settings;

– the asymptotic rejection of a certain class of interferences;

– a satisfactory transient state.

However, respecting these specifications is not sufficient to ensure a satisfying operation of the installation; it will be necessary to take into account:

– the saturations of the process;

– the level of measurement noise;

– modeling errors.

The non-compliance with these simple rules has had negative impacts on automatic control, which is then considered as a highly theoretical discipline whose industrial applications seldom exceeded the performances obtained by PIDs.

The guiding principle of the RST control is to calculate the polynomials R, S and T to obtain:

$$\frac{Y}{C} = \frac{BT}{AS + BR} = \frac{B_m}{A_m} \qquad [11.1]$$

which will be satisfied if:

$$(a) \quad BT = B_m A_o$$
$$(b) \quad AS + BR = A_m A_o$$

[11.2]

We observe that the unknown factors of the problem (R, S and T) are the solutions of polynomial equations. In particular, the latter is well-known by algebraists as a Bezout equation or Diophantus problem. That is why the following section is dedicated to some reminders on polynomial algebra. It shall be noted that this formalism was extremely well emphasized in [KUC 79] within the multi-variable and discrete context and it is the starting point of the next section.

11.1.2. *Reminders on polynomial algebra*

A certain number of traditional results on polynomials is gathered here, in order to solve the general polynomial equation:

$$AX + BY = C$$

[11.3]

We must point out that here we are interested in the single-variable case where A, B, X, Y, C are polynomials and not matrices of polynomials. Thus the multiplications can be written in a random order.

THEOREM 11.1.– *the set of the polynomials with an unknown quantity on a commutative body is a commutative unitary ring.*

The ring of polynomials on $\Re$ will be called $\Re[x]$. We note by:

– 1 the identity polynomial (the neutral element of the multiplication in $\Re[x]$);

– 0 the zero polynomial (the neutral element of the addition in $\Re[x]$),

– ∂A the degree of polynomial A.

We will assume that the concepts of polynomials division are known, as well as those of PGCD and PPCM of polynomials. If G and L are respectively the PGCD and the PPCM of A and B (A, B, G and L $\in \Re[x]$), we will write:

$$G = A \wedge B \text{ and } L = A \vee B$$

DEFINITION 11.1.– *we say that several polynomials are prime among themselves when their PGCD is of 0 degree, i.e. when their only common divisors are non-zero constants.*

THEOREM 11.2 (BEZOUT THEOREM).– *a necessary and sufficient condition for n A_i polynomials to be prime among themselves is that there are n V_i polynomials such that:*

$$\sum_{i=1}^{n} A_i V_i = 1 \qquad [11.4]$$

THEOREM 11.3 (BEZOUT EQUALITY).– *since A and B are two polynomials prime among themselves, other than constants, there is only one pair of polynomials X and Y verifying:*

$$XA + YB = C \text{ with } \partial X < \partial B \text{ and } \partial Y < \partial A \qquad [11.5]$$

THEOREM 11.4 (GENERALIZATION).– *if A and B are two polynomials of PGCD G, then there is only one pair of polynomials X and Y such that:*

$$XA + YB = G \quad \begin{cases} \partial X < \partial B - \partial G \\ \partial Y < \partial A - \partial G \end{cases} \qquad [11.6]$$

THEOREM 11.5.– *equation $XA + YB = 1$ [11.7] has a solution if and only if the PGCD of A and B divides C.*

THEOREM 11.6.– *let (X0,Y0) be a particular solution of $XA + YB = C$ [11.8] and let A_1 and B_1 be two polynomials prime among themselves such that $A = A_1 G$ [11.9] and $B = B_1 G$ [11.10] where $G = A \wedge B$; thus the general solution is given by:*

$$\begin{cases} X = X_0 - B_1 P \\ Y = Y_0 + A_1 P \end{cases} \qquad [11.11]$$

where P is any polynomial of $\mathfrak{R}[x]$.

Among all these solutions it is usual to seek a single solution which confirms a particular property. The most usual is the solution of minimum degree.

Let (X_0, Y_0) be a particular solution of [11.3]; we know (Theorem 11.6) that the general solution is written:

$$\begin{cases} X = X_0 - B_1 P \\ Y = Y_0 + A_1 P \end{cases} \tag{11.12}$$

with $A = A_1 G$ and $B = B_1 G$ where $G = A \wedge B$ and P is any polynomial of $\mathfrak{R}[x]$.

By carrying out the Euclidean division of X_0 by B_1 we obtain:

$$X_0 = B_1 U + V \quad \text{with} \quad \partial V < \partial B_1 \tag{11.13}$$

by replacing in [11.12] we obtain:

$$X = V - B_1(P - U) \tag{11.14}$$

the solution for [11.3] with minimum degree in X will be obtained for $P = U$ or:

$$\left. \begin{array}{l} X = V \\ Y = Y_0 + A_1 U \end{array} \right\} \tag{11.15}$$

Indeed, based on [11.14]:

$$\partial X \leq \max\{\partial V, \partial B_1(P - U)\} \tag{11.16}$$

If $P \neq U$, then:

$$\partial B_1(P - U) \geq \partial B_1 \tag{11.17}$$

and since by construction:

$$\partial V < \partial B_1 \tag{11.18}$$

$$\max\{\partial V, \partial B_1(P - U)\} \geq \partial B_1 \tag{11.19}$$

the hypothesis P ≠ U leads to a solution in X of a higher degree than that obtained for P = U.

❑

NOTE 11.1.– the solution of minimum degree for X does not generally coincide with the solution of minimum degree in Y.

The preceding theorems make it thus possible to calculate the solution for [11.3]. Now that the resolution tools of polynomial equations are known, it is advisable to specify in relations [11.2] the degrees of freedom available to the designer and also to equally translate the constraints of synthesis related to the nature of the problem and the specifications of the correction.

11.2. The obvious objectives of the correction

11.2.1. *Internal stability*.

It is difficult to take a final decision at this stage since the representation of the correction given in Figure 11.1 is formal and does not represent the real implementation. However, it is clear [DOY 92] that the denominator of all the transfers being $A_m A_o$, these two polynomials must be stable (besides the simplification carried out by A_o in [11.1] already supposed the stability of A_o); on the other hand, there should be no simplification of unstable root of A or B by the correctors built. On the other hand, the reverse is possible, i.e. we can choose some of the polynomials R, S and T in order to carry out such simplifications.

Thus, based on the transfer in closed loop, $\dfrac{Y}{C} = \dfrac{BT}{AS + BR}$ it is possible to hide zeros and (stable) poles of the model of the process by using S or R. Let us note, following the example of [AST 90]:

$$A = A^+ A^- \text{ and } B = B^+ B^- \tag{11.20}$$

where $P^+ P^-$ represents the spectral factorization of the polynomial P, the roots of P^+ being all stable[1], the roots of P^- being all unstable. By supposing that:

$$S = B^+ S' \text{ and } R = A^+ R' \tag{11.21}$$

---

1 Open left half-plane for the continuous systems, the open disc of unit radius for the discrete models.

we get:

$$\frac{Y}{C} = \frac{BT}{AS + BR} = \frac{B^+ B^- T}{A^+ B^+ (A^- S' + B^- R')}$$

[11.22]

The choice of $T = A^+ T'$ makes it possible to simplify by $A^+ B^+$. We are in fact brought back to the preceding problem where R, S and T are replaced by R', S' and T', and A, B by A^-, B^-. This is why subsequently, unless told otherwise, the simplifications will not be mentioned.

11.2.2. *Stationary behavior*

Since the internal stability is guaranteed, it is now possible to deal with the following stage, namely with the stationary behavior. The specifications of the correction outline the settings and interferences likely to stimulate the process. Let $e(t)$ be the error signal (not explicit in the correction structure in Figure 11.1) neglecting the supposed noise of zero mean value:

$$E = C - Y = \frac{AS + B(R - T)}{AS + BR} C + \frac{BS}{AS + BR} D$$

[11.23]

Generally, the authors [AST 90] then use [11.2 (b)] to simplify the expression of the contribution of the setting. In this case, the stationary behavior with respect to the order depends only on A_m and B_m, the asymptotic follow-up of a step function setting resulting in the choice of a reference model of unit static gain. However, as it is noticed in [COR 96, WOL 93] this supposes a perfect identification of the procedure! In fact, the relations [11.2] are only true for the model of the procedure. Let A' and B' be "the true" values of the denominator and numerator of the procedure; the real error obtained through the implementation of the RST corrector, calculated using model A, B, will in fact be:

$$E = C - Y = \frac{A'S + B'(R - T)}{A'S + B'R} C + \frac{B'S}{A'S + B'R} D$$

[11.24]

and of course $A'S + B'R \neq A_m A_o$. We suppose that:

$$C = \frac{N_c}{D_c^+ D_c^-} \quad \text{and} \quad D = \frac{N_d}{D_d^+ D_d^-}$$

[11.25]

where N_x and D_x are polynomials prime among themselves, the indices + and − having the same significance as in [11.20]. Thus, for a continuous ramp setting we will have $D_c^- = s^2$ and for a sinusoidal disturbance of angular frequency ω_o, $D_d^- = s^2 + \omega_o^2$.

By supposing that the calculated correction is sufficiently robust so that $A'S + B'R$ has all its roots stable, the stationary error will be cancelled only if D_c^- divides $AS + B(R - T)$ and D_d^- divides BS. As seen above, the values of A and B are not exact and thus it is R, S and T that will provide this function[2]. The stationary specifications thus lead to imposing the following constraints (without taking into account possible integrations of the process):

$$\begin{cases} S = D_c^- S' \\ R - T = D_c^- L \\ S = D_d^- S'' \end{cases} \text{or} \begin{cases} S = D_{dc}^- S_1 \\ R - T = D_c^- L \\ D_{dc}^- = D_c^- \vee D_d^- \end{cases} \qquad [11.26]$$

The preceding section s made it possible to set a certain number of constraints on the unknown factors of the problem and provided a general context for its solving. The following section will provide a calculation tool for polynomials R, S and T.

11.2.3. *General formulation*

We must solve [11.2] with the conditions [11.26], or:

$$\begin{aligned} BT &= B_m A_o \\ AS + BR &= A_m A_o \end{aligned} \quad \text{with} \begin{cases} S = D_{dc}^- S_1 \\ R - T = D_c^- L \\ D_{dc}^- = D_c^- \vee D_d^- \end{cases} \qquad [11.27]$$

---

2 When the process is integrator we can write $A = sA'$ despite identification errors.

Since $BT = B_m A_o$, B must divide the $B_m A_o$ product. We saw above (section 11.1.3) that polynomial A_o must be stable and thus it can share with B only stable roots. Let B_1 be the part of factorized B in $A_o{}^3$. Consequently, polynomial B_m must "become in charge" with the non-factorized part of B in A_o. Hence, let us assume that:

$$B = B_1 B_2 \quad A_o = B_1 A_o' \quad B_m = B_2 B_m'$$ [11.28]

Therefore, B_m will have to contain at least all the unstable roots of B. Taking into account these factorizations, we obtain:

$$T = B_m' A_o$$ [11.29]

On the other hand, according to [11.2(b)], since $B1$ divides A_o and B it also divides AS. However, A and B are prime between themselves by hypothesis and therefore B_1 divides S and S_1 (since B_1 is stable and not D_{dc}^-). Finally, we can write:

$$(a) \quad B = B_1 B_2 \quad B_1 \quad stable$$
$$(b) \quad D_{dc}^- = D_c^- \vee D_d^-$$
$$(c) \quad A D_{dc}^- \overline{S} + B_2 R = A_m A_o'$$
$$(d) \quad D_c^- L + B_m' A_o' = R \qquad\qquad [11.30]$$
$$(e) \quad S = D_{dc}^- B_1 \overline{S}$$
$$(f) \quad T = B_m' A_o'$$
$$(g) \quad B_m = B_2 B_m'$$

All these relations express the respect of internal stability (by supposing of course A_m and A_o' stable) and desired stationary performances. We notice that these relations require the choice of polynomials (A_m, A_o') and the factorization of B and then the solving of two Diophantus equations [11.30(c)] and [11.30(d)]. The following section is dedicated to the complete resolution of [11.30]. In particular it will be pointed out which are the degrees of freedom available to the designer in the choices mentioned above.

---

3 We will have a maximum of $B_1 = B^+$ according to the notations in section 3.1.3, equation [3.20].

11.3. Resolution

As previously seen, it is possible to develop a general solution (Theorem 11.6) by formal calculation. However, it is more usual to solve the Diophantus equations resulting from this approach by using linear algebra. This approach makes it possible to set the degrees of freedom of the designer. Indeed, if we write[4]:

$$
\begin{cases}
A = \displaystyle\sum_{i=0}^{\partial A} a_i s^i \quad & B = \displaystyle\sum_{i=0}^{\partial B} b_i s^i \quad & C = \displaystyle\sum_{i=0}^{\partial C} c_i s^i \\[2em]
X = \displaystyle\sum_{i=0}^{\partial X} x_i s^i \quad & Y = \displaystyle\sum_{i=0}^{\partial Y} y_i s^i
\end{cases}
\qquad [11.31]
$$

The resolution of equation [11.3] goes back to that of the following system:

$$
\begin{bmatrix}
a_0 & 0 & 0 & \cdots & b_0 & 0 & 0 & \cdots \\
a_1 & a_0 & 0 & \cdots & b_1 & b_0 & 0 & \ddots \\
\vdots & a_1 & a_0 & \ddots & \vdots & b_1 & b_0 & \ddots \\
 & \vdots & a_1 & \ddots & & \vdots & b_1 & \ddots \\
 & & \vdots & \ddots & b_{\partial B} & & \vdots & \ddots \\
a_{\partial A} & & & & 0 & b_{\partial B} & & \\
0 & a_{\partial A} & & & \vdots & 0 & b_{\partial B} & \\
\vdots & 0 & a_{\partial A} & & & \vdots & 0 & \ddots \\
 & \vdots & 0 & \ddots & & & \vdots & \ddots \\
 & & \vdots & \ddots & & & \vdots & \ddots
\end{bmatrix}
\begin{bmatrix}
x_0 \\ x_1 \\ \vdots \\ \\ \\ x_{\partial X} \\ y_0 \\ \vdots \\ \\ \\ y_{\partial Y}
\end{bmatrix}
=
\begin{bmatrix}
c_0 \\ c_1 \\ \vdots \\ \\ \\ \\ \\ \\ \vdots \\ c_{\partial C}
\end{bmatrix}
\qquad [11.32]
$$

Each row is obtained by equalizing the terms having the same power in [11.3]. This system is called the Sylvester system. The resolution of this system of equations requires knowing the degrees of the various polynomials and that part has not yet been set. It must be noted that in our problem, A and B are prime between themselves and, consequently, according to Theorem 11.5 [11.32] has one solution.

---

4 For discrete systems the variable would be "z" and not "s".

Before solving the general case, we will deal with a particular case (the one that is the most frequently dealt with in other works), which will enable us to show the approach used.

11.3.1 *Resolution of a particular case*

We find ourselves here in the case when no specification is made on the setting and the interference. Thus, only relations [11.2] should be solved. We know that polynomials A_m and A_o must be stable, but this information is not sufficient for the designer and we must know the degrees of these polynomials to write the Sylvester system. Can we choose these degrees randomly? The following section makes it possible to answer this question.

11.3.1.1. *Conditions on the degrees*

We suppose [DOY 92] that the model of the process is strictly proper, whereas the correctors will be supposed simply proper. Consequently:

$$\begin{cases} (a) & \partial A > \partial B \\ (b) & \partial S \geq \partial R \\ (c) & \partial S \geq \partial T \end{cases} \qquad [11.33]$$

From relations [11.33(a) and (b)] and [11.2(b)], we obtain:

$$\partial B + \partial R < \partial A + \partial S = \partial A_o + \partial A_m \qquad [11.34]$$

The uniqueness of the solution will thus be obtained by simply imposing:

$$\text{number of equations} = \text{number of unknown factors} \qquad [11.35]$$

The unknown factors in [11.2(b)] are the coefficients of the polynomials S and R and thus[5]:

$$\text{number of unknown factors} = \partial S + \partial R + 2 \qquad [11.36a]$$

---

5 A polynomial of degree n has $n + 1$ coefficients.

The number of equations is the number of rows in the Sylvester system and thus:

$$\text{number of equations} = \partial A_m + \partial A_o + 1 \qquad [11.36b]$$

Considering the uniqueness of the solution and taking into account [11.34], we obtain:

$$\partial A_m + \partial A_o + 1 = \partial S + \partial R + 2 \qquad [11.37]$$

$$\partial R = \partial A - 1 \qquad [11.38]$$

while of course always using [11.34]:

$$\partial S = \partial A_m + \partial A_o - \partial A \qquad [11.39]$$

Until this stage, polynomials A_m and A_o do not have any constraint except for stability. The conditions for the regulators to be proper will introduce the following constraints:

$$\partial S \geq \partial R \quad \Rightarrow \quad \partial A_m + \partial A_o - \partial A \geq \partial A - 1 \qquad [11.40]$$

from where we obtain the first inequality referring to the degrees of A_m and A_o:

$$\partial A_m + \partial A_o \geq 2\partial A - 1 \qquad [11.41]$$

by using [11.2(a)] and the fact that $\dfrac{T}{S}$ is proper, we obtain:

$$\partial S \geq \partial T \quad \Rightarrow \quad \partial S + \partial B \geq \partial T + \partial B = \partial B_m + \partial A_o \qquad [11.42]$$

and finally by using [11.39] we obtain a second inequality:

$$\underbrace{\partial A_m + \partial A_o - \partial A}_{\partial S} + \partial B \geq \partial B_m + \partial A_o \qquad [11.43]$$

which rearranged gives:

$$\partial A_{\mathrm{m}} - \partial B_{\mathrm{m}} \geq \partial A - \partial B \qquad [11.44]$$

This simply means that the correction can only increase the relative degree.

11.3.1.2. *Standard solution*

We must first of all choose B_{m}. By using the factorization $B = B_1 B_2$ where B_1 is stable (we can choose $B_1 = 1$ if we want to have a completely free choice of A_{o}) and consequently:

$$A_{\mathrm{o}} = B_1 A_{\mathrm{o}}' \text{ and } B_{\mathrm{m}} = B_2 B_{\mathrm{m}}' \qquad [11.45]$$

In order to minimize the complexity of the elements of the corrector, we usually choose a polynomial $B_{\mathrm{m}}' = \alpha$ the constant α being chosen in order to ensure a unit static gain for the model of reference[6]. Relations [11.41] and [11.44] thus become ($\partial B_{\mathrm{m}}' = 0$):

$$\partial A_{\mathrm{m}} - \underbrace{(\partial B_2 + \partial B_{\mathrm{m}}')}_{\partial B_{\mathrm{m}}} \geq \partial A - \underbrace{(\partial B_1 + \partial B_2)}_{\partial B}$$

$$\Rightarrow \partial A_{\mathrm{m}} \geq \partial A - \partial B_1 \qquad [11.46]$$

$$\partial A_{\mathrm{m}} + \partial A_{\mathrm{o}}' + \partial B_1 \geq 2\partial A - 1$$

$$\Rightarrow \partial A_{\mathrm{m}} + \partial A_{\mathrm{o}}' \geq 2\partial A - 1 - \partial B_1$$

We have seen in [11.30] that polynomial S is thus divisible by B_1. Figure 11.2 gives a graphic representation of the conditions [11.41] and [11.46].

---

6 It is pointed out that this choice is only a necessary condition to the asymptotic follow-up of a step function setting (see section 11.2.2).

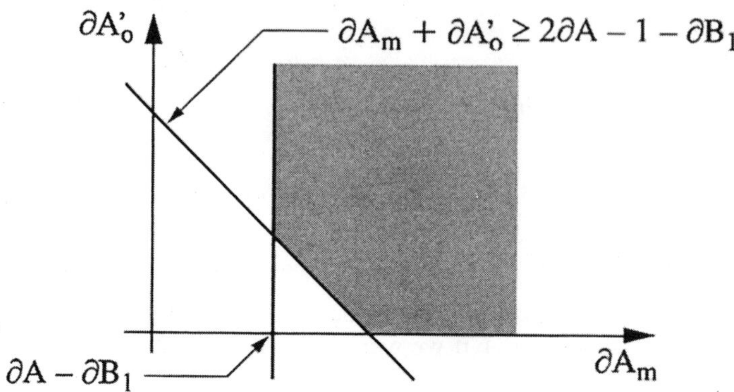

Figure 11.2. *Choice of the degrees of A_m and A_o'*

11.3.1.3. *Example*

Let us take an academic example. Let the process be described by the transfer:

$$\frac{B}{A} = \frac{s+1}{s(s^2 + 0,1s + 1)}$$

Therefore, we have the choice to make $B_1 = 1$ or $s+1$ and thus $B_2 = s + 1$ or respectively.

$$B_1 = s+1 \text{ and } B_2 = 1$$

The inequalities [11.46] thus lead to the relations:

$$\partial A_m \geq 3 - 1 = 2$$
$$\partial A_m + \partial A_o' \geq 2*3 - 1 - 1 = 4$$

If we choose the polynomials of minimum degree, we obtain:

$$\partial A_m = 2$$
$$\partial A_o' = 2$$

We can thus choose[7]:

$$A_m = A_o' = (s + 2)^2$$
$$B_m = 4$$

This choice of B_m ensures a unit static gain to the reference model. We can thus use the resolution of the Bezout equation:

$$AS + BR = A_m A_o$$

The resolution of this system of equations in this particular case gives:

$$S = (s + 1)(s + 7.9)$$
$$R = 22.21s^2 + 24.1s + 16$$
$$T = B_m A_o' = 4(s + 2)^2$$
$$B_1 = 1$$

The inequalities [11.46] lead then to the relations:

$$\partial A_m \geq 3$$
$$\partial A_m + \partial A_o' \geq 2*3 - 1 = 5$$
$$\text{here} \quad A_o' = A_o$$

If we choose the polynomials of minimum degree, we have:

$$\partial A_m = 3$$
$$\partial A_o' = 2$$

---

7 The choice of the roots of these polynomials will be seen later. For the moment the only constraint is to choose them stable.

We can thus take:

$$A_{\mathrm{m}} = (s + 2)^3$$
$$A_{\mathrm{o}} = (s + 2)^2$$
$$B_{\mathrm{m}} = 8(s + 1)$$

By using the same procedure as before we obtain:

$$S = s^2 + 7s + 7$$
$$R = 20s^2 + 41s + 32$$
$$T = B_{\mathrm{m}}'.A_{\mathrm{o}} = 8(s + 2)^2$$

11.3.2. *General case*

11.3.2.1. *Choice of degrees*

This section is dedicated to the resolution of relations [11.30]. The methodology is the same as the one used in the preceding paragraph, but here the conditions of being proper do not relate directly to the unknown polynomials. Relations [11.33], [11.34] and [11.39] are always valid since we must solve:

$$A\underbrace{(D_{\mathrm{dc}}^- B_1 \overline{S})}_{S} + \underbrace{(B_1 B_2)}_{B} R = A_{\mathrm{m}}\underbrace{(B_1 A_{\mathrm{o}}')}_{A_{\mathrm{o}}} \qquad [11.47]$$

By taking into account [11.39] and the factorization of S, the degree of $\overline{S}$ is given by:

$$\partial \overline{S} = \partial A_{\mathrm{m}} + \partial A_{\mathrm{o}} - \partial A - \partial D_{\mathrm{dc}}^- \qquad [11.48]$$

The uniqueness of the solution will be ensured if the number of equations is equal to the number of unknown factors, or:

$$\underbrace{\partial A + \partial \overline{S} + \partial D_{\mathrm{dc}}^- + 1}_{\text{number of equations}} = \underbrace{\partial \overline{S} + \partial R + 2}_{\text{number of unknown factors}} \qquad [11.49]$$

$$\Rightarrow \partial R = \partial A + \partial D_{\mathrm{dc}}^- - 1$$

To solve [11.30(c)], it is necessary to know the degree of S and thus that of B_m', which itself is solution of [11.30(d)]. For this equation, there is no constraint on being proper and we will set the uniqueness of the solution by retaining that of minimum degree in B_m'. The idea is to minimize the complexity of the transfers to be done. We consequently obtain:

$$\partial B_m{}' = \partial D_c^- - 1 \, (\partial D_c^- \geq 1)$$

[11.50]

We must now represent the property of the corrector by using [11.48] and [11.49]. The inequality $\partial S \geq \partial R$ gives:

$$\partial A_m + \partial A_o \geq 2\partial A + \partial D_{dc}^- - 1$$

[11.51]

also, the inequality $\partial S \geq \partial T$ leads to the condition:

$$\partial A_m \geq \partial A + \partial D_c^- - \partial B_1 - 1$$

[11.52]

Figure 11.3 represents these inequalities geometrically.

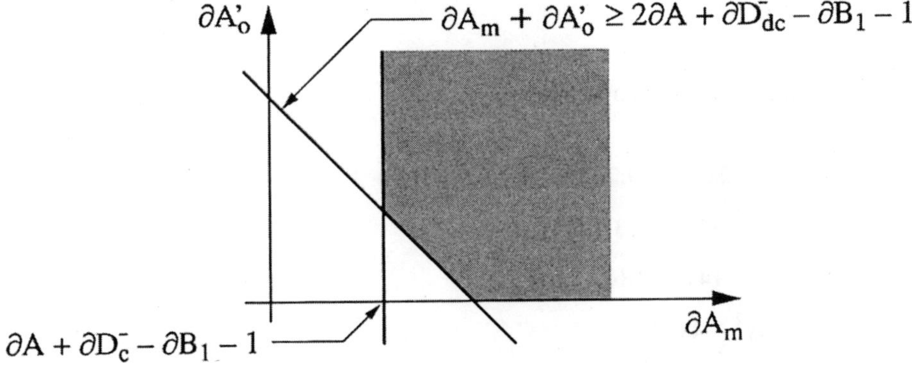

Figure 11.3. *Choice of degrees of A_m and A_o'*

11.3.2.2. *Example*

Let us take the following example:

$$A = s(s^2 + 0.1s + 1)$$

$$B = s + 0.5$$

$$\left. \begin{array}{l} D_c^- = s^2 \\ D_d^- = s \end{array} \right\} \Rightarrow D_{dc}^- = s^2$$

the inequalities [11.51] and [11.52], by choosing $B_1 = 1$, give:

$$\partial A_m + \partial A_m \geq 2\partial A + \partial D_{dc}^- - 1 = 2*3 + 2 - 1 = 7$$

$$\partial A_m \geq \partial A + \partial D_c^- - 1 = 3 + 2 - 1 = 4$$

by using the minimal degrees and by choosing identical dynamics for A_m and A_o we can take:

$$A_m = (s+1)^4$$

$$A_o = (s+1)^3$$

We obtain in this particular case:

$$R = 16.1621s^4 + 19.7042s^3 + 22s^2 + 10s + 2$$

$$S = s^2(s^2 + 6.9s + 3.1479)$$

$$T = 4s^4 + 14s^3 + 18s^2 + 10s + 2$$

$$B_m = 4s^2 + 4s + 1$$

11.4. Implementation

In section 11.1.1 it was mentioned that the structure in Figure 11.1 is formal because the represented transfers are not proper. This section makes it possible to carry out the control law.

11.4.1. *First possibility*

$$Su = Tc - Ry \qquad\qquad [11.53]$$

with physically feasible operators. A possibility [IRV 91] consists of introducing a stable auxiliary polynomial F of an equal degree to that of S into relation [11.53], which becomes:

$$\frac{S}{F}u = \frac{T}{F}c - \frac{R}{F}y \qquad\qquad [11.54]$$

the corrector is thus carried out as indicated in Figure 11.4.

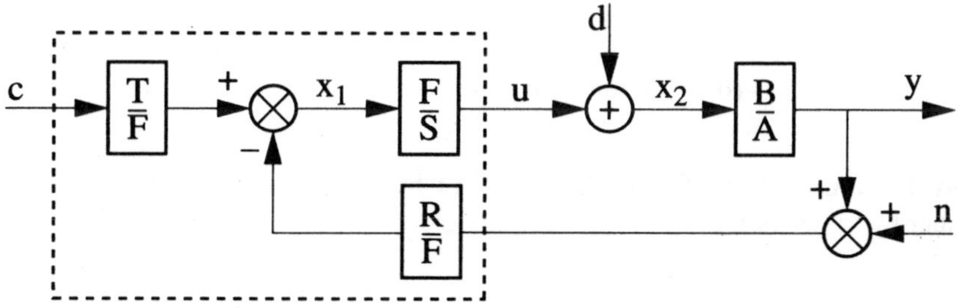

Figure 11.4. *Realization of the corrector*

This realization is not minimal because it leads to the construction of three transfers of ∂S order. The following section provides a minimal representation of the RST regulator [CHE 87].

11.4.2. *Minimal representation*

If we return to relation [11.53], we can obviously write:

$$u = \frac{T}{S}c - \frac{R}{S}y \qquad\qquad [11.55]$$

The realization of the first term leads in the majority of cases to the achievement of an unstable transfer (S always has a zero root). It is rather necessary to regard the corrector as a system having two inputs $c(t)$ and $y(t)$ and one output $u(t)$ and thus it is enough to write an equation of state verified by this system. The following example illustrates the procedure.

11.4.2.1. *Example*

For $A = s^2 + s + 1$; $B = 2$; $A_m = A_o = (s+1)^2$; $D_{\bar{c}} = D_{\bar{d}} = s$, by using the previously described procedure, we obtain:

$$R = s^2 + 0.5s + 0.5$$

$$S = s^2 + 3s$$

$$T = 0.5s^2 + s + 0.5$$

Hence, the control verifies:

$$(s^2 + 3s)U = (0.5s^2 + s + 0.5)C - (s^2 + 0.5s + 0.5)Y$$

By leaving on the left the term in U of highest degree and by dividing each member by s^2, we obtain:

$$U = 0.5C - 0.5Y + \frac{1}{s}\{-3U + C - 0.5Y + \frac{1}{s}[0.5C - 0.5Y]$$

and thus by supposing that[8]:

$$X_1 = \frac{1}{s}[0.5C - 0.5Y]$$

$$X_2 = \frac{1}{s}[-3U + C - 0.5Y + X_1]$$

---

[8] Signals x_1 and x_2 here do not have any relationship with those in Figure 11.1.

in the time field we have:

$$\dot{x}_1 = 0.5c - 0.5y$$
$$\dot{x}_2 = -3u + c - 0.5y + x_1$$
$$u = 0.5c - y + x_2$$

finally, by replacing in $\dot{x}_2$, u by its expression according to c, y and x_2:

$$\dot{x}_1 = 0.5c - 0.5y$$
$$\dot{x}_2 = x_1 - 3x_2 - 0.5c + 2.5y$$
$$u = 0.5c - y + x_2$$

and in the matrix, we have:

$$A = \begin{bmatrix} 0 & 0 \\ 1 & -3 \end{bmatrix} \quad B = \begin{bmatrix} 0.5 & -0.5 \\ -0.5 & 2.5 \end{bmatrix} \quad C = \begin{bmatrix} 0 & 1 \end{bmatrix} \quad D = \begin{bmatrix} 0.5 & -1 \end{bmatrix}$$

11.4.2.2. *Generalization*

If we write[9]:

$$S = \sum_{i=0}^{n} \sigma_i s^i$$

$$R = \sum_{i=0}^{n} \rho_i s^i \quad \text{with} \quad \sigma_n = 1 \quad \text{and} \quad n = \max(\partial S, \partial R, \partial T) = \partial S \tag{11.56}$$

$$T = \sum_{i=0}^{n} \tau_i s^i$$

---

[9] The continuous case is used here, but the approach is completely identical to the discrete case: it is enough to replace s by z in what follows.

this procedure can be generalized. Using [11.52] and [11.56], we obtain:

$$s^n.(u - \tau_n c + \rho_n y) = \sum_{i=0}^{n-1} s^i(-\sigma_i u + \tau_i c - \rho_i y)$$

either by dividing by s^n:

$$(u - \tau_n c + \rho_n y) = s^{-1}\{-\sigma_{n-1}.u + \tau_{n-1}c - \rho_{n-1}y + ...$$
$$s^{-1}\{[(-\sigma_{n-2}u + \tau_{n-2}c - \rho_{n-2}y) + ... + s^{-1}(-\sigma_0 u + \tau_0 c - \rho_0 y)]\} \}$$

or by supposing that:

$$\begin{aligned}
x_1 &= s^{-1}(-\sigma_0 u + \tau_0 c - \rho_0 y) \\
x_2 &= s^{-1}\{(-\sigma_1 u + \tau_1 c - \rho_1 y) + x_1\} \\
&\vdots \\
x_{n-1} &= s^{-1}\{(-\sigma_{n-2}u + \tau_{n-2}c - \rho_{n-2}y) + x_{n-2}\} \\
x_n &= s^{-1}\{(-\sigma_{n-1}u + \tau_{n-1}c - \rho_{n-1}y) + x_{n-1}\} \cdot \\
&= u + \tau_n c - \rho_n y
\end{aligned}$$

$$[11.57]$$

It is easy to see that the last equation makes it possible to express u according to c, y and x_n. That represents of course the output equation and thus:

$$\begin{aligned}
C &= [0 \cdots\cdots 0\ 1] \\
D &= [-\tau_n \quad \rho_n]
\end{aligned}$$

$$[11.58]$$

By replacing u by $x_n - \tau_n c + \rho_n y$ in the expressions of $x_1, x_2, \ldots x_n$ we obtain matrices $\mathbf{A}$ and $\mathbf{B}$ of the equation of state:

$$
\mathbf{A} = \begin{bmatrix}
0 & 0 & . & . & . & 0 & -\sigma_0 \\
1 & 0 & . & . & . & 0 & -\sigma_1 \\
0 & 1 & \ddots & . & . & 0 & -\sigma_2 \\
0 & 0 & \ddots & \ddots & . & \vdots & \vdots \\
\vdots & \vdots & \ddots & \ddots & \ddots & \vdots & \vdots \\
\vdots & \vdots & . & \ddots & \ddots & 0 & -\sigma_{n-2} \\
0 & 0 & . & . & 0 & 1 & -\sigma_{n-1}
\end{bmatrix}
$$

$$
\mathbf{B} = \begin{bmatrix}
\sigma_0 \tau_n + \tau_0 & -\sigma_0 \rho_n - \rho_0 \\
\sigma_1 \tau_n + \tau_1 & -\sigma_1 \rho_n - \rho_1 \\
\vdots & \vdots \\
\vdots & \vdots \\
\sigma_{n-1} \tau_n + \tau_{n-1} & -\sigma_{n-1} \rho_n - \rho_{n-1}
\end{bmatrix}
$$

$$[11.59]$$

11.4.3. *Management of saturations*

11.4.3.1. *Method*

Many failures that occurred when the so-called "advanced" techniques were applied could have been avoided if the implemented regulators had managed the inherent saturations of every industrial procedure. The RST control is no exception. The previously discussed example can be used to emphasize the problem and its solution. We will be dealing with the structure described in Figure 11.5.

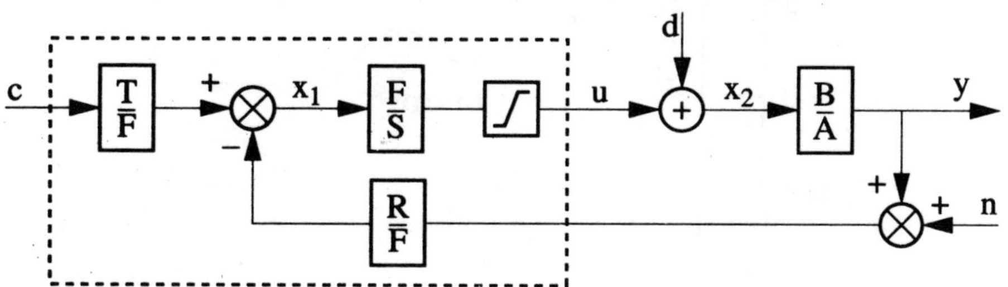

Figure 11.5. *Corrector with saturation*

The following answers show the difference in operation with and without saturation.

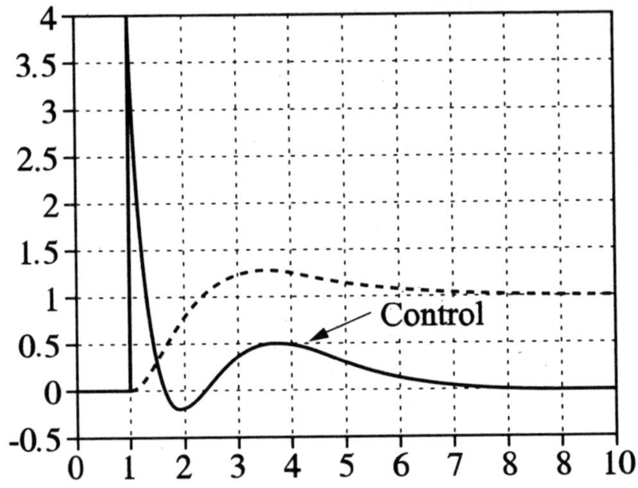

no saturation
output y(t) is confused with that of reference model B_m/A_m

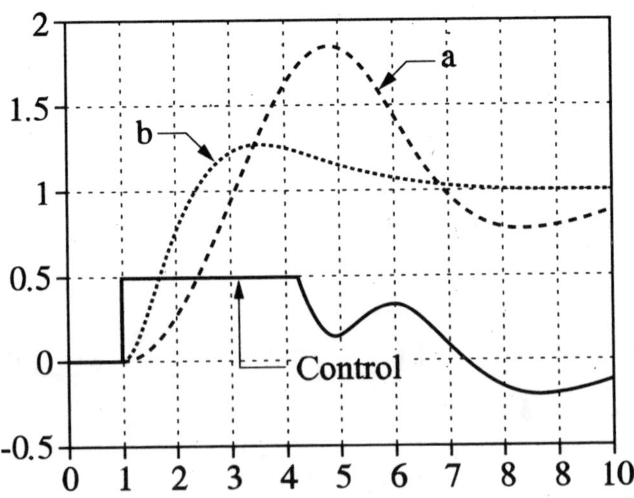

saturation at ±0.5
(a) with saturation, (b) reference model

Figure 11.6. *Comparison of behavior with and without saturation (note that the scales in ordinate are not the same!); in both cases $F(s) = (s + 1)^4$*

The use of the strategy [ÅST 90, LAR 93] described in the following figure makes it possible to considerably decrease the effect of saturation as can be seen on the simulation in Figure 11.8.

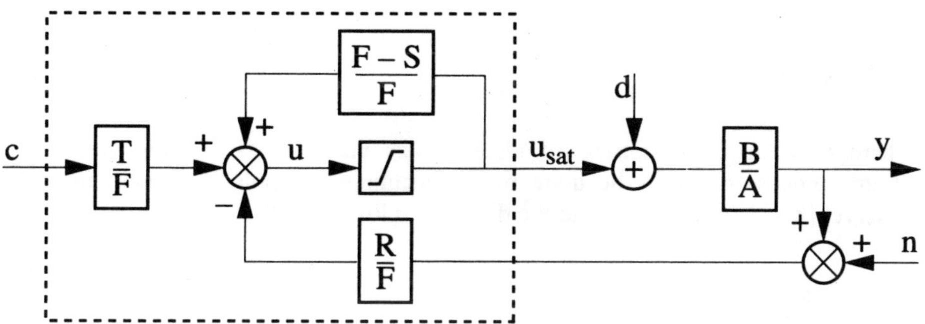

Figure 11.7. *RST correction with "anti-wind up"*

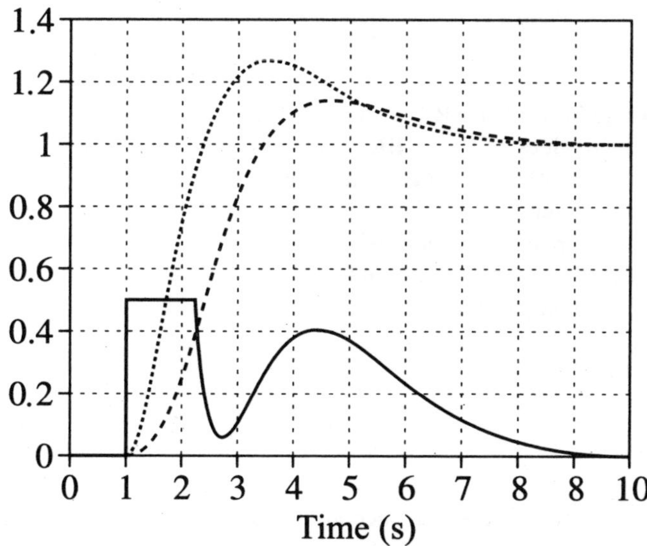

Figure 11.8. *Behavior in the presence of saturation and with "anti-wind up" with the response of the reference model represented by the dotted lines*

This structure can of course be implemented with the help of the realization presented in section 11.4.2.2.

11.4.3.2. *Justification of the method [KAI 80]*

The transfer process $\dfrac{B}{A}$ can be described by an equation of state:

$$\dot{\mathbf{x}} = \mathbf{A}\mathbf{x} + \mathbf{B}\mathbf{u}$$
$$\mathbf{y} = \mathbf{C}\mathbf{x}$$

$$[11.60]$$

where **A**, **B** and **C** are matrices of adequate size and **x** the state vector. The control of this process can thus be done by the technique of pole placement using an observer [KWA 72] according to the diagram in Figure 11.9.

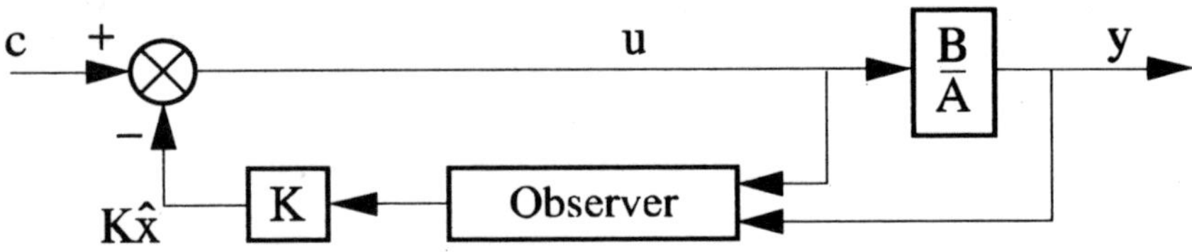

Figure 11.9. *State feedback control with observer*

It is known [KWA 72, LAR 93] that the complete system thus corrected has as its poles the eigenvalues of the **A-BK** matrix (placement of poles) and the poles of the observer (principle of separation). If we do not take into account the deterministic interferences and the characteristics of the instructions, we have, at least, 2n-1 poles to place of which n-1 come from the minimal order observer. We thus find the significance of polynomials A_m and A_o (from where we get the name of polynomial of the observer for A_o)

Let us say that:

$$v = \mathbf{K}\hat{\mathbf{x}}$$

$$[11.61]$$

Signals u, y and v are scalar; according to what was previously said, we can thus define transfers such as:

$$v = \frac{N_u}{A_o}u + \frac{N_y}{A_o}y$$

$$[11.62]$$

and thus, based on Figure 11.9, we have:

$$\left. \begin{array}{l} u = c - v \\[2mm] = c - \dfrac{N_u}{A_o}u - \dfrac{N_y}{A_o}y \\[4mm] u = \dfrac{A}{B}y \end{array} \right\} \Rightarrow y = \dfrac{BA_o}{A(A_o + N_u) + N_y B}c \qquad [11.63]$$

By approaching the expression [11.1], we note that the state feedback control with observer corresponds to an RST structure where we chose:

$$T = A_o \qquad\qquad [11.64]$$

polynomials R and S thus being:

$$\begin{array}{l} S = A_o + N_u \\[2mm] R = N_y \end{array} \qquad\qquad [11.65]$$

In the presence of saturation, there is the control described in Figure 11.11.

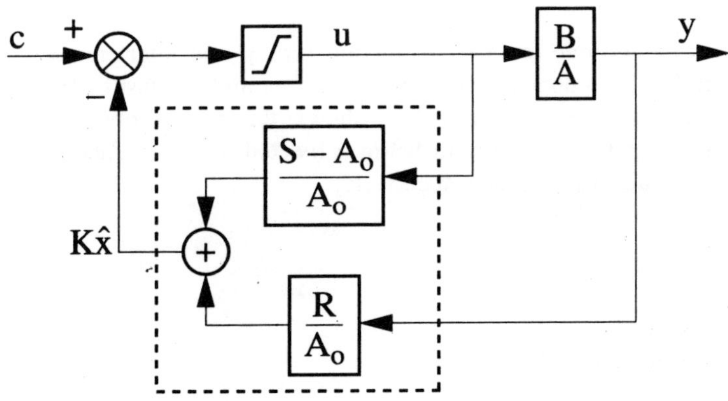

Figure 11.10. *State return control with observer in the presence of control saturation*

It is easy to make a connection between this diagram and the one in Figure 11.5. The interest in this structure is thus understood since signal v is calculated according to the "true" control of the process and its output.

11.5. Methodology

We have so far defined the various degrees of freedom of the RST control and now, to return to the objectives of section 11.1.1, we still have to define a "good transient state" by taking into account the level of noise on the control and the *robustness aspect* with respect to modeling errors.

11.5.1. *Intuitive approach*

Many mishaps can be avoided if we keep in mind the following rule: *the more important the required performances are, the more disturbed the control will be and the more sensitive the corrected system will become to modeling errors.*

The following examples illustrate this "saying" within the context of this chapter. The model of the process considered is defined by:

$$\frac{B}{A} = \frac{2}{s(s^2 + 0.1s + 1)}$$

the synthesis being done for a step function setting and interference. Taking into account what has been done above, polynomials A_m and A_o must be both of a degree at least equal to 3. The table below shows the results obtained for various polynomials A_m and A_o, the system being subjected to a step function setting and interference at 1 and 15 seconds respectively.

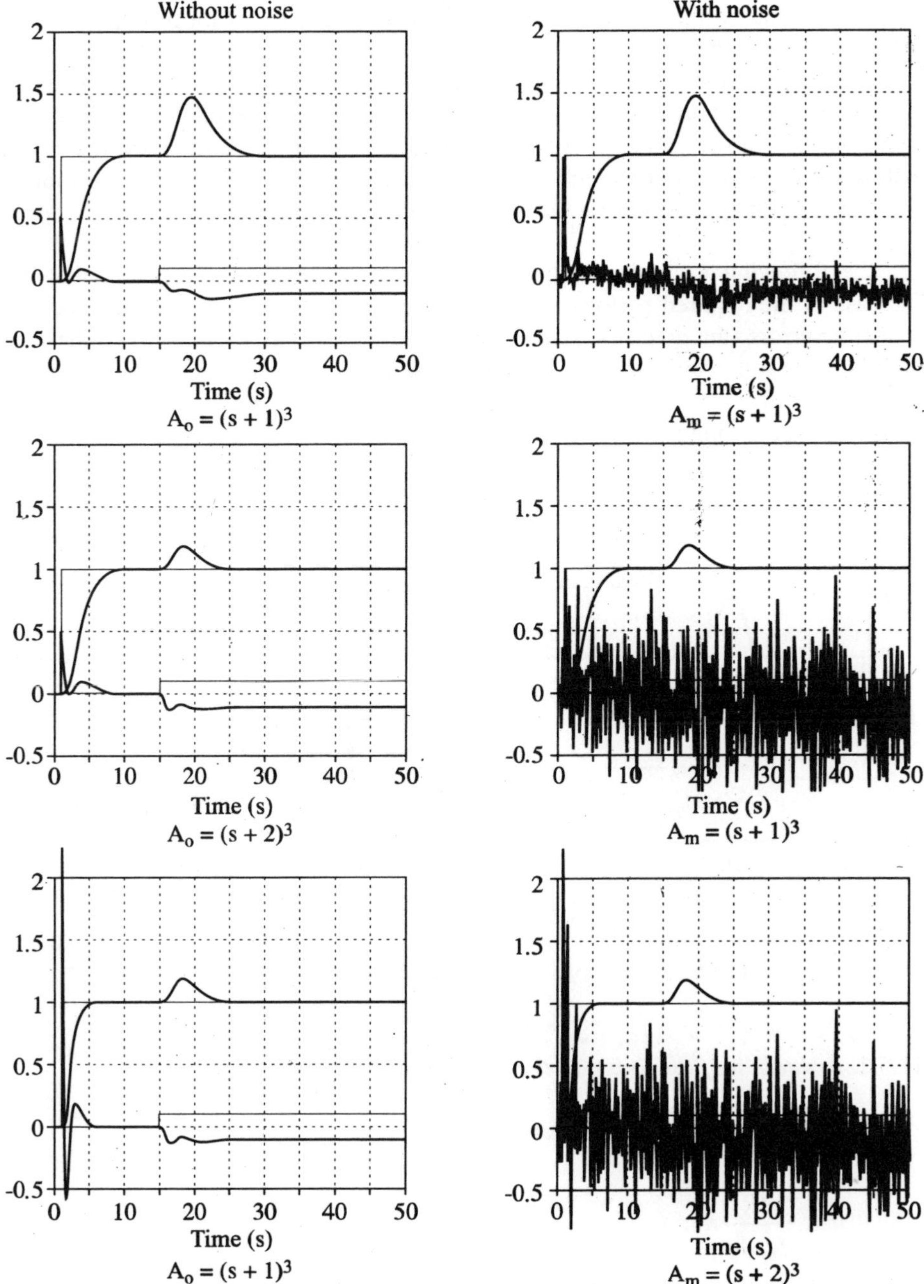

Figure 11.11. *Influence of the measurement noise according to the dynamics chosen for A_o and A_m*

The column on the left shows the results in the absence of measurement noise. We observe that the more the poles of A_m are on the left in plane s, the faster the response to the setting is and the better rejected the interference is when the poles of A_o are on the left in plane s. On the other hand, we note that the increase in dynamics by both A_m and A_o leads to an amplification of the noise on the control. It will never be pointed out enough that the designer should not be satisfied to look at the output of the process during the simulation!

The following example illustrates the finally accepted phenomenon, that the increase of the required performances is confronted with robustness with respect to modeling errors (the precision-stability *dilemma* was a beginning). The synthesis is always made following the model of the preceding section, the "real" process having a time constant of 0.5 s, which is neglected in the model. The results of simulation do not require comments.

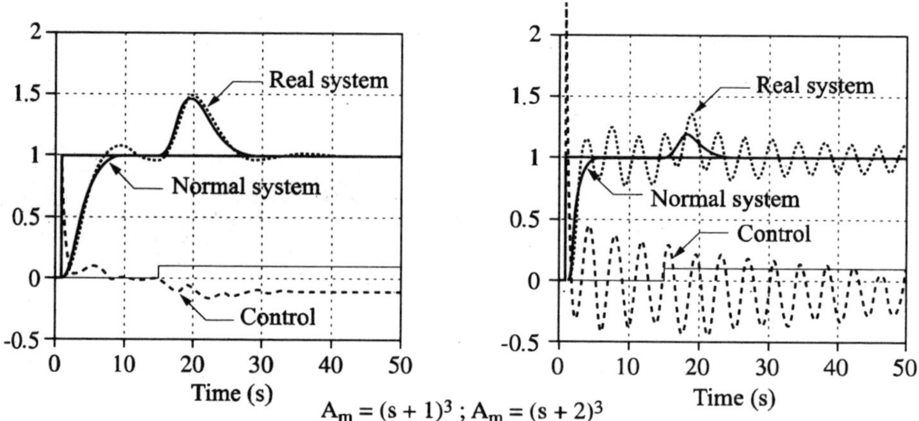

Figure 11.12. *Influence of a modeling error according to the choice of A_m*

11.5.2. *Reduction of the noise on the control by choice of degrees*

From Figure 11.1, it appears that:

$$u = \frac{AR}{A_m A_o} n \qquad [11.66]$$

So if we choose the minimum degrees for A_m and A_o, according to relations [11.49] and [11.51], we observe that the transfer between the control and the measurement noise is simply proper. This means that all the high frequencies of n will pass in u. Always according to the same relations, all we have to do in order to obtain a strictly proper transfer between u and n is to choose a polynomial A_o of degree higher than the minimal value. The following curves properly illustrate this phenomenon in the scalar field.

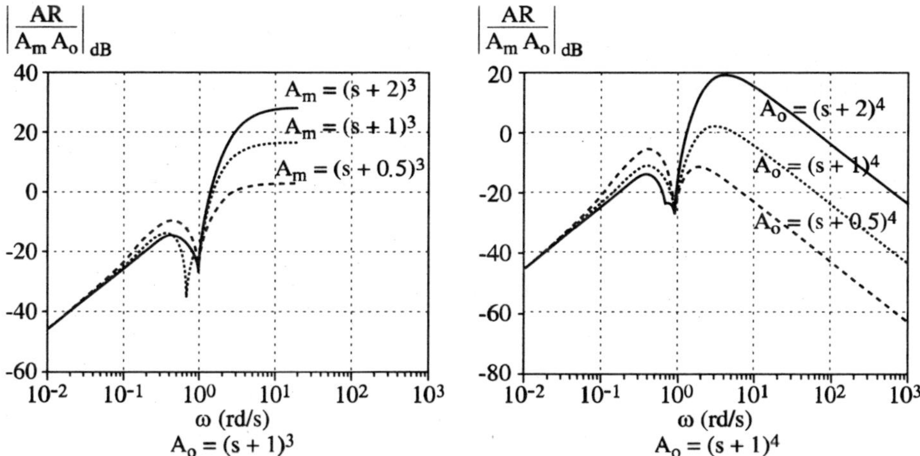

Figure 11.13. *Influence of the choice of the degree of A_o on u/n*

11.5.3. *Choice of the dynamics of A_m and A_o*

Section 11.5.1 shows the influence of the dynamics of these polynomials on the performances of corrected system. Section 11.5.2 provides the starting point of the solution by offering an additional choice criterion on their degree. We will discuss here the choice of their roots. This section is based on [LAR 99] where the author develops a methodology based on the LQG-LTR control.

In all that follows, we will suppose that the setting and the interference are step functions and that polynomial T is chosen equal to A_o, i.e.:

$$D_c^- = D_d^- = D_{dc}^- = s \qquad [11.67]$$

in this case, conditions [11.30] on the degrees become:

$$\begin{aligned}
\partial A_m &= \partial A \\
\partial A_o &= \partial A + 1 \\
\partial R &= \partial A \\
\partial S &= \partial A + 1 \quad \text{with} \quad S = sS'
\end{aligned}$$

[11.68]

S' being the solution of:

$$AsS' + BR = A_m A_o$$

[11.69]

11.5.3.1. *Optimal choice of A_m*

In the absence of interference, according to [11.23], the tracking error is given by:

$$E = C - Y = C - \frac{BT}{A_m A_o} C = C\left[1 - \frac{B}{A_m}\right]$$

[11.70]

A possible synthesis strategy can be the choice of A_m in order to minimize the H_2 standard of this transfer. A direct application of the variational calculation leads to the following stable optimal solution:

$$A_m^{\text{opt}} = B^+(s)B^-(-s)$$

[11.71]

This means that the roots of A_m^{opt} are the stable roots of B and the unstable roots of B made symmetrical with respect to the imaginary axis. Of course this solution does not satisfy the constraint on the degrees. For that we will have to add a certain number of roots, such as:

$$A_m^{\text{opt}} = B^+(s)B^-(-s)(1 + T_c s)^{\partial A - \partial B}$$

[11.72]

If the model has minimum phase difference ($B^- = 1$), we can show that:

$$\lim_{T_c \to 0} S' = B$$

[11.73]

indeed, S' is the solution of:

$$As S' + BR = B(1 + T_c s)^{\partial A - \partial B} A_o \qquad [11.74]$$

and thus S' is divisible by B. After reduction by B, we obtain:

$$R = (1 + T_c s)^{\partial A - \partial B} A_o - As S''$$
$$S' = BS'' \qquad [11.75]$$

when T_c tends toward zero, we see that S'' must tend toward a constant so that ∂R remains equal to ∂R. Thus, by considering the highest degree terms of A and A_o as equal, S'' will tend toward 1 when T_c tends toward zero.

11.5.3.2. *Optimal choice of A_o*

We have seen above that the transfer connecting the measurement noise to the control is given by:

$$\frac{U}{N} = \frac{AR}{A_m A_o} \qquad [11.76]$$

for instance, in the limit case mentioned previously:

$$\frac{U}{N} = \frac{AR}{A_m A_o} = \frac{A(A_o - sA)}{B(1 + T_c s)^{\partial A - \partial B} A_o} \qquad [11.77]$$

In order to reduce the influence of the noise on the control[10] we can choose A_o in order to minimize the H_2 standard of $1 - \dfrac{sA}{A_o}$. The use of the variational calculation leads as before to a polynomial A_o whose roots are the stable roots of sA and also symmetric with respect to the imaginary axis of the unstable roots. However, here there is the difficulty of the root at the origin. So we define the under optimal solution:

---

10 This procedure supposes seeking an observer which could minimize the restoration noise of $K\hat{x}$ (see section 11.4.3.2).

$$A_{\mathrm{o}} = A^{+}(s)A^{-}(-s)\left(s + \frac{1}{T_{\mathrm{o}}}\right) \qquad [11.78]$$

this solution tending towards the optimal solution for T_{o} tending towards infinity[11].

11.5.3.3. *Study of the sensitivity function*

Let L(s) be the transfer of the loop; we can define the sensitivity function Σ by:

$$\Sigma = \frac{1}{1 + L} = \frac{AS}{AS + BR} = \frac{AS}{A_{\mathrm{m}}A_{\mathrm{o}}} \qquad [11.79]$$

in the case of a system without unstable zero, the choice of an optimal polynomial A_{m} leads to the sensitivity function:

$$\Sigma = \frac{AsS'}{A_{\mathrm{o}}B(1 + T_{\mathrm{c}}s)^{\partial A - \partial B}} \qquad [11.80]$$

and thus for the choice of A_{o} seen above and when T_{c} tends toward zero we have:

$$\lim_{T_{\mathrm{c}} \to o} |\Sigma| = \left|\frac{sA}{A_{\mathrm{o}}}\right| = \left|\frac{sA^{+}(s)A^{-}(s)}{\left(s + \dfrac{1}{T_{\mathrm{o}}}\right)A^{+}(s)A^{-}(-s)}\right| = \left|\frac{s}{s + \dfrac{1}{T_{\mathrm{o}}}}\right| < 1 \qquad [11.81]$$

this for a process without integration ($A(0) \neq 0$).

Thus, everything occurs as if the loop transfer was $\dfrac{1}{sT_{\mathrm{o}}}$ and we obtain a phase margin of $\pi/2$ to the angular frequency $\omega_{\mathrm{p}} = \dfrac{1}{T_{\mathrm{o}}}$, which gives a limit delay margin of $M_{\mathrm{r}} = \dfrac{\pi}{2}T_{\mathrm{o}}$.

---

11 The presence of integrators in the process will be treated in a similar way.

In the case of a model with one integration, the limit sensitivity function will be:

$$\lim_{T_c \to o} |\Sigma| = \left| \frac{s^2}{\left(s + \dfrac{1}{T_o}\right)^2} \right| < 1 \qquad [11.82]$$

the equivalent loop transfer thus becomes:

$$L = \frac{1 + 2T_o s}{(T_o s)^2} \qquad [11.83]$$

and the lag margin being $M_r = 0.647\, T_o$

It should be noted that the sensitivity function is always of a module less than 1 (except for the infinite frequency).

11.5.3.4. *Some improvements*

If the poles of A (or their symmetric) are slightly damped or too slow, the choice of A_o suggested above may lead to bad dynamics of regulation. This is why it is advisable to project these undesirable poles on the abscise vertical $-1/\,T_o$.

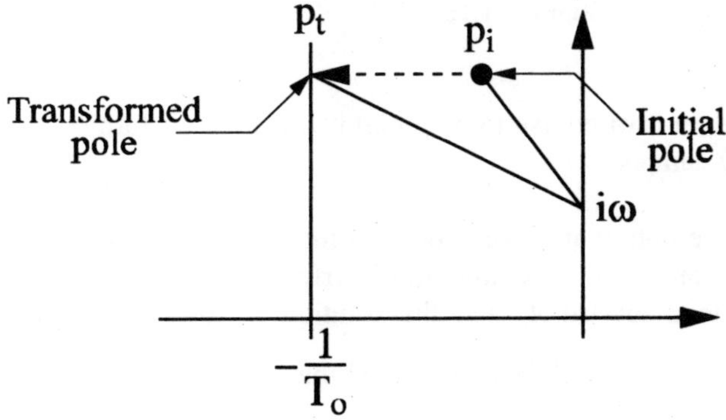

Figure 11.14. *Choice of the roots of A_o*

Thus, the contribution of this pair to the sensitivity function is:

$$\left| \frac{(s - p_i)}{(s - p_t)} \right|_{s = i\omega} < 1 \qquad\qquad [11.84]$$

and therefore this transformation does not affect the main result of these choices: the limit sensitivity function is always of a module less than 1.

Similarly, the presence of strong dynamic zeros leads, through the procedure of optimal choice of A_m, to imposing fast methods to the corrected process. Several methods are proposed in order to avoid this obstacle [LAR 99]. We will suggest here the procedure illustrated in Figure 11.15.

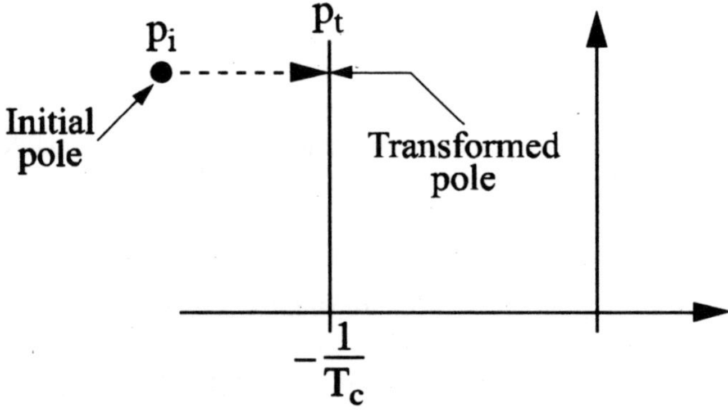

Figure 11.15. *Choice of the roots of A_m*

This procedure is of course to be prohibited if the transformed pole proves to be not sufficiently damped.

NOTE 11.2.– we note that if we choose to make A_o tend toward sA (always under the hypothesis step function setting and interference), the R solution of the Bezout equation will tend toward A_o and the solution S' toward A_m. Consequently, the sensitivity function will still have the same limit $\dfrac{sAS'}{A_o A_m} = \dfrac{sA}{A_o} = \dfrac{s}{s + \dfrac{1}{T_o}}$

We can thus reach the same limit sensitivity function either by making A_m tend toward B ($T_c \rightarrow 0$) or by making A_o tend toward sA ($T_o \rightarrow \infty$).

11.5.3.5. *Method of adjustment*

Based on the particular choices for A_m and A_o:

$$A_o = A\left(s + \frac{1}{T_o}\right)$$

$$A_m = B(1 + T_c s)^{\partial A - \partial B}$$

[11.85]

the optimal solution corresponds to $T_o \to \infty$ and $T_c \to 0$, or in practice when:

$$\frac{T_o}{T_c} \to \infty$$

[11.86]

Practically, we obtain a sensitivity close to the limit sensitivity for a 5 order ratio.

Increasing T_o increases the lag margin at the expense of the dynamics of regulation.

Decreasing T_c makes the corrected system more sensitive to measurement noises (see the example in section 11.5.2).

The initial value of T_o can be either:

– the desired lag margin;

– the dominant time constant in desired regulation;

– a value close to the dominant time constant in open loop (for a stable process of course).

11.5.4. *Examples*

In section 11.5.1 we presented an example in order to highlight the influence of the dynamics of polynomials A_m and A_o on the correction. We will reiterate this example by applying the methodology developed above. As the model of this example presents a slightly damped oscillating dynamics, we will be able to clarify the "improvements" described in section 11.5.3.3. The model is thus described by the transfer:

$$\frac{B}{A} = \frac{2}{s(s^2 + 0.1s + 1)} \qquad [11.87]$$

the setting and the interference being always steps functions. In accordance with the suggested methodology, we will have:

$$A_m = 2(T_c s + 1)^3 \qquad [11.88]$$

and:

$$A_o = (s^2 + 0.1s + 1)\left(s + \frac{1}{T_o}\right)^2 \qquad [11.89]$$

if we "hide" the oscillating poles, or:

$$A_o = \left(s + \frac{1}{T_o}\right)^4 \qquad [11.90]$$

in the opposite case. The following figures show the results obtained in the scalar field. Figure 11.16 corresponds to choice [11.89], whereas Figure 11.17 corresponds to choice [11.90]. We notice that in each case, the target sensitivity function $\dfrac{sA}{A_o}$ and the real sensitivity function $\dfrac{sAS'}{A_o A_m}$ have their modules close, thus justifying the approximation carried out in the design. In addition, in choice [11.89], we notice that the target sensitivity function is of a module less than 1.

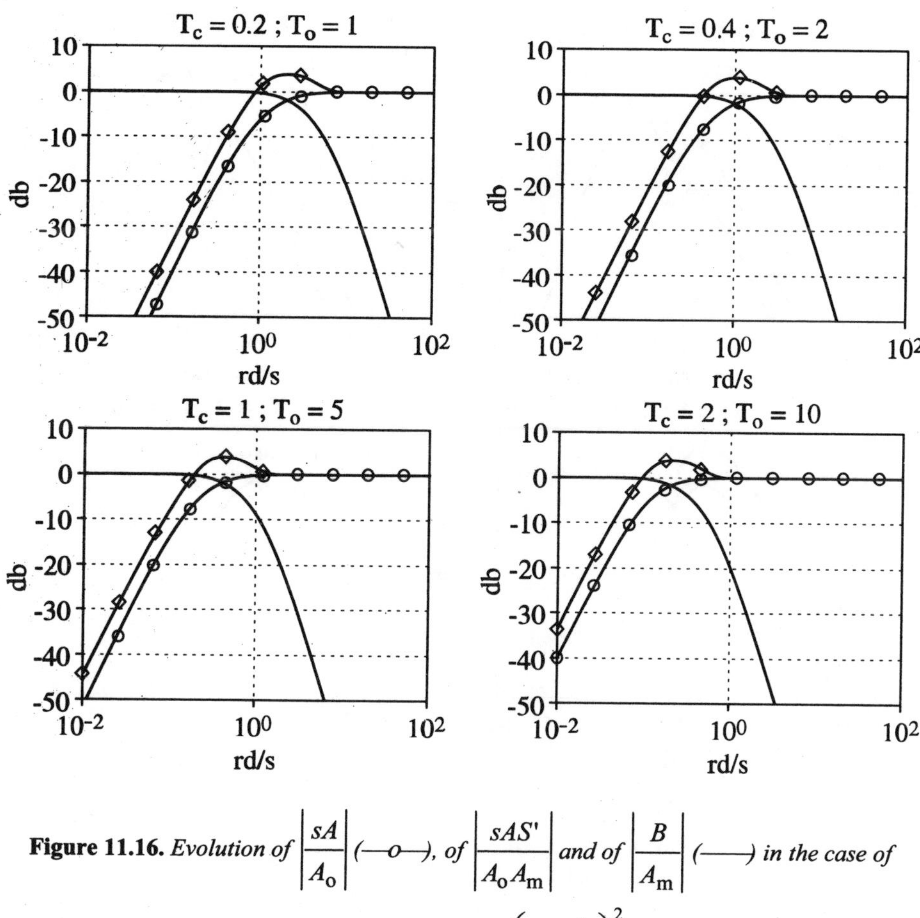

Figure 11.16. *Evolution of* $\left|\dfrac{sA}{A_o}\right|$ *(—o—), of* $\left|\dfrac{sAS'}{A_o A_m}\right|$ *and of* $\left|\dfrac{B}{A_m}\right|$ *(———) in the case of*

$$A_o = (s^2 + 0.1s + 1)\left(s + \frac{1}{T_o}\right)^2$$

The following figures give an outline of the time performances obtained in each one of these cases for various choices of T_o. Each figure represents the behavior of the corrected process following a rule:

– without measurement noise and model identical to the process;

– with measurement noise and model identical to the process;

– without measurement noise and process with a time constant of 0.5 s neglected in the model.

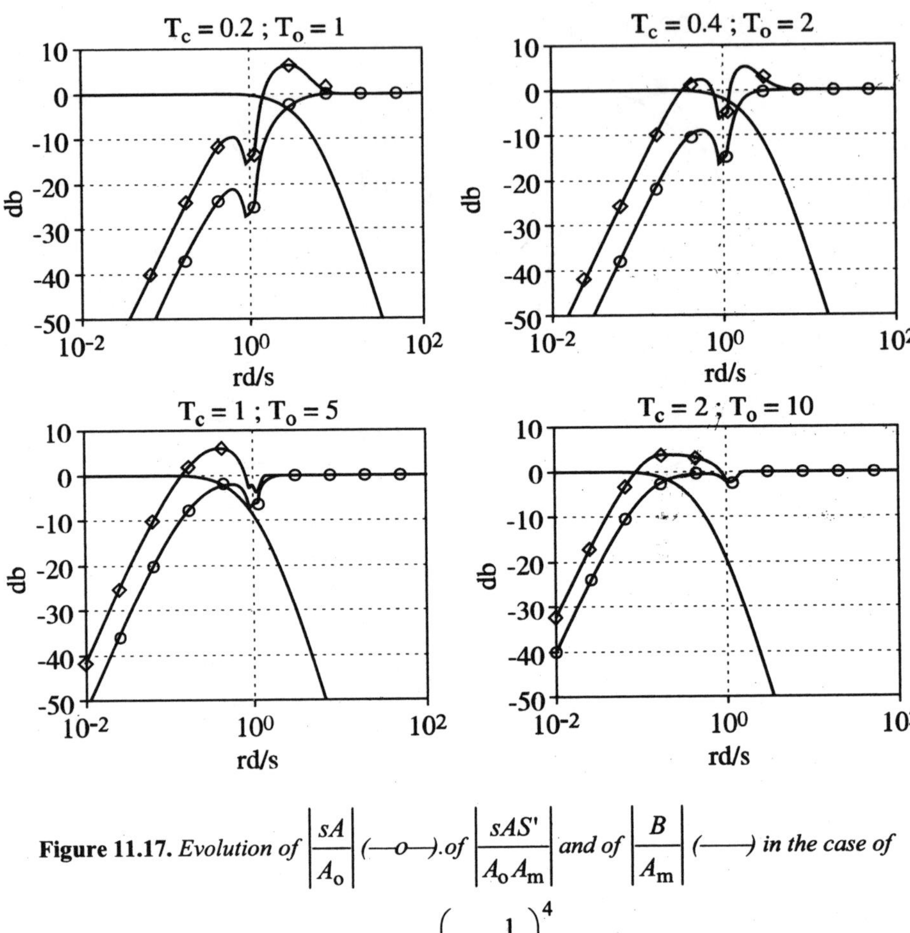

Figure 11.17. *Evolution of* $\left|\dfrac{sA}{A_o}\right|$ *(—o—).of* $\left|\dfrac{sAS'}{A_oA_m}\right|$ *and of* $\left|\dfrac{B}{A_m}\right|$ *(———) in the case of*

$$A_o = \left(s + \frac{1}{T_o} \right)^4$$

From these answers we notice that the lower T_c, the faster the feedback control response, but on the other side the control is subject to more noise. On the other hand, the higher T_o, the more robust the system will be to modeling errors, with a rejection dynamics of less competitive interference (we will have noticed in the last simulation the appearance of instability). Finally, in [11.89], the fact of hiding the oscillating poles of the process leads to a rejection of oscillating interference.

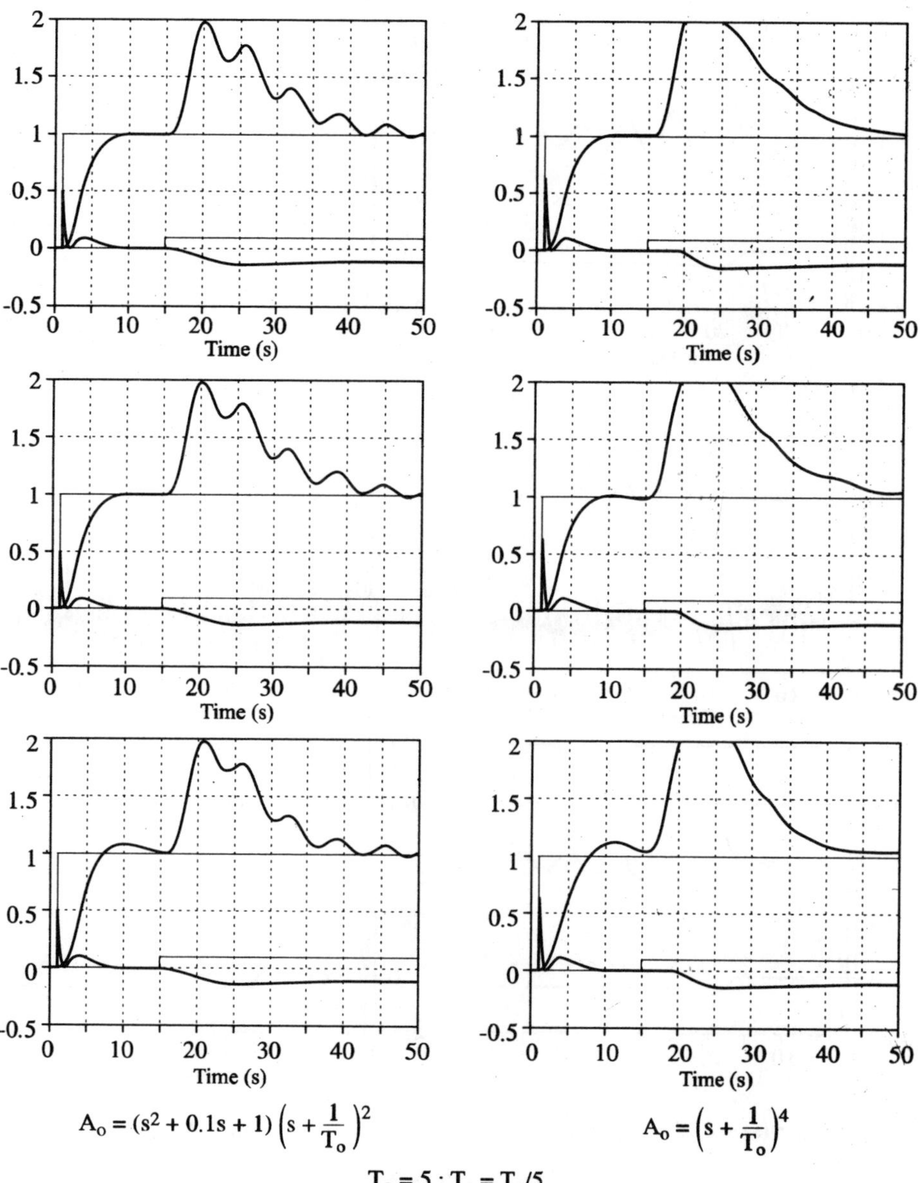

Figure 11.18. *Time responses for $T_o = 5$ seconds*

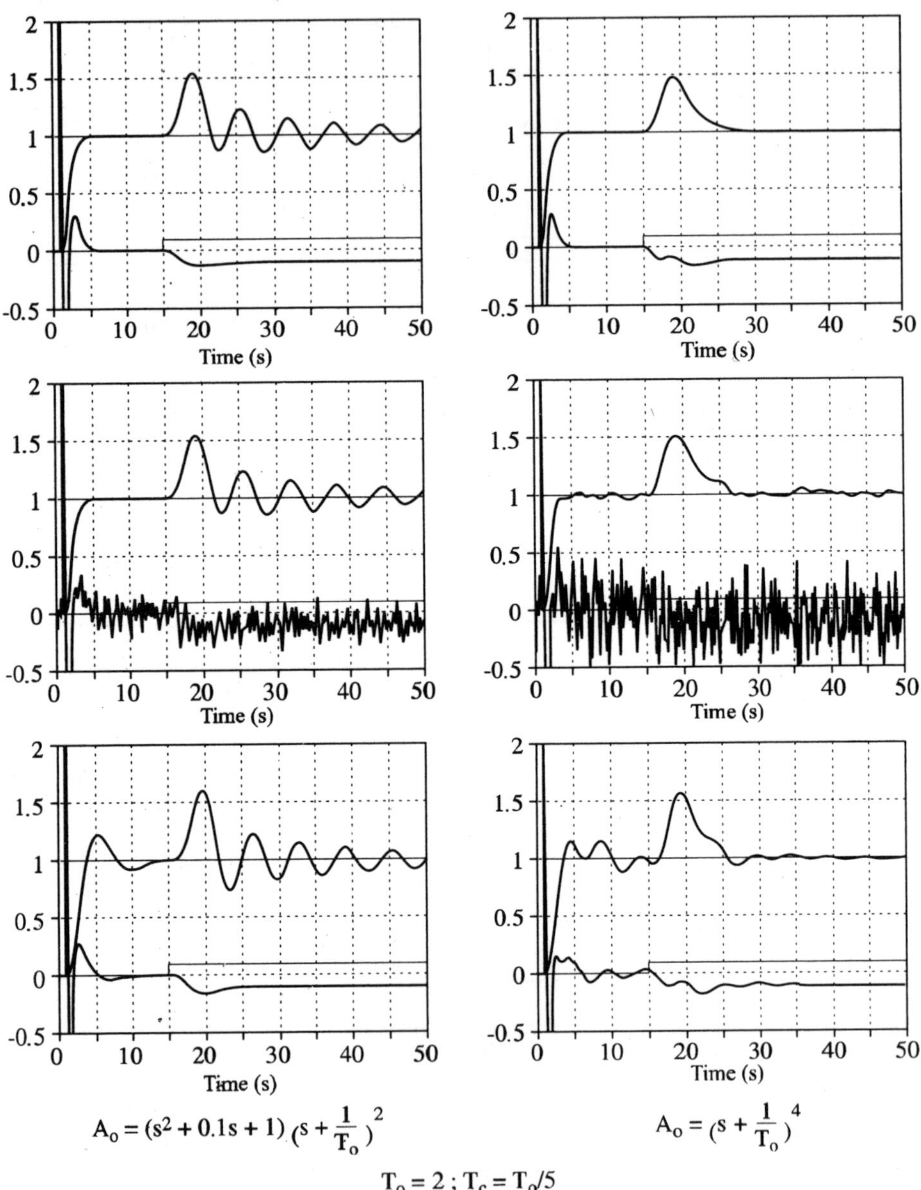

$$A_o = (s^2 + 0.1s + 1)\,(s + \tfrac{1}{T_o})^2 \qquad A_o = (s + \tfrac{1}{T_o})^4$$

$$T_o = 2 \;;\; T_c = T_o/5$$

Figure 11.19. *Time responses for $T_o = 2$ seconds*

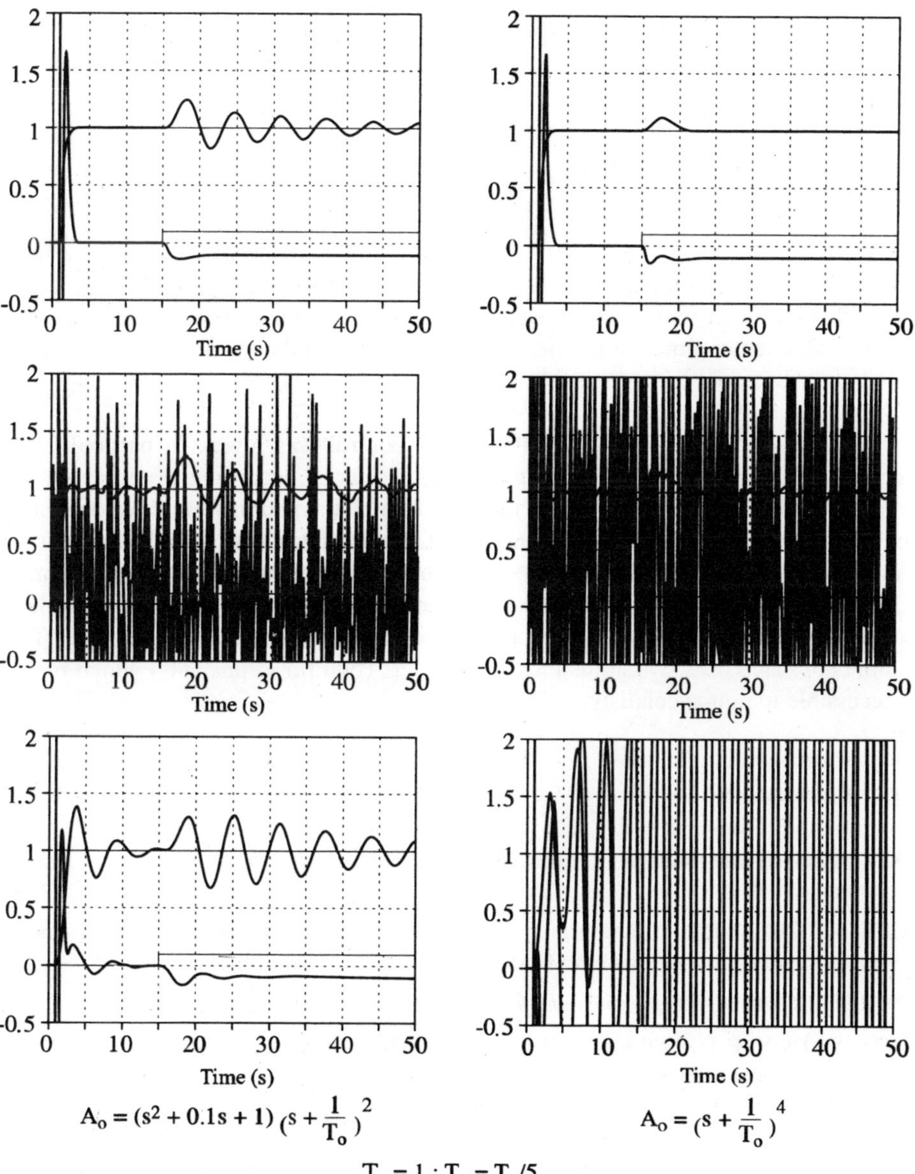

Figure 11.20. *Time responses for $T_o=1$ second*

Due to the degree of robustness when facing modeling errors, of the level of measurement noise, the designer can "easily" adjust the only two adjustment parameters T_o and T_c.

11.6. Conclusion

Many problems of automatic control can be reduced to single-variable corrections. In this context, the placement of poles performed by the RST structure, i.e. with a corrector with two degrees of freedom, has been developed these last few years. This chapter aimed at providing the designer with the maximum of degrees of freedom to carry out this synthesis.

Initially, the general solution of the RST control was given. In particular, the choice of synthesis polynomials degrees was discussed according to the classes of settings and interferences. Then the aspects of implementation were developed and, in particular, the problem of saturation of the actuator was largely dealt with. Finally, a methodology of choice of feedback control and regulation dynamics (choice of polynomials A_m and A_o) was provided. This approach makes it possible to set performances compatible with a level of robustness chosen in advance. This simple approach uses two adjustment parameters (two time constants) which makes it accessible to non-specialists.

11.7. Bibliography

[ÅST 90] ÅSTRÖM K.J., WITTENMARK B., *Computer Controlled Theory and Design*, Prentice Hall, 1990.

[CHE 87] CHEN T.C., "Introduction to the linear algebraic method for control system design", *IEEE Control Systems Magazine*, vol. 7, no. 5 October 1987.

[COR 96] CORRIOU J.P., *Commande des Procédés*, Lavoisier Tec-Doc, 1996.

[DOY 92] DOYLE J.C., FRANCIS B.A., TANNENBAUM A.R., *Feedback Control Theory*, MacMillan Publishing Company, 1992.

[FAR 86] FARGEON *et al.*, *Commande Numérique des systèmes: application aux engins mobiles et aux robots*, Masson, 1986.

[IRV 91] IRVING E., *Notes de cours EDF*, IMA, 1991.

[KAI 80] KAILATH T., *Linear Systems*, Prentice Hall, 1980.

[KUC 79] KUCERA V., *Discrete Linear Control: The Polynomial Equation Approach*, Wiley, New York 1979.

[KWA 72] KWAKERNAAK H., SIVAN R., *Linear Optimal Control Systems*, Wiley-Interscience, 1972.

[LAR 93] DE LARMINAT P., *Automatique*, Hermès, 1993.

[LAR 96] DE LARMINAT P., *Automatique*, Hermès, 1996 (2nd edition).

[LAR 99] DE LARMINAT P., PUREN S., "Robust Pole Placement", *IFAC 14th World Congress*, Beijing (China), 1999.

[LAN 93] LANDAU I.D., *Identification et Commande des Systèmes*, Hermès, 1993.

[WOL 93] WOLOVITCH W.A., *Automatic Control Systems: Basic Analysis & Design*, New York Holt, Rinehart & Winston, 1993.

Chapter 12

Predictive Control

The developments presented in this chapter aim to cover the main ideas of predictive control and then to indicate the details of the analytical minimization of the criterion for two individual structures enabling the elaboration of the equivalent polynomial regulator. The choice of adjustment parameters will also be analyzed, providing some simple rules that guarantee the corrected system good stability and robustness.

12.1. General principles of predictive control

Predictive control is based on some relatively old and intuitive ideas [RIC 78], but it has been developed as an advanced control technique mainly since the 1980s. This development was done mainly according to two privileged main lines:

– generalized predictive control (GPC) by Clarke (1985);

– functional predictive control (FPC) by Richalet (1987).

The philosophy of predictive control lies on the definition of five great ideas, common to all the methods.

12.1.1. *Anticipative aspect*

This anticipative effect is obtained by using explicit knowledge on the evolution of the trajectory to be followed in the future (necessary knowledge required at least

Chapter written by Patrick BOUCHER and Didier DUMUR.

on the horizon of some points beyond the present moment). This constraint which makes it possible to make good use of all the resources of the method, necessarily restricts the application field to the control of the systems for which the trajectory to follow is perfectly known and stored pixel by pixel in the computer. It is the case of the numerical control of machine-tools (cutting the pieces), of the control of robots arms, of monitoring the temperature profile of the applications in home automation, etc.

12.1.2. *Explicit prediction of future behavior*

The method requires the definition of a numerical model of the system, which makes it possible to predict the future behavior of the system. This discrete model results mainly from a preliminary offline identification. This feature makes it possible to classify predictive control in the big family of Model Based Control (MBC).

12.1.3. *Optimization by minimization of a quadratic criterion*

The optimization which makes it possible to obtain the control law is done by minimizing a quadratic criterion with finite horizon referring to the errors of future prediction, the variance between the predicted output of the system and the future setting or the reference trajectory inferred from this setting.

12.1.4. *Principle of the sliding horizon*

The elaboration of a sequence of future controls results from the preceding minimization, which is optimal in what the quadratic criterion is concerned, out of which only the first value is applied to the system and the model.

The preceding steps are then repeated during the following sampling period according to the principle of sliding horizon, as seen in Figure 12.1.

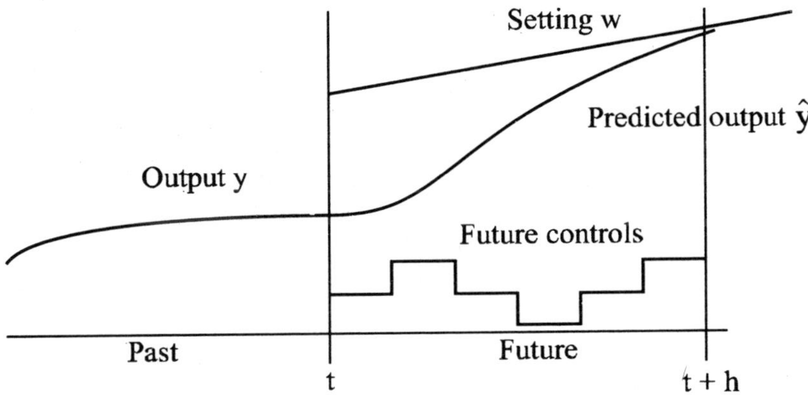

Figure 12.1. *Principle of sliding horizon*

The objective of the polynomial predictive regulator obtained by minimizing the criterion is that the predicted output joins the setting or the reference trajectory on a given prediction horizon. The principles that we have just mentioned make it possible to establish the operation diagram in Figure 12.2.

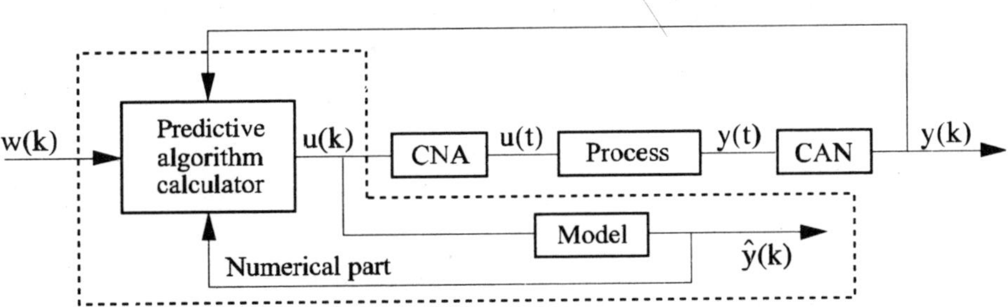

Figure 12.2. *Operation principle of a predictive algorithm*

Hence, the principle of the sliding horizon means that only the control at the present moment $u(t)$ is applied on the system. Therefore, it is possible to limit the number of estimated values of the sequence.

12.2. Generalized predictive control (GPC)

12.2.1. *Formulation of the control law*

The objective of this section is to indicate the fundamental points of the predictive structure considered [CLA 87a, CLA 87b], in the monovariable case, from the mathematical translation of the preceding general concepts up to the obtaining the equivalent polynomial regulator.

12.2.1.1. *Definition of the numerical model*

All the forms are allowable for the model but the input/output polynomial approach by transfer functions is preferred.

Traditionally the model is represented as CARIMA (*Controlled AutoRegressive Integrated Moving Average*):

$$A(q^{-1})y(t) = B(q^{-1})u(t-1) + \frac{\xi(t)}{\Delta(q^{-1})} \qquad [12.1]$$

where $\Delta(q^{-1}) = 1 - q^{-1}$, $u(t)$ and $y(t)$ are the input and output of the model, $\xi(t)$ is a centered white noise, q^{-1} is the delay operator and $A(q^{-1})$ and $B(q^{-1})$ are polynomials defined by:

$$\begin{cases} A(q^{-1}) = 1 + a_1 q^{-1} + \cdots\cdots + a_{n_a} q^{-n_a} \\ B(q^{-1}) = b_0 + b_1 q^{-1} + \cdots\cdots + b_{n_b} q^{-n_b} \end{cases} \qquad [12.2]$$

This model, which is also called incremental model, introduces an integral action and makes it possible to undo all the static errors with respect to the input or step function interference.

12.2.1.2. *Optimal predictor*

The predicted output $y(t+j/t)$ is traditionally decomposed into a free and forced response [FAV 88], including a polynomial form meant to properly conclude the final polynomial synthesis:

$$y(t+j/t) = \underbrace{F_j(q^{-1})y(t) + H_j(q^{-1})\Delta u(t-1)}_{\text{free response}} + \underbrace{G_j(q^{-1})\Delta u(t+j-1) + J_j(q^{-1})\xi(t+j)}_{\text{forced response}} \qquad [12.3]$$

The unknown polynomials F_j, G_j, H_j, J_j are single solutions of Diophantus equations, which are obtained by equality of the inputs and output of transfer functions of equations [12.1] and [12.3] and they are solved recursively:

$$\Delta(q^{-1})A(q^{-1})J_j(q^{-1}) + q^{-j}F_j(q^{-1}) = 1$$
$$G_j(q^{-1}) + q^{-j}H_j(q^{-1}) = B(q^{-1})J_j(q^{-1})$$

[12.4]

with:

$$\text{degree}\left[J_j\left(q^{-1}\right)\right] = j - 1 \qquad\qquad \text{degree}\left[F_j\left(q^{-1}\right)\right] = \text{degree}\left[A\left(q^{-1}\right)\right]$$
$$\text{degree}\left[G_j\left(q^{-1}\right)\right] = j - 1 \qquad\qquad \text{degree}\left[H_j\left(q^{-1}\right)\right] = \text{degree}\left[B\left(q^{-1}\right)\right] - 1$$

The set of calculations may be done in real-time off loop. The optimal predictor is finally defined by considering that the best noise prediction in the future is its mean (here supposed as zero), let us suppose that:

$$\hat{y}(t + j/t) = F_j(q^{-1})y(t) + H_j(q^{-1})\Delta u(t-1) + G_j(q^{-1})\Delta u(t + j - 1)$$

[12.5]

12.2.1.3. *Definition and minimization of the quadratic criterion*

The control law is obtained by minimizing a quadratic criterion pertaining to future errors with a weighting term on the control:

$$J = \sum_{j=N_1}^{N_2} [\hat{y}(t + j) - w(t + j)]^2 + \lambda \sum_{j=1}^{N_u} \Delta u^2(t + j - 1)$$

[12.6]

with: $\Delta u(t + j) \equiv 0$ for $j \geq N_u$.

The criterion requires the definition of four adjustment parameters:

– N_1 : minimal prediction horizon;

– N_2 : maximal prediction horizon;

– N_u : prediction horizon on the control;

– λ : weighting coefficient on the control.

12.2.1.4. *Synthesis of the equivalent polynomial RST regulator*

The minimization of the criterion is based on writing the prediction equation [12.5] and the cost function [12.6] in a matrix form, such as:

$$\hat{\mathbf{y}} = \mathbf{G}\tilde{\mathbf{u}} + \mathbf{if}\ y(t) + \mathbf{ih}\ \Delta u(t-1)$$

$$J = \left[\mathbf{G}\tilde{\mathbf{u}} + \mathbf{if}\ y(t) + \mathbf{ih}\ \Delta u(t-1) - \mathbf{w}\right]^{\mathrm{T}} \tag{12.7}$$
$$\left[\mathbf{G}\tilde{\mathbf{u}} + \mathbf{if}\ y(t) + \mathbf{ih}\ \Delta u(t-1) - \mathbf{w}\right] + \lambda\,\tilde{\mathbf{u}}^{\mathrm{T}}\,\tilde{\mathbf{u}}$$

$$\text{with:}\quad
\begin{aligned}
\mathbf{if} &= \left[F_{N_1}(q^{-1}) \quad \cdots \quad F_{N_2}(q^{-1})\right]^{\mathrm{T}} \\
\mathbf{ih} &= \left[H_{N_1}(q^{-1}) \quad \cdots \quad H_{N_2}(q^{-1})\right]^{\mathrm{T}} \\
\tilde{\mathbf{u}} &= \left[\Delta u(t) \quad \cdots \quad \Delta u(t + N_u - 1)\right]^{\mathrm{T}}
\end{aligned}$$

$$\mathbf{G} = \begin{bmatrix}
g_{N_1}^{N_1} & g_{N_1-1}^{N_1} & \cdots & \cdots \\
g_{N_1+1}^{N_1+1} & g_{N_1}^{N_1+1} & \cdots & \cdots \\
\cdots & \cdots & \cdots & \cdots \\
g_{N_2}^{N_2} & g_{N_2-1}^{N_2} & \cdots & g_{N_2-N_u+1}^{N_2}
\end{bmatrix}$$

The analytical minimization of the criterion leads to an optimal sequence of future controls:

$$\tilde{\mathbf{u}}_{opt} = \mathbf{N}\left[\mathbf{w} - \mathbf{if}\ y(t) - \mathbf{ih}\ \Delta u(t-1)\right] \tag{12.8}$$

$$\text{with:}\quad \mathbf{N} = \left[\mathbf{G}^{\mathrm{T}}\mathbf{G} + \lambda\,\mathbf{I}_{N_u}\right]^{-1}\mathbf{G}^{\mathrm{T}} = \left[\mathbf{n}_1^{\mathrm{T}} \quad \cdots \quad \mathbf{n}_{N_u}^{\mathrm{T}}\right]^{\mathrm{T}}$$

$$\tilde{\mathbf{u}}_{opt} = \left[\Delta u(t)_{opt} \quad \cdots \quad \Delta u(t + N_u - 1)_{opt}\right]^{\mathrm{T}}$$
$$\mathbf{w} = \left[w(t + N_1) \quad \cdots \quad w(t + N_2)\right]^{\mathrm{T}}$$

Traditionally, in a predictive control, only the first value of the sequence, equation [12.8] is applied to the system, according to the principle of the sliding horizon:

$$u_{opt}(t) = u_{opt}(t-1) - \mathbf{n}_1^T \left[\mathbf{if}\ y(t) + \mathbf{ih}\ \Delta u_{opt}(t-1) - \mathbf{w} \right] \qquad [12.9]$$

Based on the above relation, it is finally possible to obtain the polynomial representation of the equivalent regulator as indicated in Figure 12.3. This traditional RST structure enables the implementation of the control law by a simple difference equation:

$$S(q^{-1})\Delta u(t) = -R(q^{-1})y(t) + T(q)w(t) \qquad [12.10]$$

The three polynomials have the following form:

$$S(q^{-1}) = (1 + \mathbf{n}_1^T\ \mathbf{ih}\ q^{-1}) \qquad \text{degree}\left[S(q^{-1}) \right] = \text{degree}\left[B(q^{-1}) \right]$$

$$R(q^{-1}) = \mathbf{n}_1^T\ \mathbf{if} \qquad \text{degree}\left[R(q^{-1}) \right] = \text{degree}\left[A(q^{-1}) \right]$$

$$T(q) = \mathbf{n}_1^T \left[q^{N_1} \quad \cdots \quad q^{N_2} \right]^T \qquad \text{degree}\left[T(q) \right] = N_2$$

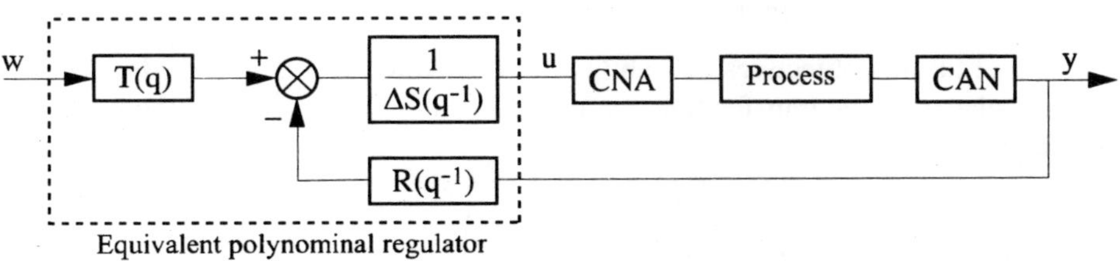

Figure 12.3. *Structure of the equivalent polynomial regulator*

We observe that polynomial $T(q)$ encloses the non-causal structure (positive power of q) inherent to the predictive control.

The interest resulted from the RST representation (actually very general because any numerical control law can be modeled this way [LAN 88]) is that, finally, the real-time loop proves to take little calculation time as the control applied to the system is calculated through a simple difference equation [12.10]. The three polynomials *R, S, T* are actually elaborated offline and uniquely defined as soon as the four adjustment parameters are chosen.

Consequently, this type of control favors the selection of short sampling periods and it proves to be well-adapted to the control of fast electro-mechanical systems (machine-tool, high-speed machining, etc.).

Another major interest in the RST structure pertains to the study of stability of the corrected loop and thus the characterization of stability of the elaborated predictive control, which is from that moment on possible for a set of parameters of the fixed criterion. This study is examined in the following section.

12.2.2. *Automatic synthesis of adjustment parameters*

The definition of the quadratic criterion [12.6] showed that the user must set four adjustment parameters. However, this choice of parameters proves to be difficult for a person who is not a specialist because there are no empirical relations which make it possible to relate these parameters to traditional "indicators" in control such as stability margins or a bandwidth.

Based on the study of a great number of single-variable systems, it is however possible to issue some "rules" based on the traditional criteria of stability and robustness [BOU 92] that we summarize.

12.2.2.1. *Criterion of stability and robustness*

First of all, the objectives of stability are related to the study in Bode, Black or Nyquist planes of the transfer function of the open loop corrected by the predictive regulator:

$$H_{bo}(q^{-1}) = \frac{q^{-1} B(q^{-1})R(q^{-1})}{A(q^{-1})S(q^{-1})\Delta(q^{-1})} \qquad [12.11]$$

It is generally agreed that a "good" adjustment is characterized by:

– a phase margin $\Delta\varphi$ higher to 45°;

– a minimal gain margin ΔG from 6 to 8 dB (decibels).

The objectives of robustness are linked to the calculation of the delay margin

$$\Delta\tau = \Delta\varphi/\omega_c \qquad (\Delta\varphi \text{ in rad, } \omega_c \text{ gap angular frequency at 0 dB}) \qquad [12.12]$$

to the study, in the scalar plane, of the direct sensitivity functions σ_d and complementary sensitivity functions σ_c :

$$\sigma_d = \frac{A(q^{-1})S(q^{-1})\Delta(q^{-1})}{A(q^{-1})S(q^{-1})\Delta(q^{-1}) + q^{-1}B(q^{-1})R(q^{-1})} \qquad [12.13]$$

$$\sigma_c = \frac{q^{-1}B(q^{-1})R(q^{-1})}{A(q^{-1})S(q^{-1})\Delta(q^{-1}) + q^{-1}B(q^{-1})R(q^{-1})} \qquad [12.14]$$

It is generally agreed that a "good" adjustment is characterized by:

– a delay margin higher than a sampling period;

– a direct sensitivity function of a module lower than 6 dB;

– a complementary sensitivity function of a module lower than 3 dB.

12.2.2.2. *Selection procedure of the criterion parameters*

From the criteria formulated above with the help of the traditional tools of scalar Automation, it is possible to choose the sets of satisfactory adjustment parameters:

– N_1 : prediction horizon lower on the output. The product $N_1 T_e$ (T_e sampling period) is chosen as equal to the pure delay of the system;

– N_2 : prediction horizon higher on the output. The product $N_2 T_e$ is limited by the value of the response time. The bigger N_2 is, the more stable and slower the corrected system becomes;

– N_u : prediction horizon on the control. Choosing N_u equal to 1 simplifies the calculation and does not penalize the stability margins (on the contrary, a higher value tends to decompose the phase margin);

– λ : weighting coefficient on the control. This parameter is related to the gain of the system, through the empirical relation:

$$\lambda_{opt} = \text{tr}(\mathbf{G}^{\mathrm{T}}\mathbf{G}) \qquad (\mathbf{G} \text{ matrix described in 12.2.1}) \qquad [12.15]$$

The choice of parameters is frequently limited to a bi-dimensional search (N_2 and λ) ending with the selection of a "good" adjustment.

12.2.3. *Extension of the basic version*

Based on the preceding easy version, several derived strategies were developed, which made it possible to recognize:

– closed loop pre-specified dynamics (structure of multiple reference models);

– several variables to control (cascade structure);

– constraints imposed on the input and output signals.

12.2.3.1. *Structure of multiple reference models*

The aim of this predictive structure of multiple reference models is double. Firstly, it makes it possible to impose a *reference trajectory* through a stable pursuit of a model determined by the user who tones down the conformity with the setting. This pursuit model imposes the dynamics of the looped system (input/output behavior) and it may be considered as a pole placement.

It is also a matter of weakening the quick control variations that we can sometimes recognize through the preceding algorithm, by trying to recreate the reasonable *reference control* that must be applied to the system in order to obtain, at the output, the reference trajectory and by creating in the criterion a minimization on the control error and not only on the control.

The digital model of prediction is defined here again as CARIMA:

$$A(q^{-1})y(t) = B(q^{-1})u(t-1) + \frac{\xi(t)}{\Delta(q^{-1})} \qquad [12.16]$$

The pursuit model chosen by the user makes it possible to specify the reference trajectory $y_r(t)$ that the output of the system will have to follow:

$$A_r(q^{-1})y_r(t) = q^{-1}B_r(q^{-1})w(t) \qquad [12.17]$$

where: $B_r(q^{-1}) = B(q^{-1})P(q^{-1})$.

$P(q^{-1})$ is conceived in such a way as to insure the asymptotical behavior: $y_r(\infty) = w(\infty)$ Thus, for a step function setting, we can choose:

$$P(q^{-1}) = \frac{A_r(1)}{B(1)} = \text{Cte}$$

$A_r(q^{-1})$ is generally a second degree polynomial making it possible to impose a desired response time as well as an adapted damping coefficient.

Coupled to the reference trajectory $y_r(t)$, a reference control $u_r(t)$, which is allowed by the system, is equally defined, the two trajectories being related by the relation:

$$A(q^{-1})y_r(t) = q^{-1}B(q^{-1})u_r(t) \tag{12.18}$$

In order to avoid the reverse of the model and the stability problems related to polynomial $B(q^{-1})$ that may result, equation [12.18] can be formulated again based on relation [12.17] by:

$$A_r(q^{-1})u_r(t) = A(q^{-1})P(q^{-1})w(t) \tag{12.19}$$

Figure 12.4 sums up the principle of this structure with reference models [IRV 86].

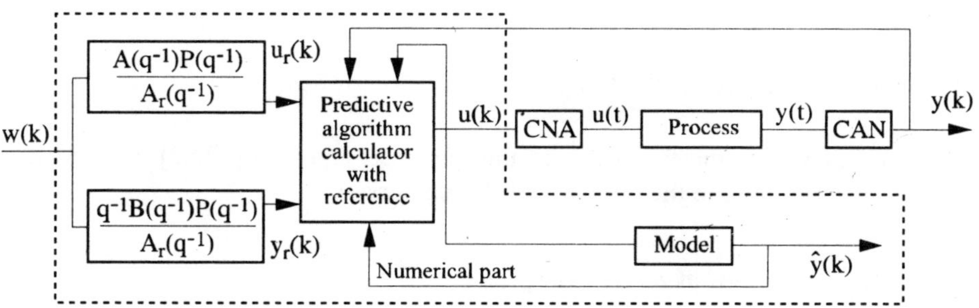

Figure 12.4. *Principle of GPC/MRM algorithm*

The cost function is henceforth a weighted sum affecting the squares of the output predicted errors and the squares of the future control error increments:

$$J = \sum_{j=N_1}^{N_2} \varepsilon_y^2(t+j) + \lambda \sum_{j=1}^{N_u} \varepsilon_u^2(t+j-1) \qquad [12.20]$$

with: $\varepsilon_u(t+j) \equiv 0$ for $j \geq N_u$ and: $\begin{cases} \varepsilon_y(t+j) = \hat{y}(t+j) - y_r(t+j) \\ \varepsilon_u(t+j) = \Delta u(t+j) - \Delta u_r(t+j) \end{cases}$.

Based on relations [12.16], [12.17] and [12.18], we notice that, on the one hand, the increments of control errors and, on the other hand, the output errors are linked by the relation:

$$A(q^{-1})\varepsilon_y(t) = B(q^{-1})\varepsilon_u(t-1) + \xi(t) \qquad [12.21]$$

which corresponds exactly to the CARIMA structure [12.16], parameterized again in terms of signals pertaining to input/output errors. The entire theory previously developed in the case of the GPC "traditional" algorithm can be preserved by replacing in the minimization process y by ε_y, Δu by ε_u and w by 0 (the system must indeed follow a zero error setting). From this moment on, the minimization of the quadratic criterion [12.20] reaches the optimal sequence:

$$\varepsilon_{\mathbf{u}\ opt} = -\mathbf{N}\left[\mathbf{if}\ \varepsilon_y(t) + \mathbf{ih}\ \varepsilon_{u\ opt}(t-1) \right] \qquad [12.22]$$

with: $\varepsilon_{\mathbf{u}\ opt} = \left[\varepsilon_u(t)_{opt} \quad \cdots \quad \varepsilon_u(t+N_u-1)_{opt} \right]^{\mathrm{T}}$.

Here again, only the first value of the sequence, equation [12.22], is applied to the system, according to the principle of sliding horizon:

$$\varepsilon_{u\ opt}(t) = -\mathbf{n}_1^{\mathrm{T}}\left[\mathbf{if}\ \varepsilon_y(t) + \mathbf{ih}\ \varepsilon_{u\ opt}(t-1) \right] \qquad [12.23]$$

We infer from it the equivalent polynomial regulator of this restated problem in terms of error signals:

$$S(q^{-1})\varepsilon_{u\ opt}(t) = -R(q^{-1})\varepsilon_y(t) \qquad [12.24]$$

with:
$$S(q^{-1}) = (1 + \mathbf{n}_1^T \ \mathbf{ih} \ q^{-1})$$

$$R(q^{-1}) = \mathbf{n}_1^T \ \mathbf{if}$$

$$\text{degree}\left[S(q^{-1})\right] = \text{degree}\left[B(q^{-1})\right]$$

$$\text{degree}\left[R(q^{-1})\right] = \text{degree}\left[A(q^{-1})\right]$$

The control applied to the system is inferred from the difference equation:

$$S(q^{-1})\Delta u(t) = -R(q^{-1})y(t) + T_r(q^{-1})w(t) \tag{12.25}$$

with, based on relations [12.17] and [12.19]:

$$T_r(q^{-1}) = \frac{P(q^{-1})}{A_r(q^{-1})} \ (A(q^{-1})\Delta(q^{-1})S(q^{-1}) + q^{-1}B(q^{-1})R(q^{-1})) \tag{12.26}$$

This control law is based again on an RST structure, with the same polynomials $R(q^{-1})$ and $S(q^{-1})$ as those obtained through the traditional algorithm; only polynomial $T(q)$ is modified, becoming a causal rational fraction and explicitly considering the pursuit model chosen by the user. Furthermore, the calculation of the input/output closed loop makes it possible to verify that the resulting dynamics is defined by the pursuit model, which is not at all the case of the transfer function between the output and the interference.

12.2.3.2. *Cascade structure*

The cascade structure suggested makes it possible, in the case of a two-loop version, to simultaneously control two variables (for instance speed and position, for the regulation of the electro-mechanical systems). In the internal loop it includes a predictive structure with multiple reference models developed above, paired to a GPC traditional algorithm for the external loop, as indicated in Figure 12.5

The synthesis of the regulator of the internal loop is considered according to the GPC/MRM strategy of the previous section, in such a way that the internal regulator R_2, S_2 and T_2 is finally implemented by the following difference equation:

$$S_2(q^{-1})\Delta u(t) = -R_2(q^{-1})y_2(t) + T_2(q^{-1})w_2(t) \tag{12.27}$$

The predictive model used for the synthesis of the external regulator consists of two terms: on the one hand the model corresponding to the asymptotical behavior of the closed internal loop and on the other hand the model issued from the external system (defined by the polynomials $A_1(q^{-1})$ and $B_1(q^{-1})$), according to the relation:

$$\frac{y_1(t)}{w_2(t)} = \frac{q^{-1}B_{1r}(q^{-1})}{A_{1r}(q^{-1})} \approx \frac{q^{-1}B_r(q^{-1})}{A_r(q^{-1})} \frac{B_1(q^{-1})}{A_1(q^{-1})}$$

[12.28]

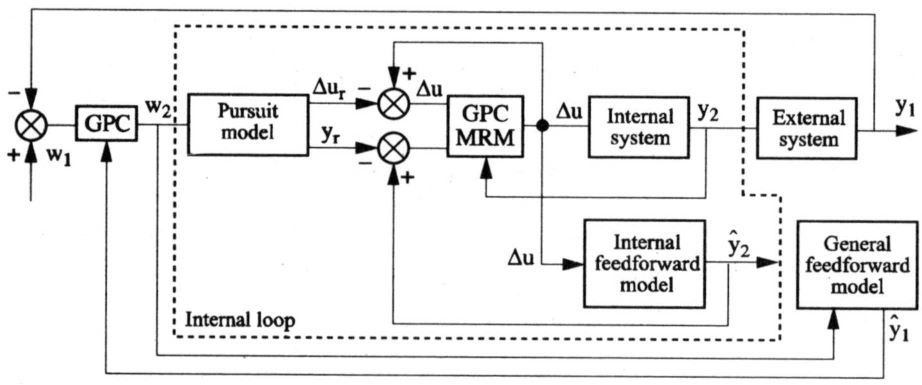

Figure 12.5. *Predictive cascade structure*

In order to obtain this cascade structure [BOU 91], a GPC "traditional" algorithm is perfected, obtaining the external regulator as an equivalent RST polynomial form, this regulator is also implemented with the help of a second difference equation:

$$S_1(q^{-1})\Delta w_2(t) = -R_1(q^{-1})y_1(t) + T_1(q)w_1(t)$$

[12.29]

The global implementation of this two-loop structure thus requires the programming of two difference equations, with weaker degree polynomials and therefore a short calculation time for the algorithm. The fact that this structure can be generalized to any number of loops proves that it is really adapted to the fast real-time loops, like the single-loop structures considered previously. It is important to indicate that only a "single rhythm" cascade has been presented so far (i.e. with one and the same sampling period for the two loops). Other cascade predictive structures were developed, either "multi-rhythm" (different sampling periods between loops), or by using operator δ when over-sampling problems appear. The reader can refer to [BOU 95] and [BOU 96] for more details.

12.2.3.3. *Recognition of equality type terminal constraints (CRHPC)*

A number of various physical problems require the recognition of constraints on the output signal as well as on the control. These constraints may be of "equality" type, when we try to impose precise values to the signals considered, or of "inequality" type, making it possible to define saturations or ranges of specific variations for a signal. The aim of this section is to examine the recognition of "equality" type constraints within the GPC algorithm.

Imposing "equality" type constraints requires the minimization of the cost criterion GPC subjected to a set of m equality constraints in the future, called terminal constraints, defined by:

$$\hat{y}(t + N_2 + j) = w(t + N_2) \text{ for } j = 1,..., m \tag{12.30}$$

where m is the number of points for which the predicted output $\hat{y}$ must coincide to the setting w after the higher prediction horizon N_2. This strategy was developed in a version called CRHPC (*Constrained Receding Horizon Predictive Control*) [CLA 91].

Equation [12.30] may be transposed in the following matrix form:

$$\mathbf{G_c}\,\tilde{\mathbf{u}} = \mathbf{w_c} - \mathbf{f_c} \tag{12.31}$$

with:
$$\mathbf{G_c} = \begin{bmatrix} g_{N_2+1} & g_{N_2} & \cdots & g_{N_2-N_u+2} \\ g_{N_2+2} & g_{N_2+1} & \cdots & g_{N_2-N_u+3} \\ \vdots & \cdots & \cdots & \vdots \\ g_{N_2+m} & g_{N_2+m-1} & \cdots & g_{N_2-N_u+m+1} \end{bmatrix}$$

$$\mathbf{w_c} = \left[w(t + N_2),..., w(t + N_2) \right]^{\mathrm{T}}$$

$$\mathbf{f_c} = \left[f_c(t + N_2 + 1),..., f_c(t + N_2 + m) \right]^{\mathrm{T}}$$

$\mathbf{f_c}$ represents the free response of the system under constraint, defined similarly for the free response of the non-constraint system (equation [12.6]), by:

$$\mathbf{f_c} = \mathbf{if_c}\, y(t) + \mathbf{ih_c}\, \Delta u(t - 1) \tag{12.32}$$

with:

$$\mathbf{if_c} = \left[F_{N_2+1}(q^{-1}), \ ..., \ F_{N_2+m}(q^{-1}) \right]^{\mathrm{T}}$$

$$\mathbf{ih_c} = \left[H_{N_2+1}(q^{-1}), \ ..., \ H_{N_2+m}(q^{-1}) \right]^{\mathrm{T}}$$

The algorithm CRHPC consists of a GPC traditional algorithm related to the concept of terminal constraints. Based on the numerical model of the system (equation [12.1]), of the optimal predictor (equation [12.5]), of the quadratic criterion (equation [12.6]) and of the terminal constraints (equation [12.30]) and with the help of Lagrange multiplier factors, the optimal solution of the problem (equation [12.6]) under the constraints (equation [12.30]) is obtained in matrix form:

$$\tilde{\mathbf{u}}_{opt} = 2\,[\mathbf{H}^{-1}\mathbf{G_c^T}(\mathbf{G_c}\mathbf{H}^{-1}\mathbf{G_c^T})^{-1}\mathbf{G_c}\mathbf{H}^{-1} - \mathbf{H}^{-1}]\ \ \mathbf{G^T}\ \mathbf{c} +$$
$$+\mathbf{H}^{-1}\mathbf{G_c^T}(\mathbf{G_c}\mathbf{H}^{-1}\mathbf{G_c^T})^{-1}\ \mathbf{b} \qquad\qquad [12.33]$$

with: $\mathbf{H} = 2(\mathbf{G^T}\mathbf{G} + \lambda \mathbf{I}_{N_u})$

$$\mathbf{c}\ \ = (\mathbf{w} - \mathbf{if}\ y(t) - \mathbf{ih}\ \Delta u(t-1))$$
$$\mathbf{b}\ \ = (\mathbf{if_c}\ y(t) + \mathbf{ih_c}\ \Delta u(t-1))$$

All the vectors and matrices defined above are made up of coefficients intervening in the j- interval predictors (equation [12.5]) for $N_1 \leq j \leq N_2 + m$.

With an approach similar to the one adopted for the simple predictive structure, only the first value of the previous sequence is applied to the system, according to the principle of the sliding horizon:

$$\Delta u_{opt}(t) = \mathbf{m_1^T}\ \mathbf{b} + \mathbf{n_1^T}\ \mathbf{c} \qquad\qquad [12.34]$$

with: $\mathbf{m_1^T}$ first row of $(\mathbf{H}^{-1}\mathbf{G_c^T}(\mathbf{G_c}\mathbf{H}^{-1}\mathbf{G_c^T})^{-1})$

$\mathbf{n_1^T}$ first row of $[2\,(\mathbf{H}^{-1} - \mathbf{H}^{-1}\mathbf{G_c^T}(\mathbf{G_c}\mathbf{H}^{-1}\mathbf{G_c^T})^{-1}\mathbf{G_c}\mathbf{H}^{-1})\ \ \mathbf{G^T})]$

Equation [12.34] corresponds to a linear corrector that can be written in an RST form, totally similar to the one obtained without constraint:

$$S(q^{-1})\Delta u(t) = -R(q^{-1})y(t) + T(q)w(t) \qquad [12.35]$$

The three polynomials R, S, T take the following form:

$$
\begin{aligned}
S(q^{-1}) &= 1 + q^{-1}(\mathbf{n}_1^{\mathrm{T}}\mathbf{ih} + \mathbf{m}_1^{\mathrm{T}}\mathbf{ih}_c) \\
R(q^{-1}) &= \mathbf{n}_1^{\mathrm{T}}\mathbf{if} + \mathbf{m}_1^{\mathrm{T}}\mathbf{if}_c \\
T(q) &= \mathbf{n}_1^{\mathrm{T}}[q^{N_1} \quad \cdots \quad q^{N_2}]^{\mathrm{T}} + \mathbf{m}_1^{\mathrm{T}}[q^{N_2} \quad \cdots \quad q^{N_2}]^{\mathrm{T}}
\end{aligned} \qquad [12.36]
$$

We can find in this structure a basic part resulting from the algorithm without traditional constraint, with elements corresponding to the recognition of constraints.

The fundamental advantage of this version under CRHPC constraints is to ensure the stability of the looped system for particular choices of adjustment parameters [LEV 93, NIC 93]:

$$
\begin{aligned}
N_1 &\approx \text{pure delay of the system} \\
N_2 &\geq \text{degree}\left[A(q^{-1})\right] + 2 \\
N_u &= N_2 + 1 \\
m &= \text{degree}\left[A(q^{-1})\right] + 1 \\
\lambda &> 0
\end{aligned}
$$

The recognition of terminal constraints was developed here only in the context of a "traditional" structure of the GPC algorithm. This formalism can also be introduced in the single- or multi-rhythm cascade structures with operator δ ... A unified version including the equality and inequality constraints was perfected, leading to the elaboration of GPC regulators by non-linear quadratic optimization. The reader can refer to [DUM 98].

12.3. Functional predictive control (FPC)

This second structure of predictive control is introduced here by indicating the big ideas of the method, starting with the form of the model, the quadratic criterion and up to the examination of adjustment parameters. The formalism and the calculation necessary to the analytical minimization of the criterion will not be dealt with so that the presentation does not become too difficult. The reader may refer to [COM 94, RIC 87] for minimization details of a simple structure and [RIC 93] in the case of a cascade structure.

12.3.1. *Definition of numerical model*

As in the case of GPC, there is no restriction for the form of the model, but the approach by state variable representation is preferred. The first versions of FPC used even a representation by convolution with the help of the coefficients of the discrete impulse response.

If we note by $u(t)$ and $s_m(t)$ respectively the input and output of the model, this model is traditionally represented by the system of equations:

$$\begin{cases} \mathbf{x}_m(t+1) = \mathbf{F}\,\mathbf{x}_m(t) + \mathbf{G}\,u(t) \\ s_m(t) = \mathbf{H}\,\mathbf{x}_m(t) \end{cases} \qquad [12.37]$$

12.3.2. *Choice of a reference trajectory*

The reference trajectory, initialized on the output of the process at instant t, specifies the way in which we want the process to relate to the setting on a given prediction horizon. All choices are possible but the easiest is to consider a first order dynamics for the variance between the setting and the reference trajectory. If we note by $s_p(t)$ the output of the system and $w(t)$ the setting to be followed, this reference trajectory $s_R(t)$ is then defined by the relation:

$$\left[w(t+j) - s_R(t+j)\right] = \alpha^j \left[w(t) - s_p(t)\right] \qquad [12.38]$$

where $0 \le \alpha \le 1$ is a parameter that conditions the speed of the conformity desired:

$$\alpha = \exp(-\frac{3T_e}{T_r}) \qquad [12.39]$$

with T_e the sampling period and T_r the response time in closed loop.

12.3.3. *Object-model difference*

FPC makes it possible to adjust the prediction of the process output obtained through the model, by taking into account the modeling and identification errors as well as possible interferences. Hence, we introduce a signal measuring the variance between the system and the model [SAN 94], called *omd* (object/model difference):

$$omd(t) = s_p(t) - s_m(t) \qquad\qquad [12.40]$$

Hence the aim is to provide a future prediction of this variance $\hat{omd}(t + j)$, so that we have:

$$\hat{s}_p(t + j) = \hat{s}_m(t + j) + \hat{omd}(t + j) \qquad\qquad [12.41]$$

The "level" prediction that consists of considering:

$$omd(t + j) = omd(t) = s_p(t) - s_m(t) \qquad\qquad [12.42]$$

corresponds to a case of a 0 degree self-compensator capable of blocking a static variance. In general, the self-compensator is written:

$$\hat{omd}(t + j) = s_p(t) - s_m(t) + \sum_{i=1}^{d_e} e_i(t)j^i \qquad\qquad [12.43]$$

where d_e is the degree of the extrapolator.

12.3.4. *Structure of the future control*

In an original way, compared to other predictive techniques, the future control is structured here in the form of a linear combination of preliminarily chosen functions, called "basic functions" and marked $\{u_{b\,k}\}, k = 1 \cdots n_b$:

$$u(t + j) = \sum_{k=1}^{n_b} \mu_k(t) u_{b\,k}(j) \qquad\qquad [12.44]$$

Hence, the calculation of the future control sequence requires the determination, at every instant t, of the unknown coefficients $\{\mu_k\}, k = 1\cdots n_b$. Furthermore, based on the strategy of sliding horizon, only the first value of the sequence is applied, which requires that the choice of basic functions must be done in such a way that at least one function verifies $u_{bk}(0) \neq 0$.

Traditionally, the basic functions are canonical functions (step function, ramp, parabola), which are chosen according to the type of the setting and the integrator character of the process. Table 12.1 provides the value of error $w - s_p$ in the case of a non-integrator system.

Functions Input	Step function	Step function + ramp	Step function + ramp + parabola
Step function	0	0	0
Ramp	Cte $\neq 0$	0	0
Parabola	∞	Cte $\neq 0$	0

Table 12.1. *Error $w - s_p$ for a non-integrator system*

12.3.5. *Structure of the optimal predictor*

The predicted output $s_m(t + j / t)$ is traditionally decomposed into a loose response and forced response:

$$\hat{s}_m(t + j / t) = s_{mL}(t + j) + s_{mF}(t + j) \tag{12.45}$$

with, taking into account the basic functions and the state model:

$$\begin{cases} s_{mF}(t + j) = \sum_{k=1}^{n_b} \mu_k(t) s_{mbk}(j) \\ s_{mL}(t + j) = \mathbf{H}\,\mathbf{F}^j \mathbf{x}_m(t) \end{cases} \tag{12.46}$$

s_{mbk} representing the forced response of the model at input u_{bk}.

If an input/output polynomial structure is chosen, the optimal predictor will have a similar form to the one developed in the case of GPC equation [12.5] [DUM 92].

12.3.6. *Definition of quadratic criterion, concept of match points*

The FPC control law is obtained by minimization of a quadratic criterion pertaining to the future errors with a weighting term on the control:

$$D = \sum_{j=1}^{n_h} [\hat{s}_p(t+h_j) - s_R(t+h_j)]^{\,2} + \lambda\; u^2(t) \tag{12.47}$$

Based on equations [12.41], [12.43] and [12.45], the criterion thus chosen minimizes the variance between the output of the predicted process and the reference trajectory in a certain number of points called match points: let h_j be these points and n_h their number. The approach followed during the minimization of the criterion, before reaching the future controls structured by relation [12.44], is summed up in Figure 12.6.

As for the generalized predictive control, the minimization (not detailed here) of the preceding criterion leads to an equivalent polynomial regulator in RST form represented in Figure 12.3.

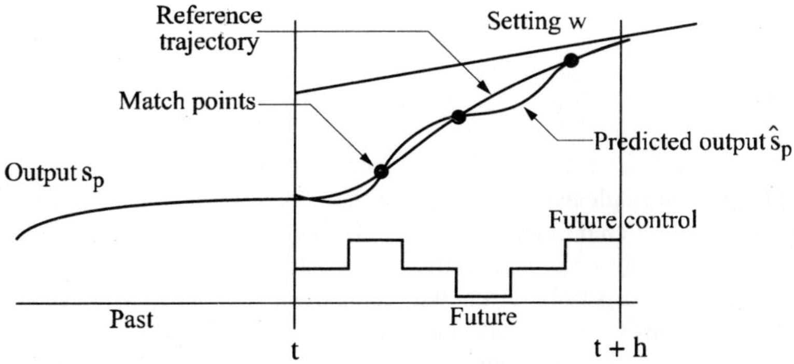

Figure 12.6. *Reference trajectory and match points*

12.3.7. *Adjustment parameters*

Based on the preceding theoretical developments, it appears that the implementation of a functional predictive control law involves the choice of the following parameters:

– T_r : desired response time. This parameter is used to indicate the sampling time if necessary. It also intervenes in the definition of coefficient α of the reference trajectory;

– n_b and $u_{b,k}$: the number of basic functions and their nature. These parameters are set as soon as the nature of the setting signal and the integrator type of the process are known;

– n_h and h_j : the number of match points and their place. The mathematical resolution requires a number of equations more than or equal to the number of unknown factors; traditionally, we choose $n_h = n_b$ and the match points are placed within the prediction horizon, which is limited by the response time desired. The more a point will be placed at the beginning of the horizon, the faster the system will most probably be;

– λ : weighting coefficient on the control. This parameter is related to the gain of the system have through a relation similar to the one defined in GPC.

The choice of parameters proves to have, in the single-variable case, a complexity equal to that observed in the case of a generalized predictive control structure, as only the match points (often restricted to one) and the weighting coefficient on the control are to be fixed, equivalent to horizon N_2 and to coefficient λ of GPC.

12.4. Conclusion

Predictive methods considered in the preceding sections showed the simplicity of their design and implementation because they are always translated, irrespective of the versions considered (simple or cascade), outside inequality constraints, by the real-time programming of several difference equations, which are generated from the RST polynomial structure of the equivalent regulator. This fundamental characteristic implies very fast real-time loops (since the polynomials are actually of low degrees, very few online operations prove to be necessary) because all the calculation of the synthesis phase are made off loop as soon as the adjustment parameters are chosen. This control structure is indicated for applications for which the specifications in terms of sampling period are more and more severe (high speed machining in machine tools, for example).

Parallel to this simple implementation, predictive techniques make it possible to satisfy the very strict specifications, in terms of stability, speed and precision (static or dynamic) but also in terms of robustness, with respect to interferences or neglected dynamics. This level of performance, that "traditional" controls cannot achieve is, however, very complex in terms of the choice of adjustment parameters, not only by a higher number, but also due to a stronger interaction between these parameters. This is why it appears more and more necessary to design, apart from the traditional synthesis of the regulator, a support module with adjustment parameters, which will make the implementation as transparent as possible for the user who is not always a specialist in advanced control laws. The design of a support module is more and more conceivable starting with traditional tools of Automation for the study and analysis of stability and robustness.

All the advantages listed above – simplicity, performance, etc. – ensure that these predictive techniques are implemented in various industrial applications, in very different fields, but preferably when the trajectory to be followed in the future is already known, in a way to profit entirely from the anticipative aspect of this control law.

Hence, a privileged field of predictive control is robotics and machine-tool, for which the elaborated versions (single or multi-rhythm cascades, with operator δ under constraints, etc.) make it possible to deal with sampling periods that can be very strict, to cover a very large range of problems that are already known and to slightly improve the outcomes that have been accessible so far [BOU 87, RIC 87]. however, these techniques were also implemented on slow processes, as the thermal systems [CLA 88], the problems of monitoring the temperature of buildings [DUM 98], or the food industry [DUM 98].

Finally, it should be noted that all these structures developed in the context of single-variable systems can be generalized to multi-variable systems without any particular theoretical difficulty [BOU 96]. The influence of adjustment parameters becomes, however, more complex since the scalar study of the stability and robustness require the use of the techniques obtained from the μ-analysis with the concept of structured and non-structured uncertainties [BOU 99, MOH 92]. One of the multi-variable applications of GPC pertains, for example, to the torque-flow control of asynchronous machines for which couplings are very important and the operation is non-linear.

The performances of predictive laws also open the possibility of implementing these techniques to adaptive structures, which make it possible to maintain an "optimal" behavior when the system presents parametrical drifts in time. The perspectives in this field prove to be very interesting because the principles of direct predictive adaptive control laws (for which the parameters of the regulator, presented in RST polynomial form, are updated directly in real-time and in a single step) are added to indirect adaptive predictive versions henceforth traditional. These methods do not impose the real-time calculation of the predictors required for the creation of the regulator and leave the hope of a gain in time for the significant calculation.

12.5. Bibliography

[BIT 90] BITMEAD R.R., GEVERS M., WERTZ V., *Adaptive Optimal Control. The Thinking Man's GPC*, Prentice Hall International, Systems and Control Engineering, 1990.

[BOU 91] BOUCHER P., DUMUR D., DAUMÜLLER S., "Predictive Cascade Control of Machine Tools Motor Drives", *Proceedings EPE'91*, vol. 2, p. 120-125, Florence, September 1991.

[BOU 92] BOUCHER P., DUMUR D., DAUMÜLLER S., "Autotuned Predictive Control", *Proceedings IFAC Workshop MICC'92*, p. 209-213, Prague, September 1992.

[BOU 95] BOUCHER P., DUMUR D., "Predictive Motion Control", *Journal of Systems Engineering. Special Issue on Motion Control Systems*, vol. 5, p. 148-162, Springer-Verlag, London, 1995.

[BOU 96] BOUCHER P., DUMUR D., *La Commande Prédictive*, Méthodes et Pratiques de l'Ingénieur collection, Editions Technip, Paris, 1996.

[BOU 99] BOUCHER P., DUMUR D., ROUGEBIEF C., RAGUENAUD P., "Predictive Multivariable Generalised Predictive control for Cognac distillation", 5th *European Control Conference*, F203, Karlsruhe, September 1999.

[CLA 87a] CLARKE D.W., MOHTADI C., TUFFS P.S., "Generalized Predictive Control, Part I "The Basic Algorithm", Part II "Extensions and Interpretation", *Automatica*, vol. 23-2, p. 137-160, mars, 1987.

[CLA 87b] CLARKE D.W., MOHTADI C., TUFFS P.S., "Properties of Generalized Predictive Control", *Proceedings 10th World Congress IFAC'87*, vol. 9, p. 63-74, Munich, July 1987.

[CLA 88] CLARKE D.W., "Application of Generalized Predictive Control to Industrial Processes", *IEEE Control Systems Magazine*, p. 49-55, April 1988.

[CLA 91] CLARKE D.W., SCATOLLINI R., "Constrained Receding Horizon Predictive Control", *Proceedings IEE-D*, vol. 138, p. 347-354, 1991.

[COM 94] COMPAS J.M., ESTIVAL J.L., FULGET N., MARTIN R., RICHALET J., "Industrial Applications of Predictive Functional Control", *Proceedings 3rd IEEE Conference on Control Applications*, vol. 3, p. 1643-1655, Glasgow, August 1994.

[DUM 92] DUMUR D., CHÊNE A., LAFABRÈGUE E., BOUCHER P., "Polynomial Predictive Functional Controller for a.c. Motors", *Proceedings IFAC Workshop MCIA'92*, p. 165-170, Perugia, October 1992.

[DUM 98] DUMUR D., BOUCHER P., "A Review Introduction to Linear GPC and Applications", *Journal A*, vol. 39, no. 4, p. 21-35, December 1998.

[FAV 88] FAVIER G., DUBOIS D., ROUGERIE C., "A Review of K-Step Ahead Predictors", *Proceedings IFAC'88 Identification and System Parameter Estimation*, Beijing, August 1988.

[IRV 86] IRVING, E., FALINOWER C.M., FONTE C., "Adaptive Generalized Predictive Control with Multiple Reference Model", *Proceedings 2nd ACASP/86*, June 1986.

[LAN 88] LANDAU I.D., *Identification et commande des systèmes*, Hermès, 1988.

[LEV 93] LEVA A., SCATOLLINI R., "Predictive Control with Terminal Constraints", *Proceedings 2nd European Control Conference*, vol. 2, p. 932-936, Groningen, June, 1993.

[MOH 86] MOHTADI C., CLARKE D.W., "Generalized Predictive Control, LQ, or Pole Placement: A United Approach", *Proceedings 25th Conference on Decision and Control*, Athens, December, 1986.

[MOH 92] MOHTADI C., SHAH S.L., FISHER D.G., "Frequency response characteristics of MIMO GPC", *International Journal of Control*, vol. 55-4, p. 877-900, 1992.

[NIC 93] DE NICOLAO G., SCATOLLINI R., "Stability and Output Terminal Constraints in Predictive Control", *Advances in Model-Based Predictive Control*, p. 105-121, Oxford Science Publications, Oxford University Press, 1993.

[RIC 78] RICHALET J., RAULT A., TESTUD J.L., PAPON J., "Model Predictive Heuristic Control: Applications to Industrial Processes", *Automatica*, vol. 14, 1978.

[RIC 87] RICHALET J., ABU EL ATA S., ARBER C., KUNTZE M.B., JACUBASCH A., SCHILL W., "Predictive Functional Control. Application to Fast and Accurate Robots", *Proceedings 10th IFAC World Congress*, Munich, July, 1987.

[RIC 93] RICHALET J., *Pratique de la Commande Prédictive*, Hermès, 1993.

[SAN 94] SANZO M., RICHALET J., PRADA C., "Matching the Uncertainty of the Model Given by Global Identification Techniques to the Robustness of Model-Based Predictive Controller", *Advances in Model-Based Predictive Control*, p. 386-401, Oxford Science Publications, Oxford University Press, 1994.

[WER 87] WERTZ V., GOREZ R., ZHU K.Y., "A New Generalized Predictive Controller Application to the Control of Process with Uncertain Dead-Time", *Proceedings of the 26th Conference on Decision and Control*, p. 2168-2173, Los Angeles, December 1987.

Chapter 13

Methodology of the State Approach Control

Designing the "autopilot" of a multivariable process, be it quasi-linear, represents a delicate thing. If the theoretical and algorithmic tools concerning the analysis and control of multivariable linear systems have largely progressed during the last 40 years, designing a control law is left to the specialist. The best engineer still has difficulties in applying his knowledge related to multivariable control acquired during his automation course. It is not a mater here to question the interest and importance of automation in the curriculum of an engineer but to stress the importance of "methodology". The teaching of a "control methodology", coherently reuniting the various fundamental automation concepts, is the *sine qua non* condition of a fertile transfer of knowledge from laboratories toward industry.

The methodological challenge has been underestimated for a long time. How else can we explain the little research effort in this field? It is, however, important to underline among others (and in France) the efforts of de Larminat [LAR 93], Bourlès [BOU 92], Duke [DUC 99], Bergeon [PRE 95] or Magni [MAG 87] pertaining to multivariable control methodology.

This chapter deals with a state-based control methodology which is largely inspired by the "standard state control" suggested by de Larminat [LAR 00].

Chapter written by Philippe CHEVREL.

13.1. Introduction

Controlling a process means using the methods available for it in order to adjust its behavior to what is needed. The control applied in time uses information (provided by the sensors) concerning the *state* of the process to react to any unforeseen evolution. Designing even a little sophisticated control law requires the data of *a behavior model of the process* but also relevant information on its *environment*. Which types of disturbances are likely to move the trajectory of the process away from the desired trajectory and which is the information available *a priori* on the desired trajectory?

Finally, a method of designing control laws must make it possible to *arbitrate* among various requirements:

– dynamic performances (which must be even better when the transitional variances between the magnitudes to be controlled and the related settings are weak);

– static performances (which must be even better when the established variances between the magnitudes to be controlled and the related settings are weak);

– weak stress on the control, low sensitivity to measurement noises (to prevent a premature wear and the saturation of the actuators, but to also limit the necessary energy and thus the associated cost);

– robustness (qualitatively invariant preceding properties despite the model errors).

Although this last requirement is not intrinsic (it depends on the model retained for the design), it deserves nevertheless to be discussed. It translates the following important fact. Since the control law is inferred from models whose validity is limited (certain parameters are not well known, idealization by preoccupation with simplicity), it will have to be robust in the sense that the good properties of control (in term of performances and stress on the control) apply to the process as well as to the model and this despite behavior variations.

This need for *arbitrating* between various control requirements leads to two types of reflection.

It is utopian to suppose that detailed specifications of these requirements can be formalized independently of the design approach of the control law. In practice, the designer is very often unaware of what he can expect of the process and an efficient control methodology will have as a primary role to help him become aware of the attainable limits. The problem of robustness can also be considered in two ways[1]. In the first instance, modeling uncertainties are assumed to be quantified in the worst case and we seek to *directly* obtain a regulator guaranteeing the expected performances despite these uncertainties. At their origin, the H_∞ control [FRA 87] and the μ-synthesis [DOY 82, SAF 82, ZHO 96] pursued this goal. A more realistic version consists of preferring a two-time approach alternating the synthesis of a corrector and the analysis of the properties which it provides to the controlled system. Hence, the methodology presented in this chapter will define a limited number of *adjustment parameters* with decoupled effects, so as to efficiently manage the various control compromises.

How can the various control compromises be better negotiated than by defining a criterion formalizing the satisfaction degree of the control considered? The compromise would be obtained by optimizing this criterion after *weighting* each requirement. Weightings would then play the part of adjustment parameters. *A priori* very tempting, this approach faces the difficulties of optimizing the control objectives and the risks of an excess of weightings which may make the approach vain. It is important in this case to define a standard construction procedure of the criterion based on meta-parameters from which the weightings will be obtained. These meta-parameters will be the adjustment parameters.

The methodology proposed here falls under the previously defined principles, i.e. it proceeds by minimization of the judiciously selected *standard* of functional calculus. When we think of *optimal control*, we initially think[2] of control H_2 or H_∞. We will prefer working in Hardy's space H_2 (see section 13.2) for the following reasons:

– the criterion, expressed by means of H_2 standard (H_2 is a Hilbert space), can break up as the sum of elementary criteria;

– control H_2 has a very fertile reinterpretation in terms of LQG control which was the subject of many research works in the past whose results can be used with benefit (robustness of LQ control, principle of separation, etc.);

– the principle of the "worst case" inherent to control H_∞ is not necessarily best adapted to the principle of arbitration between various requirements. In addition, and even if the algorithmic tools for the resolution of the problem of standard H_∞ optimization operates in the state space, the philosophy of the H_∞ approach is based more on an "input-output" principle than on the concept of state.

---

1 In [CHE 93] we used to talk of *direct* methods versus *iterative* methods.
2 For linear stationary systems.

In fact, the biggest difficulty is not in the choice of the standard used (working in H_∞ would be possible) but in the definition of the functional calculus to minimize. This functional calculus must standardize the various control requirements and be possible to parameterize based on a reduced number of coefficients. In the context of controls H_2 or H_∞, it is obtained from the construction of a *standard control model*. This model includes not only the model of the process but also information on its environment (type and direction of input of disturbances, type of settings) and on the control objectives (magnitudes to be controlled, weightings). The principle of its construction is the essence of the methodology presented in this chapter. The resolution of the optimization problem finally obtained requires to remove certain generally allowed assumptions within the framework of the optimization problem of standard H_2.

In short, the methodological principles which underline the developments of this chapter are as follows:

– to concentrate on an optimization problem so as to arbitrate between the various control requirements;

– to privilege an iterative approach alternating the design of a corrector starting from the adjustment of a reduced number of parameters up to the decoupled effects and the analysis of the controlled system;

– to express the control law based on intermediate variables having an identified physical direction and thus to privilege the state approach and the application of the separation principle in its development. The control will be obtained from the instantaneous *state* of the process and its environment.

This chapter is organized as follows. Section 13.2 presents the significant theoretical results relative to the H_2 control and optimization and carries out certain preliminary methodological choices. The minimal information necessary to develop a competitive control law is listed in section 13.3 before being used in section 13.4 for the construction of the standard control model. The methodological approach is summarized in this same section and precedes the conclusion.

13.2. H$_2$ control

The traditional results pertaining to the design of regulators by H_2 optimization and certain extensions are given in this chapter. Its aim is not to be exhaustive but to introduce all the notions and concepts which will be useful to understand the methodology suggested later on.

13.2.1. *Standards*

13.2.1.1. *Signal standard*

Let us consider the space L_2^n of the square integrable signals on $[0,\infty[$, with value in R^n. We can define in this space (which is a Hilbert space) the scalar product and the standard[3] defined below:

$$\langle x,y \rangle = \int_0^{+\infty} x(t)^T y(t)\,dt, \qquad \|x\|_2 = \left(\int_0^{+\infty} x(t)^T x(t)\,dt \right)^{\frac{1}{2}} \qquad [13.1]$$

The *Laplace transform* TL() makes the Hardy space H_2^n of analytical functions $X(s)$ in $Re(s) \geq 0$ and of integrable square correspond to L_2^n. Parseval's theorem makes it possible to connect the standard of a temporal signal of L_2^n to the standard of its Laplace transform in H_2^n:

$$\|x\|_2 = \|X\|_2 = \left(\frac{1}{2\pi} \int_{-\infty}^{+\infty} Graph\,(X^*(j\omega)X(j\omega))\,d\omega \right)^{\frac{1}{2}} \qquad [13.2]$$

13.2.1.2. *Standard induced on the systems*

Let us consider the multivariable system defined by the proper and stable (rational) transfer matrix $G(s)$ or alternatively by its impulse response $g(\cdot) = TL^{-1}(\cdot)$.

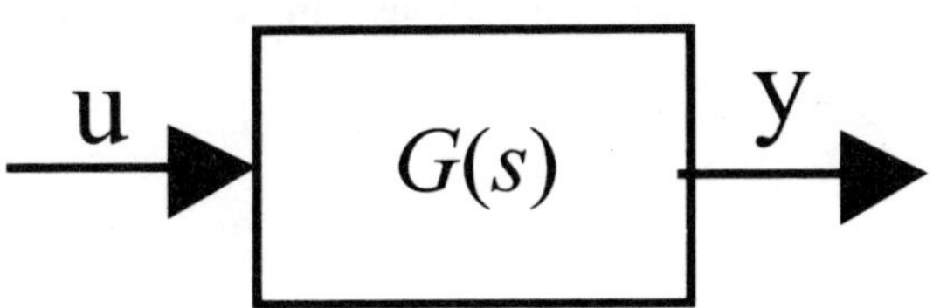

The "H_2 standard" of the input-output operator associated with this system is defined, when it exists, by:

$$\|G\|_2 = \left(\frac{1}{2\pi} \int_{-\infty}^{+\infty} Graph\,(G^*(j\omega)G(j\omega))\,d\omega \right)^{1/2} . \qquad [13.3]$$

Let us note that $u(t) \in R^m$ and $y(t) \in R^p$ respectively the input and output of the system at moment t. Let $R_{uu}(t)$, $R_{yy}(t)$ be the autocorrelation matrices and $S_{uu}(j\omega)$, $S_{yy}(j\omega)$ the associated spectral density matrices. We recall that these matrices are defined as follows. For a given u signal we have: $R_{uu}(\tau) = \lim\limits_{T \to +\infty} \frac{1}{2T} \int_{-T}^{T} u(t+\tau)\,u^T(t)\ dt$. For a centered random u signal, whose certain stochastic characteristics (in particular its 2 order momentum) are known, $R_{uu}(\cdot)$ could be also defined by the equality: $R_{uu}(\tau) = E[u(t+\tau)u^T(t)]$. The two definitions are reunited in the case of a random signal having stationarity and ergodicity properties [PIC 77]. In addition we have the relation: $S_{uu}(j\omega) = \int_{-\infty}^{+\infty} R_{uu}(\tau)e^{j\omega\tau}d\tau$. These notations enable us to give various interpretations to the H_2 standard of G. The results of Table 13.1 are easily obtained from Parseval's equality or the theorem of interferences [PIC 77, ROU 92]. They make it possible to conclude that $\|G\|_2$ is also the energy of the output signal in response to a Dirac impulse or that it characterizes the capacity of the system to *transmit* a white noise[4]. These interpretations will be important further on.

---

4. Characterized by a unitary spectral density matrix.

Characteristic of the input signal	$\|G\|_2$ Significance
$u(t) = I_m \delta(t)$	$\|G\|_2 = \|y\|_2 = \|g\|_2$
$u(\cdot) \; / \; \begin{cases} u(\cdot) \text{ is of zero mean} \\ R_{uu}(t) = I_m \delta(t) \end{cases}$	$\|G\|_2^2 = E\left[\|y(t)\|^2\right] = graph\ (R_{yy}(0))$ $= \displaystyle\int_{-\infty}^{\infty} graph\ (S_{yy}(j\omega))\ d\omega$

Table 13.1. *Several interpretations of* $\|G\|_2$

13.2.1.3. *The grammians' role in the calculation of the* H_2 *standard*

Let us consider the quadruplet $A \in R^{n \times n}, B \in R^{n \times m}, C \in R^{p \times n}, D \in R^{p \times p}$ such that:

$$G(s) = C(sI - A)^{-1}B + D \tag{13.4}$$

In other words, the state $x(t) \in R^n$ of the system Σ evolves according to:

$$\begin{pmatrix} \dot{x}(t) \\ y(t) \end{pmatrix} = \begin{pmatrix} A & B \\ C & D \end{pmatrix} \begin{pmatrix} x(t) \\ u(t) \end{pmatrix} \quad \text{with:} x(0) = x_0 \tag{13.5}$$

The partial grammians associated with this system are defined by:

$$G_c(t) = \int_0^t e^{A\tau} BB^T e^{A^T \tau}\ d\tau$$
$$\tag{13.6}$$
$$G_o(t) = \int_0^t e^{A^T \tau} C^T C e^{A\tau}\ d\tau$$

Table 13.2 presents the results emerging from these definitions.

Input signal characteristic	Significance of grammians
$u(t) = I_m \delta(t)$, $x_0 = 0$	$G_c(t) = \displaystyle\int_0^t x(\tau) \; x(\tau)^T \; d\tau$
$u(\cdot) / \begin{cases} u(\cdot) \text{ is of zero mean} \\ R_{uu}(t) = I_m \delta(t) \end{cases}$	$G_c(t) = E(x(t) \; x^T(t))$
$u(t) = 0$, $x(0) = x_0$	$x_0^T G_o(t) x_0 = \displaystyle\int_0^t y(\tau)^T \, y(\tau) \; d\tau$

Table 13.2. *Several interpretations of grammians*

$G_c(t)$ and $G_o(t)$ are respectively called partial grammians of controllability and observability. In fact, $[G_c(t)]^{-1}$ is directly connected to the minimal "control energy" necessary to transfer the system from state $x(0) = 0$ to state $x(t) = x_1$ [KWA 72]. Basically, $u(\tau) = B^T e^{A^T(t_1 - \tau)} [G_c(t_1)]^{-1} x_1 \; 0 \le \tau < t_1$ is the minimal energy control $\left(\int_0^t u^T(\tau) u(\tau) = x_1^T [G_c(t)]^{-1} x_1 \right)$ that changes the state $x(\cdot)$ from $x_0 = 0$ to $t = 0$ to x_1 to $t = t_1$.

There are also the following equivalences:

– (A, B) is controllable $\Leftrightarrow$ $\forall t > 0$, $G_c(t) > 0$;

– (C, A) is observable $\Leftrightarrow$ $\forall t > 0$, $G_o(t) > 0$.

It is shown without difficulty that $G_c(t)$ and $G_o(t)$ are solutions of Lyapunov differential equations:

$$\dot{G}_c(t) = AG_c(t) + G_c(t)A^T + BB^T \qquad G_c(0) = 0$$

$$\dot{G}_o(t) = A^T G_o(t) + G_o(t)A + C^T C \qquad G_o(0) = 0 \qquad [13.7]$$

The partial grammians can be effectively calculated by integrating this system of first order differential equations (see section 13.6.1).

The "total" grammians (this qualifier is generally omitted) result from the partial grammians by: $G_c = \lim\limits_{T\to+\infty} G_c(T)$ and $G_o = \lim\limits_{T\to+\infty} G_o(T)$. Their existence results from the stability of the system. They are the solution of Lyapunov algebraic equations obtained by canceling the derivatives $\dot{G}_c(t)$ and $\dot{G}_o(t)$:

$$AG_c + G_c A^T + BB^T = 0 \text{ and } A^T G_o + G_o A + C^T C = 0.$$

The following important property is therefore inferred. Let $G(s)$ be the transfer matrix defined by the presumed minimal realization $G(s) := \begin{pmatrix} A & B \\ C & 0 \end{pmatrix}$.

Then:

$$\left\| G(s) \right\|_2^2 = Graph\,(B^T G_o B) = Graph\,(C G_c C^T) \tag{13.8}$$

Numerically, standard H_2 of $G(s)$ could be obtained by resolution of an Lyapunov algebraic equation obtained from the state matrices A, B, C. Let us note that matrix $G(s)$ must be strictly proper for the existence of $\left\| G(s) \right\|_2$.

A last interesting interpretation of standard H_2 of $G(s) := \begin{pmatrix} A & B \\ C & 0 \end{pmatrix}$ is as follows. Let $B_{\bullet 1}, B_{\bullet 2}, \cdots B_{\bullet m}$ be the columns of B. Let y_{Li} be the free response of the system on the basis of the initial condition $x_{0i} = B_{\bullet i}$. It is verified then that the following identity is true:

$$\left\| G(s) \right\|_2^2 = \left\| y_{L1} \right\|_2^2 + \left\| y_{L2} \right\|_2^2 + L \left\| y_{Lm} \right\|_2^2 \tag{13.9}$$

Thus, standard H_2 gives, for a system whose state vector consists of internal variables easy to interpret, an energy indication on its free response for a set of initial conditions contained in $Im(B)$.

13.2.2. H_2 optimization

13.2.2.1. Definition of the standard H_2 problem [DOY 89]

Any closed loop control can be formulated in the standard form of Figure 13.1.

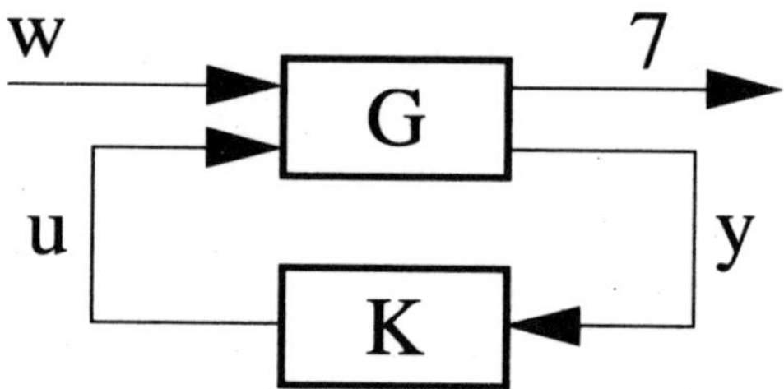

Figure 13.1. *Standard feedback diagram*

The quadripole **G**, also called a standard model, and feedback **K** are supposed to be defined as follows, by using the transfer matrices $G(s)$ and $K(s)$ and their realization in the state space:

$$G(s) = \begin{pmatrix} G_{11}(s) & G_{12}(s) \\ G_{21}(s) & G_{22}(s) \end{pmatrix} := \left[\begin{array}{c|cc} A & B_1 & B_2 \\ \hline C_1 & D_{11} & D_{12} \\ C_2 & D_{21} & D_{22} \end{array} \right] \Leftrightarrow \begin{pmatrix} \dot{x} \\ z \\ y \end{pmatrix} = \left[\begin{array}{c|cc} A & B_1 & B_2 \\ \hline C_1 & D_{11} & D_{12} \\ C_2 & D_{21} & D_{22} \end{array} \right] \begin{pmatrix} x \\ w \\ u \end{pmatrix}$$

$$K(s) = D_K + C_K (sI - A_K)^{-1} B_K \qquad\qquad [13.10]$$

NOTE 13.1.– the size of each matrix results from the size of the various signals: $w \in R^{m1}, u \in R^{m2}, x \in R^n, z \in R^{p1}, y \in R^{p2}$.

The closed loop system of input w and output z, noted by $\boldsymbol{T_{zw}}$, is obtained:

$$T_{zw}(s) = F_l(G(s), K(s)) \overset{\Delta}{=} G_{11}(s) + G_{12}(s)K(s)(I - G_{22}(s)K(s))^{-1} G_{21}(s)$$

$$= D_{bf} + C_{bf}(sI - A_{bf})^{-1} B_{bf} \qquad\qquad [13.11]$$

with:

$$A_{bf} = \begin{bmatrix} A + B_2(I - D_K D_{22})^{-1} D_K C_2 & B_2(I - D_K D_{22})^{-1} C_K \\ B_K(I - D_{22}D_K)^{-1} C_2 & A_K + B_K(I - D_{22}D_K)^{-1} D_{22}D_K \end{bmatrix},$$

$$B_{bf} = \begin{bmatrix} B_1 + B_2(I - D_K D_{22})^{-1} D_K D_{21} \\ B_K(I - D_{22}D_K)^{-1} D_{21} \end{bmatrix},$$

$$C_{bf} = \begin{bmatrix} C_1 + D_{12}(I - D_K D_{22})^{-1} D_K C_2 & , & D_{12}(I - D_K D_{22})^{-1} C_K \end{bmatrix}$$

$$D_{bf} = \begin{bmatrix} D_{11} + D_{12}(I - D_K D_{22})^{-1} D_K D_{21} \end{bmatrix}$$

It has the property of *internal stability* if and only if the eigenvalues of A_{bf} are all of negative real part.

The standard H_2 optimization problem is generally referred to as a problem consisting of finding $\boldsymbol{K_{H_2}}$ which ensures:

– the inner stability of the closed loop system $\boldsymbol{T_{zw}} = F_l(\boldsymbol{G}, \boldsymbol{K_{H_2}})$;

– the minimality of the criterion $J_{H_2}(\boldsymbol{K_{H_2}}) = \|T_{zw}\|_2$.

13.2.2.2. *Resolution of the H_2 standard optimization problem*

The solution of the problem above is well-known [ZHO 96]. To begin with, let us distinguish two elementary cases before presenting the general case.

The "state feedback" (SF) case: it is the case where $y = x$. All the state components of the standard model are accessible for feedback.

The "output injection" (OI) case: it is the case where the feedback can act independently on each component of the evolution equation. This case occurs during the design of an *observer*.

In these two cases, there are the following particular standard models:

$$G_{RE}(s) := \left[\begin{array}{c|cc} A & B_1 & B_2 \\ \hline C_1 & D_{11} & D_{12} \\ I & 0 & 0 \end{array} \right] \qquad G_{IS}(s) := \left[\begin{array}{c|cc} A & B_1 & I \\ \hline C_1 & D_{11} & 0 \\ C_2 & D_{21} & 0 \end{array} \right] \qquad [13.12]$$

The optimum of the H_2 criterion, in the case of the *state feedback*, has the characteristic that it can be obtained by a static feedback:

$$K_{RE_{H_2}}(s) = D_K = -(D_{12}{}^T D_{12})^{-1}(B_2{}^T P + D_{12}{}^T C_1) \qquad [13.13]$$

with:

$$\begin{cases} P \geq 0 \quad \text{(P positive semi-defined)} \\ A^T P + PA - (PB_2 + C_1{}^T D_{12})(D_{12}{}^T D_{12})^{-1}(B_2{}^T P + D_{12}{}^T C_1) + C_1{}^T C_1 = 0 \end{cases}$$

The optimal state feedback thus results from the resolution of this latter second order matrix equation, named the Riccati equation, which is, for the closed loop system, the Lyapunov equation:

$$(A + B_2 K_{RE_{H_2}})^T P + P(A + B_2 K_{RE_{H_2}}) + (C_1 + D_{12} K_{RE_{H_2}})^T (C_1 + D_{12} K_{RE_{H_2}}) = 0$$

P is thus the observability grammian of the looped system and it is deduced with the optimum: $\|T_{zw}\|_2^2 = \|F_l(G_{RE}, K_{RE_{H_2}})\|_2^2 = graph\,(B_1 P B_1{}^T)$. For the sake of completeness, it is necessary to specify the existence hypotheses of a solution to this problem:

– pair (A, B_2) must be stabilizable in order to enable the stability of the looped system. Let us note, however, that if the *inner* stability of the looped system is not required, the hypothesis according to which the non-stabilizable modes by u are all non-controllable by w or unobservable by z is enough. Gain $K_{RE_{H_2}}$ can then be determined from the state representation reduced to the only stabilizable states as we will see further on;

– $D_{11} = 0$ is a condition which generically ensures the strict propriety of T_{zw} and thus the existence of its H_2 standard;

– D_{12} must be of full rank (per columns) to ensure the reversibility of $D_{12}{}^T D_{12}$ in the Riccati equation. Similarly, the zero invariants of $G_{12}(s) := \begin{pmatrix} A & B_2 \\ C_1 & D_{12} \end{pmatrix}$ must not be on the imaginary axis.

The H_2 solution, which is optimal in the case of output injection, is obtained directly from what precedes by application from the *duality principle* (see section 13.6.2). Under the dual assumptions of those stated previously, we obtain:

$$K_{IS_{H_2}}(s) = D_K = -(\Sigma C_2^T + B_1 D_{21}^T)(D_{21}^T D_{21})^{-1} \qquad [13.14]$$

with:
$$
\begin{cases}
\Sigma \geq 0 \quad (\Sigma \text{ positive semi-defined}) \\
A\Sigma + \Sigma A^T - (\Sigma C_2^T + B_1 D_{21}^T)(D_{21}^T D_{21})^{-1}(C_2 \Sigma + D_{21} B_1^T) + B_1 B_1^T = 0
\end{cases}
$$

At optimum, $\left\| T_{zw} \right\|_2^2 = \left\| F_l(G_{IS}, K) \right\|_2^2 = Graph(C_1^T \Sigma C_1)$. The existence hypotheses of a solution to this problem are themselves dual of those of problem (RE).

The H_2 solution – which is optimal in the general case, is this time a dynamic system of the same size as the standard model. It is obtained from the two preceding elementary cases by applying the *separation principle* [AND 89]:

$$K_{H_2}(s) := \left(\begin{array}{c|c} A + B_2 K_{RE_{H_2}} + K_{IS_{H_2}} C_2 + K_{IS_{H_2}} D_{22} K_{RE_{H_2}} & K_{IS_{H_2}} \\ \hline -K_{RE_{H_2}} & 0 \end{array} \right) \qquad [13.15]$$

Moreover:

$$\| T_{zw}(s) \|_2^2 = \left\| F_l(G(s), K_{H_2}(s)) \right\|_2^2 = \left\| F_l\left(G_{RE}(s), K_{RE_{H_2}}\right) \right\|_2^2 + \left\| F_l\left(G_{IS}(s), K_{IS_{H_2}}\right) \right\|_2^2$$

Let us sum up the existence conditions of this solution to the standard H_2 problem (A, B_2) stabilizable and (C_2, A) detectable.

$$\forall \omega \in R, \begin{pmatrix} A - j\omega I & B_2 \\ C_1 & D_{12} \end{pmatrix} \text{ and } D_{12} \text{ are of full rank per column.}$$

$$\forall \omega \in R, \begin{pmatrix} A - j\omega I & B_1 \\ C_2 & D_{21} \end{pmatrix} \text{ and } D_{21} \text{ are of full rank per row. These hypotheses}$$

are easily understood if it is known that at optimum, the poles of $T_{zw}(s)$ tend toward the zeros of transmission of $G_{12}(s)$ and $G_{21}(s)$. In addition, the remaining invariant zero are non-controllable modes by B_1 or non-detectable modes by C_1 which would be preserved in closed loop. Hence, the absence of infinite zeros or on the imaginary axis is imposed.

$$D_{11} = 0.$$

13.2.3. $H_2 - LQG$

Various interpretations of the H_2 standard provided in the preceding section enable us to establish the link with Kalman theory and LQG control (see Chapter 6). If w is a centered, stationary, unit spectrum white noise, and if the standard model is that in Figure 13.2 [STE 87], we obtain:

$$\|T_{zw}\|_2^2 = \lim_{T \to \infty} E\left[\frac{1}{T}\int_0^T \|z(t)\|^2 \, dt\right] = \lim_{T \to \infty} E\left[\frac{1}{T}\int_0^T \left[x(t)^T \; u(t)^T\right]\begin{bmatrix} Q & N_c \\ N_c{}^T & R \end{bmatrix}\begin{bmatrix} x(t) \\ u(t) \end{bmatrix} dt\right] = J_{LQG}$$

$$\text{and} \quad E\left\{\begin{bmatrix} w_x(t) \\ w_y(t) \end{bmatrix}\left[w_x(\tau) \; w_y(\tau)\right]^T\right\} = \begin{bmatrix} V & N_f \\ N_f{}^T & W \end{bmatrix}\delta(t-\tau) \qquad\qquad [13.16]$$

The two elementary cases previously discussed in relation to H_2 correspond to the case of LQ control and the design of the Kalman filter. We have $K_{LQ} = -K_{RE_{H_2}}$ and the control law by state feedback $u = -K_{LQ}x$ minimizes

$$J_{LQ} = \lim_{T \to \infty} \frac{1}{T}\int_0^T \left[x(t)^T \; u(t)^T\right]\begin{bmatrix} Q & N_c \\ N_c{}^T & R \end{bmatrix}\begin{bmatrix} x(t) \\ u(t) \end{bmatrix} dt \,.$$ In addition, for $L_{FK} = K_{IS_{H_2}}$,

the observer $\dot{\hat{x}} = A\hat{x} + B_2 u + L_{FK}(y - C_2\hat{x})$ is precisely the Kalman filter minimizing $E\left(\|C_1(x-\hat{x})\|^2\right)$ under the hypotheses of evolution noise w_x and measurement noise w_y previously defined.

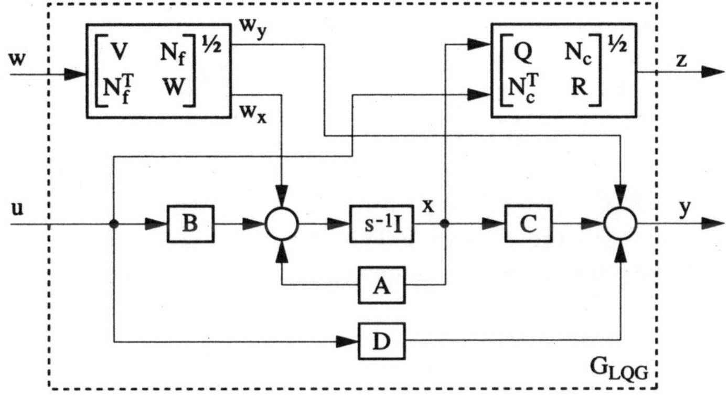

Figure 13.2. *Standard form for LQG control*

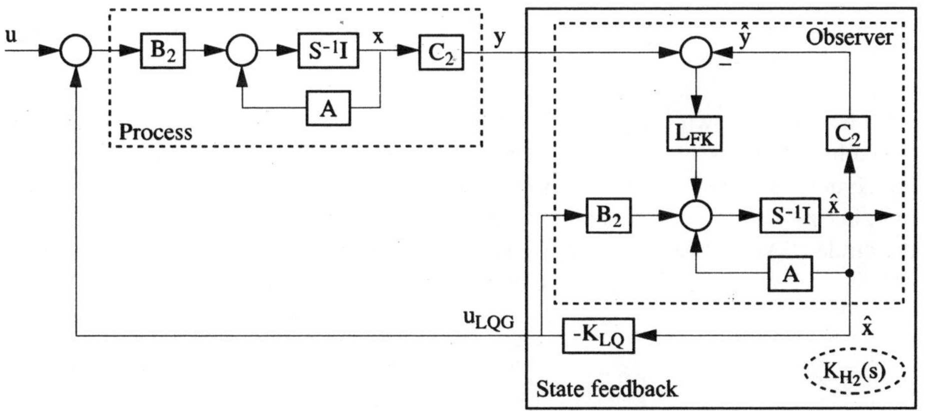

Figure 13.3. *LQG structure (state feedback/observer)*

The resulting control law illustrated in Figure 13.3 has the structure of the state feedback/observer:

$$\begin{cases} u = -K_{LQ}\hat{x} \\ \dot{\hat{x}} = (A - L_{FK}C_2)\hat{x} + B_2 u + L_{FK}(y - C_2\hat{x}) \end{cases}$$

$$\Updownarrow \qquad\qquad [13.17]$$

$$K_{H_2}(s) = -K_{LQ}(sI - A - B_2 K_{LQ} + K_{FK}C_2)^{-1}K_{FK}$$

Finally, the equivalence between the standard H_2 problem and LQG problem is obtained for: $Q = C_1^T C_1$, $R = D_{12}^T D_{12}$, $N_c = C_1^T D_{12}$, $V = B_1 B_1^T$, $W = D_{21} D_{21}^T$, $N_f = B_1 D_{21}^T$.

However, for H_2, matrices Q, R, N_c, V, W and N_f can be officially considered weighting matrices. In order to be able to wisely choose these weightings, the designer must make use of methodological rules like the ones suggested in section 13.4.

13.2.4. $H_2 - LTR$

According to what was said above, the plethoric works (see [CHE 93] and the references included) on the LQ control, LQ with frequency weightings and LQG can be useful in the context of H_2 control. This is true in particular for the results relating to robustness.

It has been known for a long time that the LQ control gives to the looped system enviable properties of robustness (see Chapter 6 and [SAF 77]) The exteriority of the Nyquist place with respect to the Kalman circle guarantees good gain and phase margins, as well as good robustness with respect to static non-linearities (criterion of the circle [SAF 80]) and a certain type of dynamic uncertainties[5]. These properties are obtained at the beginning of the process.

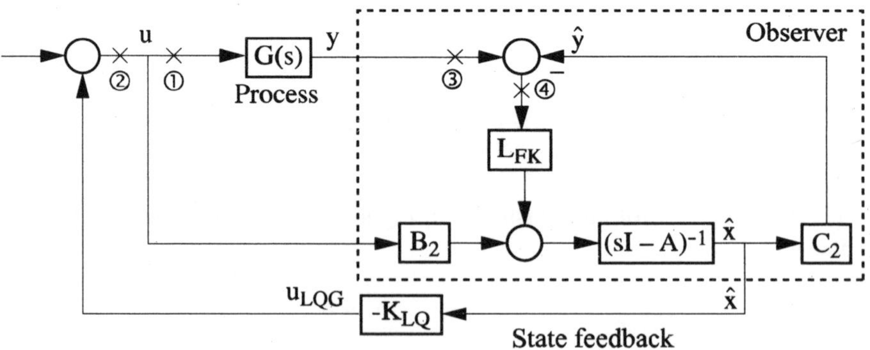

Figure 13.4. *Analysis of robustness of the LQG control*

---

5 $\Delta(s) = (L_{pu}(s) - L_u(s))L_u(s)^{-1}$ relative uncertainty on the input loop transfer if L_u and L_{pu} represent the nominal and disturbed loop transfers.

The robustness properties of the LQ control (or of H_{2RE} regulator) can be lost in the general case, i.e. when the state of the system is inaccessible. The addition of an observer, be it the Kalman observer, in fact modifies the loop transfer $L_{u_{LQ}}(s) = K_{LQ}(sI - A)^{-1}B_2$ obtained in the case of state feedback. As an example we will verify that $L_{u_{LQ}}(s)$ is also the loop transfer of control LQG if we open the loop of Figure 13.4 at point ②. Unfortunately, the need for robustness is felt at point ① and not at point ② (uncertainties due to the actuators). The LTR technique (Loop Transfer Recovery according to the Anglo-Saxon terminology [STE 87, MAC 89]) consists of choosing for problem H_2 / LQG, a particular set of weightings, allowing the restoration of the loop transfer $L_{u_{LQ}}(s)$ in point ① this time. In the diagram of Figure 13.5 it appears obvious that this will be at least closely obtained on the only condition that the transfer matrix $K_{LQ}(sI - A + L_{FK}C_2)^{-1}B_2$ is small in terms of a certain standard. This will be the case for the following particular choice of weightings (for the Kalman filter):

$$B_1 = B_2, \quad D_{21} \to 0 \quad \Leftrightarrow \quad V = B_2 B_2^{T}, \quad W \to 0, \quad N_f \to 0$$

This result is formalized by the following proposition.

Proposition (primal LTR)

$$B_1 = B_2, \quad D_{21} \to 0 \quad \Rightarrow \quad L_{FK} \text{ minimizes } \left\| K_{LQ}(sI - A + L_{FK}C_2)^{-1}B_2 \right\|_2.$$

Moreover, if the process is at phase minimum and reversible on the left $\left\| K_{LQ}(sI - A + L_{FK}C_2)^{-1}B_2 \right\|_2 \to 0$ and the robustness of LQ[6] [AND 89], [SAF 80] is recovered for regulator $H_2 - LQG$ at the *beginning* of the process (at point ①).

NOTE 13.2.– the demonstration of this result, omitted for lack of space, uses the separation principle presented in section 13.2.2.

We obtain by duality the following proposition.

Proposition (dual LTR)

$$C_1 = C_2, \quad D_{12} \to 0 \quad \Rightarrow \quad K_{LQ} \text{ minimizes } \left\| C_2 (sI - A + B_2 K_{LQ})^{-1} L_{FK} \right\|_2.$$

---

6 $S_u(s) = (I + L_{u_{LQ}}(s))^{-1}$ satisfies the equality $\left\| D_{12} S_u D_{12}^{\#} \right\|_\infty = 1$, if $C_1^T D_{12} = 0$. Note:

$D_{12}^{\#}$ represents the reverse on the left of D_{12} : $D_{12}^{\#} \overset{\Delta}{=} (D_{12}^T D_{12})^{-1} D_{12}^T$.

If moreover the process is at phase minimum and reversible on the right $\left\| C_2 \left(sI - A + B_2 K_{LQ} \right)^{-1} L_{FK} \right\|_2 \to 0$ and the robustness[7] is recovered for regulator $H_2 - LQG$ at the *end* of the process (at point ①).

Hence, the dual *LTR* makes it possible to obtain good robustness margins with respect to uncertainties at the output of the system resulting in particular from sensors.

Let us note that "input robustness" and "output robustness" are not necessarily antagonistic and that in the majority of the encountered practical cases, these properties converge. It is at least the bet of the standard state control presented in [LAR 00].

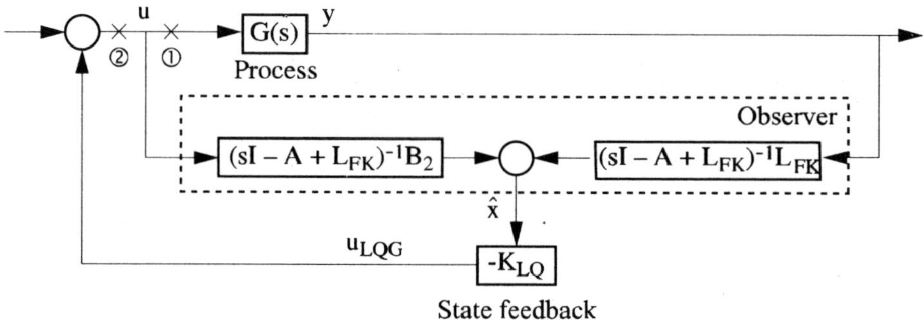

Figure 13.5. *Equivalent LQG diagram*

13.2.5. *Generalization of the H_2 standard problem*

The required (and commonly approved) hypotheses in the formulation of the standard H_2 problem are too restrictive to be able to rigorously solve the majority of control problems, at least by adopting the methodology recommended in this chapter. If the hypothesis "D_{12} and D_{21} of full rank" can be made less strict by preferring a resolution of the problem based on the latest developments regarding the optimization by positive semi-definite programming[8] [GAH 94, IWA 91], the internal stability of the relooped standard model $F_l(G, K)$ always appears as a constraint. As underlined in [CHE 93], this is restrictive in the context of the design

---

7 That of the Kalman filter this time.

8 The problem is formulated as an optimization problem under the constraint of Linked Matrix Inequalities (LMI). The numerical tools related to this type of optimization are from then on entirely competitive.

of a regulator because the constraint should only relate to the internal stability of the process and not of the standard model which potentially includes dynamic weightings. For this reason, we consider it useful to present a generalized version of the standard problem for its use in the context of the methodology of the control suggested further on.

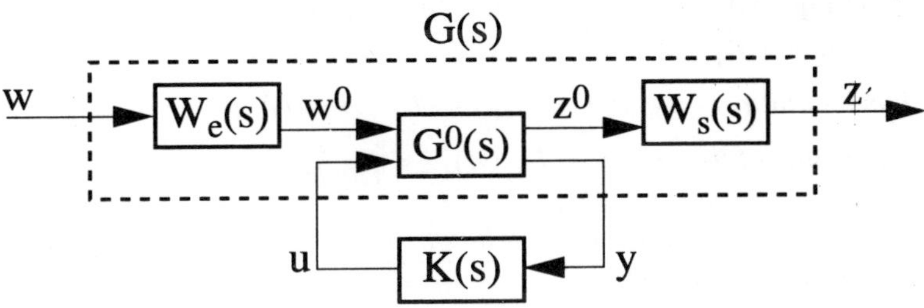

Figure 13.6. *Toward defining the H_2 generalized problem*

The generalized H_2 problem can be formalized as follows. Let us consider the looped system in Figure 13.6 with $w \in R^{m1}$, $u \in R^{m2}$, $x \in R^n$, $z \in R^{p1}$, $y \in R^{p2}$. A realization in the state space of the standard model $G(s)$ can be directly deduced from those presumed minimal of $W_e(s)$, $G^0(s)$ and $W_s(s)$:

$$G_0(s) := \left[\begin{array}{c|cc} A^0 & B_1^0 & B_2^0 \\ \hline C_1^0 & D_{11}^0 & D_{12}^0 \\ C_2^0 & D_{21}^0 & 0 \end{array} \right] \quad W_e(s) := \left[\begin{array}{c|c} A_{W_e} & B_{W_e} \\ \hline C_{W_e} & D_{W_e} \end{array} \right] \quad W_s(s) := \left[\begin{array}{c|c} A_{W_s} & B_{W_s} \\ \hline C_{W_s} & D_{W_s} \end{array} \right]$$

$$\Downarrow$$

$$G(s) := \left[\begin{array}{ccc|cc} A_{W_e} & 0 & 0 & B_{W_e} & 0 \\ B_1^0 C_{W_e} & A^0 & 0 & B_1^0 D_{W_e} & B_2^0 \\ B_{W_s} D_{11}^0 C_{W_e} & B_{W_s} C_1^0 & A_{W_s} & B_{W_s} D_{11}^0 D_{W_e} & B_{W_s} D_{12}^0 \\ \hline D_{W_s} D_{11}^0 C_{W_e} & D_{W_s} C_1^0 & C_{W_s} & D_{W_s} D_{11}^0 D_{W_e} & D_{W_s} D_{12}^0 \\ D_{21}^0 C_{W_e} & C_2^0 & 0 & D_{21}^0 D_{W_e} & 0 \end{array} \right] := \left[\begin{array}{c|cc} A & B_1 & B_2 \\ \hline C_1 & D_{11} & D_{12} \\ C_2 & D_{21} & D_{22} \end{array} \right]$$

$$[13.19]$$

We will assume $D_{11} = D_{W_s} D_{11}^0 D_{W_e} = 0$. By construction, the modes of $W_s(s)$ are unobservable by y, whereas the modes of $W_e(s)$ are non-controllable by u. If these modes are unstable, the standard model $G(s)$ is non-stabilizable by u and non-detectable by y. The standard H_2 problem cannot be solved (we are outside its context of hypothesis).

DEFINITION OF THE H_2 OPTIMIZATION PROBLEM GENERALIZED.– *it is a question of finding* $\mathbf{K_{H_2}}$ *which ensures:*

– *the inner stability of the looped process* $T_{z^0 w^0} = F_l(G^0, K_{H_2})$;

– *the minimality of the criterion* $J_{H_2}(K_{H_2}) = \|T_{zw}\|_2$.

LEMMA 13.1.– *the existence of a solution to the problem above requires the hypotheses (H0) to (H3) below.*

H0. *The poles* $W_e(s)$ *and* $W_s(s)$ *are of non-negative real part. If this were not the case, it would be enough to incorporate the stable parts of* $W_e(s)$ *and* $W_s(s)$ *into the model* $G^0(s)$.

H1. *The pairs* $\left(\begin{bmatrix} A^0 & 0 \\ B_{W_s} C_1^0 & A_{W_s} \end{bmatrix}, \begin{bmatrix} B_2^0 \\ B_{W_s} D_{12}^0 \end{bmatrix} \right)$ *and* $\left([D_{21}^0 C_{W_e} \quad C_2^0], \begin{bmatrix} A_{W_e} & 0 \\ B_1^0 C_{W_e} & A^0 \end{bmatrix} \right)$ *are*

respectively stabilizable and detectable. If these hypotheses were not satisfied but if (C_2^0, A^0, B_2^0) *is stabilizable and detectable, we will reduce beforehand the standard model (see [FRA 77]) so that it satisfies (H1).*

H2. D_{12} *(respectively* D_{21}*) is of full rank per column (resp. per row).*

H3. *The realizations of* $G_{12}(s)$ *and* $G_{21}(s)$, *obtained from equation [13.19], have no other invariant zeros on the imaginary axis that belong respectively to the spectra of* A_{W_e} *and* A_{W_s}. *Precisely:*

$$\text{H3.1} \quad \begin{pmatrix} A^0 - j\omega I & 0 & B_2^0 \\ B_{W_s} C_1^0 & A_{W_s} & B_{W_s} D_{12}^0 \\ D_{W_s} C_1^0 & C_{W_s} & D_{W_s} D_{12}^0 \end{pmatrix} \text{ is of full rank per columns } \forall \omega \in R$$

$$H3.2 \quad \begin{pmatrix} A_{W_e} - j\omega I & 0 & B_{W_e} \\ B_1^0 C_{W_e} & A^0 & B_1^0 D_{W_e} \\ D_{21}^0 C_{W_e} & C_2^0 & D_{21}^0 D_{W_e} \end{pmatrix} \quad \text{is of full rank per rows} \quad \forall \omega \in R$$

$H4.\ D_{11} = 0$.

THEOREM 13.1 (SOLUTION TO THE GENERALIZED H_2 PROBLEM).– *under the hypotheses (H0) to (H3) of lemma 13.1, we can show that the generalized H_2 problem admits a solution if and only if:*

$$\exists P \geq 0 / \ A^T P + PA - (PB_2 + C_1^T D_{12})(D_{12}^T D_{12})^{-1}(B_2^T P + D_{12}^T C_1) + C_1^T C_1 = 0$$

$$\exists \Sigma \geq 0 / \ A\Sigma + \Sigma A^T - (\Sigma C_2^T + B_1 D_{21}^T)(D_{21}^T D_{21})^{-1}(C_2 \Sigma + D_{21} B_1^T) + B_1 B_1^T = 0$$

$$\text{Hence } K = -(D_{12}^T D_{12})^{-1}(B_2^T P + D_{12}^T C_1)$$

$$\text{and } L = -(\Sigma C_2 + B_1^T D_{21})\,(D_{21}^T D_{21})^{-1} 1.$$

It is shown that:

– the only unstable modes $A + B_2 K$ are the unstable modes of A_{W_e} ;

– the only unstable modes $A + LC_2$ are the unstable modes of A_{W_s} ;

– the optimal regulator has the same size as the standard model and is given by:

$$K_{\mathbf{H_{2g}}}(s) := \left(\begin{array}{c|c} A + B_2 K + LC_2 & L \\ \hline K & 0 \end{array} \right)$$

Note that the separation principle continues to apply.

We can also show the following original result which establishes the link with the well-known Regulation Problem with Internal Stability (RPIS) introduced by Wonham [WON 85].

THEOREM 13.2 (HIDDEN EQUATIONS).– *under the same hypotheses as previously, properties 1 and 2 are equivalent as well as properties 3 and 4.*

$$\exists P \geq 0 / A^T P + PA - (PB_2 + C_1^T D_{12})(D_{12}^T D_{12})^{-1}(B_2^T P + D_{12}^T C_1) + C_1^T C_1 = 0$$

$$\exists (T_a, K_a) \; / \; \left\{ \begin{array}{l} T_a A_{W_e} - \begin{pmatrix} A^0 & 0 \\ B_{W_s} C_1^0 & A_{W_s} \end{pmatrix} T_a - \begin{pmatrix} B_2^0 \\ B_{W_s} D_{12}^0 \end{pmatrix} K_a + \begin{pmatrix} B_1^0 C_{W_e} \\ 0 \end{pmatrix} = 0 \\[3mm] (D_{W_s} C_1^0 \quad C_{W_s}) T_a + D_{W_s} D_{12}^0 K_a - D_{W_s} D_{11}^0 C_{W_e} = 0 \end{array} \right.$$

$$\exists \Sigma \geq 0 / \; A\Sigma + \Sigma A^T - (\Sigma C_2^T + B_1 D_{21}^T)(D_{21}^T D_{21})^{-1}(C_2 \Sigma + D_{21} B_1^T) + B_1 B_1^T = 0$$

$$\exists (S_a, L_a) \; / \; \left\{ \begin{array}{l} A_{W_s} S_a - S_a \begin{pmatrix} A_{W_e} & 0 \\ B_1^0 C_{W_e} & A^0 \end{pmatrix} - L_a (D_{21}^0 C_{W_e} \quad C_2^0) + (0 \quad B_{W_s} C_1^0) = 0 \\[3mm] S_a \begin{pmatrix} B_{W_e} \\ B_1^0 D_{W_e} \end{pmatrix} + L_a D_{21}^0 D_{W_e} - B_{W_s} D_{11}^0 D_{W_e} = 0 \end{array} \right.$$

Furthermore:

$$P = \begin{pmatrix} T_a^T P_3 T_a & T_a^T P_3 \\ P_3 T_a & P_3 \end{pmatrix}$$

where solution P_3 of the Riccati equation reduced to the controllable part by u is solution of 1.

$$\Sigma = \begin{pmatrix} \Sigma_1 & \Sigma_1 S_a^T \\ S_a \Sigma_1 & S_a \Sigma_1 S_a^T \end{pmatrix}$$

where solution Σ_1 of the Riccati equation reduced to the observable part by y is solution of 3.

Let us give the idea of the equivalence proof of properties 1. and 2., the equivalence of properties 3. and 4. resulting by duality.

We show that $2. \Rightarrow 1.$ *by partitioning the solution of the Riccati equation according to* $P = \begin{pmatrix} P_1 & P_2 \\ P_2^T & P_3 \end{pmatrix}$, *with* P_1 *matrix of the same size as* A_{W_e} *and then by*

verifying that $P = \begin{pmatrix} T_a^T P_3 T_a & T_a^T P_3 \\ P_3 T_a & P_3 \end{pmatrix}$ *is solution if we choose solution* P_3 *of the Riccati equation reduced to the controllable part by u and* T_a *solution of 2.*

Reciprocally, we can deduce that $1. \Rightarrow 2.$ *as follows. Equation 1 can be "seen" as the Lyapunov equation associated with the observability grammian by z of the looped system if* $u = Kx$, *with K defined in Theorem 13.1. The existence of a solution* $P \geq 0$ *leads, according to lemma 3.19 of [ZHO 96], to the conclusion that* $(A + BK)$ *is stable even since the looped system is detectable by z. The non-stability of the pair* (A, B) *leads to the conclusion that the looped system must necessarily be undetectable by z and, consequently, that equation 2. admits one solution.* $\square$

NOTE 13.3.– Theorem 13.2 generalizes the former reflections [LAR 93], [LAR 00] in the case of output frequency weightings. It introduces the dual problem of the regulator [DAV 76, FRA 77, WON 85]. Speaking of hidden problems would be more general. The problem of the regulator consists in fact of hiding, by a proper feedback, the non-stabilizable modes by u (interpreted as disturbances) in order to make them unobservable by z. The dual problem seeks to hide the non-detectable modes by y so as to make them non-controllable by w. It is clear that the existence of a solution for the H_2 problem is subordinated to the existence of a solution for each one of these sub-problems.

When they exist, the solutions to the hidden equations are not necessarily single. Equation 2 of Theorem 13.2 is a necessary and sufficient condition to the *Regulation Problem with Internal Stability (RPIS)* which is well-known in other works [WON 85]. The uniqueness of (T_a, K_a) is acquired as soon as $G_{12}^0(s)$ is reversible on the left and does not have zeros among the eigenvalues of A_{W_e} [STO 00]. In a dual way, the solution (S_a, L_a) in (4) will be unique if $G_{21}^0(s)$ is reversible on the right and does not have zeros among the eigenvalues of A_{W_s}.

13.2.6. *Generalized H_2 problem and robust RPIS*

Let us consider here the case of a standard model that does not have unstable modes unobservable by y. This restrictive and simplifying hypothesis will not block the "State Standard Control" type methodological developments. From the H_2 generalized problem, we can establish the following result which shows the presence of an *internal model* [WON 85] within the regulator.

Theorem 13.3 (H_2 REGULATOR AND INTERNAL MODEL).– *let us suppose satisfied the hypotheses of the generalized H_2 problem (degenerated hypotheses if the standard model does not have unstable modes unobservable by y). Let us suppose moreover that there is a solution to equation 2 of Theorem 13.2 (section 13.2.5). Then, the H_2 optimal regulator (see Theorem 13.1) contains a copy of unobservable dynamics of the pair* $(C_2 T_a + D_{21}, A_{W_e})$.

The demonstration results from corollary 3.3 of [STO 00].   □

COROLLARY 13.1.– the duplicate within the regulator of unobservable dynamics of the pair $(C_2 T_a + D_{21}, A_{W_e})$ is basically Wonham internal model. The H_2 regulator thus obtained satisfies the principle of the internal model.

In what follows, we will seek to specify the conditions in which one will have a robust H_2 regulator where property $\lim\limits_{t \to \infty}(t) = 0$ (condition of existence of $\|T_{zw}\|_2$) is verified despite the arbitrarily small uncertainties on $G^0(s)$. Because if we know (see [HAU 83]) that the stabilizing and detectability properties are preserved for small disturbances on $G(s)$ (or the state matrices which characterize it), it is not the same for the hidden properties.

Besides the hypotheses of the H_2 problem, we will suppose that the $W_e(s)$ modes are perfectly known and that there is a matrix M_z of appropriate size such as $C_1 = M_z C_2$. In addition, the result statement will be facilitated by the introduction of the following notation:

$$a^{q'}_{\lambda(\sigma,v_1,v_2)} \stackrel{\Delta}{=} \begin{pmatrix} \sigma & v_1 & 0 & 0 & 0 \\ v_2 & \sigma & 1 & 0 & 0 \\ 0 & v_2 & \sigma & v_1 & 0 \\ 0 & 0 & v_2 & \sigma & \ddots \\ 0 & 0 & 0 & \ddots & \ddots \end{pmatrix} \in R^{q \times q'}$$

THEOREM 13.4 (ROBUST RPIS).– *let* $m(t) = (p_0 + p_1 t + \cdots + p_q t^q)\, e^{\lambda t}$ *be a mode of the exosystem* $W_e(s)$. *Hence,* A_{W_e} *is similar to matrix* $\begin{pmatrix} a^{q'}_{\lambda} & 0 \\ 0 & * \end{pmatrix}$ *with*

$q' = q, \sigma = \lambda, v_1 = 1, v_2 = 0$ *if* $\lambda \in R$ *and* $q' = 2q,\ \sigma = re\,(\lambda), v_1 = im(\lambda),$ $v_2 = -im(\lambda)$ *if* $\lambda \in C$. *The* H_2 *optimal regulator contains a strong internal model [WON 85] associated with the mode* $m(t)$ *and consequently the rejection of the* *mode $m(t)$ on $z(t)$ will be robust if A_{W_e} is similar to* $\overline{A}_{W_e} = \begin{pmatrix} I_{p_2} \otimes a_\lambda^{q'} & 0 \\ 0 & * \end{pmatrix}$. *In*

other words, the mode $m(t)$ must be observable p_2 times by y if $y(t) \in R^{p_2}$.

If the presentation of this result is original and in particular the relation with the H_2 problem generalized, its demonstration can result from the traditional results on the regulator problem (see [ABE 00, HAU 83] and the references included).   □

The result of Theorem 13.4 is important as it provides a key for the weighting choice $W_e(s)$ when we wish to reject a robust disturbance. It reiterates the "principle of sufficient duplication" introduced in [LAR 00].

13.2.7. *Discretization of the H_2 problem*

Let us consider that we must implement on the computer a regulator H_2 designed beforehand in continuous-time [GEV 93, WIL 91]. A slightly clumsy way would be to approximate *a posteriori* the continuous-time regulator. We recommend the following way which proceeds by discretization of the H_2 problem and direct calculation of the optimal discrete regulator.

Hence, the problem consists of determining the discrete-time regulator $K_{dH_2}(z)$ which will give to the numerical control[9] in Figure 13.7 a behavior close to that of the analogical control resulted from feedback $K_{H_2}(s)$. Therefore, let us try to define the H_2 standard discrete problem for which $K_{dH_2}(z)$ would be the solution.

---

9 B_0 and E_T represent respectively the *0 order blocker* and the *sampling operator* in accordance with Chapter 3.

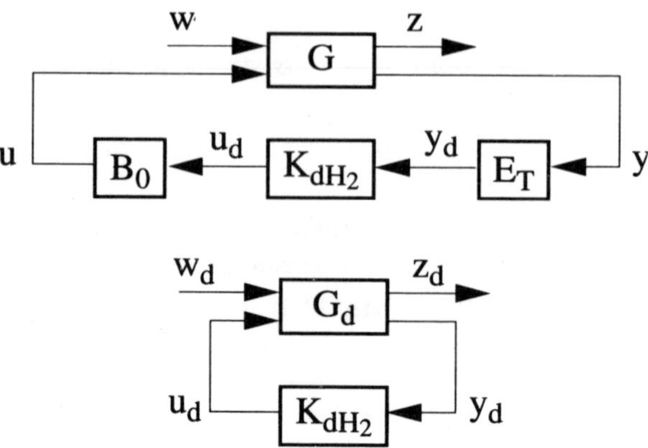

Figure 13.7. *Discretization of the standard H_2 problem*

Discretizing problem H_2 does not imply discretizing the standard model by writing $\mathbf{G_d} = E_T \circ \mathbf{G} \circ B_0$. By doing this, the behavior of the system in between two samplings would be neglected. Similarly, w would implicitly receive spectral properties that it does not have ($w(t)$ cannot be assumed as constant between two sampling instances). The following lemma makes it possible to obtain $\mathbf{G_d}$ from hypotheses 1 to 3 below:

1) $G(s) := \left[\begin{array}{c|cc} A & B_1 & B_2 \\ \hline C_1 & 0 & D_{12} \\ C_2 & D_{21} & 0 \end{array} \right]$ with[10]: $D_{21}B_1^{\ T} = 0$;

2) the sampler E_T incorporates an ideal spectrum anti-aliasing filter (band-pass filter of unitary gain on $\left[-\dfrac{f_e}{2}, \dfrac{f_e}{2} \right]$ if f_e is the sampling frequency). At this level we could also introduce a delay so as to take into account the delay of calculation. We will not do it here in order to avoid complicating the writing of the discrete model;

3) the input of control $u(\cdot)$ remains constant between two sampling instances.

---

10 This hypothesis is not at all compulsory. It just simplifies the writing of the resulting discrete model.

LEMMA 13.2 (OBTAINING THE STANDARD DISCRETE MODEL).– *under the preceding hypotheses, the discrete standard model* $\mathbf{G_d}$ *results from the continuous model by:*

$$G_d(z) := \left(\begin{array}{c|cc} A_d & B_{d1} & B_{d2} \\ \hline C_{d1} & 0 & D_{d12} \\ C_{d2} & D_{d21} & 0 \end{array} \right) \qquad [13.19]$$

with:

$$A_d = e^{AT_e}, \quad B_{d2} = \int_0^{T_e} e^{A\tau} B_2 d\tau$$

$$B_{d1} = [G_{c1}(T_e)]^{1/2}, \text{ i.e. } B_{d1}B_{d1}^T = \int_0^{T_e} e^{A\tau} B_1 B_1^T e^{A^T\tau} d\tau$$

$$[C_{d1} \quad D_{d12}] = [G_{ao}(T_e)]^{1/2}, \text{ i.e.}$$

$$\begin{bmatrix} C_{d1}^T \\ D_{d12}^T \end{bmatrix}[C_{d1} \quad D_{d12}] = \int_0^t e^{\begin{pmatrix} A & B_2 \\ 0 & 0 \end{pmatrix}^T \tau} \begin{bmatrix} C_1^T \\ D_{12}^T \end{bmatrix}[C_1 \quad D_{12}] e^{\begin{pmatrix} A & B_2 \\ 0 & 0 \end{pmatrix} \tau} d\tau$$

$$C_{d2} = C_2 \text{ and } D_{d21} = \frac{D_{21}}{\sqrt{T_e}}$$

The model thus discretized makes it possible to write Theorem 13.5.

THEOREM 13.5 (DISCRETIZATION OF H_2 PROBLEM).– $K_{dH_2}(z)$ *minimizing* $\left\| T_{z_d w_d} \right\|_2 = \left\| F_l(G_{dH_2}, K_{dH_2}) \right\|_2$ *also minimizes* $\left\| T_{zw} \right\|_2 = \left\| F_l(\mathbf{G}_{H_2}, B_0 \circ \mathbf{K}_{dH_2} \circ E_T) \right\|_2$.

We once more neglect to give the proof of the theorem. At least let us give the idea. It is shown initially that the state x_{dk} of the model G_{dH_2} has the same dynamic and stochastic properties as the discretized state $x(kT_e)$ of G_{H_2}. It is *shown then that* $\|z_d\|_2 \overset{\Delta}{=} \sum_k z_{dk}{}^T z_{dk} = \|z\|_2$. *Finally,* y_{dk} *is obtained by discretization of the white noise* $D_{21}w$. $\square$

NOTE 13.4.– we have, by definition of the H_2 standard in discrete time,

$$\|T_{z_d w_d}\|_2 = \left(\frac{1}{2\pi} \int_{-\pi/T_e}^{+\pi/T_e} Graph\ (T^*_{z_d w_d}(e^{j\omega T_e})T_{z_d w_d}(e^{j\omega T_e}))\ d\omega \right)^{1/2}.\ \text{Standard}\ \|T_{zw}\|_2$$

cannot be interpreted this way. T_{zw} transfer uses continuous and discrete-time signals at the same time and $\|T_{zw}\|_2$ is defined only through its interpretation in terms of induced standard.

NOTE 13.5.– the way the H_2 discrete problem is solved is completely similar to the way the H_2 problem in continuous-time is solved. Only the formal expressions of Lyapunov and Riccati equations change [ZHO 96]. Alternatively, we can use the operator $\delta_{T_e} - \gamma_{T_e}$ (see section 3.3) and the complex variable $w = \dfrac{2}{T_e}\dfrac{z-1}{z+1}$. In this case, the H_2 standard of the $G(w)$ transfer is defined by

$$\|G\|_2 = \left(\frac{1}{2\pi} \int_{-\infty}^{+\infty} \begin{matrix} Graph \\ Trace \end{matrix} (G^*(w)G(w))\ dw \right)^{1/2}\ \text{with}\ w = \frac{2}{T_e}\frac{e^{j\omega T_e}-1}{e^{j\omega T_e}+1}\ \text{and we find}$$

the same results as for continuous time.

13.3. Data of a feedback control problem

The goal of a methodological guide is to offer support to the designer of a control law throughout the design chain. Before being able to formalize the control problem through an H_2 optimization problem, since it is the part included here, it is important to proceed in a systematic way. Which are the "contours" of the system to be regulated? What do we know of its environment? Which are the means of action? Which is the available information in real-time? Which are the control objectives, at least from a qualitative point of view? These questions and the formalization of their answer belong to the methodological approach. We will use as much as possible the notations adopted in [LAR 00].

13.3.1. *Model of the process*

The first stage consists, without any doubt, of proceeding on the basis of functional reasoning. How should we define *the system to be adjusted* and which function must it have? From the nature of its function, we will infer the *magnitudes that need to be controlled,* denoted as y_c, and at least qualitatively the control objectives. We will define on this occasion the *setting or reference magnitudes r.* Secondly, it is a question of defining the means of action necessary or useful for the achievement of the control objectives. The action magnitudes are reunited into a *control* vector noted by *u.* Any other means of action that is not used for the feedback control considered will be labeled as disturbance. The other sources of disturbances result from the *environment* of the system to adjust. The *disturbance inputs* are gathered in vector *d.* It is finally necessary to keep track of the measurements or *observations* which are likely to be used to carry out the control law. y_o defines the vector of the noticed outputs.

The *contour* of the system to be adjusted is "drawn" at the end of these first two stages through the data of its inputs-outputs d, u, y_c, y_o. It is then important to *model* the relations of cause to effect between these various magnitudes, starting from the equations of physics which govern its behavior, or directly by minimizing the distance between the inputs-outputs of the system and that of a mathematical model. We make the assumption here that this model is *linear* and defined by

$$G_{SAR}(s) := \left[\begin{array}{c|cc} A & B_d & B_u \\ \hline C_{y_c} & D_{y_c d} & D_{y_c u} \\ C_{y_o} & D_{y_c d} & 0 \end{array} \right] \quad \text{according to the diagram in Figure 13.8.}$$

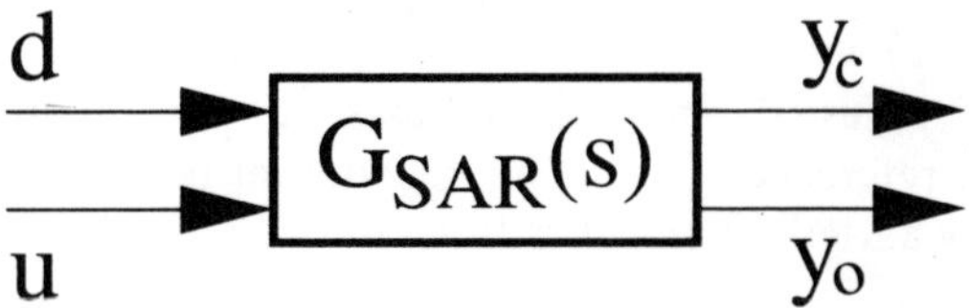

Figure 13.8. *Model of the system to be adjusted*

The model can then be simulated for various input signals in order to well understand the behavior of the system and the evolution of its internal variables x_{SAR}. It is important from now on to specify the operation *environment* of the system.

13.3.2. *Modeling the environment of the system to adjust*

Controlling any system consists of controlling its actuators so that it can fulfill the required function. The control law naturally depends on the function to be accomplished and on the conditions under which it must be accomplished. The magnitudes to be controlled were already defined in section 13.3.1. It is still necessary to specify the requirements imposed on them. Similarly, macroscopic information on the type of disturbances to which the system is subjected can be validly used during the design of the control.

Let us specify these aspects by using the example of an electric machine. The same machine can fulfill very different functions. Does it have to function with an engine or a generator? We implicitly answer this question while defining which are the control magnitudes and the magnitudes to control. Typically, position or speed control pertains to an "engine" operation. Let us suppose that we are interested in a position control. The formulation of an efficient control law requires more information on the operation conditions of the engine. Which position profile does it have to follow? Is the position reference likely to vary in an abrupt way, with stages, or on the contrary its evolution is linear, as it can be the case sometimes in robotics? It is the same with specific applications where the reference is quasi-periodic, even sinusoidal [DET 99]. In addition, does the disturbance, which is here a resistive torque, have a characteristic "signature"?

A regulator, be it a little sophisticated, will use this information to *predict* and *anticipate* the future evolution of the system. It is clear that the quality of this anticipation depends on the capacity of the regulator to predict the evolution of references and disturbances on the basis of their past evolution. For this reason, the development of a control law necessarily supposes (sometimes in an implicit way) the definition of *predictor models* which specify and formalize the environment of the system to be adjusted. In what follows, ξ will indifferently define the disturbance d, the reference r, or the aggregation of both. Generally, the signal ξ will be described by a Markov model as follows:

$$\begin{pmatrix} \dot{x}_\xi \\ \xi \end{pmatrix} = \begin{pmatrix} A_\xi & B_\xi \\ C_\xi & 0 \end{pmatrix} \begin{pmatrix} x_\xi \\ w \end{pmatrix}$$

Let us define $w(t) = \sum_i q_i \delta(t - t_i)$ with δ as a Dirac distribution, $(q_i)_{i \in N}$ and $(t_i)_{i \in N}$ as two independent random sequences with value in R. $(t_i)_{i \in N}$ is strictly increasing.

This choice enables the adequate description of many standard signals. Consider the following examples:

$$\begin{pmatrix} \dot{x}_\xi(t) \\ \xi(t) \end{pmatrix} = \begin{pmatrix} 0 & 1 \\ 1 & 0 \end{pmatrix} \begin{pmatrix} x_\xi(t) \\ w(t) \end{pmatrix} \Leftrightarrow \xi(t) = \sum_i q_i \Gamma(t - t_i)$$

enables the description of a Wiener noise.

$$\begin{pmatrix} \dot{x}_\xi(t) \\ \xi(t) \end{pmatrix} = \left[\begin{array}{cc|c} 0 & 1 & 0 \\ 0 & -\omega_0 & \omega_0 \\ \hline 1 & 0 & 0 \end{array} \right] \begin{pmatrix} x_\xi(t) \\ w(t) \end{pmatrix} \Leftrightarrow \begin{pmatrix} x_{\xi 1}(t) \\ x_{\xi 2}(t) \end{pmatrix} = \sum_i q_i \begin{pmatrix} 1 - e^{-\omega_0(t-t_i)} \\ \omega_0 e^{-\omega_0(t-t_i)} \end{pmatrix} \Gamma(t - t_i)$$

describes the evolution of the standard signal represented in Figure 13.9. For $\omega_0 = 0$, ξ evolves in ramps piece by piece.

$$\begin{pmatrix} \dot{x}_\xi(t) \\ \xi(t) \end{pmatrix} = \left[\begin{array}{cc|cc} 0 & \omega_0 & 1 & 0 \\ -\omega_0 & 0 & 0 & 1 \\ \hline 1 & 0 & 0 & 0 \end{array} \right] \begin{pmatrix} x_\xi(t) \\ w(t) \end{pmatrix} \Leftrightarrow$$

$$\xi(t) = \sum_i \left[q_i^1 \cos \omega(t - t_i) + q_i^2 \sin \omega(t - t_i) \right] \Gamma(t - t_i), \;\; si \;\; q_i = \begin{pmatrix} q_i^1 \\ q_i^2 \end{pmatrix}$$

describes the evolution of a harmonic signal.

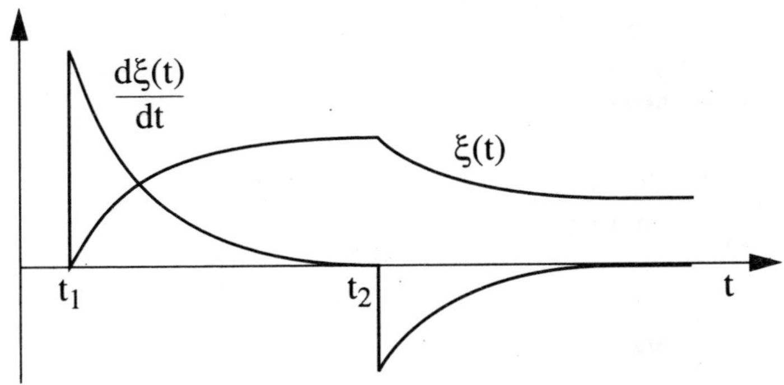

Figure 13.9. *Model of a 2nd order aperiodic signal*

Under certain hypotheses on the random variables q_i and t_i, the signal $\xi(t)$ has second order statistical properties equivalent to the signal $\xi'(t)$ obtained by injecting at the input of the model a centered white noise $w'(t)$ [BRO 92, LAR 93]. This second form is besides better adapted to take into account a disturbance on the signal measured which is often connected to a colored noise.

Any input signal w or w' is irreducible in both cases and the best prediction of $x_\xi(t + \Delta t)$ that can be done at instant t is given by $x_\xi(t + \Delta t) = e^{A_\xi \Delta t} x_\xi(t)$. This point is illustrated in Figure 13.10. $x_\xi(t)$, being internal to the signal model ξ, incorporates everything that we know on the past evolution of ξ.

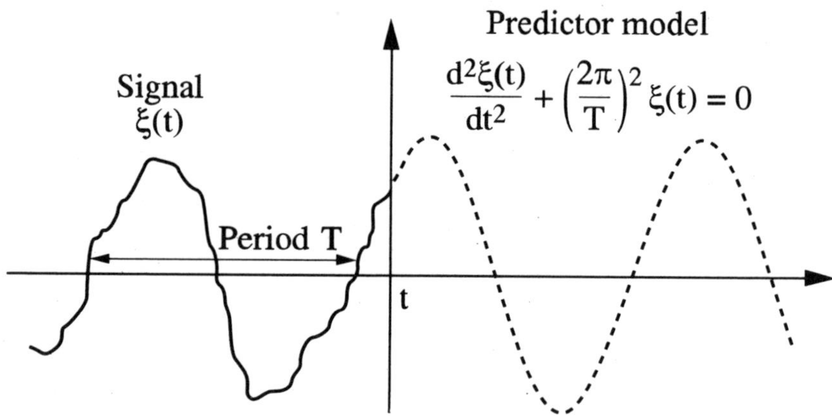

$$\frac{d^2\xi(t)}{dt^2} + \left(\frac{2\pi}{T}\right)^2 \xi(t) = 0$$

Figure 13.10. *Sinusoidal prediction*

Finally, the designer task consists of providing the type of the prediction to be carried out (constant, in ramp, sinusoidal prediction). This information is formalized through the data of predictor models of disturbances and settings:

$$\begin{pmatrix} \dot{x}_r \\ r \end{pmatrix} = \begin{pmatrix} A_r & B_r \\ C_r & 0 \end{pmatrix} \begin{pmatrix} x_r \\ 0 \end{pmatrix} \quad \text{and} \quad \begin{pmatrix} \dot{x}_d \\ d \end{pmatrix} = \begin{pmatrix} A_d & B_d \\ C_d & 0 \end{pmatrix} \begin{pmatrix} x_d \\ 0 \end{pmatrix} \qquad [13.21]$$

13.3.3. *Additional data*

The designer may specify certain statistical information that he knows such as the average frequency of a disturbance appearance, the characteristics of a measurement noise, etc. A H_2 criterion may be inferred in its "LQG" interpretation (see section 13.2.3). This approach is not viable for the design of a control law. Even if we exclude the fact that obtaining this information is a problem in itself, minimizing the hope for the output signal of the standard model would not guarantee in any way the dynamic performances expected. The interpretation of the H_2 standard as "energy" of the impulse response is better adapted here.

Matrices Q, R, N_c, V, W, N_f that intervene in the criterion $H_2 - LQG$ must be interpreted as weighting matrices. They cannot be left as such in the hands of the designer as he may give up in front of the difficulty of the task or he may settle for an average adjustment. Better than a few empirical rules, the question of choosing the weightings deserves a true "methodological hat", which makes it possible to obtain the weightings of a limited number of *adjustment parameters* with well understood effects. The designer will be able then "to adjust" these control parameters to best manage the control compromises. The definition of such "a methodological hat" is the subject of the following section.

13.4. Standard H₂ optimization problem

The relation between an H_2 *optimization problem* and the *synthesis problem of a regulator* based on the information collected in the preceding section seems still fine at this stage. We will describe the various construction stages of the standard model for the H_2 optimization, based on the information available in section 13.3 on the one hand and on the calculation of the weight matrices from the parameters left at the designer's free choice on the other hand.

13.4.1. *Construction of the conceptual control model*

The magnitudes penalized in the H_2 criterion must be selected in a coherent way to lead to a clear optimization problem. This is what is carried out by the following stages and is summed up by the diagram in Figure 13.11.

Once the signals d, u, y_c, y_o are defined as indicated in section 13.3, we establish the process model $G_{SAR}(s)$ defined by:

$$\begin{pmatrix} \dot{x}_{SAR} \\ \hline y_c \\ y_o \end{pmatrix} = \begin{bmatrix} A_{SAR} & B_d & B_u \\ \hline C_{y_c} & D_{y_c d} & D_{y_c u} \\ C_{y_o} & D_{y_c d} & 0 \end{bmatrix} \begin{pmatrix} x_{SAR} \\ \hline d \\ u \end{pmatrix}$$

We associate an additional input which will enable the additional incorporation of a noise on the evolution equation while defining $G_{SARW_x}(s)$ by:

$$\begin{pmatrix} \dot{x}_{SAR} \\ \hline y_c \\ y_o \end{pmatrix} = \left[\begin{array}{c|ccc} A_{SAR} & B_d & [0 \quad I] & B_u \\ \hline C_{y_c} & D_{y_c d} & 0 & D_{y_c u} \\ C_{y_o} & D_{y_o d} & 0 & 0 \end{array} \right] \begin{pmatrix} x_{SAR} \\ \hline d \\ w_x \\ u \end{pmatrix} \qquad [13.22]$$

We define the predictor model of disturbances and settings with which we associate an evolution noise whose intensity will be fixed later on according to the control objectives: $\begin{pmatrix} \dot{x}_\xi \\ r \\ d \end{pmatrix} = \left(\begin{array}{c|c} A_\xi & B_\xi \\ \hline C_{\xi_1} & 0 \\ C_{\xi_2} & \end{array} \right) \begin{pmatrix} x_\xi \\ w_x \end{pmatrix}$. We define the setting variance

$e = r - y_c$ which will have to be brought back to 0.

We seek the one trajectory of the process $(\cdot)(u_a(\cdot), x_a(\cdot))$ which ensures the nullity of the setting variance and thus $y_c \equiv r$. It results from the state of the predictor model of disturbances and settings according to relations $x_a = T_a x_\xi$ and $u_a = K_a x_\xi$, in which T_a and K_a are solutions of Sylvester equation:

$$\begin{cases} T_a A_\xi - A_{SAR} T_a - B_u K_a + B_d C_{\xi_2} = 0 \\ C_{y_c} T_a + C_{\xi_1} = 0 \end{cases}$$

To conclude, we define the control variance $e_u = u - u_a$ and vector $y = \begin{pmatrix} r \\ y_c \end{pmatrix} + w_y$ of the magnitudes accessible to the regulator.

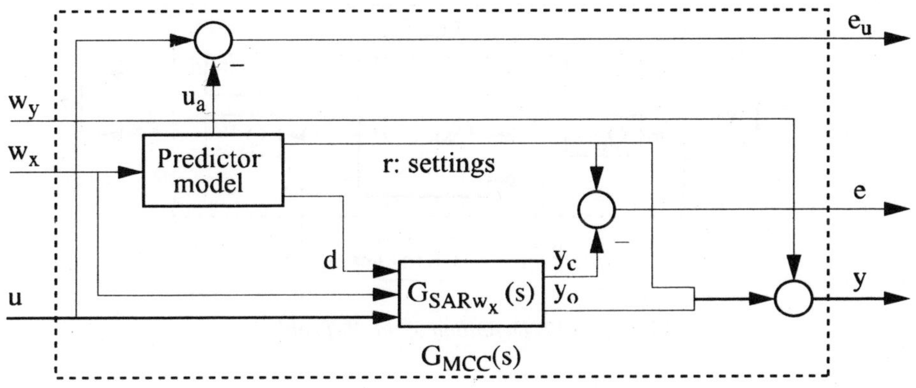

Figure 13.11. *Conceptual control model (CCM)*

We easily obtain a minimal realization of the CCM from what precedes. We note:

$$\begin{pmatrix} e_u \\ \dfrac{e}{y} \end{pmatrix} = \underbrace{\begin{pmatrix} G_{MCC_{11}} & G_{MCC_{12}} \\ G_{MCC_{21}} & G_{MCC_{22}} \end{pmatrix}}_{G_{MCC}(s)} \begin{pmatrix} w_y \\ \dfrac{w_x}{u} \end{pmatrix} \qquad [13.22]$$

We are able from now on to associate with the control problem a *clear* and relevant H_2 optimization problem. This will be done in the following section.

13.4.2. *Definition of the H₂ optimization problem*

On the basis of CCM and weighting matrices $Q_c = Q_c^{\frac{1}{2}T} Q_c^{\frac{1}{2}}, R_c = R_c^{\frac{1}{2}T} R_c^{\frac{1}{2}}$
and $Q_o = Q_o^{\frac{1}{2}} Q_o^{\frac{1}{2}T}, R_o = R_o^{\frac{1}{2}} R_o^{\frac{1}{2}T}$, we build the standard model $G_{H_2}(s)$ defining the H_2 problem (see Figure 13.12):

$$G_{H_2}(s) \overset{\Delta}{=} \begin{pmatrix} R_c^{\frac{1}{2}} & 0 & 0 \\ 0 & Q_c^{\frac{1}{2}} & 0 \\ 0 & 0 & I \end{pmatrix} G_{MCC}(s) \begin{pmatrix} R_o^{\frac{1}{2}} & 0 & 0 \\ 0 & Q_o^{\frac{1}{2}} & 0 \\ 0 & 0 & I \end{pmatrix} := \left(\begin{array}{c|cc} A & B_1 & B_2 \\ \hline C_1 & 0 & D_{12} \\ C_2 & D_{21} & 0 \end{array} \right)$$

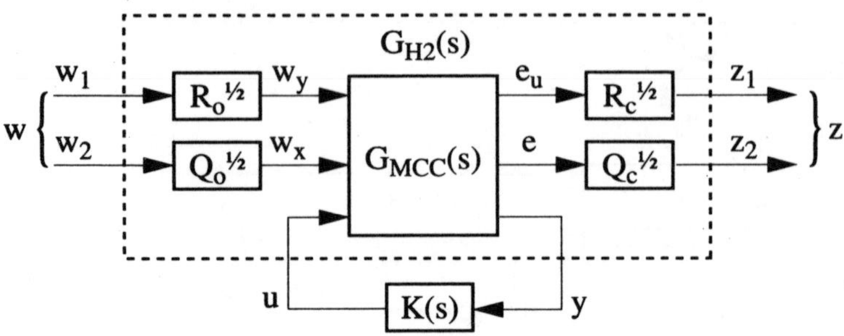

Figure 13.12. *Definition of the H_2 problem*

The H_2 problem that consists of calculating $K(s)$ ensuring the internal stability of $G_{SAR}(s)$ and minimizing $\left\| F_l(G_{H_2}(s), K(s)) \right\|_2$ is a generalized *clear H_2* problem (section 13.2.5) considering the previously taken precautions (penalization of variances e_u and e with respect to the asymptotic trajectory). It remains to be entirely defined. Choosing the weighting matrices is potentially choosing $n_{MCC} \times \dfrac{(n_{MCC} + 1)}{2}$ coefficients if n_{MCC} is the order of the CCM. It is too much to be able to properly handle them. Other works unanimously refers to Bryson's rule [BRY 69] which suggests using the diagonal matrices in order to simplify the problem. Hence, each coefficient is applied to a state variable or particular control and the "rule" stipulates choosing it in accordance with the variation range (existing or desired) of the variable considered. It is implicitly supposed that each variable has an obvious physical direction.

This *standardization*[11] approach is good but still it is too approximate and often delicate. De Larminat in [LAR 93] recommends standardizing the system according to the controllability or observability on a determined horizon, based on the partial grammians (see section 13.1). This approach seems to be the proper one (we will justify this) and is what we will present here.

Matrices Q_c and R_c will be inferred from the choice of a *control horizon T_c*:

$$R_c = T_c \int_0^{T_c} (C_{y_c} e^{A_{SAR}t} B_u)^T (C_{y_c} e^{A_{SAR}t} B_u)\,dt, \qquad Q_c = I \qquad [13.23]$$

---

11 The interest was always in standardizing the units of descriptive physical magnitudes of the process considered by expressing them in percentage of the nominal value.

Symmetrically, matrices Q_o and R_o will be inferred from the choice of an *observation horizon* T_o :

$$Q_o = [T_o \int_0^{T_o} e^{A^T t} C_2^T C_2 e^{At} dt]^{-1}, \qquad R_o = I \qquad\qquad [13.24]$$

Let us note various possible rewritings for matrices R_c and Q_o :

$$R_c = T_c \int_0^{T_c} (C_1 e^{At} B_2)^T (C_1 e^{At} B_2) dt = T_c B_2^T G_{o1}(T_c) B_2$$

$$Q_o = [T_o \int_0^{T_o} e^{A^T t} C_2^T C_2 e^{At} dt]^{-1} = [T_o G_{o2}(T_o)]^{-1}$$

Understanding these choices implies returning to the various interpretations of the grammians given in section 13.2.1. We leave this exercise to the reader who can refer to [LAR 00] for more details. We will mostly deal with the consequences of such a choice.

13.4.3. *The interest in standardization*

The previous choices are of undeniable methodological interest. From the standardization approach it results that the horizons T_c and T_o respectively adjust the dynamics of the control modes and of filtering modes (see separation principle). It is noted in [LAR 00] that the control poles, for example, are in general at the left of the real axis $-\dfrac{1}{T_c}$. Our experiment corroborates this result. A second result, equally significant, is stated in the following proposition.

Proposition: "standard" character of SSC

The properties of the feedback control system obtained according to the standard state control (SSC) procedure are independent of the state representation of the model of the system to be adjusted and of that of the predictor models.

Let us provide the demonstration idea of this original result. It results directly from the two following notes.

The corrector $K_{opt} = \underset{\min K(s) \ \textit{stabilizing}}{\arg} \left\| F_l(G_{H_2}(s), K(s)) \right\|_2$ is independent of the state representation chosen for $G_{H_2}(s)$.

Let us consider the models $G_{H_2}^1(s)$ and $G_{H_2}^2(s)$ obtained from two distinct creations of $G_{MCC}(s)$ in the state space. Let T be the passage matrix from one creation to another. In order to have $G_{H_2}^1(s) = G_{H_2}^2(s)$, it is sufficient that:

$Q_{c2} = T^{-T} Q_{c1} T^{-1}$, $R_{c2} = R_{c1}$, $Q_{o2} = T Q_o T^T$, $R_{o2} = R_{o1}$. These equalities are automatically satisfied by the weighting matrices calculated as indicated above from grammians.

Thus, SSC has the essential property for a control methodology to lead to a result which is independent of the choice of internal variables of the process and exogenous signals.

13.4.4. *Management of the control compromises*

The control parameters left to the user's free choice are two: T_c and T_o. How should they be used? T_o controls the dynamics of the observer. The higher it is, the slower the rebuild dynamics of the observer and the less its sensitivity to the measurement noises becomes. Generally, T_o, selected as higher than T_c, will fix the dominant modes of control. The control horizon will be sufficiently small to guarantee good robustness margins in terms of module margin. $T_c \to 0$ involves indeed $R_c^{1/2} \to 0$. Consequently, based on Figures 13.11 and 13.12, it is obvious that the state representation of $G_{H_2}(s)$ (see section 13.4.2) is such that $D_{12} \to 0$. The conditions of the dual *LTR* defined in section 13.2.4 apply in theory only if the condition $C_1 = C_2$ is satisfied. This is not necessarily true if the SSC rules are applied. [LAR 00]. The "LTR effect" is, however, often noted[12]. To be systematic, in H_2 criterion we will have to penalize vector $z = (e_u \ \ e \ \ e_y)^T$ with $e_y = y_o - y_a$ if y_a is defined by $y_a = C_o x_a$ ((u_a, x_a) is the asymptotic trajectory defined in section 13.4.1). Hence, if the process is at *phase minimum* in the bandwidth of the control, we have the guarantee of obtaining "good" robustness properties (module margin) at the level of sensors (system output).

---

12   Minimizing $\left\| C_1(sI - A + B_2 K_{LQ})^{-1} L_{FK} \right\|_2$ does not necessary lead to make $\left\| C_2(sI - A + B_2 K_{LQ})^{-1} L_{FK} \right\|_2$ small unless $\mathrm{Im}(C_2) \in \mathrm{Im}(C_1)$.

Finally, the compromises related to the adjustment of the process are managed from these two parameters. Nothing prevents after all to define a horizon T_r, thus making it possible to adjust the dynamics of the pursuit. This can be carried out in a very systematic way without questioning the structure state feedback/observer, which results from the separation principle and which is advisable to be used during the implementation of the regulator on the computer. In addition, the discrete-time regulator is easily obtained by applying the result given in section 13.2.7.

Finally, if we seek to obtain a "robust" setting follow-up, during the construction of the CCM (see section 13.4.1), we could to add on y_o a vector of disturbances d_o of the same size as y_o and such that:

$$\begin{pmatrix} \dot{x}_{do} \\ d_o \end{pmatrix} = \begin{pmatrix} I_{p2} \otimes A_\xi & I & 0 \\ C_{do} & 0 & I_{p2} \end{pmatrix} \begin{pmatrix} x_{do} \\ w_{xo} \\ w_{yo} \end{pmatrix}$$

$(C_{do}, I_{p2} \otimes A_\xi)$ observable

According to Theorem 13.4, this addition will guarantee obtaining a robust setting follow-up.

13.5. Conclusion

This chapter does not describe in detail SSC but it only outlines its main principles. General results, which may be original for some of us, were presented so as to lay down the theoretical bases of a control methodology using the state approach. Hence it should be noted the atypical presentation of $H_2 - LTR$, the definition and the resolution of the generalized H_2 problem, the relation between its resolution and the so-called "regulator" problem, the discretization of the standard H_2 problem, etc. The generality of these developments makes it possible to consider various extensions to SSC.

If the methodological contribution of SSC is undeniable in our eyes, it is possible to still enrich it in various ways. It may thus be interesting in certain cases to be able to obtain good robustness margins at the beginning of the process rather than at its end and to still widen the class of systems which can be approached by SSC. It is also possible to introduce frequency weightings without the state approach (in the deep sense of the term) which is the basis of SSC. Finally, it is possible to specifically reduce the sensitivity of certain transfers of the looped system with respect to uncertain parameters and thus to increase to a certain extent the robustness of the regulation. For that, we enrich the initial H_2 criterion of the parametric sensitivity of significant transfers. The optimization problem which results from this is in this case more complicated numerically as shown in [CHE 01]. A sufficiently efficient algorithm could be developed despite everything (see [YAG 01]). It remains to polish the methodological aspect that makes these tools usable by the designer.

13.6. Appendices

13.6.1. *Resolution of the Lyapunov equations*

Preliminary definitions

Let be M and N be two matrices of size $m \times n$ and $p \times q$ respectively.

$$M \otimes N = \begin{pmatrix} m_{11}N & \cdots & m_{1n}N \\ \vdots & \ddots & \vdots \\ m_{m1}N & \cdots & m_{mn}N \end{pmatrix}. \ \otimes \text{ defines the Kronecker product.}$$

$$Vec(M) = \begin{pmatrix} m_{11} & \cdots & m_{m1} & m_{12} & \cdots & \dot{m}_{m2} & \cdots & m_{1n} & \cdots & m_{mn} \end{pmatrix}^T$$

PROPOSITION 13.1.– the equivalence below is verified.

$$\dot{P} = F^T P + PF + Q \quad \Leftrightarrow \quad Vec(\dot{P}) = -[I \otimes A^T + A^T \otimes I] \ Vec(P) - Vec(Q)$$

This equivalence can be used with gain to calculate $P(t)$ starting from the data of matrices F and Q. We obtain the partial grammian of controllability for $F = A^T$ and $Q = BB^T$ and the partial grammian of observability for $F = A$ and $Q = C^T C$.

PROPOSITION 13.2.– equation $F^T P + PF + Q = 0$ with the unknown factor P and for which Q is symmetrical, has the following properties:

– F does not have eigenvalues on the imaginary axis and is a sufficient condition for the existence of a solution;

– let us suppose that: $\exists C / Q = C^T C$ and $(C, \ A)$ is detectable, then $[P \geq 0 \Rightarrow F$ *is stable*];

$$-F \text{ is stable} \Rightarrow P = \int_{0}^{+\infty} e^{A^{T}\tau} Q e^{A\tau}\, d\tau \ .$$

13.6.2. *Duality principle*

System $G(s)^{T} := \left[\begin{array}{c|c} A^{T} & C^{T} \\ \hline B^{T} & D^{T} \end{array}\right]$ defines the dual system of $G(s) := \left[\begin{array}{c|c} A & B \\ \hline C & D \end{array}\right]$. It appears trivial that a system and its dual have the same modes. We also have, without difficulty, the following relations of duality:

$$(A,B) \quad controllable \quad \Leftrightarrow \quad (B^{T}, A^{T}) \ observable$$

$$(C,A) \quad observable \quad \Leftrightarrow \quad (A^{T}, C^{T}) \ controllable$$

Let us also consider the diagram of standard feedback in Figure 13.1 pointed out below. We verify without difficulty the following equalities:

$$T_{zw}{}^{T} \triangleq \left[F_{L}\left(G,K\right)\right]^{T} = \left[F_{L}(G^{T}, K^{T})\right] \triangleq T_{\overline{z}\,\overline{w}}$$

$$\left\|T_{zw}\right\|_{2} = \left\|T_{\overline{z}\,\overline{w}}\right\|_{2}$$

Consequently, "finding K which minimizes $\left\|F_{L}(G,K)\right\|_{2}$" or "finding L which minimizes $\left\|F_{L}(G^{T}, L)\right\|_{2}$" are two *dual* problems and we can infer the solution of one from the other by the relation of duality $L = K^{T}$.

Figure 13.13. *Standard feedback diagram*

13.6.3. *Another useful interpretation of grammians*

Grammians are more than mere intermediaries for calculation of the H_2 standard. In robust control, they are used for model analysis and reduction. The controllability grammian contains useful information pertaining to the areas or directions of the state space which are the most "excited" by the inputs. The observability grammian contains similar information pertaining to the state space areas where the outputs are the most "sensitive". The construction of these areas (contours of an ellipsoid) can be obtained on the basis of a decomposition of grammians into singular value (semi-axes of the ellipsoid of R^n deduced from the values and singular vectors of the grammian concerned). Precisely, the controllability grammian G_c makes it possible to define the area of the state space reachable by inputs of given energy, whereas the observability grammian G_o defines the sub-space of initial conditions x_0 producing outputs of given energy. This interesting geometrical interpretation of grammians was provided by Moore [MOO 81].

13.6.4. *Zeros of a multivariable system*

Let $\left[\begin{array}{c|c} A & B \\ \hline C & D \end{array}\right]$ be a *minimal realization* of the transfer matrix $G(s)$ to m inputs and p outputs. Let us assume that $z_0 \in C$ is not a pole of $G(s)$. In the case mono input-mono output ($m = p = 1$), z_0 is a $G(s)$ zero (finite) if and only if $G(z_0) = 0$. In the multivariable case, the concept of zero is richer and its characterization more delicate. Before pointing out some results enabling this characterization, let us give a dynamic interpretation to the concept of zero.

PROPOSITION 13.3.– z_0 is a zero of transmission $G(s)$ if there is an input vector of the form $u(t) = u_0 e^{z_0 t}$ and an initial state $x(t_0) = x_0$ so that the output of the system is identically zero.

THEOREM 13.6.– *z_0 is a zero of transmission $G(s)$ if and only if one of the two following propositions is verified:*

1) $\exists s \in C\,/\,rank(G(z_0)) < rank(G(s))$

2) $\exists s \in C\,/\,rank\begin{pmatrix} A - z_0 I & B \\ C & D \end{pmatrix} < rank\begin{pmatrix} A - s I & B \\ C & D \end{pmatrix}$

The zero of transmission z_0 is called blocking zero if $rank(G(z_0)) = 0$ or,

equally, rank $\begin{pmatrix} A - z_0 I & B \\ C & D \end{pmatrix} = n$.

If we tone down the minimal hypothesis of realizing $G(s)$, the proposition enables the characterization of the *invariant zeros* which include, in addition to the transmission zeros, decoupling zeros (uncontrollable or/and unobservable modes of the realization) (see Chapter 4).

13.6.5. *Standardization of a system*

Let us assume the system $G(s) := \left[\begin{array}{c|c} A & B \\ \hline C & D \end{array} \right]$. Let us note by $u \in R^m$, $x \in R^n$,

$y \in R^p$ the input, state and output vectors of the system. Let N_u, N_x and N_y be diagonal matrices of suitable size enabling the standardization of the inputs, states and outputs of the system: $u_N = N_u u$, $x_N = N_x x$, $y_N = N_y y$. The standardized model $G_N(s)$ admits for realization in the state space:

$$G_N(s) := \left[\begin{array}{c|c} N_x A N_x^{-1} & N_x B N_u \\ \hline N_y^{-1} C N_x^{-1} & N_y^{-1} D N_u^{-1} \end{array} \right] .$$

13.7. Bibliography

[ABE 00] ABEDOR J., NAGPAL K., KHARGONEKAR P.P., POOLA K., "Robust Regulation in the presence of Norm Bounded Uncertainty", *IEEE Trans. Auto. Control*, vol. AC-40, no. 1, p. 147-153, 2000.

[AND 89] ANDERSON B.D.O., MOORE J.B., *Optimal Control*, Prentice Hall, 1989.

[BOU 92] BOURLES H., Contribution à la commande et à la stabilité des systèmes, Thesis, Paris XI University, June 1992.

[BRO 92] BROWN R.G., HWANG P.Y.C., *Introduction to random signal and applied Kalman Filtering*, John Wiley & Sons, 1992.

[BRY 69] BRYSON A.E., HO Y.C., *Applied Control*, Waltham (Massachusetts) Ginn and Co., 1969.

[CHE 93] CHEVREL P., Commande Robuste: application à la régulation d'un groupe turbo-alternateur, PhD Thesis, Paris XI University, 1993.

[CHE 99] CHEVREL P., Méthodologie de commande multivariable, Lecture notes, Ecole des Mines, Nantes, 1999.

[CHE 01] CHEVREL P., YAGOUBI M., DE LARMINAT P., "On Insensitive H_2 Control Problem Complexity", *European Control Conference*, 4-7 September, Portugal, 2001.

[DAV 76] DAVISON E.J., "The Robust Control of a Servomechanism Problem for Linear Time Invariant Multivariable Systems", *IEEE Trans. on Autom. Control*, vol. AC-21, no. 1, p. 25-34, 1976.

[DET 99] DETTORI M., SCHERER C.W., "Design and implementation of a gain-scheduled controller for a compact disc player", *European Control Conference*, Karlsruhe, Germany, 1999.

[DOY 82] DOYLE J.C., "Analysis of feedback systems with structured uncertainties", *IEEE Proc.*, vol. 129, Part D, p. 242-250, 1982.

[DOY 89] DOYLE J.C., GLOVER K., KHARGONEKAR P.P., FRANCIS B.A., "State-space solutions to standard H_2 and H_∞ control problems", *IEEE Trans. Auto. Control*, vol. AC-34, no. 8, p. 831-847, 1989.

[DUC 99] DUC G., FONT S., *Commande H_∞ et μ-analyse*, Hermès, 1999.

[FRA 77] FRANCIS B.A., "The linear multivariable regulator problem", *SIAM Journal on Control and Optimization*, vol. 15, no. 3, p. 486-505, 1977.

[FRA 87] FRANCIS B.A., *A course in H_∞ control theory*, Berlin, Springer, 1987.

[GAH 94] GAHINET P., APKARIAN P., "A Linear Matrix Inequality Approach to H_∞ Control", *Int. J. of Robust & Non-linear Control*, vol. 4, p. 421-448, 1994.

[GEV 93] GEVERS M., LI G., *Parametrizations in Control, Estimation and Filtering Problems*, Springer, 1993.

[HAU 83] HAUTUS M.L.J., "Linear matrix equations with applications to the regulator problem", in *Outils et modèles mathématiques pour l'automatique, l'analyse de systèmes et le traitement du signal*, vol. 3, CNRS Editions, 1983.

[IWA 91] IWASAKI T., SKELTON R.E., "All controllers for the general H_∞ control", *Int. J. Control*, vol. 54, p. 1031-1076, 1991.

[KWA 72] KWAKERNAAK H., SIVAN R., *Linear Optimal Control Systems*, Wiley, 1972

[LAN 90] LANE DAILEY R., "Lecture Notes for the Workshop on H_∞ and μ methods for Robust Control", *American Control Conference*, p. 31-39, San Diego, California, 1990.

[LAR 93] DE LARMINAT P., *Automatique*, Hermès, 1993.

[LAR 99] DE LARMINAT P., LEBRET G., PUREN S., "About some interconnections between LTR and RPIS", *7th Mediterranean Conference on Control & Automation*, Haifa, Israel, 1999.

[LAR 00] DE LARMINAT P., *Le contrôle d'état standard*, Hermès, 2000.

[LAR 02] DE LARMINAT P. (ed.), *Analyse des systèmes linéaires*, Hermès, IC2 series, Paris, 2002.

[MAG 87] MAGNI J.F., Commande modale des systèmes multivariables, PhD Thesis, Paul Sabatier University, Toulouse, 1987.

[MIT 00] MITA T., XIN X., ANDERSON B.D.O., "Extended H_∞ control with unstable weights", *Automatica*, vol. 36, p. 735-741, 2000.

[MOO 81] MOORE B.C., "Principal Component analysis in linear systems: controllability, observability and model reduction", *IEEE Trans. Auto. Control*, vol. AC-26, p. 17-32, 1981.

[PIC 77] PICINBONO B., *Eléments de théorie du signal*, Dunod, 1977.

[PRE 95] PREMPAIN E., BERGEON B., "Méthodologie R2M2 Multivariable", *APII*, vol. 29, no. 6, p. 655-678, 1995.

[ROU 92] ROUBINE E., *Distributions signal*, Eyrolles, 1992.

[SAF 82] SAFONOV M.G., "Stability margins of diagonally perturbed multivariable feedback systems", *IEEE Proc.*, vol. 129, Part D, p. 251-256, 1982.

[STE 87] STEIN G., ATHANS M., "The LQG/LTR Procedure for Multivariable Feedback Control Design", *IEEE Trans. on Autom. Control*, AC-32, no. 2, 1987.

[STO 00] STOORVOGEL A.A., SABERI A., SANNUTI P., "Performance with regulation constraints", *Automatica*, vol. 36, p. 1443-1456, 2000.

[WIL 91] WILLIAMSON D., *Digital Control and Implementation*, Prentice Hall, 1991.

[WON 85] WONHAM W.M., *Linear Multivariable Control: A Geometric Approach*, Springer, 1985.

[YAG 01] YAGOUBI M., CHEVREL P., "An ILMI Approach to Structure-Constrained LTI Controller Design", *European Control Conference*, 4-7 September, Portugal, 2001.

[ZHO 96] ZHOU K., DOYLE J.C., GLOVER K., *Robust and Optimal Control*, Prentice Hall, 1996.

Complementary references relative to LQG/LTR control

[BOU 91] BOURLES H., IRVING E., "La méthode LQG/LTR: une interprétation polynomiale temps continu/temps discret", *RAIRO APII*, 25, p. 545-592, 1991.

[DOY 78] DOYLE J.C., "Guaranteed margins for LQG regulators", *IEEE Trans. on Autom. Control*, vol. AC-2,3 p 756-757, 1978.

[DOY 81] DOYLE J., STEIN G., "Multivariable Feedback Design: Concepts for a Traditionalal/Modern Synthesis", *IEEE Trans. on Automat. Contr.*, AC-26, p. 4-16, 1981.

[LEH 81] LEHTOMAKI N.A., SANDELL N.R., ATHANS M., "Robustness Results in Linear-Quadratic Gaussian Based Multivariable Control Designs", *IEEE Trans. on Autom. Control*, vol. AC-26, p. 75-93

[LEW 92] LEWIS F.L., *Applied Optimal Control & Estimation*, Prentice Hall, 1992.

[MAC 89] MACIEJOWSKI J.M., *Multivariable Feedback Design*, Addison-Wesley, 1989.

[SAF 77] SAFONOV M.G., ATHANS M., "Gain and phase margin of multiloop LQG regulators", *IEEE Trans. on Autom. Control*, AC-22, April 1977.

[SAF 81] SAFONOV M.G., LAUB A.J., HARTMANN G., "Feedback Properties of Multivariable Systems: The Role and Use of Return Difference Matrix", *IEEE Trans. of Automat. Contr.*, AC-26, p. 47-65, 1981.

[STE 87] STEIN G., ATHANS M., "The LQG/LTR procedure for multivariable feedback control design", *IEEE Trans. on Autom. Control*, vol. AC-32, p. 105-114.

Complementary References relative to the LQGPF/H2 control

[COP 92] COPELAND B.R., SAFONOV M.G., "A generalized eigenproblem approach to singular control problems – Part II: H∞ problems", *Proc. 31st IEEE CDC*, Tucson, Arizona, 1992.

[GUP 80] GUPTA N.K., "Frequency-shaped cost functionals: extension of linear quadratic-gaussian design methods", *J. Guidance and Control*, vol. 3, p. 529-535.

[MAC 89] MACIEJOWSKI J.M., *Multivariable Feedback Design*, Addison-Wesley, 1989.

[MAM 91] MAMMAR S., DUC G., "Commande LQG/LTR à pondérations fréquentielles: application à la stabilisation d'un hélicoptère", *European Control Conference*, p. 1169-1174, Grenoble, 1991.

[MOO 81] MOORE J.B., GANGSAAS D., BLIGHT J.D., "Performance and robustness trades in LQG regulator design", *20th IEEE Conf. on Decision and Control*, p. 1191-1199, 1981.

[MOO 87] MOORE J.B., MINGORI D.L., "Robust Frequency Shaped LQ Control", *Automatica*, vol. 23, No5, pp. 641-646, 1987.

[SAF 77] SAFONOV M.G., ATHANS M., "Gain and phase margin of multiloop LQG regulators", *IEEE Trans. on Autom. Control*, AC-22, April 1977.

[SAF 80] SAFONOV M.G., *Stability and Robustness of Multivariable Feedback Systems*, MIT Press, 1980.

[STE 87] STEIN G., ATHANS M., "The LQG/LTR procedure for multivariable feedback control design", *IEEE Trans. on Autom. Control*, vol. AC-32, p. 105-114, 1987.

References relative to the regulation problem under internal stability constraint

[DAV 75] DAVISON E.J., GOLDENBERG A., "Robust Control of a General Servomechanism Problem: The Servo Compensator", *Automatica*, vol. 11, p. 461-471, 1975.

Chapter 14

Multi-variable Modal Control

14.1. Introduction

The concept of eigenstructure placement was born in the 1970s with the works of Kimura [KIM 75] and Moore [MOO 76a]. Since then, the eigenstructure placement has undergone continuous development, in particular due to its potential applications in aeronautics. In fact, the control of couplings through these techniques makes them very appropriate for this type of application. Moore's works led to numerous studies on the decoupling eigenstructure placement. The principle consists of setting the dominant eigenvalues of the system while guaranteeing, through a proper choice of related closed loop eigenvectors, certain decoupling, non-reactivity, insensitivity, etc. Within the same orientation, Harvey [HAR 78] interprets the asymptotic LQ in terms of eigenstructure placement. Alongside this type of approach, Kimura's works on pole placement through output feedback have been supported by several researchers. In these more theoretical approaches, the exact pole placement is generalized during the output feedback. The degrees of freedom of eigenvectors are no longer used in order to ensure decoupling – as in Moore's approach – but in order to set supplementary eigenvalues. Recently, research in automatics has been particularly oriented towards robustness objectives (through methods such as the H_∞ synthesis, the μ-synthesis, etc.), the control through eigenstructure placement being limited to the aim of ensuring the insensitivity of the eigenvalues placed (insensitivity to the first order) by a particular choice of eigenvectors [APK, 89, CHO 94, FAL 97, MUD 88]. It was only recently that the modal approach was adjusted to the control resisting to parametric uncertainties. This adaptation, proposed in [LEG 98b, MAG 98], is based on the alternation between the μ-analysis and the multi-model modal synthesis (technique of

---

Chapter written by Yann LE GORREC and Jean-François MAGNI.

μ-Mu iteration) and makes it possible to ensure, with a minimum of conservatism, the robustness in front of parametric uncertainties (structured real uncertainties).

In this chapter, we will describe only the traditional eigenstructure placement. We will see how to ensure certain input/output decoupling or how to minimize the sensitivity of the eigenvalues to parametric variations. These basic concepts will help whoever is interested in the robust approach [MAG 02b] to understand the problem while keeping in mind the philosophy of the standard eigenstructure placement. The implementation of the techniques previously mentioned is facilitated by the use of the tool box [MAG 02a] dedicated to the eigenstructure placement (single-model and multi-model case).

The first part of this chapter will enable us to formulate a set of definitions and properties pertaining to the eigenstructure of a system: concept of mode and relations existing between the input, output and disturbance signals and the eigenvectors of the closed loop. We will see what type of constraints on the eigenvectors of the closed loop make the desired decouplings possible. Then we will describe how to characterize the modal behavior of a system with the help of two techniques: the modal simulation and the analysis of controllability. This information will allow to choose which eigenvalues to place by output feedback. This synthesis of the output feedback will be described in detail in the second part of this chapter. Finally, the last part is dedicated to the synthesis of observers and to the eigenstructure placement with observer.

14.2. The eigenstructure

In this section we will reiterate the results formulated in [MAG 90].

14.2.1. *Notations*

14.2.1.1. *System considered*

In this part, the multi-variable linear system considered has the following form:

$$\dot{x} = Ax + Bu$$
$$y = Cx + Du$$

[14.1]

where x is the state vector, u the input vector and y the output vector. The sizes of the system will be as follows:

$$n \text{ states} \qquad x \in \mathbb{R}^n$$

$$m \text{ inputs} \qquad u \in \mathbb{R}^m$$

$$p \text{ outputs} \qquad y \in \mathbb{R}^p$$

The equivalent transfer matrix is noted $G(s)$:

$$G(s) = C(sI - A)^{-1}B + D$$

14.2.1.2. *Corrector*

In what follows, the system is corrected by an output static feedback and the inputs v (settings) are modeled with the help of a pre-control H. Therefore, the control law is:

$$u = Ky + Hv \qquad [14.2]$$

where v has the role of reference input.

If $D = 0$:

$$\dot{x} = (A + BKC)x + BHv$$

If $D \neq 0$, the expressions of y in [14.1] and of u in [14.2] make the following relation possible:

$$u = (I - KD)^{-1}KCx + (I - KD)^{-1}Hv$$

By substituting u in the relation $\dot{x} = Ax + Bu$, we obtain:

$$\dot{x} = (A + B(I - KD)^{-1}KC)x + B(I - KD)^{-1}Hv$$

By noticing that $K(I - DK)^{-1} = (I - DK)^{-1}K$, we get:

$$\dot{x} = (A + BK(I - DK)^{-1}C)x + B(I - KD)^{-1}Hv$$

14.2.1.3. *Eigenstructure*[1]

The *eigenvalues* of the state matrix of the looped system $A + BK(I - DK)^{-1}C$ are noted:

$$\lambda_1, \ldots, \lambda_n$$

the *right eigenvectors*:

$$v_1, \ldots, v_n$$

and the *input directions*:

$$w_1, \ldots, w_n$$

where (by definition):

$$w_i = (I - KD)^{-1}KC\,v_i \quad \Leftrightarrow \quad w_i = K(Cv_i + Dw_i) \qquad [14.3]$$

---

1. In this chapter, it is supposed that the eigenvalues are always distinct.

The *left eigenvectors* of matrix $A + BK(I - DK)^{-1}C$ are noted:

$$u_1, \ldots, u_n$$

and the *output directions*:

$$t_1, \ldots, t_n$$

where (by definition):

$$t_i = u_i BK(I - DK)^{-1} \quad \Leftrightarrow \quad t_i = (u_i B + t_i D)K \qquad [14.4]$$

14.2.1.4. *Matrix notations*

Let us take q vectors (generally $q = p$ or $q = n$); the scalar notations $\lambda_i, v_i, w_i, u_i, t_i$ become:

$$\Lambda = \begin{bmatrix} \lambda_1 & & 0 \\ & \ddots & \\ 0 & & \lambda_q \end{bmatrix} \qquad [14.5a]$$

$$V = \begin{bmatrix} v_1 & \cdots & v_q \end{bmatrix}, \quad W = \begin{bmatrix} w_1 & \cdots & w_q \end{bmatrix} \qquad [14.5b]$$

$$U = \begin{bmatrix} u_1 \\ \vdots \\ u_q \end{bmatrix}, \quad T = \begin{bmatrix} t_1 \\ \vdots \\ t_q \end{bmatrix} \qquad [14.5c]$$

If λ_i is not real, it is admitted that there is an index i' for which $\lambda_{i'} = \bar{\lambda}_i$. Thus, in matrices V and W, $v_{i'} = \bar{v}_i$, $w_{i'} = \bar{w}_i$ and in matrices U and T, $u_{i'} = \bar{u}_i$, $t_{i'} = \bar{t}_i$. In addition, when it is a question of *placement*, we will consider that if λ_i is *placed*, then $\lambda_{i'}$ is placed too. Vectors u_i and v_i are standardized such that:

$$UV = I \quad \text{and} \quad U(A + BK(I - DK)^{-1}C)V = \Lambda \qquad [14.6]$$

14.2.2. *Relations among signals, modes and eigenvectors*

Apart from the definition of the concept of mode, the objective of this section is to study the relations between excitations, modes and outputs in terms of the eigenstructure. *Knowing this makes it possible to consider the decoupling specifications as constraints on the right and left eigenvectors of the looped system* (constraints that could be considered during the synthesis). This knowledge is the basis of the traditional techniques of eigenstructure placement. However, in many cases, the decoupling specifications are not primordial. In fact, it would be often be preferable to place the eigenvectors of the closed loop by an orthogonal projection, this approach enabling us to better preserve the natural behavior of the system.

In this section, for reasons of clarity, we will consider a strict eigensystem (with no direct transmission $(D = 0)$). The different vectors considered are:

- the vector of regular outputs z;

- the vector of reference inputs v;

- the vector of disturbances d. These disturbances are distributed on the states and outputs of the system, respectively by E' and F';

- the vector of initial conditions x_0.

System [14.1] becomes:

$$\dot{x} = Ax + Bu + E'd$$
$$y = Cx + F'd \qquad\qquad [14.7]$$
$$z = Ex + Fu$$

14.2.2.1. *Definition of modes*

Let us take the state basis change where U corresponds to the matrix of n left closed loop eigenvectors (see [14.5]):

$$\boldsymbol{\xi} = Ux \qquad\qquad [14.8]$$

where:

$$\boldsymbol{\xi} = \begin{bmatrix} \xi_1 \\ \vdots \\ \xi_n \end{bmatrix}$$

The various components of this vector will be called the *modes of the system.*

In [14.8] there was an obvious relation between state and mode of the system. Identically, the relations between excitations, modes and outputs of the system will be detailed, which will enable us to interpret the various specifications of decoupling in terms of constraints on the eigenstructure of the system.

14.2.2.2. *Relations between excitations and modes*

The input u of [14.7] is of the form [14.2]. The effect of the initial condition is modeled by a Dirac function $x_0\delta$, hence:

$$\dot{x} = (A + BKC)x + BHv + (E' + BKF')d + x_0\delta$$

or:

$$\dot{x} = (A + BKC)x + f$$

where f corresponds to all excitations acting on the system ($f = BHv + (E' + BF'K)d + x_0\delta$). After having applied the basis change ($\xi = Ux$):

$$\dot{\xi} = \Lambda\xi + Uf$$

We obtain:

$$\xi(t) = e^{\Lambda t} * Uf(t)$$

where "$*$" is the convolution integral and $e^{\Lambda t}$ the diagonal matrix:

$$e^{\Lambda t} = \operatorname{diag}\left(e^{\lambda_1 t}, \ldots, e^{\lambda_n t}\right)$$

In addition:

$$\xi_i(t) = e^{\lambda_i t} * u_i f(t) = \int_0^t e^{\lambda_i(t-\tau)} u_i f(\tau)\, \mathrm{d}\tau \qquad [14.9]$$

14.2.2.3. *Relations between modes and states*

By returning to the original basis, we obtain:

$$x = V\xi = \sum_{i=1}^{n} \xi_i v_i \qquad [14.10]$$

This relation shows that the right eigenvectors of the system control the modes on the states.

14.2.2.4. *Relations between reference inputs and controlled outputs*

Here, $f = BHv$. Instead of considering the state vector as above, we consider the controlled output $z = Ex + Fu$. The term Ex can be written $EV\xi$ and the term Fu:

$$Fu = FKCx + FHv = FKCV\xi + FHv = FW\xi + FHv$$

The mode transmission becomes:

$$\xi_i(t) = e^{\lambda_i t} * u_i BHv \quad \text{and} \quad z = \sum_{i=1}^{n} \begin{bmatrix} E & F \end{bmatrix} \begin{bmatrix} v_i \\ w_i \end{bmatrix} \xi_i(t) + FHv \qquad [14.11]$$

The transfers between v and ξ and between the modes and z (by omitting the term that does not make the eigenvectors appear, FHv) can be written:

$$\xrightarrow{v} \boxed{UBH} \longrightarrow \boxed{(sI - \Lambda)^{-1}} \xrightarrow{\xi} \boxed{\begin{bmatrix} E & F \end{bmatrix} \begin{bmatrix} V \\ W \end{bmatrix}} \xrightarrow{z}$$

We note:

- E_k, F_k the kth rows of E, F;
- z_k, v_k the kth inputs of z, v ;
- H_k the kth columns of H.

The open loop relation between the inputs and the controlled output (by omitting the term that does not make the eigenvectors appear, FHv) is given by (see [14.9] and [14.11] by considering $W = 0$):

$$z_k(t) = \sum_{i=1}^{n} E_k v_i \int_0^t e^{\lambda_i(t-\tau)} u_i B H v(t) \, d\tau \qquad [14.12]$$

The conditions that the eigenvectors must satisfy so that there is decoupling are immediate:

$$u_i B H_k = 0 \quad \Rightarrow \quad v_k \text{ does not have any effect on the mode } \boldsymbol{\xi}_i(t)$$

$$E_k v_i + F_k w_i = 0 \quad \Rightarrow \quad \text{the mode } \boldsymbol{\xi}_i(t) \text{ does not have any effect on } z_k$$

14.2.2.5. *Relations between initial conditions and controlled outputs*

The transfers between $x_0\boldsymbol{\delta}$ and $\boldsymbol{\xi}$ and between the modes and z can be written:

$$\xrightarrow{x_0\boldsymbol{\delta}} \boxed{U} \longrightarrow \boxed{(sI - \Lambda)^{-1}} \xrightarrow{\boldsymbol{\xi}} \boxed{[E \quad F] \begin{bmatrix} V \\ W \end{bmatrix}} \xrightarrow{z}$$

Based on the notations previously mentioned, the equivalent constraints on the eigenstructure are:

$$u_i x_0 = 0 \quad \Rightarrow \quad \text{the initial condition does not have any effect on the mode } \boldsymbol{\xi}_i(t)$$

$$E_k v_i + F_k w_i = 0 \quad \Rightarrow \quad \text{the mode } \boldsymbol{\xi}_i(t) \text{ does not have any effect on } z_k$$

14.2.2.6. *Relations between disturbances and controlled outputs ($F = 0$ or $F' = 0$)*

The transfers between $\boldsymbol{d}$ and $\boldsymbol{\xi}$ and between the modes and z can be written:

$$\xrightarrow{\boldsymbol{d}} \boxed{[U \quad T] \begin{bmatrix} E' \\ F' \end{bmatrix}} \longrightarrow \boxed{(sI - \Lambda)^{-1}} \xrightarrow{\boldsymbol{\xi}} \boxed{[E \quad F] \begin{bmatrix} V \\ W \end{bmatrix}} \xrightarrow{z}$$

The equivalent constraints on the eigenstructure are:

$$u_i E'_k + t_i F'_k = 0 \quad \Rightarrow \quad \boldsymbol{d}_k \text{ does not have any effect on the mode } \boldsymbol{\xi}_i(t)$$

$$E_k v_i + F_k w_i = 0 \quad \Rightarrow \quad \text{the mode } \boldsymbol{\xi}_i(t) \text{ does not have any effect on } z_k$$

14.2.2.7. *Summarization*

The analysis of the time behavior of a controlled system was done in the modal basis. Each mode is associated to an eigenvalue λ_i of the system in the form $e^{\lambda_i t}$. We have shown that:

– the excitations act on the modes through the left eigenvectors U and the output directions T;

– the modes are distributed on the controlled outputs through the right eigenvectors V and the input directions W:

$$\boxed{\text{excitations}} \xrightarrow{U,T} \boxed{\text{modes}} \xrightarrow{V,W} \boxed{\text{controlled outputs}}$$

We have also showed that the decoupling on the controlled outputs have the form:

$$E_k v_i + F_k w_i = 0$$

EXAMPLE 14.1. The graph in Figure 14.1 is used in order to illustrate the decoupling properties accessible through this method. The system considered here is of the 3rd order and has two inputs and three outputs.

The relations linking the modes and the controlled outputs are:

$$\begin{bmatrix} z_1 \\ z_2 \\ z_3 \end{bmatrix} = \begin{bmatrix} E_1 \\ E_2 \\ E_3 \end{bmatrix} x = \begin{bmatrix} E_1 \\ E_2 \\ E_3 \end{bmatrix} \begin{bmatrix} v_1 & v_2 & v_3 \end{bmatrix} \begin{bmatrix} \xi_1 \\ \xi_2 \\ \xi_3 \end{bmatrix}$$

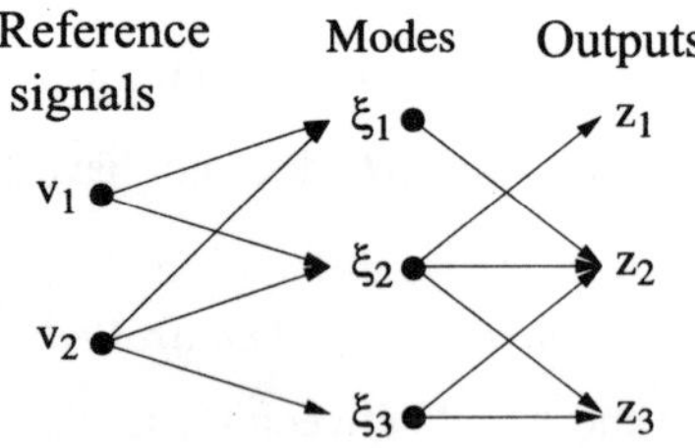

Figure 14.1. *Example of desired decoupling between inputs/modes and modes/outputs*

The decoupling constraints in Figure 14.1 are:

– the first mode must not have any effect on z_1 and z_3;

– the third mode must not have any effect on z_1;

– the reference input v_1 must not have any effect on the third mode.

Hence, we obtain the following constraints:

$$\begin{cases} E_1 v_1 = 0 \\ E_3 v_1 = 0 \\ E_1 v_3 = 0 \\ u_3 B H_1 = 0 \end{cases}$$

The first two equations will be considered during the synthesis of the corrector as constraints on the output feedback (K), whereas the third constraint pertains to the pre-control (H).

14.3. Modal analysis

14.3.1. *Introduction*

The modal synthesis consists of placing the eigenstructure of the closed loop system (see section 14.4). In order to achieve this, it is very important to know very well the modal behavior of the open loop system and the difficulties related to its modification. As for all synthesis methods, those that we will use in what follows are even more efficient if the designer has a good understanding of the system he is trying to control. The analysis described in this section will help him avoid in the future trying to impose unnatural constraints on the control law.

More precisely, *modal simulation* makes it possible to generate an answer to the following questions: what is the influence of each mode on the input-output behavior of the system? Consequently, on which models is it necessary to act in order to modify a given output? By considering afterwards synthesis-oriented objectives, we will seek to have information on the difficulty of placing certain poles. This relative measure will be obtained by using a technique of controllability analysis. A more complete study on this type of analysis can be found in [LEG 98a].

14.3.2. *Modal simulation*

This refers to the analysis of the *modal behavior* of a system. This type of technique is used when we want to know the couplings between inputs, modes and outputs, overflows, etc. It makes it possible to evaluate the contribution of each mode on a given output.

Let us consider a signal decomposed according to equation [14.12]. In this equation, we decompose the controlled outputs z. The modal simulation can also be relative to the measured outputs y; in this case, this analysis also makes it possible to detect the dominant modes (good degree of controllability/observability, etc; see also section 14.3.3). For the outputs measured, we will have:

$$\boldsymbol{y}_k(t) = C_k v_1 \int_0^t e^{\lambda_1(t-\tau)} u_1 BH\boldsymbol{v}(\tau)\,\mathrm{d}\tau$$

$$+ \cdots + C_k v_n \int_0^t e^{\lambda_n(t-\tau)} u_n BH\boldsymbol{v}(\tau)\,\mathrm{d}\tau \qquad [14.13]$$

where $\boldsymbol{y}_k$ corresponds to the kth input of $\boldsymbol{y}$. The *modal simulation* consists of simulating each component:

$$C_k v_i \int_0^t e^{\lambda_i(t-\tau)} u_i BH\boldsymbol{v}(\tau)\,\mathrm{d}\tau \qquad [14.14]$$

of the signal $\boldsymbol{y}_k(t)$ separately. This evaluation provides information on the contribution of modes λ_i to the outputs. It also makes it possible to evaluate the nature – oscillating or damped – of this contribution.

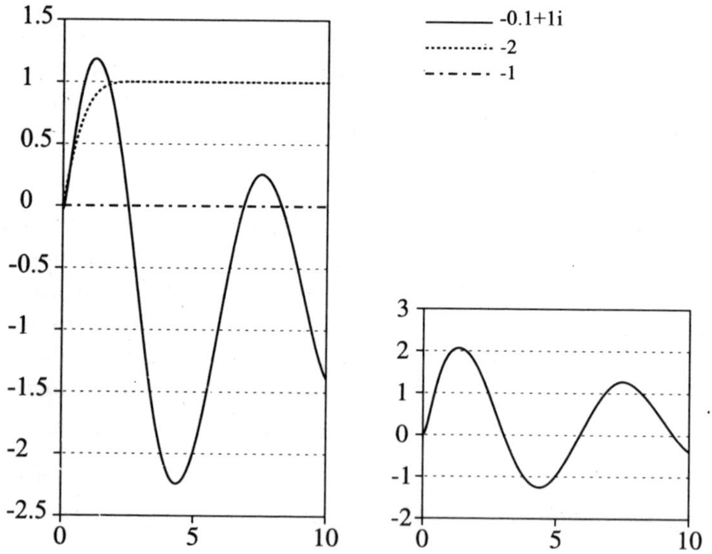

Figure 14.2. *Example of modal simulation.*
On the left: contributions of each mode;
on the right: overall contribution

EXAMPLE 14.2. An example of modal simulation is given in Figure 14.2. This example of modal simulation is taken from *Robust Modal Control Toolbox* [MAG 02a]. The simulation is meant to illustrate the modal participation of the modes of the system to a given output. A step function excitation is sent at input. On the left of Figure 14.2 are traced the different components of the form [14.14] and on the right is traced the sum of these components. On this figure, we can notice that the mode -1 does not have any influence on the output considered, thus it will not be necessary to act on this mode in order to modify the behavior of this output. However, the modes in $-0.1 \pm i$ and in -2 are very important and they will have to be considered during the synthesis. In addition, the modal simulation provides information on the type of contribution of these two modes (transient state and permanent state). The former is very oscillating whereas the latter is damped. This information visually (and thus obviously) illustrates the fact that the modes are associated to very different eigenvalues. Based on this analysis, the designer has a precise idea of the modal behavior of the system and can decide which models to modify in order to influence the outputs to control.

DEFINITION 14.1 (DOMINANT EIGENSTRUCTURE). *We call a dominant eigenstructure the set of pairs of eigenvalues and eigenvectors having a preponderant influence in terms of input-output transfer. The modal simulation makes it possible to determine the influence of each mode on the system's outputs and hence to isolate the pairs of eigenvalues and eigenvectors with a preponderant influence. This technique could be used in order to determine, among the set of eigenvalues of the system, which ones to place by output feedback. This concept of dominant mode is even more important in the context of multi-model techniques discussed in [MAG 02b].*

After dealing with the input-output modal contribution, we will now present the input-output controllability of each mode (corresponding to the difficulty of placing the modes of the system through an output feedback).

14.3.3. *Controllability*

The study of controllability is a subject that generated a lot of interest and many investigations were undertaken by researchers [HAM 89, LIM 93, MOO 81, SKE 81]. After having sought to determine if a state was or was not controllable (Kalman, Popov-Belevich and Hautus' traditional approaches (PBH), Grammian technique), the research has rapidly turned towards the study of the difficulty associated with controlling a state. That is the point of origin for the concept of controllability degree. Numerous researchers have explored this field by adapting the traditional concepts of controllability (PBH test, Grammian method, etc.). In the majority of cases, these techniques are based on a study pertaining to the open loop and are not relevant for our situation. For example, the Grammian measurement of an unstable pole is infinite (zero controllability) and does not reflect the fact that this pole can be controllable by output feedback. Through a continuity argument, the controllability measurement of a pole in terms of stability is erroneous due to the nature itself of this pole. This statement makes this type of method unusable in the context of our approaches. The technique that we choose, in order to efficiently apply the methods of eigenstructure placement, is the technique of modal residuals analysis, which provides an instantaneous criterion independent of the type of eigenvalues analyzed. Other possibilities are proposed in [LEG 98a].

Modal residuals

The modal decomposition can be evaluated by considering the time responses at a given instant and for a given input. Generally, responses to an input impulse (high frequencies) or to a step function on the state (low frequencies) are considered. Let us take equation [14.13] where $BHv(t)$ is replaced by $B_l\delta$ (impulse response) and where the measured outputs are considered; the following result is obtained.

Behavior at high frequencies: impulse response at instant $t = 0$

We have:

$$y_k(t = 0) = C_k v_1 u_1 B_l + \cdots + C_k v_n u_n B_l$$

The quantities $C_k v_i u_i B_l$, $i = 1, \ldots, n$ are called *residuals* between input number l and output number k. The evaluation of residuals $C_k v_i u_i B_l$ makes it possible to find the controllability degree of mode i.

EXAMPLE 14.3. A relative controllability analysis through the graph of modal residuals is given in Figure 14.3. The impulse residuals of each mode are represented in this figure as a bar chart.

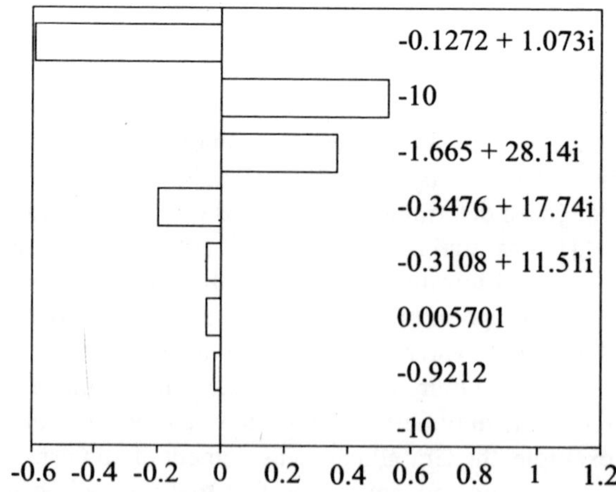

Figure 14.3. *Example of analysis of input-output controllability*

14.4. Traditional methods for eigenstructure placement

Based on the definitions of input and output directions ([14.3] and [14.4]), the following lemmas can be easily demonstrated.

LEMMA 14.1 ([MOO 76a]). *Let us take $\lambda_i \in \mathbb{C}$ and $v_i \in \mathbb{C}^n$. Vector v_i is said to be placed as the right eigenvector associated to the eigenvalue λ_i if and only if there is a vector $w_i \in \mathbb{C}$ such that:*

$$\begin{bmatrix} A - \lambda_i I & B \end{bmatrix} \begin{bmatrix} v_i \\ w_i \end{bmatrix} = 0 \qquad [14.15]$$

Any proportional gain K that makes it possible to carry out this placement satisfies:

$$K\left(Cv_i + Dw_i\right) = w_i \qquad [14.16]$$

Vectors w_i correspond to the input directions defined by [14.3].

Demonstration. If [14.15] and [14.16] are verified:

$$Av_i + Bw_i = \lambda_i v_i$$

and:

$$w_i = (I - KD)^{-1} KCv_i$$

By combining these two equations, we obtain:

$$(A + B(I - KD)^{-1}KC)v_i = \lambda_i v_i$$

which justifies the "if" part of the lemma. As for the part "only if", let us consider the last equation written as follows:

$$\begin{bmatrix} A - \lambda_i I & B \end{bmatrix} \begin{bmatrix} v_i \\ (I - KD)^{-1}KCv_i \end{bmatrix} = 0$$

By defining $w_i = (I - KD)^{-1}KCv_i$, we have $K\left(Cv_i + Dw_i\right) = w_i$, which concludes the demonstration of the lemma. $\qquad \square$

By duality, we also have the following result.

LEMMA 14.2. *Let us take $\lambda_i \in \mathbb{C}$ and $u_i^* \in \mathbb{C}^n$. Vector u_i is said to be placed as the left eigenvector associated to the eigenvalue λ_i if and only if there is a vector $t_i^* \in \mathbb{C}^p$ such that:*

$$\begin{bmatrix} u_i & t_i \end{bmatrix} \begin{bmatrix} A - \lambda_i I \\ C \end{bmatrix} = 0 \qquad [14.17]$$

Any proportional gain K that makes it possible to carry out this placement satisfies:

$$(u_i B + t_i D)\, K = t_i \qquad [14.18]$$

Vectors t_i^* correspond to the output directions defined by [14.4].

Parameterization of placeable eigenvectors

The vectors satisfying [14.15] can be easily parameterized by a set of vectors $\eta_i \in \mathbb{R}^m$. In fact, based on [14.15], the eigenvectors of the right solutions belong to the space defined by the columns of $V(\lambda_i) \in \mathbb{R}^{n \times m}$ which are obtained after resolving:

$$[A - \lambda_i I \quad B] \begin{bmatrix} V(\lambda_i) \\ W(\lambda_i) \end{bmatrix} = 0 \qquad [14.19]$$

Therefore, for a column vector $\eta_i \in \mathbb{C}^m$:

$$v_i = V(\lambda_i)\eta_i$$

Based on [14.17], the eigenvectors of the left solutions belong to the space defined by the rows of $U(\lambda_i) \in \mathbb{R}^{p \times n}$ given by:

$$[U(\lambda_i) \quad T(\lambda_i)] \begin{bmatrix} A - \lambda_i I \\ C \end{bmatrix} = 0 \qquad [14.20]$$

Thus, for a row vector $\eta_i \in \mathbb{C}^p$:

$$u_i = \eta_i U(\lambda_i)$$

14.4.1. *Modal specifications*

For any type of control, one of the main objectives is to stabilize the system, if it is unstable, or to increase its degree of stability, if poorly damped oscillations appear during the transient states. Alongside this, we can try to improve the speed of the system without deteriorating its damping. These specifications are interpreted directly in terms of eigenvalue placement. As we saw in section 14.2, a system can be dissociated into modes. Each mode corresponds to a first order (real number eigenvalue) or to a second order (self-conjugated complex number eigenvalues). These modes have different contributions evaluated due to the *modal simulation* presented in section 14.3.2, hence we will have the concept of dominant modes (see note 14.1). For these dominant modes, it is possible to formulate the following rules: for a desired response time τ_d and a desired damping ξ_d, the dominant closed loop eigenvalues must verify:

$$\mathcal{R}e(\lambda) < 0 \quad \text{for stability}$$

$$|\mathcal{R}e(\lambda)| \geqslant \frac{3}{\tau_d}$$

$$\frac{|\mathcal{R}e(\lambda)|}{|\lambda|} \geqslant \xi_d$$

These constraints define an area of the complex plane (Figure 14.4) where the eigenvalues must be placed.

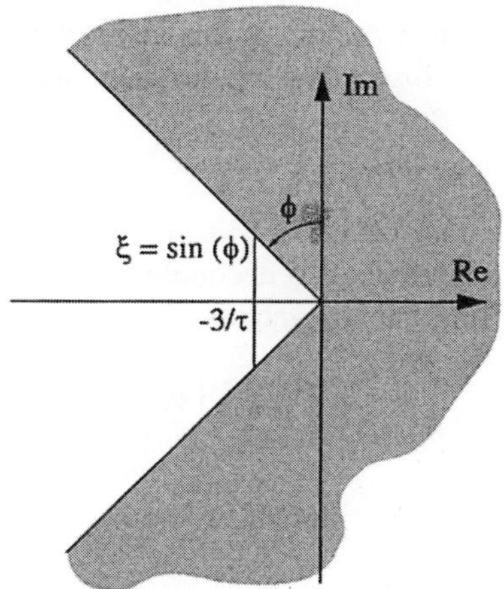

Figure 14.4. *Area of the complex plane corresponding to the desired time performances*

Let us note that – since the control is done through power systems (*closed loop controls*) with limited bandwidths – a supplementary constraint is imposed by the closed loop modes which must be placed within the same bandwidth. Hence, it is recommended to close this field by imposing a bound superior to $|\mathcal{R}e(\lambda)|$ (see Figure 14.5).

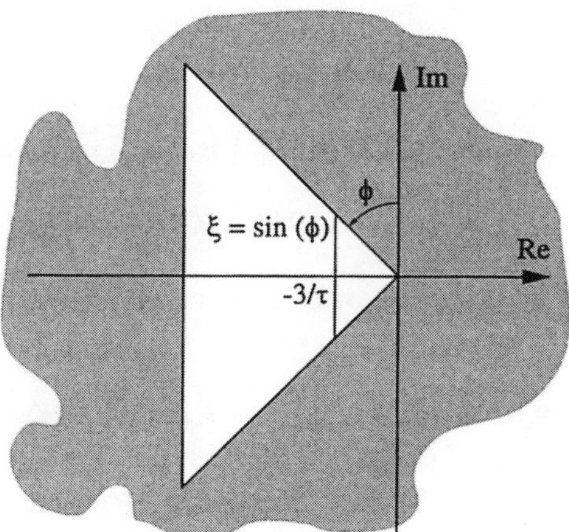

Figure 14.5. *Area of the complex plane corresponding to the desired performances and to the constraints on the bandwidth*

14.4.2. *Choice of eigenvectors of the closed loop*

The solution sub-space of [14.5] or [14.9] is of size[2] m. Hence, it is necessary to make an *a priori* choice of eigenvectors in this sub-space. Several strategies can be used in order to choose these closed loop eigenvectors (see below).

14.4.2.1. *Considering decouplings*

We seek here to reduce the size of $V(\lambda_i)$ to 1. The right eigenvectors will satisfy [14.15] and the conditions pertaining to decouplings (see section 14.2.2) are of the form $E_0 v_i + F_0 w_i = 0$. Consequently, vectors v_i and w_i are calculated by resolving:

$$\begin{bmatrix} A - \lambda_i I & B \\ E_0 & F_0 \end{bmatrix} \begin{bmatrix} v_i \\ w_i \end{bmatrix} = 0 \qquad [14.21]$$

Since matrix B is of size m, it is possible to impose $m - 1$ decoupling constraints (number of rows of E_0 and F_0).

14.4.2.2. *Considering the insensitivity of eigenvalues*

The concept of insensitivity consists of quantifying the variation of the eigenvalues of a system subjected to parametric variations. This quantification is given by lemma 14.3.

LEMMA 14.3. *Let us consider the system [14.1] corrected by the output static feedback K. For a variation of the state closed loop matrix $\hat{A} = A + B(I - KD)^{-1}KC$, we have for first order:*

$$\Delta \lambda_i = u_i \Delta \hat{A} v_i \qquad [14.22]$$

In addition, if the variation $\Delta \hat{A}$ of matrix $\hat{A}$ is due to variation ΔK of matrix K:

$$\Delta \lambda_i = (u_i B + t_i D)\Delta K(C v_i + D w_i) \qquad [14.23]$$

and if variation $\Delta \hat{A}$ is due to the respective variations ΔA, ΔB, ΔC, ΔD of A, B, C, D:

$$\Delta \lambda_i = u_i \Delta A v_i + u_i \Delta B w_i + t_i \Delta C v_i + t_i \Delta D w_i \qquad [14.24]$$

---

2. It is easily shown that the size of this sub-space is equal to m if and only if λ_i is not a non-controllable eigenvalue. In case of non-controllability, the size is superior to m. Thus, the degree of freedom which is lost when an eigenvalue is not movable due to its non-controllability, is recovered at the level of eigenvector placement which offers more degrees of freedom.

Demonstration. By definition:

$$(\hat{A} + \Delta\hat{A})(v_i + \Delta v_i) = (\lambda_i + \Delta\lambda_i)(v_i + \Delta v_i)$$

When we multiply on the left by u_i and when we simplify the equal terms while taking into consideration that $u_i v_i = 1$, we have:

$$u_i \Delta\hat{A} v_i + u_i \Delta\hat{A}\Delta v_i = \Delta\lambda_i + u_i \Delta\lambda_i \Delta v_i$$

Let us take, by neglecting the second order terms:

$$u_i \Delta\hat{A} v_i = \Delta\lambda_i$$

which corresponds to equation [14.22]. When the variation of the state closed loop matrix is due to an output feedback variation, we can replace $\Delta\hat{A}$ with:

$$\Delta\hat{A} = B(I - KD)^{-1}\Delta KC + B(I - KD)^{-1}\Delta KD(I - KD)^{-1}KC$$

Hence, equation [14.22] becomes:

$$\Delta\lambda_i = u_i B(I - KD)^{-1}\Delta K\left(Cv_i + D(I - KD)^{-1}KCv_i\right)$$

Based on the matrix identity $(I - KD)^{-1} = I + K(I - DK)^{-1}D$ and the definitions [14.3] of w_i and [14.4] of t_i, i.e. $w_i = (I - KD)^{-1}KCv_i$ and $t_i = u_i BK(I - DK)^{-1}$, equation [14.22] becomes:

$$\Delta\lambda_i = (u_i B + t_i D)\Delta K(Cv_i + Dw_i)$$

which corresponds to expression [14.23]. Let us consider now that the variations of the closed loop dynamics are due to the variations of state matrices ΔA, ΔB, ΔC, ΔD. We have:

$$\Delta\hat{A} = \Delta A + \Delta B(I - KD)^{-1}KC + B(I - KD)^{-1}K\Delta C$$
$$+ B(I - KD)^{-1}K\Delta D(I - KD)^{-1}KC$$

By using definitions [14.3] of w_i and [14.4] of t_i as before, we will immediately have:

$$\Delta\lambda_i = u_i \Delta A v_i + u_i \Delta B w_i + t_i \Delta C v_i + t_i \Delta D w_i$$

which proves expression [14.24]. $\qquad\square$

Based on equation [14.22], the variation of the eigenvalue λ_i is increased as follows:

$$|\Delta\lambda_i| \leqslant \|\Delta\hat{A}\|\|u_i\|\|v_i\|$$

In order to minimize the sensitivity of eigenvalues, we can thus minimize the criterion:

$$J = \sum_{i=1}^{n} \|u_i\| \|v_i\|$$

Let us consider all the eigenvalues and the associated eigenvectors as being real. Let $J_i = \|u_i\| \|v_i\|$. Based on the relation $u_i v_i = 1$, we obtain:

$$J_i = \frac{1}{\cos(u_i, v_i)}$$

In addition[3], $\langle u_i \rangle = \langle v_1, \ldots, v_{i-1}, v_{i+1}, \ldots, v_n \rangle^T$. J_i is thus the reverse of the *sinus* of the angle between v_i and the space generated by the other eigenvectors. *Minimizing J thus implies maximizing the angle between the eigenvectors.*

Case of state feedback

The objective is to calculate the state feedback K_e by placing the poles $\{\lambda_1, \ldots, \lambda_n\}$ while minimizing the sensitivity criterion:

$$J = \sum_{i=1}^{n} \|u_i\| \|v_i\|$$

General methods of non-linear optimization (gradient, conjugated gradient, etc.) can be used in order to carry out the optimization of the criterion. In the case of state feedback, it is possible to use an entirely algebraic method [CHU 85, KAU 90, MOO 76b]. It is based on the interpretation of insensitivity in terms of angles between the eigenvectors.

14.4.2.3. *Use of the orthogonal projection of eigenvectors*

In many applications, decouplings are not primordial. In this case, it is preferable to choose the closed loop eigenvectors as orthogonal projections of the open loop eigenvectors.

DEFINITION 14.2. *Let us consider that the open loop eigenvalue λ_{i0} is moved into λ_i. Based on the notations defined by [14.19], the open loop eigenvector v_{i0} (associated with the eigenvalue λ_{i0}) is projected as follows:*

$$\eta_i = (V^*(\lambda_i)V(\lambda_i))^{-1} V^*(\lambda_i) v_{i0} \qquad [14.25]$$

---

3. The notation $\langle \rangle$ designates the sub-space generated.

The eigenvector and the input direction of the closed loop are thus chosen as being the orthogonal projections of the open loop eigenvector with the help of relations [14.26] and [14.27]:

$$v_i = V(\lambda_i)\eta_i \qquad\qquad [14.26]$$

$$w_i = W(\lambda_i)\eta_i \qquad\qquad [14.27]$$

Properties of the orthogonal projection

The choice of closed loop eigenvectors by orthogonal projection of the open loop eigenvectors makes it possible to:

– minimize the control leading to a desired pole placement (with a minimization of secondary effects such as the destabilization of non-placed poles);

– maintain the parametric behavior of the open loop. In fact, dispersion of the open loop poles – when the system is subjected to disturbances – is often acceptable. By considering that this hypothesis is verified and by keeping in mind that the dispersion of poles is closely related to the eigenvectors (see developments at first order [14.22]), it is natural to consider using the degrees of freedom related to the choice of the eigenvectors of the closed loop in order to maintain this good dispersion [MAG 94a], as shown in Figure 14.6. Ideally, based on [14.22], we would have to choose the closed loop eigenvectors that are co-linear to those of the open loop, in order to have a $\Delta\lambda_i$ identical in open loop and in closed loop, but these eigenvectors are constrained to evolve within a space defined by equation [14.19]. Therefore, we suggest choosing them as being orthogonal projections of the open loop eigenvectors on the eigenspace solution of [14.19] in order to minimize the distance between the closed loop eigenvector and the open loop eigenvector in the sense of the Euclidian standard. This projection is done through relations [14.25], [14.26] and [14.27];

– proceed by continuity. We can *continually* move a pole towards the left by projecting the corresponding eigenvector.

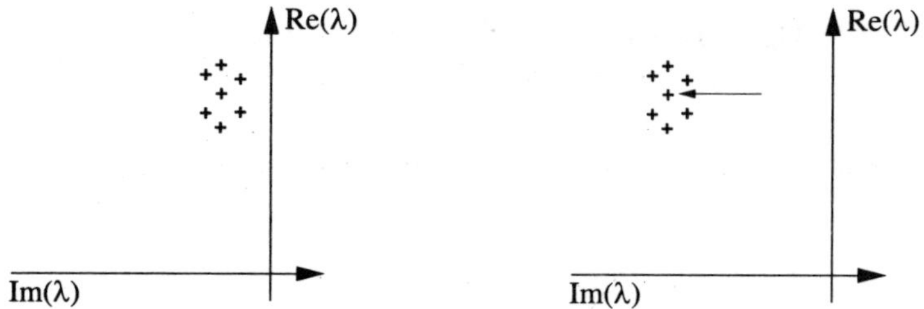

Figure 14.6. *Shift of a set of poles with minimum dispersion*

EXAMPLE 14.4. To illustrate these points, let us take a set of models pertaining to the lateral side of a jumbo jet (RCAM problem taken from [DOL 97]). The poles of the open loop are represented in Figure 14.7.

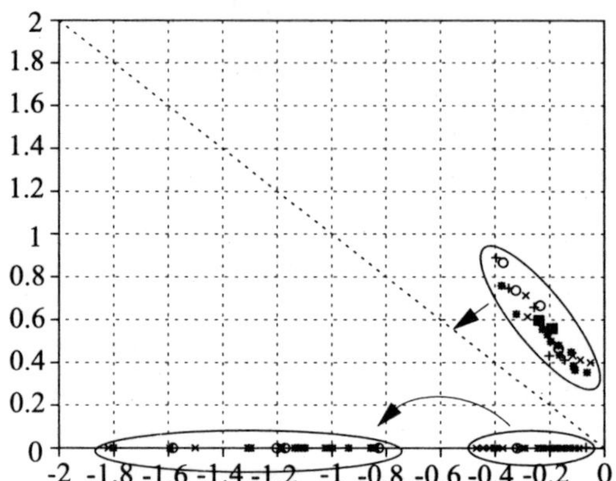

Figure 14.7. *Poles of the open loop of the lateral side of the RCAM*

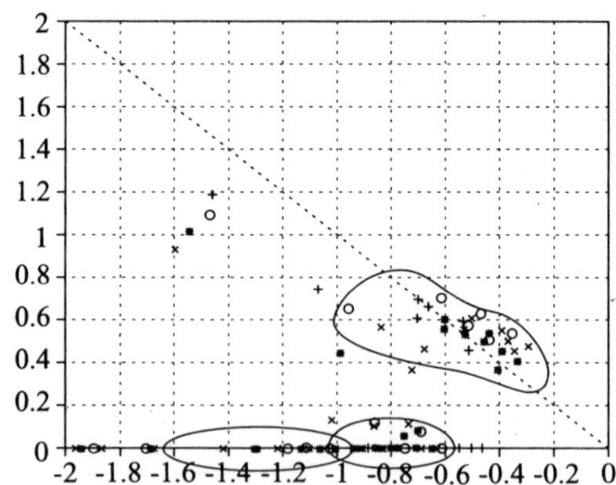

Figure 14.8. *Poles of the closed loop with orthogonal projection of the eigenvectors of the open loop*

On the nominal model, we carry out the following placement:

$$-0.23 + 0.59\,i \rightarrow -0.6 + 0.6\,i$$

$$-1.30 \rightarrow -1.30$$

$$-0.18 \rightarrow -0.8$$

by considering, for each eigenvalue placed, the orthogonal projection of the open loop eigenvector (associated to the open loop eigenvalue) on the solution closed loop eigenspace. In practice, for each eigenvalue $\lambda_i \in \{-0.23 + 0.59\,i, -1.30, -0.18\}$ placed respectively in $\lambda_i' \in \{-0.6 + 0.6\,i, -1.30, -0.8\}$, we calculate the sub-space $V(\lambda_i'), W(\lambda_i')$ solution of:

$$[A - \lambda_i' I \quad B] \begin{bmatrix} V(\lambda_i') \\ W(\lambda_i') \end{bmatrix} = 0$$

Then we place the closed loop eigenvectors v_i' associated to λ_i' by projection of the open loop eigenvectors v_i associated to λ_i on the sub-space $\langle V(\lambda_i')^T, W(\lambda_i')^T \rangle^T$. This projection is done as follows:

$$v_i' = V(\lambda_i')(V^*(\lambda_i')V(\lambda_i'))^{-1}V^*(\lambda_i')v_i$$

$$w_i' = W(\lambda_i')(V^*(\lambda_i')V(\lambda_i'))^{-1}V^*(\lambda_i')v_i$$

Figure 14.8 represents the poles of the closed loop. We can notice that the group of poles pertaining to each eigenvalue placed has shifted with a minimum of dispersion (the isolated eigenvalues correspond to the eigenvalues not dealt with).

14.4.3. *State feedback and output elementary static feedback*

When even n or p triplets (λ_i, v_i, w_i) are placed, it is not necessary to distinguish the two syntheses because the procedures are the same. The formula to use in both cases is equation [14.16]. *Usually, p (or n) triplets are placed.* The calculation of K is done as follows:

$$K = [w_1 \ \cdots \ w_p](C[v_1 \ \cdots \ v_p] + D[w_1 \ \cdots \ w_p])^{-1}$$

If $p < n$, the $n - p$ are not disturbing if they correspond to dynamics sufficiently fast or negligible in the sense of section 14.3. In the contrary case, it is necessary to use the exact pole placement techniques or to increase the number of eigenvalues placed by using an observer (see section 14.5).

In the context of our work, the result previously mentioned can be used in order to define procedure 14.1.

PROCEDURE 14.1 (EIGENSTRUCTURE PLACEMENT BY STATE FEEDBACK OR OUTPUT ELEMENTARY FEEDBACK). The procedure is decomposed as follows:

1) choosing a self-conjugated group of $q \leqslant p$ complex numbers $\lambda_1, \ldots, \lambda_q$ to place as closed loop eigenvalues;

2) for each λ_i, $i = 1, \ldots, q$, choosing a par of vectors (v_i, w_i) satisfying [14.15] i.e.:

$$\begin{bmatrix} A - \lambda_i I & B \end{bmatrix} \begin{bmatrix} v_i \\ w_i \end{bmatrix} = 0$$

with $\bar{v}_i = v_j$ for i, j, such that $\bar{\lambda}_i = \lambda_j$;

3) finding the real solution of:

$$K\left(Cv_i + Dw_i\right) = w_i \quad i = 1, \ldots, q$$

making it possible to place the eigenvalues $\{\lambda_1, \ldots, \lambda_q\}$ and the related right eigenvectors $\{v_1, \ldots, v_q\}$. If the problem is sub-specified (the number q of eigenvalues placed is inferior to the number p of outputs of the system), this solution can be obtained by a least squares resolution ($\alpha_i = (Cv_i + Dw_i)$):

$$K = [w_1, \ldots, w_r]([\alpha_1, \ldots, \alpha_r]^T[\alpha_1, \ldots, \alpha_r])^{-1}[\alpha_1, \ldots, \alpha_r]^T$$

NOTE 14.1 (GAIN COEFFICIENTS BELONGING TO $\mathbb{R}$). In the second phase, the condition on the conjugated term is necessary in order to be able to find a real solution. This is easily explained by the fact that:

$$K\left(C[v_i \quad \bar{v}_i] + D[w_i \quad \bar{w}_i]\right) = [w_i \quad \bar{w}_i]$$

which can be written (after multiplication on the right by an *arbitrary* matrix):

$$K\left(C[\Re(v_i) \quad \Im(v_i)] + D[\Re(w_i) \quad \Im(w_i)]\right) = [\Re(w_i) \quad \Im(w_i)]$$

where only real numbers are used. The group of linear constraints of the third stage can also be written (see [14.5]):

$$W = K\left(CV + DW\right) \tag{14.28}$$

NOTE 14.2 (NON-CONTROLLED EIGENVALUES). Gain K given by this algorithm makes it possible to place only p triplets. This method is thus used when the $n - p$ other eigenvalues correspond, in open loop, to low controllable modes or outside the bandwidth of the corrector, and thus will not be too disturbed by the corrector during looping.

EXAMPLE 14.5. Let us consider again the lateral model of the jumbo jet described in [DOL 97]. We are interested in the traditional measurements of β, p, r and ϕ. In particular we wish to decouple the requests in β (respectively ϕ) (noted by β_c (respectively ϕ_c)) of ϕ (respectively β) (couplings inferior to one degree, for requests in β_c and ϕ_c, of two and 20 degrees). These decouplings are illustrated in Figure 14.9.

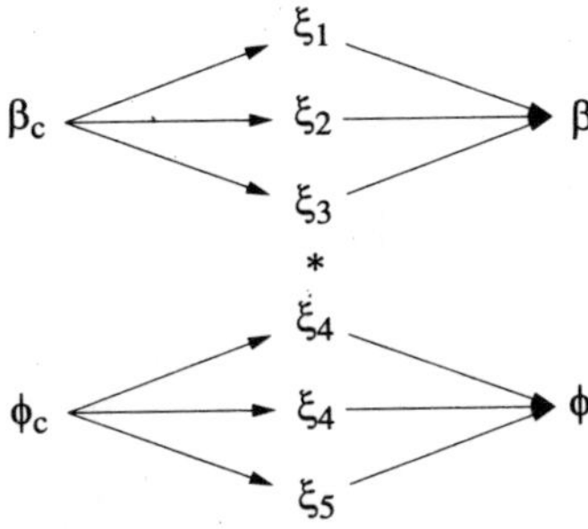

Figure 14.9. *Preferred decouplings between β and ϕ*

The system has 10 states and six outputs, hence it is possible to place only six pairs of eigenvalues and eigenvectors. Based on the relations obtained from the flight dynamics, we associate three real number modes to ϕ (modes ξ_1, ξ_2, ξ_3 associated to the eigenvalues $\lambda_1, \lambda_2, \lambda_3$) and a complex number mode as well as a real number mode to β (modes ξ_4, ξ_4^*, ξ_5 associated to the eigenvalues $\lambda_4, \lambda_4^*, \lambda_5$). Since the system has two inputs, it is possible to impose, for each eigenvalue placed, a decoupling constraint. Therefore, the output feedback will be synthesized in such a way that the three modes associated with ϕ are each decoupled from β (first output) and the three modes associated with β are each decoupled from ϕ (fourth output). The eigenvectors associated with the eigenvalues of ϕ will thus satisfy:

$$\begin{bmatrix} A - \lambda_i I & B \\ 1\ 0\ 0\ 0\ 0\ 0 & 0\ 0 \end{bmatrix} \begin{bmatrix} v_i \\ w_i \end{bmatrix} = \begin{bmatrix} 0 \\ 0 \end{bmatrix}$$

The eigenvectors associated with the eigenvalues of β satisfy:

$$\begin{bmatrix} A - \lambda_i I & B \\ 0\ 0\ 0\ 1\ 0\ 0 & 0\ 0 \end{bmatrix} \begin{bmatrix} v_i \\ w_i \end{bmatrix} = \begin{bmatrix} 0 \\ 0 \end{bmatrix}$$

These constraints make it possible to ensure decoupling between modes and outputs. The permanent state decouplings between the settings and outputs are ensured by integrators. After calculating the output feedback, we note that the modes dealt with are correctly placed and the modes that were not dealt with are fast and not disturbed by the output feedback gain thus calculated. The responses of outputs β and ϕ to settings in β_c and ϕ_c are traced in Figure 14.10 (respectively equal to a two degree step function and to a 20 degree step function).

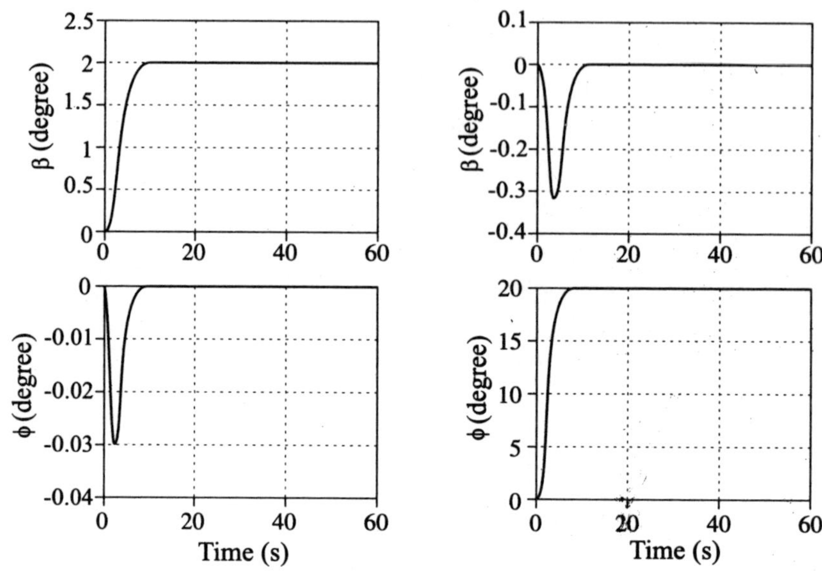

Figure 14.10. *Decouplings between β and ϕ by output static feedback*

We can notice in this figure that couplings (anti-diagonal faces) remain, in both cases, inferior to the degree. Hence, the required decouplings have been correctly considered during the synthesis.

14.5. Eigenstructure placement as observer

14.5.1. *Elementary observers*

A modal approach of the observers' synthesis is proposed in [MAG 91, MAG 94b, MAG 96]. This modal approach is based on the following lemma.

LEMMA 14.4. *The system defined by (see Figure 14.11):*

$$\dot{\hat{z}}_i = \pi_i \hat{z}_i - t_i \boldsymbol{y} + u_i B \boldsymbol{u} + t_i D \boldsymbol{u} \qquad [14.29]$$

where $u_i \in \mathbb{C}^n$, $t_i \in \mathbb{C}^p$ and $\pi_i \in \mathbb{C}$ satisfy:

$$u_i A + t_i C = \pi_i u_i \qquad [14.30]$$

is an observer of the variable $z_i = u_i \boldsymbol{x}$. The observation error is given by $\epsilon_i = \hat{z}_i - u_i \boldsymbol{x}$ satisfying:

$$\dot{\epsilon}_i = \pi_i \epsilon_i$$

Demonstration. Based on [14.1] and [14.29] we have:

$$\dot{\hat{z}}_i - u_i\dot{x} = \pi_i\hat{z}_i - t_iCx + u_iBu - u_iAx - u_iBu$$

Based on [14.30], we have:

$$\dot{\hat{z}}_i - u_i\dot{x} = \pi_i(\hat{z}_i - u_ix) \qquad \square$$

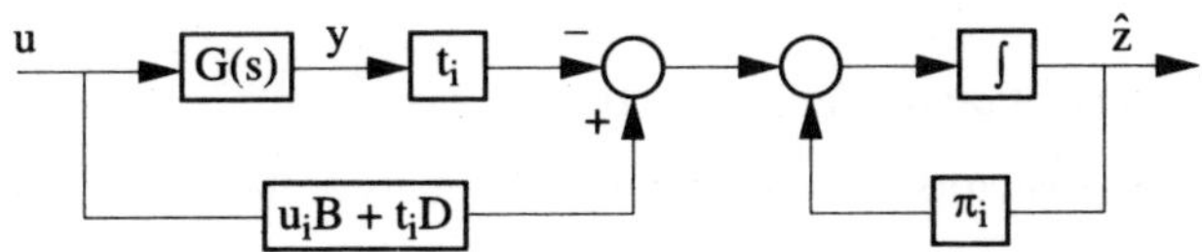

Figure 14.11. *Elementary observer of the variable* $z = u_ix$

14.5.2. *Observer synthesis*

The previous lemma establishes that a linear combination of states u_ix can be estimated by a mono-dimensional observer and this observer is obtained with the help of vector u_i satisfying [14.30] for a given vector t_i and a complex number π_i. This equation can be written (see equation [14.15]):

$$\begin{bmatrix} u_i & t_i \end{bmatrix} \begin{bmatrix} A - \pi_iI \\ C \end{bmatrix} = 0 \qquad [14.31]$$

If q elementary observers are used in parallel, it is possible to represent the overall observer as in Figure 14.11, but by replacing u_i, t_i and π_i with their matrix notations U, T and Π where:

$$U = \begin{bmatrix} u_1 \\ \vdots \\ u_{n_c} \end{bmatrix}, \quad T = \begin{bmatrix} t_1 \\ \vdots \\ t_{n_c} \end{bmatrix}. \quad \Pi = \text{Diag}\{\pi_1 \cdots \pi_{n_c}\} \qquad [14.32]$$

and where each triplet (π_i, u_i, t_i) satisfies [14.29]. These n_c equations can overall be described as:

$$UA + TC = \Pi U \qquad [14.33]$$

Here, z becomes a vector of size n_c. This structure (including the output feedback) is described in Figure 14.12.

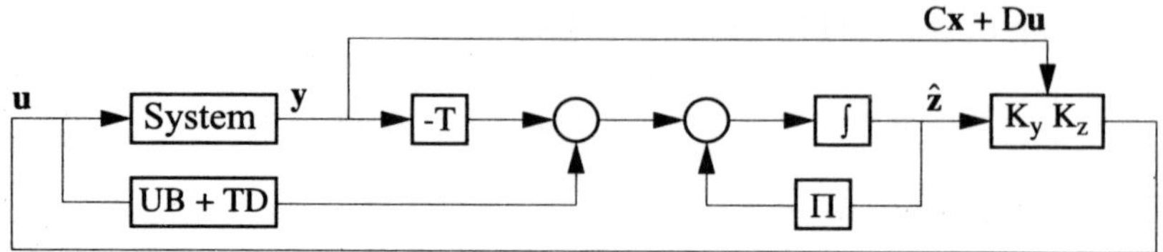

Figure 14.12. *Closed loop observer*

14.5.2.1. *Parameterization of elementary observations*

Based on [14.31], vectors u_i correspond to elements of the eigen sub-space defined by:

$$\begin{bmatrix} U(\pi_i) & T(\pi_i) \end{bmatrix} \begin{bmatrix} A - \pi_i I \\ C \end{bmatrix} = 0$$

Hence, the elementary observers can be parameterized by a vector $\eta_i^* \in \mathbb{C}^{n-p}$ as follows:

$$u_i = \eta_i^* U(\pi_i)$$

14.5.2.2. *Choice of elementary observations*

Decoupling

In order to choose vectors u_i, it is possible to consider decoupling objectives (for details, see [MAG 91], dual problem of the control issue).

Projection

Let us consider λ_{i0} an open loop eigenvalue and u_{i0} its related left eigenvector. It is possible to dualize the projection proposed in section 14.4.2. Hence, we obtain:

$$u_i = \eta_i U(\pi_i) \qquad\qquad [14.34]$$

$$t_i = \eta_i T(\pi_i) \qquad\qquad [14.35]$$

where:

$$\eta_i = u_{i0}(U(\pi_i)U(\pi_i)^T)^{-1}U(\pi_i) \qquad\qquad [14.36]$$

The choice of the projection can be justified by considering that u_{i0} is exactly placed. In this case, the elementary observer that corresponds to this *placement* is characterized by the open loop triplet $(\lambda_{i0}, u_{i0}, t_{i0})$ with $t_{i0} = 0$. It is obvious that this triplet satisfies [14.31]. The corresponding observer is represented in Figure 14.13. From this figure it results that the measurement vector y is not used for the observation of the variable $z_i = u_i x$. Consequently, when the increased output feedback is used as indicated in Figure 14.12, a dynamic pre-control structure is obtained. More generally, when a projection is used, through a continuity argument, the pre-control structure becomes "dominant". The effects of the output feedback are thus minimized.

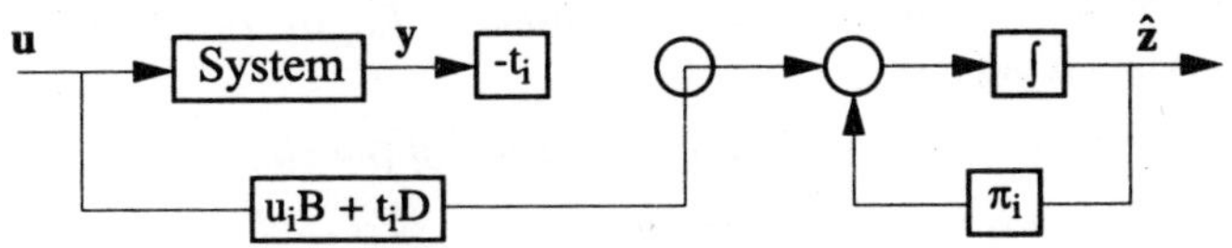

Figure 14.13. *Elementary observer of*
$z = u_i x$ *where* u_i *is a left eigenvector of* A

14.5.3. *Synthesis of output dynamic feedback in the form of observer*

When a set of elementary observers is considered, the "measurements" available to us are provided by y and $\hat{z}$. The problem of synthesis supposes seeking matrices K_y and K_z so that the system:

$$\begin{cases} \dot{x} = Ax + Bu \\ \dot{\hat{z}} = \Pi\hat{z} + (UB + TD)u - Ty \\ y = Cx + Du \end{cases} \qquad [14.37]$$

corrected by:

$$u = K_y y + K_z \hat{z}$$

has the dynamics hoped for. Due to the separation principle, this synthesis is usually divided into two sub-problems.

The separation principle, mentioned below, establishes that the syntheses of the observer and the output feedback can be done independently:

1) the observer is synthesized by the choice of matrices U, T, Π according to equation [14.33];

2) instead of synthesizing the output feedback as $u = K_y y + K_z \hat{z}$ specific to system [14.37], we synthesize it as $u = K_y y + K_z z$ (it should be noted that the estimation $\hat{z}$ of z is replaced by signal z) of the following system:

$$\begin{cases} \dot{x} = Ax + Bu \\ \begin{bmatrix} y \\ z \end{bmatrix} = \begin{bmatrix} C \\ U \end{bmatrix} x + \begin{bmatrix} D \\ 0 \end{bmatrix} u \end{cases} \qquad [14.38]$$

THEOREM 14.1 (SEPARATION PRINCIPLE). *Let us consider that:*

1) an observer of order n_c is synthesized, i.e. that three matrices $U \in \mathbb{R}^{n_c \times n}$, $T \in \mathbb{R}^{n_c \times p}$ and $\Pi \in \mathbb{R}^{n_c \times n_c}$ satisfying [14.33] are synthesized;

2) two gain matrices $K_y \in \mathbb{R}^{n_c \times n_c}$ and $K_z \in \mathbb{R}^{m \times n_c}$ specific to [14.38] are calculated, thus making it possible to place $p + n_c$ eigenvalues and eigenvectors.

Then, if U, T, Π, K_y and K_z are used based on Figure 14.12, the corresponding closed loop system is such that:

1) the eigenvalues of Π belong to the spectrum of the closed loop system;

2) the $p + n_c$ eigenvalues of the system [14.38] corrected by K_y and K_z belong to the spectrum of the closed loop.

Demonstration. After simplifying the expression of $\hat{z}$ (see [14.37]):

$$\begin{cases} \dot{x} = Ax + Bu \\ \dot{\hat{z}} = \Pi\hat{z} + UBu - TCx \\ y = Cx + Du \end{cases}$$

Let us consider the *feedback* in the conceivable form ($\hat{z}$ replaces z) $u = K_y y + K_z \hat{z}$. This will control the increased system:

$$\begin{cases} \begin{bmatrix} \dot{x} \\ \dot{\hat{z}} \end{bmatrix} = \begin{bmatrix} A & 0 \\ -TC & \Pi \end{bmatrix} \begin{bmatrix} x \\ \hat{z} \end{bmatrix} + \begin{bmatrix} B \\ UB \end{bmatrix} u \\ \begin{bmatrix} y \\ \hat{z} \end{bmatrix} = \begin{bmatrix} C & 0 \\ 0 & I \end{bmatrix} \begin{bmatrix} x \\ \hat{z} \end{bmatrix} + \begin{bmatrix} D \\ 0 \end{bmatrix} u \end{cases}$$

by the *feedback* $u = \tilde{K}\begin{bmatrix} y \\ \hat{z} \end{bmatrix}$ where:

$$\tilde{K} = (I - K_y D)^{-1} \begin{bmatrix} K_y & K_z \end{bmatrix}$$

The equation of the closed loop system is thus:

$$\begin{bmatrix} \dot{x} \\ \dot{\hat{z}} \end{bmatrix} = \left(\begin{bmatrix} A & 0 \\ -TC & \Pi \end{bmatrix} + \begin{bmatrix} B \\ UB \end{bmatrix} \tilde{K} \begin{bmatrix} C & 0 \\ 0 & I \end{bmatrix} \right) \begin{bmatrix} x \\ \hat{z} \end{bmatrix} \qquad [14.39]$$

Let us use the estimation error variable $\varepsilon = \hat{z} - Ux$ by considering the following variable change:

$$\begin{bmatrix} x \\ \varepsilon \end{bmatrix} = \begin{bmatrix} I & 0 \\ -U & I \end{bmatrix} \begin{bmatrix} x \\ \hat{z} \end{bmatrix} \qquad \Leftrightarrow \qquad \begin{bmatrix} x \\ \hat{z} \end{bmatrix} = \begin{bmatrix} I & 0 \\ U & I \end{bmatrix} \begin{bmatrix} x \\ \varepsilon \end{bmatrix}$$

System [14.39] becomes, by using the equality $UA + TC = \Pi U$:

$$\begin{bmatrix} \dot{x} \\ \dot{\varepsilon} \end{bmatrix} = \left(\begin{bmatrix} A & 0 \\ 0 & \Pi \end{bmatrix} + \begin{bmatrix} B \\ 0 \end{bmatrix} \tilde{K} \begin{bmatrix} C & 0 \\ U & I \end{bmatrix} \right) \begin{bmatrix} x \\ \varepsilon \end{bmatrix}$$

or:

$$\begin{bmatrix} \dot{x} \\ \dot{\varepsilon} \end{bmatrix} = \begin{bmatrix} A + B\tilde{K} \begin{bmatrix} C \\ U \end{bmatrix} & B\tilde{K} \begin{bmatrix} 0 \\ I \end{bmatrix} \\ 0 & \Pi \end{bmatrix} \begin{bmatrix} x \\ \varepsilon \end{bmatrix} \qquad [14.40]$$

The block triangular structure demonstrates the lemma: the closed loop spectrum will consist of the observer spectrum and the system spectrum [14.38] looped according to $u = K_y y + K_z z$. $\qquad \square$

Let us consider the problem of the equivalent output feedback described by the system [14.38]. Procedure 14.1 can be applied to this system (see procedure 14.2). The supplementary rows of the "new matrix C" of the system make it possible to place $p + n_c$ eigenvectors instead of p (if $n_c + p \geqslant n$, n eigenvalues and eigenvectors are placed). The separation principle makes it possible to consider the looping $u = K_y y + K_z z$ for the synthesis, whereas for the implementation we use $u = K_y y + K_z \hat{z}$. For example, we could use this type of an observer (by choosing an order equal to $n - p$) in order to place all the eigenvalues of the system (but it is preferable to place only the dominant modes) or to artificially increase the number of outputs in order to deal with robustness problems (see [MAG 97]).

NOTE 14.3 (NON-CONTROLLABILITY OF OBSERVER MODES). Based on equation [14.40] we note that the left eigenvectors associated to modes π_i of the system looped by the observer have the form $[0 \quad u_{ci}]$ (for a certain vector u_{ci} of size n_c). Consequently, the product "left eigenvector by matrix B" is zero ($[0 \quad u_{ci}][B^T \quad 0]^T = 0$), which immediately translates the non-controllability.

Now we will sum up the above discussions.

PROCEDURE 14.2 (EIGENSTRUCTURE PLACEMENT THROUGH AN OBSERVER). Let us suppose that we want to place q (dominant) poles and that $q > p$. The procedure is divided as follows:

1) choosing matrices U, T, Π satisfying equation [14.33]. The size n_c of these matrices is equal to the size of the dynamic extension used for placing the $q - p$ pairs of eigenvalues and eigenvectors that cannot be placed by output static feedback, i.e. $n_c = q - p$ (in the case of placement of all eigenvalues of the system, we have $n_c = p - q$);

2) choosing a self-conjugated set of q complex numbers $\lambda_1, \ldots, \lambda_q$ to place as closed loop eigenvalues;

3) for each λ_i, $i = 1, \ldots, q$, choosing a pair of vectors (v_i, w_i) satisfying [14.15], i.e.:

$$\begin{bmatrix} A - \lambda_i I & B \end{bmatrix} \begin{bmatrix} v_i \\ w_i \end{bmatrix} = 0$$

with $\bar{v}_i = v_j$ for i, j such that $\bar{\lambda}_i = \lambda_j$;

4) finding the real number solution of:

$$\begin{bmatrix} K_c & K_z \end{bmatrix} \left(\begin{bmatrix} C \\ U \end{bmatrix} v_i + \begin{bmatrix} D \\ 0 \end{bmatrix} w_i \right) = w_i \quad i = 1, \ldots, q \qquad [14.41]$$

If the problem is sub-specified ($q < n_c + p$), this solution can be obtained in the least squares sense:

$$\begin{bmatrix} K_c & K_z \end{bmatrix} = [w_1, \ldots, w_q](\Gamma^T \Gamma)^{-1} \Gamma^T$$

where:

$$\Gamma = \begin{bmatrix} Cv_1 + Dw_1 & \cdots & Cv_q + Dw_q \\ Uv_1 & \cdots & Uv_q \end{bmatrix}$$

NOTE 14.4 (EQUIVALENT DYNAMIC EQUALIZER). We can show that the *feedback* calculated by the above procedure corresponds to the dynamic gain:

$$\begin{cases} A_c = \Pi + (UB + TD)K_z \\ B_c = -T + (UB + TD)K_c \\ C_c = K_z \\ D_c = K_c \end{cases} \qquad [14.42]$$

and to the dynamic pre-control:

$$\begin{cases} A_p = A_c \\ B_p = UB + TD \\ C_p = C_c \\ D_p = I \end{cases} \qquad [14.43]$$

NOTE 14.5 (EFFECT OF THE EIGENSTRUCTURE OF THE OBSERVER). Based on the demonstration of the separation principle, the dynamics of the observer is non-controllable. Let us consider $\dot{x} = Ax + B(u + v)$. It is possible to show that:

$$\begin{bmatrix} x \\ x_c \end{bmatrix} = \Sigma_i^{n+q} \int_0^t \begin{bmatrix} v_i \\ v_{ic} \end{bmatrix} e^{\lambda_i(t-\tau)} [u_i \ u_{ci}] \begin{bmatrix} B \\ UB \end{bmatrix} v(\tau) \, d\tau \qquad [14.44]$$

Note 14.3 underlines the fact that the non-controllability of the observer's modes implies $u_i B + u_{ci} U B = 0$; hence, the term corresponding to π_i in [14.44] disappears. The eigenstructure of the observer does not have *any* effect on that of the closed loop system.

EXAMPLE 14.6. Let us take again Example 14.4. We consider here that the only available outputs are the measurements of the slide-slip angle β and of the bank angle ϕ. These two measurements enable us to place only two pairs of eigenvalues and eigenvectors through the output static feedback. That is the reason why we will use the procedure of eigenstructure placement in the form of an observer. This eigenstructure – by choosing an observer of a sufficient order – will enable us to set a part or all the dynamics of the system.

The system considered has $p = 2$ outputs and we will place the same eigenvalues as in Example 14.4, i.e. $q = 4$ eigenvalues. The observer necessary to such a placement must thus be of order $n_c = q - p = 2$. We choose the dynamics of the observer according to objectives specific to the bandwidth of the corrector. For example:

$$\pi_{1.2} = -0.8 \pm 0.8\,i$$

Matrices U and T are obtained through the equation of Example 14.4 by orthogonal projection.

We will place the same eigenvalues as previously, i.e.:

$$-0.2360 + 0.5954\,i \rightarrow -0.6 + 0.6\,i$$

$$-1.3017 \rightarrow -1.317$$

$$-0.1837 \rightarrow -0.8$$

with the same decoupling constraints.

We calculate matrices K_z and K_c of the observer by resolving a set of four equations similar to [14.41].

Finally, in the spectrum of the closed loop, we find the modes placed by the corrector in the form of observer as well as the modes of the observer itself (separation principle). The non-controlled modes have a correct behavior. We note that these modes could have been placed in the same way by increasing the size of the observer. Figure 14.14 shows the four modal simulations between β_c, Φ_c and β, Φ. We note that the decouplings were taken into consideration.

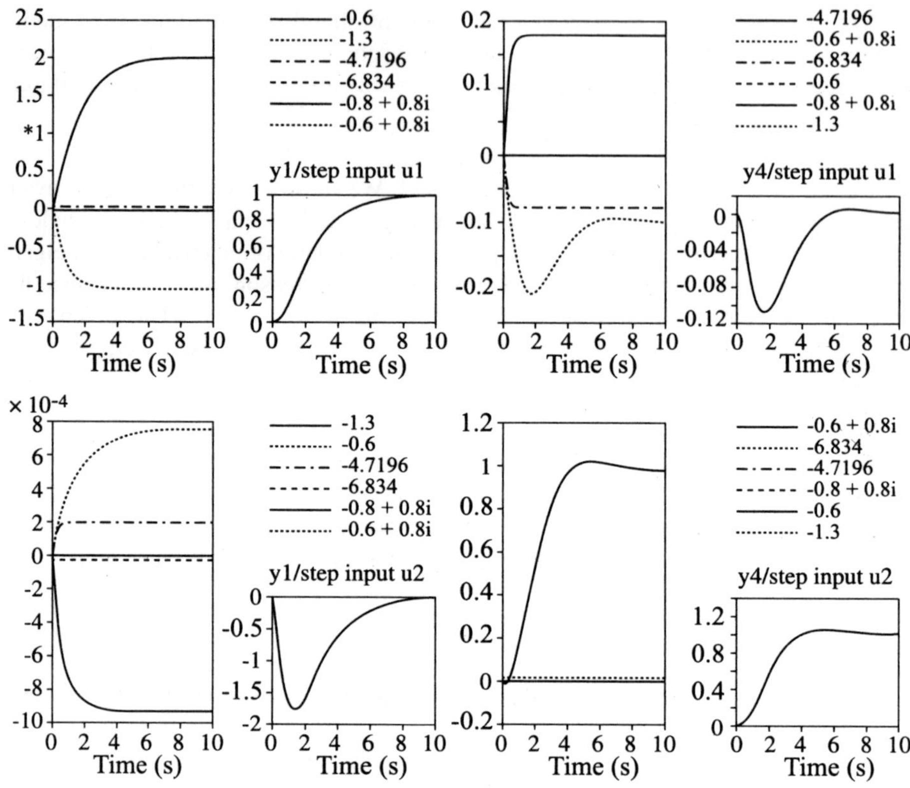

Figure 14.14. *Inputs-outputs modal simulation*

14.6. Conclusion

In this chapter, we have defined and studied the modal behavior of a system. We have seen how to choose the eigenvalues and eigenvectors of the closed loop according to the objectives:

– on the dynamics of time responses;

– on the couplings-inputs, outputs or modes-outputs;

– on the local robustness to parametric variations.

Once this choice is made, we have seen how to synthesize a static corrector making the placement of this eigenstructure possible. In the case of the output feedback, it is possible to place p pairs of eigenvalues and eigenvectors with a constant gain. In order to remedy this inconvenient, we have presented the synthesis of observers and the technique that makes it possible to directly carry out an eigenstructure placement in the form of an observer. Therefore, it is possible to place the totality of pairs of eigenvalues and eigenvectors of the system.

As we have said in our introduction, our intention was to describe here a synthesis technique in which we choose *a priori* the eigenvalues and eigenvectors of the closed loop, which leads us to a linear problem. Other techniques do not make this choice and adopt other policies calling upon more complex resolutions or non-convex optimization problems. Those interested in finding out more may refer to [APK 89, CHO 94, MUD 88].

Finally, the traditional approaches that we described are the basis of more complete syntheses (detailed in [MAG 02b]), that are meant, in particular, to deal with problems of parametric robustness.

14.7. Bibliography

[APK 89] APKARIAN P., CHAMPETIER C., MAGNI J.F., "Design of an helicopter output feedback control law using modal and structured-robustness techniques", *International Journal of Control*, vol. 50, no. 4, p. 1195–1215, 1989.

[CHO 94] CHOUAIB I., PRADIN B., "On mode decoupling and minimum sensitivity by eigen structure assignment", in *Proceedings of MELECON'94 (Antalya, Turkey)*, p. 663–666, 12–14 April 1994.

[CHU 85] CHU E.K.W., NICHOLS N.K., KAUTSKY J., "Robust pole assignment by output feedback", in Coock P. (ed.), *Proceedings of the Fourth IMA Conference on Control Theory*, Academic Press, p. 137–146, 1985.

[DOL 97] DÖLL C., MAGNI J.F., LE GORREC Y., "A modal multi-model approach, RCAM part", in *Robust Flight Control*, Springer-Verlag, Lecture Notes in Control and Information Sciences 224, p. 258–277, 1997.

[FAL 97] FALEIRO L.F., MAGNI J.F., DE LA CRUZ J., SCALA S., "Eigenstructure assignment, Tutorial part", in *Robust Flight Control*, Springer-Verlag, Lecture Notes in Control and Information Sciences 224, p. 22–32, 1997.

[HAM 89] HAMDAN A.M.A., NAYFEH A.H., "Measures of modal controllability and observability for first and second-order linear systems", *Journal of Guidance, Control, and Dynamics*, vol. 12, no. 3, p. 421–428, 1989.

[HAR 78] HARVEY C.A., STEIN G., "Quadratic weight for asymptotic regulator properties", *IEEE Transactions on Automatic Control*, vol. AC–23, p. 378–387, 1978.

[KAU 90] KAUSTKY J., NICHOLS N.K., "Robust pole assignment in systems subject to structured perturbations", *Systems and Control Letters*, vol. 15, p. 373–380, 1990.

[KIM 75] KIMURA H., "Pole assignment by gain output feedback", *IEEE Transactions on Automatic Control*, vol. AC–20, p. 509–516, 1975.

[LEG 98a] LE GORREC Y., Commande modale robuste, synthèse de gains autoséquencés: approche multimodèle, PhD Thesis, Ecole nationale supérieure de l'aéronautique et de l'espace (SUPAERO), Toulouse, France, December 1998.

[LEG 98b] LE GORREC Y., MAGNI J.F., DÖLL C., CHIAPPA C., "A modal multimodel control design approach applied to aircraft autopilot design", *AIAA Journal of Guidance, Control, and Dynamics*, vol. 21, no. 1, p. 77–83, 1998.

[LIM 93] LIM K.B., GAWRONSKI W., "Modal grammian approach to actuator and sensor placement for flexible structures", in *Proceedings of AIAA Guidance, Navigation, and Control Conference (Scottsdale, Arizona)*, August 1993.

[MAG 90] MAGNI J.F., CHAMPETIER C., Commande modale des systèmes multivariables, Polycopié, Ecole nationale supérieure de l'aéronautique et de l'espace (SUPAERO), Toulouse, France, 1990.

[MAG 91] MAGNI J.F., MOUYON P., "A tutorial approach to observer design", in *Proceedings of AIAA Conference on Guidance, Navigation, and Control (New Orleans)*, vol. III, p. 1748–1755, August 1991.

[MAG 94a] MAGNI J.F., MANOUAN A., "Robust flight control design by eigen structure assignment", in *Proceedings of the IFAC Symposium on Robust Control (Rio de Janeiro, Brazil)*, p. 388–393, September 1994.

[MAG 94b] MAGNI J.F., MOUYON P., "On residual generation by observer and parity space approaches", *IEEE Transactions on Automatic Control*, vol. 39, no. 2, p. 441–447, 1994.

[MAG 96] MAGNI J.F., "Continuous time parameter identification by using observers", *IEEE Transactions on Automatic Control*, vol. AC–40, no. 10, p. 1789–1792, 1996.

[MAG 97] MAGNI J.F., LE GORREC Y., CHIAPPA C., "An observer based multimodel control design approach", in *Proceedings of the Asian Control Conference (Seoul, Korea)*, vol. I, p. 863–866, July 1997.

[MAG 98] MAGNI J.F., LE GORREC Y., CHIAPPA C., "A multimodel-based approach to robust and self-scheduled control design", in *Proceedings of the Thirty-seventh IEEE Conference on Decision Control (Tampa, Florida)*, p. 3009–3014, 1998.

[MAG 02a] MAGNI J.F., *Robust Modal Control with a Toolbox for Use with Matlab*, Kluwer Academic/Plenum Publishers, New York, 2002.

[MAG 02b] MAGNI J.F., LE GORREC Y., "La commande multimodèle", in Bernussou J., Oustaloup A. (ed.), *Conception de commandes robustes*, Chapter 5, p. 155–187, Hermès, IC2 series, 2002.

[MOO 76a] MOORE B.C., "On the flexibility offered by state feedback in multivariable system beyond closed loop eigenvalue assignment", *IEEE Transactions on Automatic Control*, vol. AC–21, p. 659–692, 1976.

[MOO 76b] MOORE B.C., KLEIN G., "Eigenvalue selection in the linear regulator combining modal and optimal control", in *Proceedings of the Fifteenth IEEE Conference on Decision and Control*, p. 214–215, December 1976.

[MOO 81] MOORE B.C., "Principal component analysis in linear systems: Controllability, observability, and model reduction", *IEEE Transactions on Automatic Control*, vol. AC–26, p. 17–32, 1981.

[MUD 88] MUDGE S.K., PATTON R.J., "Analysis of the technique of robust eigenstructure assignment with application to aircraft control", *IEEE Proceedings, Part D: Control Theory and Applications*, vol. 135, no. 4, p. 275–281, 1988.

[SKE 81] SKELTON R.E., "Cost decomposition of linear systems with application to model reduction", *International Journal of Control*, vol. 6, no. 32, 1981.

Chapter 15

Robust H_∞/LMI Control

The synthesis of a control law passes through the utilization of patterns which are nothing other than an imperfect representation of reality: besides the fact that the laws of physics provide only a global representation of phenomena, valid only in a certain range, there are always the uncertainties of pattern establishment because the behavior of the physical process cannot be exactly described using a mathematical pattern.

Even if we work with patterns whose validity is limited, we have to take into account the *robustness* of the control law, i.e. we have to be able to guarantee not only the stability but also certain performances related to incertitude patterns. This last issue requires completing the pattern establishment work with a precise description of pattern uncertainties, to include them in a general formalism enabling us to take them into account and to reach certain conclusions.

The synthesis of a control law is hence articulated around two stages which are being alternatively repeated until the designer reaches satisfactory results:

– controller calculation: during this stage only certain performance objectives and certain robustness objectives can be taken into account;

– analysis of the controlled system properties, from the perspective of its performances as well as their robustness.

<hr>

Chapter written by Gilles DUC.

The approaches presented in this chapter are articulated around these two concepts.

15.1. The H$_\infty$ approach

The preoccupation for robustness, which is inherent among the methods used by traditional automatic control engineering, reappears around the end of the 1970s after having been so widely obscured due to the development of state methods. It is at the root of the development of H_∞ approaches.

15.1.1. *The H$_\infty$ standard problem*

Within this approach, the designer considers a synthesis scheme whose general form is presented in Figure 15.1: vector u represents the controls and vector y the available measurements; vector w reunites the considered exterior inputs (i.e. reference signals, disturbances, noises), which can be the inputs of the shaper filters chosen by the designer. Finally, vector e reunites the signals chosen to characterize the good functioning of the feedback control system, which are generally obtained from the signals existing in the feedback control loop with the help of the filters chosen there also by the designer.

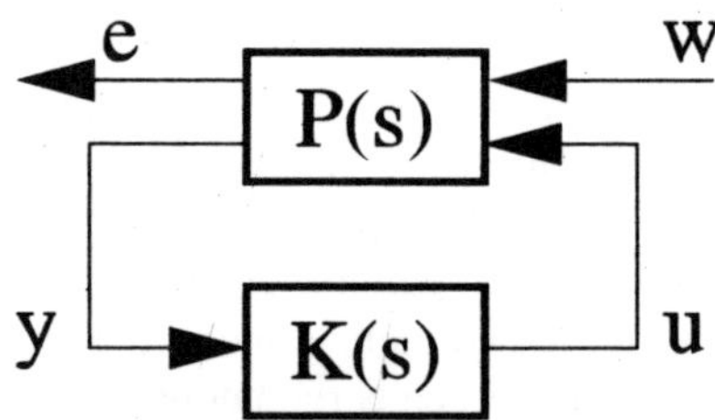

Figure 15.1. *H$_\infty$ standard problem*

The objective of the problem considered is thus to determine a corrector $K(s)$ that ensures the stability of the closed loop control system in Figure 15.1, conferring to the transfer $T_{ew}(s)$ between w and e a norm H_∞ less than a given level γ. This can be defined as follows:

$$\left\| \, T_{ew}(s) \, \right\|_\infty := \sup_{\omega \in \mathbf{R}} \sqrt{\overline{\lambda}\left(T_{ew}(j\omega)\; T_{ew}(-j\omega)^T\right)}$$

$$= \sup_{\omega \in \mathbf{R}} \sqrt{\overline{\lambda}\left(T_{ew}(-j\omega)^T\; T_{ew}(j\omega)\right)}$$

[15.1]

where $\overline{\lambda}$ designates the highest eigenvalue.

Let us suppose that the level γ has been reached. Then, by using the properties of norm H$_\infty$ [DUC 99], we can establish that:

– each transfer $T_{e_i w_j}(s)$ between a component w_j irrespective of w and a component e_i irrespective of e verifies:

$$\forall\, \omega \in \mathbf{R} \quad \left| T_{e_i w_j}(j\omega) \right| < \gamma$$

[15.2]

– the system remains stable for any uncertainty of the pattern that would introduce a looping of e over w in the form $w(s) = \Delta(s)\, e(s)$, $\Delta(s)$ being a stable transfer matrix irrespective of the norm H_∞ less than $1/\gamma$.

We can therefore use these results in different manners:

– to impose templates to certain transfers by choosing the signals e and w, in an appropriate manner; if, for example, $e(s) = W_1(s)\, z(s)$, where z is the output to be controlled and w is a disturbance, we obtain:

$$\forall\, \omega \in \mathbf{R} \quad \left| T_{zw}(j\omega) \right| < \frac{\gamma}{\left| W_1(j\omega) \right|}$$

[15.3]

so that the filter $W_1(s)$ makes it possible to impose a template to the transfer $T_{zw}(s)$ between the disturbance and the output;

– to perform the synthesis of a corrector which ensures the robustness related to the incertitude of $\Delta(s)$ pattern marked by norm (in this case, the signals e and w do not correspond to the feedback control inputs and outputs but they are the results of an appropriated pattern establishment);

– to adopt a combination of these two approaches.

It is worth mentioning that, historically, the second approach is the root of the H_∞ syntheses development and gathering all the patterns uncertainties in a single transfer matrix $\Delta(s)$ is a very poor representation which leads in most of the practical cases to limited results. The synthesis H_∞ must then be seen, according to the first approach, as a way to impose templates to nominal patterns of the feedback control without being able to take into account all the robustness objectives from the synthesis.

15.1.2. *Example*

Let us consider a system with the input y and the control u, where the nominal pattern is:

$$Y(s) \;=\; G(s)\,U(s) \;=\; \frac{1}{(s+1)(s+2)}\,U(s) \tag{15.4}$$

We want to create a feedback control in accordance with the block diagram in Figure 15.2, where the corrector $K(s)$ must ensure the following objectives:

i) the output y must be controlled over a constant reference r, with a static error less than 0.01;

ii) the gain of the feedback control[1] must contain all the angular frequencies between 0 and 1 rd/s at least;

iii) the module gain[2] must be at least equal to 0.7;

iv) the gain of the transfer function between r and u must be less than 10 for all angular frequencies and it must decrease following a gradient of –20 dB/decade beyond 10 rd/s;

v) the gain of the transfer function between r and y must be less than 0.5 beyond 10 rd/s.

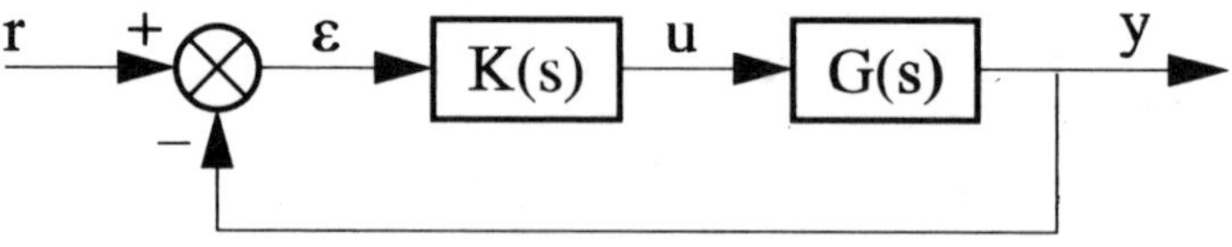

Figure 15.2. *Block diagram of the feedback control*

---

1 Conventionally defined as the set of angular frequencies for which the gain between the reference r and the error ε is less than 1.
2 Defined as the minimum distance between a point of Nyquist plot of the equalized system and the critical point –1.

Points i) to iii) can be translated through stresses on the transfer function $T_{\varepsilon r}(s) = \left(1 + G(s)\ K(s)\right)^{-1}$, where the gain must be:

– less than 0.01 in steady regime;

– less than 1 below 1 rd/s;

– less than 1/0.7 above.

Point iv) explicitly concerns the transfer $T_{ur}(s) = K(s)\left(1 + G(s)\ K(s)\right)^{-1}$.

Finally, point v) concerns the transfer $T_{yr}(s) = G(s)\,K(s)\left(1 + G(s)\,K(s)\right)^{-1}$.

This brings us to construct the block scheme in Figure 15.3, where the filters $W_i(s)$ are chosen in accordance to these specifications.

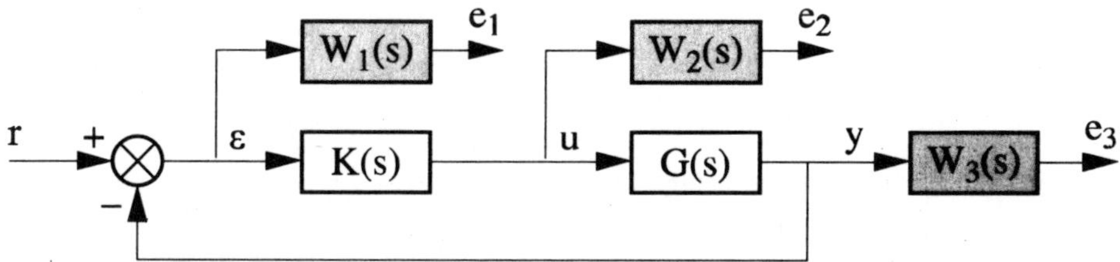

Figure 15.3. *Diagram used by the synthesis*

$$W_1(s) = \left(\frac{1}{0.7}\frac{s+0.01}{s+1/0.7}\right)^{-1} = \frac{0.7s+1}{s+0.01}$$

$$W_2(s) = \left(10\,\frac{1+s/1{,}000}{1+s/10}\right)^{-1} = 10\,\frac{s+10}{s+1{,}000} \qquad [15.5]$$

$$W_3(s) = \left(0.5\,\frac{s+10}{s}\right)^{-1} = \frac{2s}{s+10}$$

It must be noted that denominator $W_2(s)$ does not result from specifications but it is introduced in order to make this filter an eigenfilter: this condition is generally required by the resolution algorithms.

The scheme in Figure 15.3 is presented in the general form in Figure 15.1 by choosing $w = r$, $y = \varepsilon$ and $e = \begin{pmatrix} e_1 & e_2 & e_3 \end{pmatrix}^T$. We are then going to search for a corrector $K(s)$ solution of the following problem:

$$\left\| \begin{pmatrix} W_1(s)\, T_{\varepsilon r}(s) \\ W_2(s)\, T_{ur}(s) \\ W_3(s)\, T_{yr}(s) \end{pmatrix} \right\|_\infty < \gamma \qquad [15.6]$$

If this problem accepts a solution, we shall then have:

$$\forall \omega \quad \left| W_1(j\omega)\, T_{\varepsilon r}(j\omega) \right|^2 + \left| W_2(j\omega)\, T_{ur}(j\omega) \right|^2 + \left| W_3(j\omega)\, T_{yr}(j\omega) \right|^2 < \gamma^2$$

which implies:

$$\forall \omega \quad \left\{ \begin{array}{l} \left| W_1(j\omega)\, T_{\varepsilon r}(j\omega) \right| < \gamma \;\; \Leftrightarrow \;\; \left| T_{\varepsilon r}(j\omega) \right| < \dfrac{\gamma}{\left| W_1(j\omega) \right|} \\[2ex] \left| W_2(j\omega)\, T_{ur}(j\omega) \right| < \gamma \;\; \Leftrightarrow \;\; \left| T_{ur}(j\omega) \right| < \dfrac{\gamma}{\left| W_2(j\omega) \right|} \\[2ex] \left| W_3(j\omega)\, T_{yr}(j\omega) \right| < \gamma \;\; \Leftrightarrow \;\; \left| T_{yr}(j\omega) \right| < \dfrac{\gamma}{\left| W_3(j\omega) \right|} \end{array} \right. \qquad [15.7]$$

so that the objectives will be reached if the value of γ is less than 1 (or at the most close to 1).

By applying one of the resolution methods which are to be subsequently presented, we obtain a corrector corresponding to the value $\gamma = 1.029$ whose equation is the following, after an order reduction that makes it possible to eliminate the useless terms (a pole and a zero in high frequency and an almost exact compensation between a pole and a zero):

$$K(s) = 71 \frac{(s+1)(s+2)}{(s+0.01)(s^2+15.7s+73)} \qquad [15.8]$$

The transfer functions obtained for the feedback control are written:

$$T_{\varepsilon r}(s) = \frac{(s+0.01)(s^2+15.7s+73)}{s^3+15.7s^2+73.2s+71.7}$$

$$T_{ur}(s) = \frac{(s+1)(s+2)}{s^3+15.7s^2+73.2s+71.7}$$

$$T_{\varepsilon r}(s) = \frac{71}{s^3+15.7s^2+73.2s+71.7}$$

[15.9]

Figure 15.4 shows the Bode diagram for each of these functions compared to that of the inverse of the filters: it makes it possible to verify that the inequalities [15.7] are satisfied and hence that the synthesis objectives are reached.

In terms of robustness, the last of the inequalities [15.7] introduces a bound over the transfer bandwidth between the reference and the regulated magnitude: this ensures that the closed loop control system can tolerate high frequency dynamics which are not taken into account by the pattern [15.7] without risk for stability. In order to illustrate this idea, we suppose as an example that the pattern [15.7] does not consider an additional first order term at the denominator, so that a more precise pattern would be:

$$G'(s) = \frac{1}{(s+1)(s+2)(1+\tau s)}$$

[15.10]

Acknowledging that:

$$G'(s) = G(s)\left(1 - \frac{\tau s}{1+\tau s}\right)$$

[15.11]

The closed loop control system is presented in Figure 15.5a, which is equivalent to that in Figure 15.5b. In this latter figure, the transfer from r' to y' verifies the third inequality [15.7]:

$$\forall \omega \quad |T_{y'r'}(j\omega)| = \left|\frac{K(j\omega)G(j\omega)}{1+K(j\omega)G(j\omega)}\right| < \frac{\gamma}{|W_3(j\omega)|}$$

[15.12]

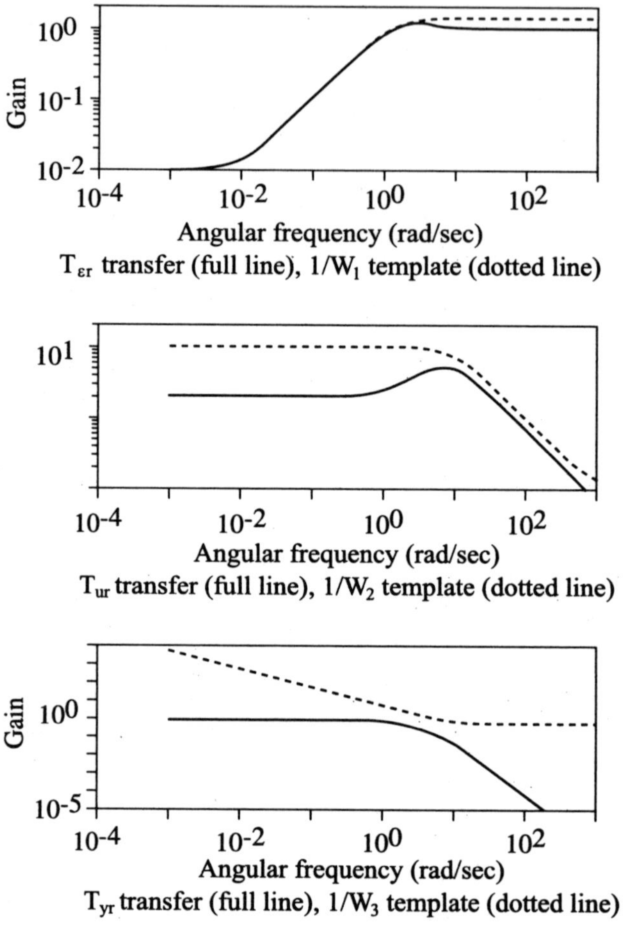

$T_{\varepsilon r}$ transfer (full line), $1/W_1$ template (dotted line)

T_{ur} transfer (full line), $1/W_2$ template (dotted line)

T_{yr} transfer (full line), $1/W_3$ template (dotted line)

Figure 15.4. *Bode diagrams for different transfers (full lines) and for their templates (dotted lines)*

We therefore infer that the closed loop control system in Figure 15.5b is stable for any value of τ such that:

$$\forall \omega \quad \left| T_{y'r'}(j\omega) \, \frac{\tau \, j\omega}{1 + \tau \, j\omega} \right| < 1 \quad \Leftrightarrow \quad \forall \omega \quad \left| \frac{\tau \, j\omega}{1 + \tau \, j\omega} \right| < \frac{|W_3(j\omega)|}{\gamma} \qquad [15.13]$$

because Figure 15.5b then corresponds to a system where the open loop (in y'') is stable, with a gain always less than 1: from Nyquist criterion, the close loop is then also stable.

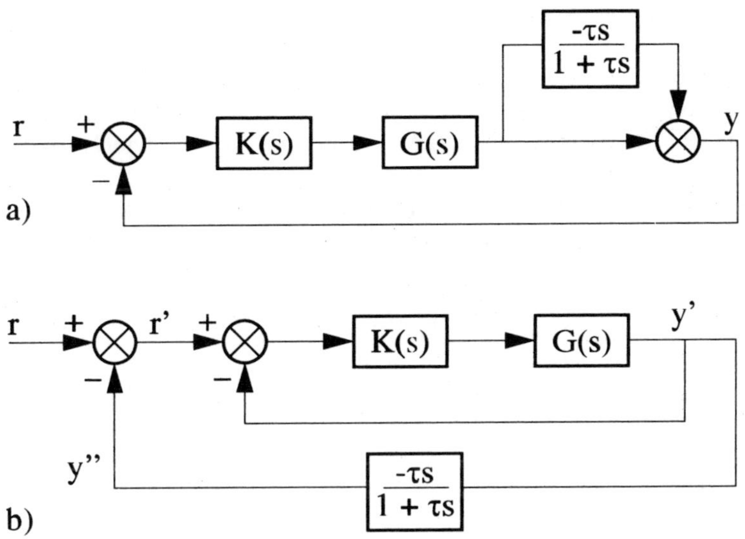

Figure 15.5. *Study of the neglected dynamics robustness*

Figure 15.6 makes it possible to compare the two functions' Bode diagrams which appear in the second inequality [15.13] (W_3/γ with full line and graphs for three different values of τ in dotted line): we see that stability is ensured for any value of τ less than 0.2.

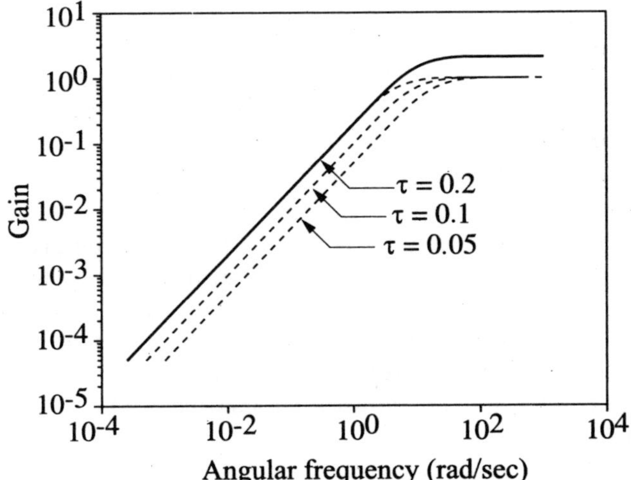

Figure 15.6. *Determination of a bound value of the neglected time constant*

15.1.3. *Resolution methods*

We can consider different methods in order to solve the H_∞ standard problem. We therefore present the approach through the Riccati equations and the approach through Linear Matrix Inequalities (LMI), which are the most widely used.

These two methods use a state representation of the interconnection matrix $P(s)$ which is written in the following form:

$$\begin{pmatrix} \dot{x}(t) \\ e(t) \\ y(t) \end{pmatrix} = \left(\begin{array}{c|cc} A & B_w & B_u \\ \hline C_e & D_{ew} & D_{eu} \\ C_y & D_{yw} & D_{yu} \end{array} \right) \begin{pmatrix} x(t) \\ w(t) \\ u(t) \end{pmatrix} \qquad [15.14]$$

with $x \in \mathbf{R}^n$; $w \in \mathbf{R}^{n_w}$; $u \in \mathbf{R}^{n_u}$; $e \in \mathbf{R}^{n_e}$; $y \in \mathbf{R}^{n_y}$.

15.1.4. *Resolution of H_∞ standard problem through the Riccati equations*

To solve the H_∞ standard problem, we suppose the following hypotheses as being satisfied:

H1) (A, B_u) can be stabilized and (C_y, A) can be detected;

H2) $\mathrm{rank}\,(D_{eu}) = n_u$ and $\mathrm{rank}\,(D_{yw}) = n_y$;

H3) $\forall\,\omega \in \mathbf{R}$ $\mathrm{rank}\begin{pmatrix} A - j\omega I_n & B_u \\ C_e & D_{eu} \end{pmatrix} = n + n_u$;

H4) $\forall\,\omega \in \mathbf{R}$ $\mathrm{rank}\begin{pmatrix} A - j\omega I_n & B_w \\ C_y & D_{yw} \end{pmatrix} = n + n_y$.

From a practical point of view, hypothesis H1 forces the user to choose the stable filters $W_i(s)$: placed outside the loop, these are actually non-controllable by u and non-observable by ε. In order to be verified, hypothesis H2 supposes the presence of direct transmissions between the controls u and the regulated variables e on the one hand, and between external inputs w and the measures y on the other hand. Hypotheses H3 and H4 are verified when the transfers $P_{eu}(s)$ and $P_{yw}(s)$ are not zero on the imaginary axis.

We present below the solution of a simplified case, which is characterized by the following relations:

$$D_{ew} = 0 \qquad D_{eu}^T \begin{pmatrix} C_e & D_{eu} \end{pmatrix} = \begin{pmatrix} 0 & I_{n_u} \end{pmatrix}$$

$$D_{yu} = 0 \qquad \begin{pmatrix} Bw \\ Dyw \end{pmatrix} D_{yw}^T = \begin{pmatrix} 0 \\ I_{n_y} \end{pmatrix} \qquad [15.15]$$

The general case is presented in [GLO 88]. We can also bring this general case back to the simplified one by using variable changes [ZHO 96].

The following theorem makes it possible in the first place to test the feasibility of the standard problem.

THEOREM 15.1.– *having the hypotheses H1-H4 and the conditions [15.15], the H_∞ standard problem has a solution if and only if the following 5 conditions are satisfied:*

i) matrix $H_\infty = \begin{pmatrix} A & \gamma^{-2} B_w B_w^T - B_u B_u^T \\ - C_e^T C_e & - A^T \end{pmatrix}$ *has no eigenvalue on the imaginary axis;*

ii) there is a symmetrical matrix $X_\infty \geq 0$ *solution of the Riccati equation:*

$$X_\infty A + A^T X_\infty + X_\infty (\gamma^{-2} B_w B_w^T - B_u B_u^T) X_\infty + C_e^T C_e = 0 \qquad [15.16]$$

iii) matrix $J_\infty = \begin{pmatrix} A^T & \gamma^{-2} C_e^T C_e - C_y^T C_y \\ - B_w B_w^T & - A \end{pmatrix}$ *has no eigenvalue on the imaginary axis;*

iv) there is a symmetrical matrix $Y_\infty \geq 0$ *solution of the Riccati equation:*

$$Y_\infty A^T + A Y_\infty + Y_\infty (\gamma^{-2} C_e^T C_e - C_y^T C_y) Y_\infty + B_w B_w^T = 0 \qquad [15.17]$$

v) $\rho (X_\infty Y_\infty) < \gamma^2$

where $\rho (\ )$ *designates the module of the highest eigenvalue.*   ∎

Finally, a solution for the standard problem is given by the following theorem.

THEOREM 15.2.– *based on the conditions of Theorem 15.1, a corrector $K(s)$ stabilizing the system and accomplishing $\| T_{ew}(s) \|_\infty < \gamma$ is described by the state representation [15.18]:*

$$\begin{pmatrix} \dot{x}_c(t) \\ u(t) \end{pmatrix} = \left(\begin{array}{c|c} \hat{A}_\infty & Z_\infty\, Y_\infty\, C_y^T \\ \hline -B_u^T X_\infty & 0 \end{array} \right) \begin{pmatrix} x_c(t) \\ y(t) \end{pmatrix}$$

$$\hat{A}_\infty = A + \gamma^{-2} B_w\, B_w^T\, X_\infty - B_u\, B_u^T X_\infty - Z_\infty\, Y_\infty\, C_y^T C \qquad\qquad [15.18]$$

$$Z_\infty = (I_n - \gamma^{-2} Y_\infty\, X_\infty)^{-1} \qquad\qquad\qquad\qquad \blacksquare$$

Thus the application of this solution consists of using firstly the results of Theorem 15.1 to find an admissible value of γ (we can use iterations on y by exploring through dichotomy a range of values previously chosen). Afterwards, we calculate a corrector by applying Theorem 15.2.

15.1.5. *Resolution of the H_∞ standard problem by LMI*

Synthesis by LMI provides another way to solve the standard problem. It is more general, since it requires only the hypothesis H1. We shall limit the exposition to the case when the condition [15.19] is verified:

$$D_{yu} = 0 \qquad\qquad\qquad\qquad [15.19]$$

In the opposite case, we firstly solve the problem by considering fictional measure units $\hat{y}$ corresponding to this case and we modify *a posteriori* the corrector obtained by carrying out the change of the variable $y = \hat{y} - D_{yu} u$ within its state equations.

The feasibility of the standard problem is tested using the following theorem [GAH 94].

THEOREM 15.3.– *having the hypothesis H1 and condition [15.19], the problem H_∞ standard has a solution if and only if there are 2 symmetric matrices R and S, verifying the following 3 matrix inequalities:*

$$\begin{pmatrix} \mathbf{N}_R & 0 \\ 0 & I_{n_w} \end{pmatrix}^T \begin{pmatrix} AR + RA^T & RC_e^T & B_w \\ C_e R & -\gamma I_{n_e} & D_{ew} \\ B_w^T & D_{ew}^T & -\gamma I_{n_w} \end{pmatrix} \begin{pmatrix} \mathbf{N}_R & 0 \\ 0 & I_{n_w} \end{pmatrix} < 0 \qquad [15.20a]$$

$$\begin{pmatrix} \mathbf{N}_S & 0 \\ 0 & I_{n_e} \end{pmatrix}^T \begin{pmatrix} A^T S + SA & SB_w & C_e^T \\ B_w^T S & -\gamma I_{n_w} & D_{ew}^T \\ C_e & D_{ew} & -\gamma I_{n_e} \end{pmatrix} \begin{pmatrix} \mathbf{N}_S & 0 \\ 0 & I_{n_e} \end{pmatrix} < 0 \qquad [15.20b]$$

$$\begin{pmatrix} R & I_n \\ I_n & S \end{pmatrix} \geq 0 \qquad [15.20c]$$

where $\mathbf{N}_R$ and $\mathbf{N}_S$ form a core basis of $(B_u^T \ D_{eu}^T)$ and $(C_y \ D_{yw})$, respectively.

Additionally, the $r < n$ order correctors exist if and only if the inequalities [15.20a, b, c] are verified by the matrices R and S which satisfy the additional condition:

$$rank \begin{pmatrix} R & I_n \\ I_n & S \end{pmatrix} \leq n + r \iff rank\left(I_n - R S\right) \leq r \qquad [15.20d] \ \blacksquare$$

The matrix inequalities [15.20a, b, c], which replace Theorem 15.1 conditions from i) to v), are closely connected to the unknown parameters R and S: they are usually designated by LMI. It is easy to verify that the set of matrices satisfying one or several LMIs is a convex set. Specific solvers are dedicated to this kind of problems [GAH 95].

We can additionally seek the optimal value of γ by solving the following problem, which is a convex optimization problem:

$$\min_{R=R^T,\ S=S^T} \gamma \quad \text{under [15.20a, b, c]} \tag{15.21}$$

From the solutions of matrices R and S in the previous problems, we can consider various procedures to form a corrector: explicitly formulae are given especially in [IWA 94], whereas [GAH 94] proposes a resolution by LMI, which can be summed up as follows.

Let:

$$\begin{cases} \dot{x}_c(t) = A_c x_c(t) + B_c y(t) \\ u(t) = C_c x_c(t) + D_c y(t) \end{cases} \tag{15.22}$$

with $x_c \in \mathbf{R}^r$ being a state representation of the corrector of order $r \leq n$ sought. The closed loop control system in Figure 15.1 has as a state representation:

$$\begin{pmatrix} \dot{x} \\ \dot{x}_c \\ \hline e \end{pmatrix} = \left(\begin{array}{cc|c} A + B_u D_c C_y & B_u C_c & B_w + B_u D_c D_{yw} \\ B_c C_y & A_c & B_c D_{yw} \\ \hline C_e + D_{eu} D_c C_y & D_{eu} C_c & D_{ew} + D_{eu} D_c D_{yw} \end{array} \right) \begin{pmatrix} x \\ x_c \\ \hline w \end{pmatrix} = \left(\begin{array}{c|c} A_f & B_f \\ \hline C_f & D_f \end{array} \right) \begin{pmatrix} x \\ x_c \\ \hline w \end{pmatrix} \tag{15.23}$$

and, based on the "Bounded Real Lemma" [BOY 94], its norm H_∞ is less than y if and only if there is a matrix $X = X^T > 0$ that verifies:

$$\begin{pmatrix} A_f^T X + X A_f & X B_f & C_f^T \\ B_f^T X & -\gamma I_{n_w} & D_f^T \\ C_f & D_f & -\gamma I_{n_e} \end{pmatrix} < 0 \tag{15.24a}$$

(which is a bilinear matrix inequality in X, A_c, B_c, C_c, D_c). A suitable matrix X can be obtained by performing a decomposition into singular values of $I_n - R S$, from where we can infer 2 full rank matrices $M, N \in \mathbf{R}^{n \times r}$ verifying:

$$M N^T = I_n - R S \tag{15.24b}$$

which make it possible to determine:

$$X = \begin{pmatrix} S & N \\ N^T & -M^+ R N \end{pmatrix} \qquad [15.24c]$$

where M^+ designates the pseudo-reciprocal of M ($M^+ M = I_r$). The inequality [15.24a] is therefore an LMI in A_c, B_c, C_c, D_c, where the resolution then provides a corrector.

15.1.6. *Restricted synthesis on the corrector order*

The two resolution methods presented in the previous sections lead to correctors with an order equal to that of the matrix $P(s)$, which contains the pattern of the regulation system increased by the filters expressing the synthesis objectives. However, we easily understand that this order, which can be very high, is not inevitably necessary to obtain a satisfactory control policy.

The LMI formulation makes it possible to consider the synthesis H_∞ with a restricted order. Let $r < n$ be the order of the corrector sought. It is necessary to establish matrices R and S, which are solutions of LMI [15.20a, b, c] and satisfying at the same time the restriction [15.20d] (about which we can say that it is always verified for $r \geq n$): this restriction leads to the loss of convexity of the set of matrices solutions, but heuristic methods dedicated to this type of problem can be efficiently used [DAV 94, ELG 97, VAL 99].

15.2. The μ-analysis

The μ-analysis is a technique which makes it possible to study system properties in the presence of different uncertainties of the pattern establishment. It should be noted that it is no longer a matter of calculating a corrector but, a corrector being given, it is about characterizing the robustness it provides to the closed loop control system. This technique, which appeared at the beginning of the 1980s, represented a major progress, perceptible especially through the change in the judging manner: it enables in fact the description and analysis of the properties on a patterns family and no longer on an unique pattern about which we know that it is not capable to represent the set of possible behaviors of a process.

15.2.1. *Analysis diagram and structured single value*

The μ-analysis uses the general diagram in Figure 15.7 (where we can observe the relationship with the one used in synthesis H_∞): all the pattern uncertainties are reunited in the matrix $\Delta(s)$; the transfer matrix $H(s)$ – which, in the case of a feedback system obviously depends on the corrector – establishes a pattern for the interconnections between the inputs w, the objectives e and the signals v and z which make the uncertainties possible.

If the transfer matrix $H(s)$ can be anything, the situation is not the same for the matrix $\Delta(s)$, which generally has a particular structure. Typically, this matrix will be block diagonal and consist of, on the one hand real diagonals blocks (representing the parametrical uncertainties) and on the other hand, transfer functions (or matrices) (representing neglected or uncertain dynamic phenomena):

$$\Delta(s) = \mathrm{diag}\,\{\delta_1 I_{r_1},\ \dots\ ,\ \delta_r I_{r_r},\ \Delta_1(s),\ \dots\ ,\ \Delta_q(s)\}$$

$$\delta_i \in \mathbf{R}\ ;\quad \Delta_i(s) \in \mathbf{R}H_\infty^{n_i \times n_i} \tag{15.25}$$

where $\mathbf{R}H_\infty^{n \times n}$ conventionally designates the set of stable transfer matrices of size $n \times n$. Further on, we shall name $\mathbf{S}$ the set of all complex matrices having size and structure identical to those of $\Delta(s)$:

$$\mathbf{S} = \left\{\Delta = \mathrm{diag}\left\{\delta_1 I_{r_1},\ \dots\ ,\ \delta_r I_{r_r}, \Delta_1,\ \dots\ ,\Delta_q\right\}\right\}$$

$$\delta_i \in \mathbf{R}\ ;\ \Delta_i \in \mathbf{C}^{n_i \times n_i} \tag{15.26}$$

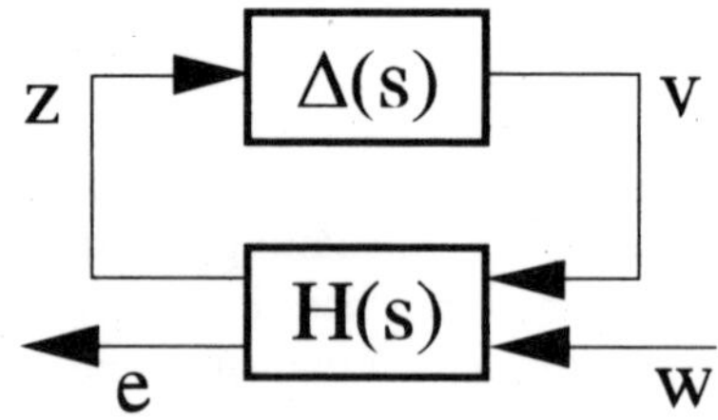

Figure 15.7. *Robustness analysis diagram*

In other terms, $\Delta(s) \in \mathbf{S}$ for all the values of s.

Let M be a square complex matrix having the same size as $\Delta(s)$. We note by Δ^* the transpose-conjugated of Δ. *The structured single value of M, of the set* **S**, *is defined by:*

$$\mu_{\mathbf{S}}(M) := \left(\inf_{\Delta \in \mathbf{S}} \left(\overline{\lambda}(\Delta^*\Delta) : \det\left(I - \Delta\ M\right) = 0 \right) \right)^{-1}$$

$$:= 0 \ \text{if} \ \forall \Delta \in \mathbf{S} \ \det\left(I - \Delta\ M\right) \neq 0$$

[15.27]

15.2.2. *Main results of robustness*

The structured single value makes it possible to establish different results [ZHO 96]. Further on, we divide the transfer matrix $H(s)$ in Figure 15.7 into:

$$\begin{pmatrix} z(s) \\ e(s) \end{pmatrix} = \begin{pmatrix} H_{11}(s) & H_{12}(s) \\ H_{21}(s) & H_{22}(s) \end{pmatrix} \begin{pmatrix} v(s) \\ w(s) \end{pmatrix}$$

[15.28]

with $\dim(z) = \dim(v) = n_1$ and $\dim(e) = \dim(w) = n_2$.

THEOREM 15.4.– *if $H(s)$ is stable, the system in Figure 15.7 is stable for any matrix $\Delta(s)$ of type [15.25] so that $\left\| \Delta(s) \right\|_\infty < 1/\alpha$ if and only if:*

$$\forall\ \omega \in \mathbf{R},\ \mu_{\mathbf{S}}\left(H_{11}(j\omega)\right) \leq \alpha$$

[15.29]

If, in addition, $\left\| H_{22}(s) \right\|_\infty < \beta$, the system in Figure 15.7 has a norm H_∞ less than β for any matrix $\Delta(s)$ of type [15.25] so that $\left\| \Delta(s) \right\|_\infty < 1/\beta$ if and only if:

$$\forall\ \omega \in \mathbf{R},\ \mu_{\mathbf{S}'}\left(H(j\omega)\right) \leq \beta$$

[15.30]

where **S'** *is obtained by completing* **S** *by any complex matrices having the same size as $H_{22}(s)$:*

$$\mathbf{S'} = \{\Delta' = \mathrm{diag}\{\Delta\ ,\ \Delta_{22}\}\ ;\ \Delta \in \mathbf{S}\ ;\ \Delta_{22} \in \mathbf{C}^{n_2 \times n_2}\}$$

[15.31] ∎

The first result of Theorem 15.4 is clearly a result of the stability robustness with pattern establishment uncertainties. The second one is the result of performance robustness because it guarantees that each transfer function $T_{e_i w_j}(s)$ has a gain less than 1 for all frequencies.

15.2.3. *Example*

We consider the closed loop control system in Figure 15.8, with a constant corrector $K(s) = 2$. The system to be controlled is characterized by the transfer function $G(s)$, whose nominal expression is:

$$G(s) = \frac{1}{(s + a)^2} \; ; \; 1 < a < 3 \qquad [15.32]$$

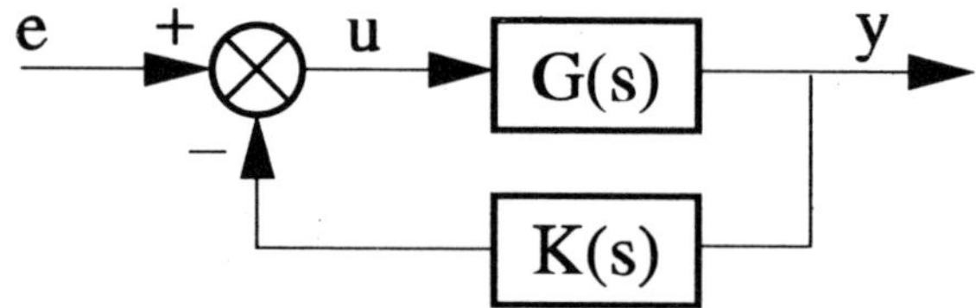

Figure 15.8. *Studied system*

In addition, this pattern neglects the high frequency dynamics, which are globally represented by a time constant with a maximal value equal to 0.5 s.

To characterize the parametrical uncertainty on the transfer function $G(s)$, we firstly suppose that $a = 2 + \delta$, with $-1 < \delta < 1$. The transfer function corresponds then to the differential equations:

$$\begin{cases} \dfrac{d\,y_1}{d\,t} + 2\,y_1 + \delta\,y_1 = u \\[2mm] \dfrac{d\,y_2}{d\,t} + 2\,y_2 + \delta\,y_2 = y_1 \end{cases} \qquad [15.33]$$

(having for instant $y_2 = y$). In order to separate the uncertainty δ from the rest of the system, in accordance with the general diagram in Figure 15.7, we have:

$$\begin{cases} z_1 = y_1 ; & v_1 = \delta z_1 \\ z_2 = y_2 ; & v_2 = \delta z_2 \end{cases} \qquad [15.34]$$

Equations [15.33] are then written:

$$\begin{cases} \dfrac{d\,y_1}{d\,t} + 2\,y_1 = -v_1 + u \\[2mm] \dfrac{d\,y_2}{d\,t} + 2\,y_2 = -v_2 + y_1 \end{cases} \qquad [15.35]$$

In order to represent the neglected dynamics, by reiterating the approach presented in section 15.1.2, we note that a possible pattern of the system is the following:

$$G(s) = \frac{1}{(s+a)^2 (1+\tau s)} = \frac{1}{(s+a)^2}\left(1 - \frac{\tau s}{1+\tau s}\right) \qquad [15.36]$$

with $0 < \tau < 0.5$. We can contain this type of patterns within the set of transfer functions in the following form:

$$G(s) = \frac{1}{(s+a)^2}\left(1 + W_d(s)\,\Delta_d(s)\right) \qquad [15.37]$$

where the filter $W_d(s)$ is chosen according to the previous knowledge of neglected dynamics and where $\Delta_d(s)$ is a restricted norm stable transfer function:

$$W_d(s) := \frac{0.5\,s}{1+0.5\,s} \qquad [15.38]$$

$$\left\|\Delta_d(s)\right\|_\infty = \sup_\omega \left|\Delta_d(j\omega)\right| < 1$$

By reuniting these two pattern establishments, we can redraw the block diagram of the closed loop control system in the form given in Figure 15.9 (always with

$K(s) = 2$). We easily identify the matrices $\Delta(s)$ and $H(s)$ of the general diagram in Figure 15.7 by considering $v = \begin{pmatrix} v_1 & v_2 & v_d \end{pmatrix}^T$ and $z = \begin{pmatrix} z_1 & z_2 & z_d \end{pmatrix}^T$:

$$\Delta(s) = \begin{pmatrix} \delta & 0 & 0 \\ 0 & \delta & 0 \\ 0 & 0 & \Delta_d(s) \end{pmatrix} \qquad [15.39]$$

$$H(s) = \frac{1}{s^2 + 4s + 6} \left(\begin{array}{ccc|c} -(s+2) & 2 & -2s & s+2 \\ -1 & -(s+2) & \dfrac{-2s}{s+2} & 1 \\ -1 & -(s+2) & \dfrac{-2s}{s+2} & 1 \\ \hline -1 & -(s+2) & s(s+2) & 1 \end{array} \right) \qquad [15.40]$$

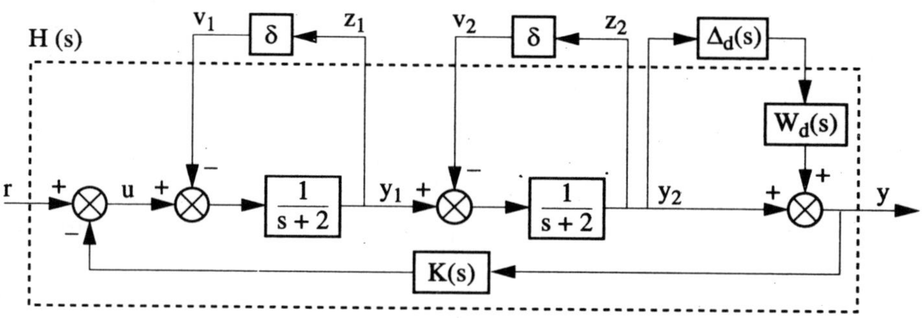

Figure 15.9. *Diagram used for the robustness analysis*

We can verify that $H(s)$ corresponds to a stable system and that:

$$\left\| H_{22}(s) \right\|_\infty = \sup_\omega \left| H_{22}(j\omega) \right| = 1/6 \qquad [15.41]$$

In accordance with the above results, we define the sets **S** and **S'** by:

$$\mathbf{S} = \left\{ \Delta = \mathrm{diag}\left\{ \delta I_2, \Delta_d \right\}; \delta \in \mathbf{R}; \Delta_d \in \mathbf{C} \right\} \qquad [15.42a]$$

$$\mathbf{S'} = \left\{ \Delta' = \mathrm{diag}\left\{ \delta I_2, \Delta_d, \Delta_{22} \right\}; \delta \in \mathbf{R}; \Delta_d \in \mathbf{C}; \Delta_{22} \in \mathbf{C} \right\} \qquad [15.42b]$$

Figure 15.10 shows an upper bound of $\mu_S(H_{11}(j\omega))$ according to ω (obtained following an approach which will be presented in the next section). Its value remains less than 0.7 for any ω. We infer from this that the closed loop control system is stable for any $\Delta(s)$ with the structure [15.39] such that:

$$\left\|\Delta(s)\right\|_\infty < 1/0.7 \ \Leftrightarrow \ \begin{cases} \left|\delta\right| < 1/0.7 \\ \left\|\Delta_d(s)\right\|_\infty < 1/0.7 \end{cases} \qquad [15.43]$$

The first condition is equivalent to a condition on a, which is presented below. From the second condition we can infer a maximal value for τ by noticing that:

$$\left\|\Delta_d(s)\right\|_\infty < 1/0.7 \ \Leftrightarrow \ \forall \omega \ \left|W_d(j\omega)\Delta_d(j\omega)\right| < \frac{1}{0.7}\left|\frac{j\omega}{j\omega+2}\right| \qquad [15.44]$$

If we apply this inequality to our particular case, i.e.:

$$W_d(s)\Delta_d(s) = \frac{-\tau s}{\tau s + 1} \qquad [15.45]$$

we infer from it a maximal admissible value for τ :

$$\forall \omega \ \left|\frac{-j\omega}{j\omega+1/\tau}\right| < \frac{1}{0.7}\left|\frac{j\omega}{j\omega+2}\right| \ \Leftrightarrow \ \left|\tau\right| < 0.5/0.7 \qquad [15.46]$$

We can finally state that the closed loop control system is stable if:

$$\begin{cases} 0.572 < a < 3.428 \\ \ \ 0 < \tau < 0.714 \end{cases} \qquad [15.47]$$

Figure 15.10b shows an upper bound of $\mu_{S'}\left(H(j\omega)\right)$ according to ω . Its value remains less than 0.89 for any ω. We infer from here that the closed loop control system preserves a norm H_∞ less than 0.89 for any $\Delta(s)$ of structure [15.39] such that $\left\|\Delta(s)\right\|_\infty < 1/0.89 = 1.12$, which is a condition accomplished for:

$$\begin{cases} 0.88 < a < 3.12 \\ \ \ 0 < \tau < 0.56 \end{cases} \qquad [15.48]$$

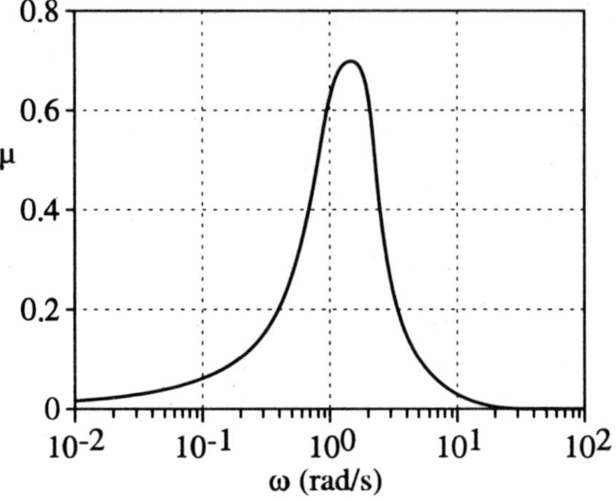

a) Robustness of stability

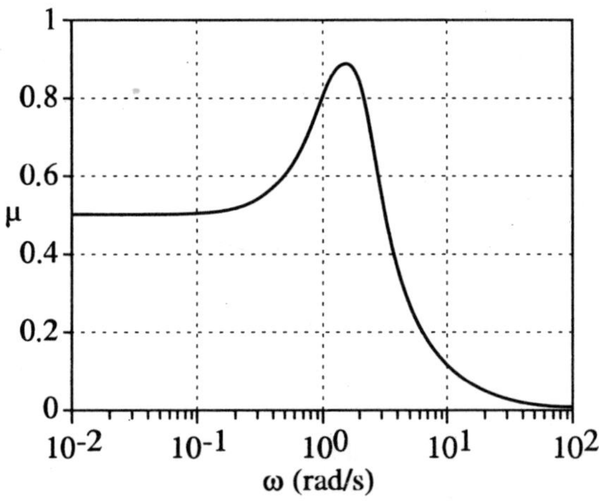

b) Robustness of performance

Figure 15.10. *Upper bounds of the structured single value*

15.2.4. *Evaluation of structured single value*

The calculation of the structured single value is recognized since the beginning as a difficult mathematical problem (except for simple cases, it is part of the problems with non-polynomial complexity). Nevertheless, we know how to search various upper bounds by solving optimization problems under LMI constraints. This means that the following sets of matrices, whose structure is inferred from that of the set S:

$$\mathbf{D} = \left\{ \begin{array}{c} D = diag\left\{ D_1, \cdots, D_r, d_1 I_{n_1}, \cdots, d_q I_{n_q} \right\} \\ D_i \in \mathbf{C}^{r_i \times r_i} \ ; \ d_i \in \mathbf{R} \end{array} \right\}$$
[15.49a]

$$\mathbf{D}_H = \{ D \in \mathbf{D} \ ; \ D = D^* > 0 \}$$
[15.49b]

$$\mathbf{G} = \left\{ \begin{array}{c} G = diag\left\{ G_1, \cdots G_r, O_{n_1}, \cdots, O_{n_q} \right\} \\ G_i \in \mathbf{C}^{r_i \times r_i} \ ; \ G_i = G_i^* \end{array} \right\}$$
[15.49c]

The set $\mathbf{D}$ consists of matrices which can be substituted by any matrix of S: $D\Delta = \Delta D$; the set $\mathbf{D}_H$ is formed of Hermitian matrices positively defined by $\mathbf{D}$; the set $\mathbf{G}$ is formed of Hermitian matrices (not necessarily defined) such that for any matrix of S: $G\Delta = \Delta^* G$: in fact the single non-zero blocks of matrices G correspond to real blocks of Δ .

We then demonstrate the following results [YOU 95, ZHO 96]. A first upper bound is obtained by using only the matrices D:

$$\mu_S(M) \leq \gamma_1^* = \min_{D \in \mathbf{D}} \sqrt{\overline{\lambda}\left(D^{-1} M^* D^2 M D^{-1} \right)}$$
[15.50a]

This first upper bound can be calculated by solving the following optimization problem:

$$\gamma_1^* = \min_{D \in \mathbf{D}} \gamma_1 = \min_{D \in \mathbf{D}_H} \gamma_1 \text{ under the constraints:}$$
[15.50b]

$$\gamma_1 \geq 0$$
[15.50c]

$$M^* D M - \gamma_1^2 D \leq 0$$
[15.50d]

If the matrices $\Delta(s)$ contain real blocks, a more precise upper bound is obtained by using in conjunction the matrices D and G:

$$\mu_S(M) \leq \gamma_2^* \text{ with} \tag{15.51a}$$

$$\gamma_2^* = \min_{\substack{D \in \mathbf{D} \\ G \in \mathbf{G}}} \gamma_2 = \min_{\substack{D \in \mathbf{D}_H \\ G \in \mathbf{G}^H}} \gamma_2 \text{ under the constraints:} \tag{15.51b}$$

$$\gamma_2 \geq 0 \tag{15.51c}$$

$$M^* D M + j(G M - M^* G) - \gamma_2^2 D \leq 0 \tag{15.51d}$$

The interest of these formulations is that, with fixed γ_1 and γ_2, the inequalities [15.50d] and [15.51d] are LMIs, in D or in D and G respectively. The calculation of γ_1^* and γ_2^* can be performed by using regulators dedicated to this type of problem [BAL 93, GAH 95].

The approach usually used to perform a μ-analysis consists of searching an upper bound of $\mu_S(H_{11}(j\omega))$ or $\mu_{S'}(H(j\omega))$ for a previously chosen set of values for ω. If these functions are regular enough (which unfortunately is not always the case [PAC 93]), we obtain quite easily upper bounds of their maximum: based on the results in section 15.2.2, this is the information we need in order to conclude on the issues related to stability robustness or the performances robustness. This approach has been used in the example of section 15.2.3.

15.3. The μ-synthesis

15.3.1. *A H_∞ robust synthesis*

In an interesting manner, the μ-synthesis combines the two previous approaches by searching an answer to the following problem: can we determine a corrector which guarantees that the norm H_∞ of a closed loop control system remains less than a given γ level, *this system being submitted to different pattern uncertainties?*

In order to approach this problem, we consider the block diagram in Figure 15.11, which combines the diagram of H_∞ standard synthesis (Figure 15.1) and that of μ-analysis (Figure 15.7). As in the previous section, the uncertainties $\Delta(s)$ have the general structure:

$$\Delta(s) = \text{diag}\left\{\delta_1 I_{r_1}, \, \ldots \,, \delta_r I_{r_r}, \Delta_1(s), \, \ldots \,, \Delta_q(s)\right\}$$
$$\delta_i \in \mathbf{R} \; ; \; \Delta_i(s) \in \mathbf{R}H_\infty^{\,n_i \times n_i} \tag{15.52}$$

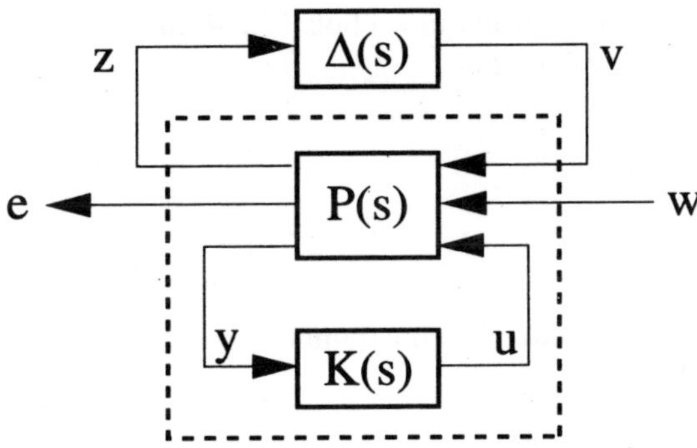

Figure 15.11. *The problem of robust synthesis*

and we shall suppose that each uncertainty has a norm bounded by 1:

$$\begin{cases} \delta_i \in \,]-1\,;+1\,[\,, i = 1, \cdots r \\ \left\|\Delta_i(s)\right\|_\infty < 1, i = 1, \cdots q \end{cases} \Leftrightarrow \left\|\Delta(s)\right\|_\infty < 1 \tag{15.53}$$

Moreover, by supposing that the level γ to be satisfied is equal to 1 (we can always return to this case by integrating the value of γ in the matrix $P(s)$), the problem is the following: to establish a corrector $K(s)$ so that the norm H_∞ of the transfer of w toward e should be less than 1 for any $\Delta(s)$ of type [15.52] such that $\left\|\Delta(s)\right\|_\infty < 1$.

Let $T(s)$ be the transfer between $(v\ w)^T$ and $(z\ e)^T$ of the closed loop system through $K(s)$ (Figure 15.11). Based on the results in section 15.2.2, this property is verified if and only if:

$$\forall\, \omega \in \mathbf{R}\,,\; \mu_{S'}(T(j\omega)) \leq 1 \tag{15.54}$$

where the set $\mathbf{S}'$ is defined as in [15.31]. Establishing a corrector verifying [15.54] is called a *μ-synthesis problem* [ZHO 96]. Unfortunately, except for simple cases, it has no solution known nowadays. Therefore, we have to try to solve it by using an alternative method.

15.3.2. *Approach by D-K iterations*

However, based on the results in section 15.2.4, an inequality of the following type is verified for each value of ω:

$$\mu_{\mathbf{S}'}\big(T(j\omega)\big) \leq \gamma_\omega^* \quad \text{with} \tag{15.55a}$$

$$\gamma_\omega^* = \min_{D_\omega \in \mathbf{D}'} \gamma_\omega \quad \text{under the constraints:} \tag{15.55b}$$

$$\gamma_\omega \geq 0 \tag{15.55c}$$

$$T(-j\omega)^T D_\omega T(j\omega) - \gamma_\omega^2 D_\omega \leq 0 \tag{15.55d}$$

$$\text{and } \mathbf{D}' = \left\{ \begin{array}{c} D = diag\left\{ D_1, \cdots, D_r, d_1 I_{n_1}, \cdots, d_q I_{n_q}, d_{q+1} I_{n_2} \right\} \\ D_i \in \mathbf{C}^{r_i \times r_i} \ ; \ d_i \in \mathbf{R} \end{array} \right\} \tag{15.55e}$$

The set $\mathbf{D}'$ is formed of matrices which can be substituted with any matrix of $\mathbf{S}'$. A more realistic problem consists of searching a corrector $K(s)$ and a stable transfer matrix $D(s)$ so that its reverse $D(s)^{-1}$ is stable and can be substituted with any matrix of $\mathbf{S}'$, such that:

$$\left\| D(s)\, T(s)\, D(s)^{-1} \right\|_\infty < 1 \tag{15.56}$$

In fact, condition [15.56] will ensure that:

$$\forall \omega \;\; \mu_{\mathbf{S}'}\left(T(j\omega)\right) \le \gamma_\omega^* = \min_{D_\omega \in \mathbf{D}'} \; \sqrt{\bar{\lambda}(D_\omega^{-1}\, T(-j\omega)^T\, D_\omega^{\;2}\, T(j\omega)\, D_\omega^{-1})}$$

$$\le \sqrt{\bar{\lambda}(D(j\omega)^{-1}\, T(-j\omega)^T\, D(j\omega)^2\, T(j\omega)\, D(j\omega)^{-1})}$$

$$\le \left\| D(s)\, T(s)\, D(s)^{-1} \right\|_\infty < 1$$

Let us note that the constraint of having $D(s)$ and $D(s)^{-1}$ as stable is necessary in order to have a corrector which stabilizes the transfer matrix which appears in [15.56].

Once more, this problem has no generally known solution, but we can try to solve it by searching alternative matrices $K(s)$ and $D(s)$. In fact:

– calculate fixed $K(s)$ to $D(s)$ is nothing else than a problem of synthesis H_∞, corresponding to the block diagram in Figure 15.12;

– with fixed $K(s)$, the search for $D(s)$ can be conducted by calculating the upper bound [15.55] for a set of previously chosen values of ω and then by interpolating the matrices D_ω obtained by a stable and inversely stable transfer matrix.

These two steps are repeated until the convergences of matrices D_ω or the fulfillment of condition [15.54]. This procedure is named *D-K iteration*.

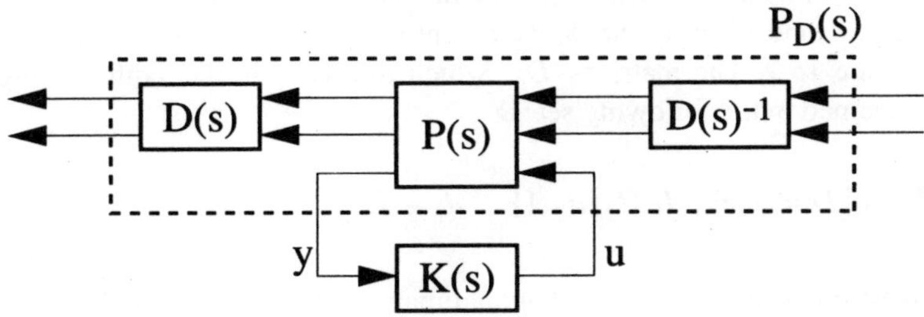

Figure 15.12. *H_∞ standard problem solved during D-K iterations*

We note that if the calculation of $K(s)$ on the one hand and the calculation of each matrix D_ω on the other hand are perfectly solved problems, the procedure convergence is ensured only if $D(s)$ perfectly interpolates the matrices D_ω. In practice, we are limited to a rough interpolation, meaning that we lose any guarantee of the convergence of the *D-K* iteration.

We should also note that the H_∞ synthesis which provides $K(s)$ uses an interconnection matrix $P_D(s)$ which contains the matrices $D(s)$ and $D(s)^{-1}$ (Figure 15.12): the corrector order obtained by the Riccati equations (section 15.1.4) or by LMI (section 15.1.5) will then be equal to the $P_D(s)$ order, which increases along with the order chosen for $D(s)$ during the interpolation of matrices D_ω.

This procedure is obviously heavier insofar as the number of uncertainties taken into account in $\Delta(s)$ is significant. An intelligent use of this technique consists of using it with a limited number of uncertainties and taking care to choose those which are the most penalizing for the synthesis. After that, we shall perform a much more refined analysis of robustness, the μ-analysis procedure developed within section 15.2 not presenting the same inconvenient.

15.3.3. *Example*

Let us recall the example presented in section 15.2.3 in order to illustrate the μ-analysis. We had established that having a constant corrector $K(s) = 2$, the robustness analysis tracked by using the diagram in Figure 15.9, made it possible to guarantee an H_∞ norm less than 0.89 for any value of the parameters verifying inequalities [15.48].

We want to enhance this result by calculating the corrector $K(s)$ by μ-synthesis. For that, we will apply the *D-K* iterations procedure by identifying the transfer $T(s)$ which used to appear during the development of the previous section at the transfer $H(s)$ in Figure 15.9. The matrices D_ω which intervene in the synthesis procedure will be contained in the following set $\mathbf{D}'$:

$$\mathbf{D}' = \left\{ D = diag\left\{ d_1, d_2, d_3, 1 \right\} \; ; \; d_i \in \mathbf{C} \right\} \qquad [15.57]$$

Let us note that we choose the diagonal matrices to limit the number of transfers to determine at the moment of interpolation (whereas the presence of a repeated uncertainty δI_2 would authorize to replace the first two elements with a plain matrix 2×2). Additionally, the procedure does not make a difference between real and unreal uncertainties. These two remarks show that the family of the patterns to be considered forms an over-set of those in which we are directly interested.

The *D-K* iterations have been conducted by choosing to interpolate each element of the matrices D_ω by a first order transfer function and by using the software

[BAL 93]. The maximum of the $\mu_{S'}\left(H(j\omega)\right)$ upper bound, calculated in accordance with the selection performed in [15.57] by considering four complex uncertainties, is established at the end of four iterations on an order value of 0.72.

By recalculating $\mu_{S'}\left(H(j\omega)\right)$ as in section 15.2.3 (i.e. with one real repeated uncertainty and two complex uncertainties), we obtain a hardly different value, namely 0.697 (Figure 15.13.a). We infer from this that the closed loop control system preserves an H_∞ norm less than 0.697 for any value of the parameters verifying the inequalities:

$$\begin{cases} 0.57 < a < 3.43 \\ 0 < \tau < 0.72 \end{cases} \qquad [15.58]$$

The matrix $D(s)$ obtained after the last iteration is given by:

$$D(s) = diag\left\{ \frac{0.685\,s + 1.038}{s + 0.860} ; \frac{19.79\,s + 36.37}{s + 40.72} ; \frac{30.18\,s + 8.36}{s + 41.68} ; 1 \right\} \qquad [15.59]$$

The obtained corrector is of 9th order (equal to the sum of orders of $G(s)$, $W_d(s)$, $D(s)$ and $D(s)^{-1}$), but it can be easily reduced to a 2nd order transfer function whose expression is:

$$K(s) = 1.14 \frac{1 + s/1.062}{\left(1 + s/68.13\right)\left(1 + s/100\right)} \qquad [15.60]$$

The Bode diagram for this corrector is given in Figure 15.13b.

15.4. Synthesis of a corrector depending on varying parameters

15.4.1. *Problem considered and L_2 gain*

The use of a fixed corrector, even "robust", is not conceivable for a process whose parameters vary strongly or rapidly. In this section, we shall suppose that the process to be controlled can be described by a *linear system with variable parameters* (LVP system), whose general form is the following.

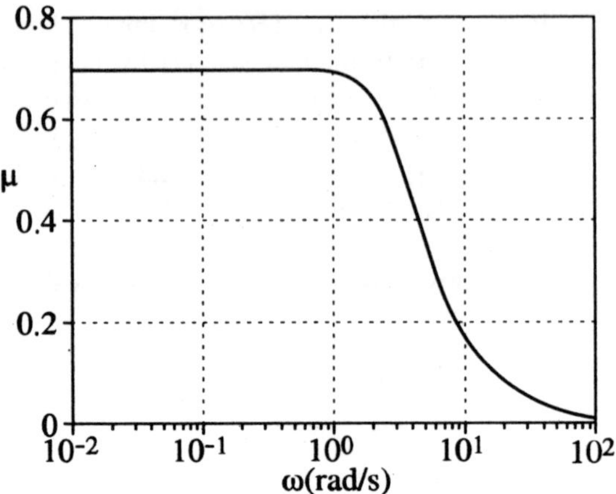

b) Performance robustness

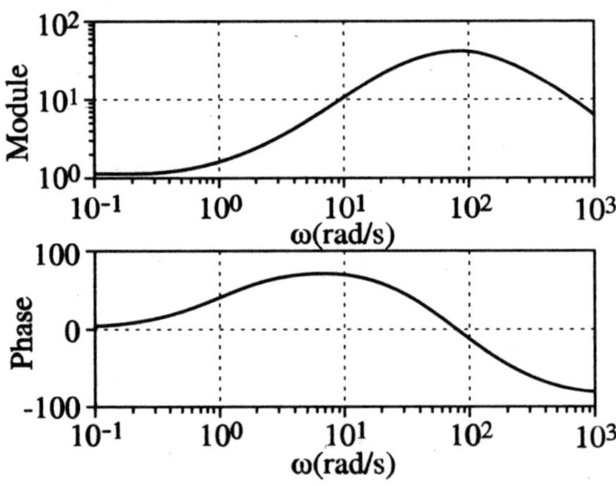

b) Bode diagram of the corrector

Figure 15.13. *Results of μ-synthesis*

$$\mathbf{P}(\theta): \begin{pmatrix} \dot{x}(t) \\ e(t) \\ y(t) \end{pmatrix} = \left(\begin{array}{c|cc} A(\theta) & B_w(\theta) & B_u(\theta) \\ \hline C_e(\theta) & D_{ew}(\theta) & D_{eu}(\theta) \\ C_y(\theta) & D_{yw}(\theta) & D_{yu}(\theta) \end{array} \right) \begin{pmatrix} x(t) \\ w(t) \\ u(t) \end{pmatrix} \qquad [15.61]$$

where $\theta(t) = (\theta_1(t), \theta_2(t), ..., \theta_p(t))^T$ is a vector with time depending parameters, each component of $\theta(t)$ having the possibility to be measured in real-time. This last hypothesis enables us to search for a corrector $\mathbf{K}(\theta)$ in the same form (Figure 15.14).

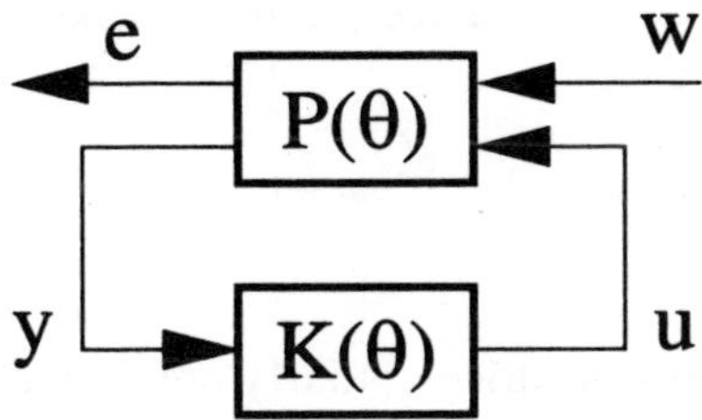

Figure 15.14. *System and corrector depending on parameters*

It is suitable first of all to define a measure for the performances to be reached. For this, we can generalize the concept of H_∞ norm (which is applicable only to invariant systems) in the following way; let $L_2(\mathbf{R}^n)$ be the set of signals $s(t)$ with $\mathbf{R}_+$ in $\mathbf{R}^n$ whose L_2 norm:

$$\| s \|_2 := \sqrt{\int_0^\infty s(t)^T s(t)\, dt} \qquad [15.62]$$

is limited. The L_2 gain of the system in Figure 15.14 is then defined by:

$$\gamma = \sup_{w(t) \in L_2} \frac{\| e \|_2}{\| w \|_2} \qquad [15.63]$$

The synthesis problem we are going to consider is to establish a corrector $\mathbf{K}(\theta)$ depending on the parameters which ensure the stability of the closed loop control system in Figure 15.14 for any possible evolution of $\theta(t)$, while providing to the closed loop control system a L_2 gain between w and e less than a given value y. Further on, we consider two versions of this problem corresponding to two pattern establishments different from the process and from the corrector.

15.4.2. *Polytopic approach*

Let us suppose that each parameter $\theta_i(t)$ could have any value within a range of type $[\underline{\theta}_i \, ; \overline{\theta}_i]$. The vector $\theta(t)$ can then have any value within a section of $\mathbf{R}^P$. We shall note by π_i, $i = 1, \cdots, 2^P$ the peaks of this section. If equations [15.61] of the system to be controlled are connected in $\theta(t)$, each matrix of its state representation evolves within a "polytope" whose peaks are successively obtained by considering $\theta(t) = \pi_i$, $i = 1, \cdots, 2^p$. More specifically, at any moment t, the vector $\theta(t)$ can be expressed as a barycenter of the peaks π_i:

$$\theta(t) = \sum_{i=1}^{2^p} \alpha_i \, \pi_i \ ; \ \ \alpha_i \geq 0 \ ; \ \ \sum_{i=1}^{2^p} \alpha_i = 1 \tag{15.64}$$

and the state representation matrices [15.61] are expressed based on the same coefficients α_i:

$$\left(\begin{array}{c|cc} A(\theta) & B_w(\theta) & B_u(\theta) \\ \hline C_e(\theta) & D_{ew}(\theta) & D_{eu}(\theta) \\ C_y(\theta) & D_{yw}(\theta) & D_{yu}(\theta) \end{array} \right) = \sum_{i=1}^{2^p} \alpha_i \left(\begin{array}{c|cc} A(\pi_i) & B_w(\pi_i) & B_u(\pi_i) \\ \hline C_e(\pi_i) & D_{ew}(\pi_i) & D_{eu}(\pi_i) \\ C_y(\pi_i) & D_{yw}(\pi_i) & D_{yu}(\pi_i) \end{array} \right) \tag{15.65}$$

It then seems possible to search for a corrector:

$$\mathbf{K}(\theta): \begin{pmatrix} \dot{x}_K(t) \\ u(t) \end{pmatrix} = \left(\begin{array}{c|c} A_K(\theta) & B_K(\theta) \\ \hline C_K(\theta) & D_K(\theta) \end{array} \right) \begin{pmatrix} x_K(t) \\ y(t) \end{pmatrix} . \tag{15.66}$$

whose state representation matrices are expressed once more based on the same linear combination:

$$\left(\begin{array}{c|c} A_K(\theta) & B_K(\theta) \\ \hline C_K(\theta) & D_K(\theta) \end{array} \right) = \sum_{i=1}^{2^p} \alpha_i \left(\begin{array}{c|c} A_K(\pi_i) & B_K(\pi_i) \\ \hline C_K(\pi_i) & D_K(\pi_i) \end{array} \right) \tag{15.67}$$

Further on, we shall suppose that the following hypotheses are verified:

H5) $D_{yu}(\theta) = 0$ or in an equivalent manner $D_{yu}(\pi_i) = 0$ for $i = 1, \cdots, 2^p$;

H6) $B_u(\theta)$, $C_y(\theta)$, $D_{eu}(\theta)$ and $D_{yw}(\theta)$ are independent of θ, or in an equivalent manner:

$$B_u(\pi_i) = B_u , \ C_y(\pi_i) = C_y , \ D_{eu}(\pi_i) = D_{eu} , \ D_{yw}(\pi_i) = D_{yw} , i = 1, \cdots, 2^p \tag{15.68}$$

H7) for any possible evolution of $\theta(t)$ within the top peaks π_i, $\left(A(\theta), B_u\right)$ can be quadratically stabilized and $\left(C_y, A(\theta)\right)$ can be quadratically detected, or in an equivalent manner there are real symmetric matrices $X > 0$ and $Y > 0$ satisfying respectively the followings LMIs:

$$\mathbf{N}_u^T \, (A_i^T X + X A_i) \mathbf{N}_u < 0 \, , \ i = 1, \cdots, 2^P \tag{15.69a}$$

$$\mathbf{N}_y^T \, (A_i Y + Y A_i^T) \mathbf{N}_y < 0 \, , \ i = 1, \cdots, 2^P \tag{15.69b}$$

where $\mathbf{N}_u$ and $\mathbf{N}_y$ form a base of the cores of B_u^T and C_y respectively.

The feasibility of the problem presented is tested by using the following theorem [APK 95a].

THEOREM 15.5.– *having the hypotheses H5, H6, H7, there is a corrector stabilizing the system in Figure 15.14 and ensuring an L_2 gain less than y for any possible evolution of $\theta(t)$ within the top peaks π_i, if and only if there are two symmetric matrices R and S verifying the following three matrix inequalities:*

$$\begin{pmatrix} \mathbf{N}_R & 0 \\ 0 & I_{n_w} \end{pmatrix}^T \begin{pmatrix} A(\pi_i)R + RA(\pi_i)^T & R\,C_e(\pi_i)^T & B_w(\pi_i) \\ C_e(\pi_i)R & -\gamma I_{n_e} & D_{ew}(\pi_i) \\ B_w(\pi_i)^T & D_{ew}(\pi_i)^T & -\gamma I_{n_w} \end{pmatrix} \begin{pmatrix} \mathbf{N}_R & 0 \\ 0 & I_{n_w} \end{pmatrix} < 0$$

$$i = 1, \cdots, 2^P \tag{15.70a}$$

$$\begin{pmatrix} \mathbf{N}_S & 0 \\ 0 & I_{n_e} \end{pmatrix}^T \begin{pmatrix} A(\pi_i)^T S + S A(\pi_i) & S B_w(\pi_i) & C_e(\pi_i)^T \\ B_w(\pi_i)^T S & -\gamma I_{n_w} & D_{ew}(\pi_i)^T \\ C_e(\pi_i) & D_{ew}(\pi_i) & -\gamma I_{n_e} \end{pmatrix} \begin{pmatrix} \mathbf{N}_S & 0 \\ 0 & I_{n_e} \end{pmatrix} < 0$$

$$i = 1, \cdots, 2^P \tag{15.70b}$$

$$\begin{pmatrix} R & I_n \\ I_n & S \end{pmatrix} \geq 0 \tag{15.70c}$$

where $\mathbf{N}_R$ *and* $\mathbf{N}_S$ *form a base of cores of* $(B_u{}^T \ D_{eu}{}^T)$ *and* $(C_y \ D_{yw})$ *respectively.* ■

Inequalities [15.70], calculated on inequalities [15.20.a, b, c] of the H_∞ synthesis problem are LMIs in R and S. We note that, unlike the H_∞ problem, the first two inequalities are both replaced by a system of 2^P inequalities which must be simultaneously verified by the same matrix R or S.

After the resolution of this inequality system, the construction of matrices $A_K(\pi_i)$, $B_K(\pi_i)$, $C_K(\pi_i)$, $D_K(\pi_i)$ can be done, for every peak π_i, by following the approach presented in section 15.1.5 for the construction of the H_∞ corrector. The final corrector is hence obtained by the formula [15.67]. It is actually dependant on the evolution of $\theta(t)$ because, at the moment I, we have to infer from $\theta(t)$ the values of the coefficients α_i, from where we infer the corrector matrices. We understand that this operation can become tedious if the number of variable parameters taken into account within the synthesis is increased.

15.4.3. *A more general approach*

Let us take again as departing point equations [15.61]. When the different matrices of this state representation depend on $\theta(t)$ in a rational manner, this system can be presented like in Figure 15.15, where an invariable system described by the matrix $P(s)$ is looped by a matrix $\Theta(t)$ isolating the parameters:

$$\Theta(t) = diag\left(\theta_1(t)I_{n_1}, \ \theta_2(t)I_{n_2}, \ ..., \ \theta_p(t)I_{n_p}\right) \tag{15.71}$$

This operation is similar to the one necessary in order to perform a μ-analysis. We can search for this system a corrector in the same form, i.e. presented as the looping of an invariant system of transfer matrix $K(s)$ and of a response of matrix $\Theta(t)$ (Figure 15.15).

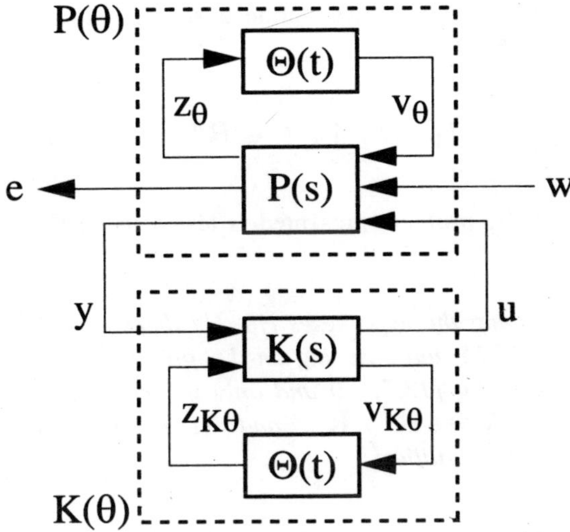

Figure 15.15. *Structures of the system and of the corrector*

Without being any less general, we can always suppose that the elements contained into $\Theta(t)$ have been standardized, so that:

$$\left|\theta_i(t)\right| \le 1 \quad \forall\, i = 1, \cdots, p \tag{15.72}$$

Further on, we shall note the state representation of $P(s)$ in the form:

$$P(s): \begin{pmatrix} \dot{x}(t) \\ z_\theta(t) \\ e(t) \\ y(t) \end{pmatrix} = \left(\begin{array}{c|ccc} A & B_\theta & B_w & B_u \\ \hline C_\theta & D_{\theta\theta} & D_{\theta w} & D_{\theta u} \\ C_e & D_{e\theta} & D_{ew} & D_{eu} \\ C_y & D_{y\theta} & D_{yw} & D_{yu} \end{array}\right) \begin{pmatrix} x(t) \\ v_\theta(t) \\ w(t) \\ u(t) \end{pmatrix} \tag{15.73}$$

and we shall suppose as verified the following hypotheses:

H8) $D_{yu} = 0$;

H9) $\left(A, B_u\right)$ can be stabilized and (C_y, A) can be detected.

As in the case of μ-analysis, we define a set of matrices calculated on the structure of the matrix $\Theta(t)$:

$$\mathbf{L}_H = \left\{ L = diag \left\{ L_1, \cdots, L_p \right\} ; \; L_i \in \mathbf{R}^{n_i \times n_i} ; \; L = L^T > 0 \right\} \qquad [15.74]$$

The feasibility of the problem presented is tested using the following theorem [APK 95b].

THEOREM 15.6.– *having the hypotheses H8, H9 there is an corrector stabilizing the system in Figure 15.15 and ensuring an L_2 gain less than y for any possible evolution of $\theta(t)$ verifying [15.72] if and only if there are two pairs of existing symmetric matrices $(R,S) \in \mathbf{R}^n \times \mathbf{R}^n$ and $(L,J) \in \mathbf{L}_H \times \mathbf{L}_H$, verifying the following three matrix inequalities:*

$$\begin{pmatrix} N_R & 0 \\ 0 & I \end{pmatrix}^T \begin{pmatrix} R\,A^T + A\,R & R\,C_\theta^T & R\,C_e^T & B_\theta\,J & B_w \\ C_\theta\,R & -J & 0 & D_{\theta\theta}\,J & D_{\theta w} \\ C_\theta\,R & 0 & -\gamma I & D_{e\theta}\,J & D_{ew} \\ J\,B_\theta^T & J\,D_{\theta\theta}^T & J\,D_{e\theta}^T & -J & 0 \\ B_w^T & D_{\theta w}^T & D_{ew}^T & 0 & -\gamma I \end{pmatrix} \begin{pmatrix} N_R & 0 \\ 0 & I \end{pmatrix} < 0$$

$$[15.75a]$$

$$\begin{pmatrix} N_S & 0 \\ 0 & I \end{pmatrix}^T \begin{pmatrix} S\,A + A^T S & S\,B_\theta & S\,B_w & C_\theta^T L & C_e^T \\ B_\theta^T S & -L & 0 & D_{\theta\theta}^T L & D_{e\theta}^T \\ B_w^T S & 0 & -\gamma I & D_{\theta w}^T L & D_{ew}^T \\ L\,C_\theta & L\,D_{\theta\theta} & L\,D_{\theta w} & -L & 0 \\ C_e & D_{e\theta} & D_{ew} & 0 & -\gamma I \end{pmatrix} \begin{pmatrix} N_S & 0 \\ 0 & I \end{pmatrix} < 0$$

$$[15.75b]$$

$$\begin{pmatrix} R & I \\ I & S \end{pmatrix} \geq 0 \qquad\qquad [15.75c]$$

$$\begin{pmatrix} L & I \\ I & J \end{pmatrix} \geq 0 \qquad\qquad [15.75d]$$

where $\mathbf{N}_R$ *and* $\mathbf{N}_S$ *form a base of cores of* $(B_2^T \; D_{\theta u}^T \; D_{eu}^T)$ *and* $(C_y \; D_{y\theta} \; D_{yw})$ *respectively.*

Based on matrices R, S, L and J, a corrector responding to the problem can be built following an approach similar to that presented in section 15.1.5 for the H_∞ corrector [APK 95b].

This second approach covers a range of systems wider than that of the polytopic systems because the dependence in $\theta(t)$ is supposed to be rational (and not necessarily connected). Moreover, it does not require expressing $\theta(t)$ at each moment as the baric center of different peaks of its evolution field. On the other hand, it requires putting the system in the form of Figure 15.15. The resolution uses two additional matrices, but the number of LMIs to be verified is smaller.

15.4.4. *Example*

We consider a system described by the following state equations:

$$\begin{cases} \dot{x}(t) = \begin{pmatrix} 0 & 4/7 \\ 0 & -\delta/0.015 \end{pmatrix} x(t) + \begin{pmatrix} 0 \\ 420\,\delta/0.015 \end{pmatrix} v(t) \quad ; \ 0.1 \le \delta \le 2 \\ z(t) = \begin{pmatrix} 1 & 0 \end{pmatrix} x(t) \end{cases} \qquad [15.76]$$

This system corresponds to an integrator followed by a first order transfer whose time constant varies between 7.5 and 150 ms.

Taking into account the range of values of δ given below, we are going to search for a corrector which minimizes the L_2 gain of the system described in the block diagram in Figure 15.16. On this appear two exogenous inputs w_1, w_2 which act, one in the system's output (and which represents, for example, a setting input) and the other at the input and two outputs e_1, e_2 which correspond to the feedback error and to the control (this type of problem is rather usual within H_∞ synthesis [DUC 99]).

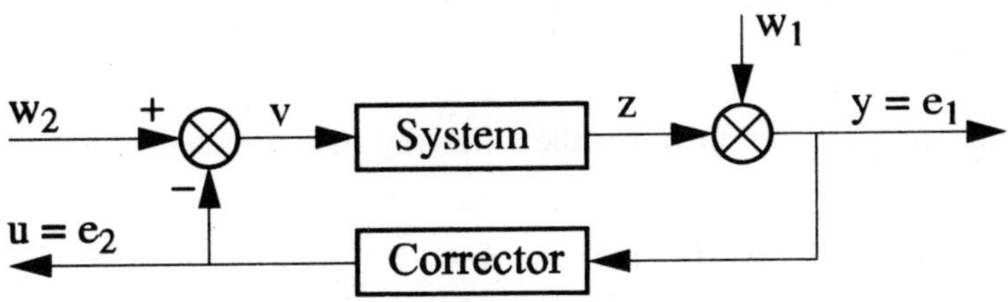

Figure 15.16. *Synthesis problem considered*

The system being polytopic, we can apply the approach of section 15.4.2, by defining the peaks π_1 and π_2 from two extreme values of δ. Certainly, it does not verify hypotheses [15.68] but we can modify the problem by adding at the system input a filter of negligible constant time T_f:

$$\begin{cases} \dot{x}(t) = \begin{pmatrix} 0 & 4/7 & 0 \\ 0 & -\delta/0.015 & 420\,\delta/0.015 \\ 0 & 0 & -1/T_f \end{pmatrix} x(t) + \begin{pmatrix} 0 \\ 0 \\ 1/T_f \end{pmatrix} v(t) \\ z(t) = \begin{pmatrix} 1 & 0 & 0 \end{pmatrix} x(t) \end{cases} \qquad [15.77]$$

Further on, we have considered $T_f = 0.1\,\text{ms}$. The minimization of γ under constraints [15.70], based on the software [GAH 95], leads to the value $\gamma = 9.37$. The correctors corresponding to each peak are given by·

$$\left(\begin{array}{c|c} A_{K1} & B_{K1} \\ \hline C_{K1} & D_{K1} \end{array} \right) = \left(\begin{array}{ccc|c} 3.34\,10^2 & -3.37\,10^2 & 1.84\,10^3 & -2.25\,10^3 \\ -6.06\,10^3 & -3.46\,10^4 & -3.37\,10^5 & -37.8 \\ -7.82\,10^4 & -2.17\,10^5 & -2.44\,10^6 & 2.34\,10^5 \\ \hline -0.600 & -2.41 & -32.8 & 0 \end{array} \right)$$

$$\left(\begin{array}{c|c} A_{K2} & B_{K2} \\ \hline C_{K2} & D_{K2} \end{array} \right) = \left(\begin{array}{ccc|c} 3.93\,10^2 & -6.65\,10^3 & 9.55\,10^3 & -2.25\,10^3 \\ -5.74\,10^3 & -3.47\,10^4 & -3.18\,10^5 & -37.8 \\ -8.08\,10^4 & 1.96\,10^5 & -2.87\,10^6 & 2.34\,10^5 \\ \hline -0.600 & -2.41 & -32.8 & 0 \end{array} \right)$$

$$[15.78]$$

The parameter δ is expressed according to the peaks 0.1 and 2 by the expression:

$$\delta = 0.1\alpha + 2(1-\alpha) \;\Leftrightarrow\; \alpha = \frac{2-\delta}{1.9} \qquad [15.79]$$

the parameterized corrector in δ has the following state representation:

$$\left(\begin{array}{c|c} A_K & B_K \\ \hline C_K & D_K \end{array} \right) = \frac{2-\delta}{1.9} \left(\begin{array}{c|c} A_{K1} & B_{K1} \\ \hline C_{K1} & D_{K1} \end{array} \right) + \frac{\delta-0.1}{1.9} \left(\begin{array}{c|c} A_{K2} & B_{K2} \\ \hline C_{K2} & D_{K2} \end{array} \right) \qquad [15.80]$$

Figure 15.17a shows the response of the output $z(t)$ and of the control $v(t)$ to an amplitude interval -1 on the input w_1, the parameter δ evolving throughout this response in accordance with the relation $\delta(t) = 1.05 + 0.95\cos(50t)$. We note that, despite a very brutal variation of δ during this transient, the response is very well damped and the control is very smooth.

Figure 15.17b shows the output in response to the same input signal, this time for constant values of δ, which are contained between 0.1 and 2. We note that the various responses are relatively homogenous despite this strong parametrical dispersion. A synthesis of a fixed corrector, even with all issues identical, does not make it possible to obtain this result.

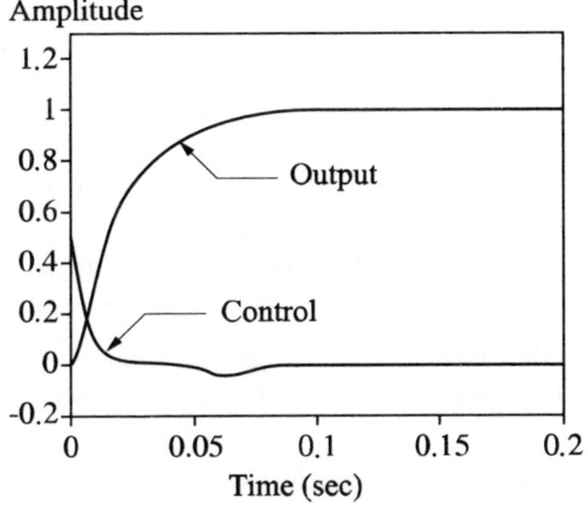

a) Response with parametric variation

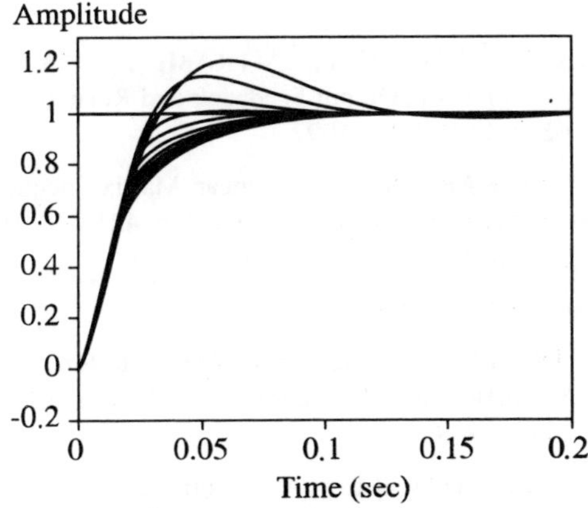

b) Fixed parameter response

Figure 15.17. *Unit-step responses with LVP corrector*

15.5. Conclusion

In this chapter, we noted that the concepts inferred from the H_∞ norm connected with the matrix inequalities make it possible to implement very strong analysis and synthesis tools. The developments presented here have many extensions within the contemporary works. See [DUC 98, DUC 99] for a more complete bibliography on this subject.

15.6. Bibliography

[APK 95a] APKARIAN P., GAHINET P., BECKER G., "Self-scheduled H_∞ Control of Linear Parameter-varying Systems: A Design Example", *Automatica*, vol. 31, p. 1251-1261, 1995.

[APK 95b] APKARIAN P., GAHINET P., "A Convex Characterization of Gain-scheduled H_∞ Controllers", *IEEE Trans. Autom. Control*, AC 40, p. 853-864, 1995.

[BAL 93] BALAS G., DOYLE J.C., GLOVER K., PACKARD A., SMITH R., *μ-Analysis and Synthesis Toolbox*, The Math Works Inc., 1993.

[BOY 94] BOYD S., EL GHAOUI L., FÉRON E., BALAKRISHNAN V., *Linear Matrix Inequalities in Systems and Control Theory*, SIAM Publications, 1994.

[DAV 94] DAVID J., DE MOOR B., "Designing Reduced Order Output Feedback Controllers Using a Potential Reduction Method", *American Control Conf.*, p. 845-849, Baltimore, 1994.

[DUC 98] DUC G., "Panorama des Principales Approches Relevant du Cadre H_∞", *Traitement du Signal*, vol. 15, p. 627-631, 1998.

[DUC 99] DUC G., FONT S., *Commande H_∞ et μ-analyse: des outils pour la robustesse*, Hermès, 1999.

[ELG 97] EL GHAOUI L., OUSTRY F., AITRAMI M., "A Cone Complementary Linearization Algorithm for Static Output Feedback and Related Problems", *IEEE Trans. Autom. Control*, AC 42, p. 1171-1175, 1997.

[GAH 94] GAHINET P., APKARIAN P., "A Linear Matrix Inequality Approach to H_∞ Control", *Int. J. of Robust & Nonlinear Contr.*, vol. 4, p. 421-448, 1994.

[GAH 95] GAHINET P., NEMIROVSKI A., LAUB A.J., CHILALI M., *LMI Control Toolbox*, The Math Works Inc., 1995.

[GLO 88] GLOVER K., DOYLE J.C., "State-Space Formulae for all Stabilizing Controllers That Satisfy an H_∞-Norm Bound and Relations to Risk Sensitivity", *Systems & Control Letters*, vol. 11, p. 77-172, 1988.

[IWA 94] IWASAKI T., SKELTON R.E., "All Controllers for the General H_∞ Control Problem: LMI Existence Conditions and State-Space Formulas", *Automatica*, vol. 30, p. 1307-1317, 1994.

[PAC 93] PACKARD A., PANDEY P., "Continuity Properties of the Real/Complex Structured Singular Value", *IEEE Trans. Autom. Control*, AC 38, p. 415-428, 1993.

[VAL 99] VALENTIN-CHARBONNEL C., DUC G., LE BALLOIS S., "Low-Order Robust Attitude Control of an Earth Observation Satellite", *Control Engineering Practice*, vol. 7, p. 493-506, 1999.

[YOU 95] YOUNG P.M., NEWLIN M.P., DOYLE J.C., "Computing Bounds for the Mixed μ Problem", *Int. J. of Robust & Nonlinear Control*, vol. 5, p. 573-590, 1995.

[ZHO 96] ZHOU K., DOYLE J.C., GLOVER K., *Robust and Optimal Control*, Prentice Hall, 1996.

Chapter 16

Linear Time-Variant Systems

The complexity of the physical phenomena studied cannot be reduced to only one modeling by linear dynamic systems with constant coefficients. These models are sometimes poorly adapted because, for example, they can only deal with magnitudes having an exponentially decreasing correlation. However, in fields full of variety such as hydrology [HUR 65], electronics [VAN 88], traffic [RIE 97, WIL 95], electrical engineering [CHA 81] and mechanics [CLE 98, ZHU 96], there are many situations that generate behaviors that do not obey these quite simple models. Therefore, in the last 30 years, new analysis models and tools have appeared. In this perspective, one of the goals of research is to extend the class of linear dynamic systems by including those for which the coefficients vary in time. These variations can be divided into two classes. The first class concerns the sudden non-stationarities or "failures" characterized by time intervals where the coefficients are constant. The non-stationarity is due only to the presence of instantaneous shifts in their values. This modeling is found, for example, in the field of monitoring and diagnostic [BAS 93]. As such, the problem is to essentially detect the instants of change as well as the amplitude of the parametric shifts. The second class pertains to the systems where the coefficients are functions of time. When these dynamics are "slow" with respect to those of the system, they can be dealt with through adaptive techniques.

However, there are also many cases in which the evolution of parameters is "fast" (T-periodic systems [RAB 92], auto-similar systems [GUG 01], etc.). This last category makes the development and the implementation of specific methods in both the control and identification fields indispensable. Firstly, it is important to have the basic mathematical tools indispensable to their analysis.

This chapter is dedicated to the analysis of the dynamic systems required by the linear differential equations with time-variant coefficients. The approach presented consists of an approach parallel to that adopted for the constant coefficient systems. The Laplace transform, even if it can be always applied to the input/output magnitudes of the system, can no longer be used in order to define the transfer function of these systems.

However, this transfer concept can, despite everything, be extended to the non-stationary linear differential systems provided they operate on the non-commutative body of rational fractions. This body is isomorphic to the group generated by the non-stationary linear dynamic systems. Hence, we can elaborate the composition rules of these systems with the help of the algebraic rules applied to the transfer functions. The results obtained can be used in order to solve control or/and identification problems.

This chapter deals, in the first place, with the construction of the non-commutative polynomial ring and of the body of related rational fractions. As in the traditional case, the relation between the basic properties of dynamic systems (stability, etc.) and the characteristics of the elements of the body of fractions (poles, etc.) can be established. The second part pertains to the construction of the systems: serialization or/and parallelization. It is possible, for each association diagram, to write the transfer function of the system composed with the help of simple algebraic rules. Finally, based on these results, two applications illustrate the use of the results obtained. The first one concerns the modeling of multi-component polynomial phase signals and the second is dedicated to the design of a pole placement control law.

16.1. Ring of non-commutative polynomials

Let $\Pi(\lambda)$ be the set of polynomials of degree n:

$$\Pi(\lambda) = \left\{ P(\lambda) = a_n(t)\lambda^n + a_{n-1}(t)\lambda^{n-1} + \ldots + a_0(t) \big| a_i(t) \in K \; \forall i \in [1, n] \right\}$$

Let K be a differential body, i.e. a body on which is defined a derivation operator $da/dt \; \forall a(t) \in K$ (noted from now by $\dot{a}$) which will satisfy the traditional derivation properties:

$$(a+b)^{\cdot} = \dot{a} + \dot{b} \quad \text{and} \quad \dot{ab} = \dot{a}\,b + a\,\dot{b}$$

when coefficient $a_n(t)$ is equal to 1, polynomial $P(\lambda)$ is standardized.

The set $\Pi(\lambda)$ including addition and multiplication which satisfies:

$$\forall a(t) \in K, \ \lambda a(t) = a(t)\lambda + \dot{a}(t)$$

has a non-commutative ring structure [ORE 33].

16.1.1. *Division and the right highest divisor (RHD)*

$$\forall \ P_1(\lambda), P_2(\lambda) \in \Pi(\lambda) \otimes \Pi(\lambda)$$

insofar as $n_{P_1} \geq n_{P_2}$ (where n_X represents the degree of $X(\lambda)$) there is a unique pair of polynomials $[Q(\lambda), R(\lambda)]$ so that:

$$P_1(\lambda) = Q(\lambda)P_2(\lambda) + R(\lambda) \text{ with } n_R < n_{P_2}$$

We can infer Euclid's division algorithm:

$$P_1(\lambda) = Q_1(\lambda)P_2(\lambda) + P_3(\lambda)$$

$$\dots\dots \qquad \dots\ \dots$$

$$P_{n-1}(\lambda) = Q_{n-1}(\lambda)P_{n-2}(\lambda) + P_n(\lambda)$$

The RHD of $P_1(\lambda)$, $P_2(\lambda)$ is then defined as the standardized polynomial resulted from $P_n(\lambda)$.

16.1.2. *Right least common multiple (RLCM)*

It is then possible to define the RLCM of $P_1(\lambda), P_2(\lambda)$ as the lowest degree standardized polynomial divisible on the right by both $P_1(\lambda)$ and $P_2(\lambda)$.

$$M(\lambda) = Q_1(\lambda)P_1(\lambda) = Q_2(\lambda)P_2(\lambda)$$

Generally, the existence of the Euclidian division implies the existence of the RHD [ORE 33].

16.1.3. *Explicit formulation of RLCM*

Based on all the factors of the Euclidian division, it is possible to express the RLCM:

$$M(\lambda) = \alpha[P_{n-1}(\lambda)]\,[P_n(\lambda)]^{-1}\,[P_{n-2}(\lambda)]\,[P_{n-1}(\lambda)]^{-1}....\,[P_2(\lambda)][P_3(\lambda)]^{-1}[P_1(\lambda)]$$

The constant $\alpha(t)$ (which does not depend on λ) is such that polynomial $M(\lambda)$ is normalized.

NOTE.– the product of two polynomials $P_1(\lambda)P_2(\lambda)$ cannot generally be divided on the right by $P_1(\lambda)$ which makes the use of all the terms of the Euclidian division compulsory in the expression above.

$M(\lambda)$ is a polynomial and writing it in the inverse form is basically a useful notation.

By applying the same approach, we can define the left highest divisor and the left least common multiple.

16.1.4. *Factoring, roots, relations with the coefficients*

Any polynomial $P(\lambda)$ can be factorized in the general form:

$$P(\lambda) = \alpha(t)(\lambda - p_1(t))(\lambda - p_2(t))....(\lambda - p_{n-1}(t))(\lambda - p_n(t))$$

The roots of $P(\lambda)$ are provided by the solutions of equation $P(\lambda) = 0$.

The relation between the roots and the coefficients of $P(\lambda)$ is a non-linear differential equation [KAM 88]:

$$S^{n-1}p_n(t) + \sum_{2}^{n-1} a_i(t)S^{i-1}p_n(t) + a_1(t)S^{n-1}p_n(t) + a_0(t) = 0$$

where S is the operator defined by:

$$Sp_n(t) = p_n^2(t) + \dot{p}_n(t)$$
$$S^i p_n(t) = p_n(t)S^{i-1} + d(S^{i-1}p)/dt \quad \forall i \geq 2$$

EXAMPLE 16.1.– let us consider the second degree polynomial:

$$P(\lambda) = \lambda^2 + a_1(t)\lambda + a_0(t) = (\lambda - p_1(t))(\lambda - p_2(t))$$

From the following relations:

$$p_1(t) + p_2(t) = -a_1(t) \text{ and } p_1(t)p_2(t) - \dot{p}_2(t) = a_0(t)$$

we infer that the roots are solutions of:

$$p_2^2(t) + \dot{p}_2(t) + a_1(t)p_2(t) + a_0(t) = 0$$

Particular case: $a_1(t) = a_0(t) = 0 \Rightarrow P(\lambda) = \lambda^2 = 0$

$$\Rightarrow p_{21}(t) = p_{22}(t) = 0 \quad \text{and} \quad p_{21}(t) = -\frac{1}{t+k}, p_{22}(t) = \frac{1}{t+k} \quad \forall k \in \Re$$

Hence, the factoring of $P(\lambda)$:

$$P(\lambda) = \lambda^2 = (\lambda + \frac{1}{t+k})(\lambda - \frac{1}{t+k}) \quad \forall k \in \Re \quad .$$

16.2. Body of rational fractions

In general it is not possible to define a body of rational functions from polynomials for which the unknown factor and the parameters cannot be switched (*skew*). However, if the polynomials verify the two following conditions called ORE [AMI 54]:

$$\forall\ P_1(\lambda), P_2(\lambda) \in \Pi(\lambda) \otimes \Pi(\lambda)$$
$$\exists\ \hat{P}_1(\lambda), \hat{P}_2(\lambda) \quad \text{so that } \hat{P}_2(\lambda)P_1(\lambda) = \hat{P}_1(\lambda)P_2(\lambda)$$

(condition on the left)

and:

$$\forall\ P_1(\lambda), P_2(\lambda) \in \Pi(\lambda) \otimes \Pi(\lambda)$$
$$\exists\ \tilde{P}_1(\lambda), \tilde{P}_2(\lambda) \ \text{ so that }\ P_1(\lambda)\tilde{P}_1(\lambda) = P_2(\lambda)\tilde{P}_2(\lambda)$$

(condition on the right)

it is possible to consider the set:

$$F(\lambda) = \left\{ P(\lambda)Q(\lambda)^{-1}, \forall\, P(\lambda) \in \Pi(\lambda),\ \forall Q(\lambda) \in \Pi^*(\lambda)\ \right\} =$$
$$\left\{ Q(\lambda)^{-1}P(\lambda), \forall\, P(\lambda) \in \Pi(\lambda),\ \forall Q(\lambda) \in \Pi^*(\lambda)\ \right\}$$

where $\Pi^*(\lambda) = \Pi(\lambda) - \{0\}$ which has a body structure [ZHU 89].

NOTE.– the inverse of polynomial $P(\lambda)$ is unique and verifies:

$$P(\lambda)P^{-1}(\lambda) = P(\lambda)P^{-1}(\lambda) = 1$$

16.3. Transfer function

In the case of linear systems with constant coefficients, the transfer function is defined as the ratio between the Laplace transforms of the output and those of the input. When the coefficients are time functions, it is possible to extend this concept of transfer function which preserves certain properties obtained in the traditional case even if it does not represent any longer the ratio between the Laplace transforms of the pair input/output.

Let $\sum(\frac{d}{dt})$ be the set of n degree single-variable systems described by the linear differential equation with variable coefficients belonging to a derivable body K:

$$y^{(n)}(t)+a_1(t)y^{(n-1)}(t)+\cdots+a_n(t)y^{(0)}(t) = b_0 u^{(n)}(t)+b_1(t)u^{(n-1)}(t)+\cdots+b_n(t)u^{(0)}(t)$$

with $y^{(i)}(t) = d^i y(t)/dt^i$, $u^{(i)}(t) = d^i u(t)/dt^i$

It is easy to show that this set, consisting of two internal operations (serialization and parallelization) is a body. Hence, the application of $\sum(\frac{d}{dt})$ [1] on $F(\lambda)$ is an isomorphism.

Therefore, system Σ can be formally described by its transfer function:

$$H(\lambda) = Q(\lambda)^{-1} P(\lambda)$$

where:

$$Q(\lambda)=\lambda^n +a_1(t)\lambda^{n-1} +\cdots+ a_n(t)\ ,\ P(\lambda) =b_0\lambda^n +b_1(t)\lambda^{n-1} +\cdots+ b_n(t)$$

16.3.1. *Properties of transfer functions*

THEOREM 16.1.– *two transfer functions $H_1(\lambda)$ and $H_2(\lambda)$ are equivalent if and only if there is a polynomial $D(\lambda)$ such that [KAM 88]:*

$$P_1(\lambda) = D(\lambda)P_2(\lambda) \text{ and } Q_1(\lambda) = D(\lambda)Q_2(\lambda)$$

The transfer function $H_1(\lambda)$ is said to be minimal if the numerator and denominator are first on the right.

---

1 Henceforth, $\Sigma(\frac{d}{dt})$ will be simply noted by Σ.

16.3.2. *Normal modes*

If we make the homogenous equation:

$$y^{(n)}(t)+a_1(t)y^{(n-1)}(t)+\cdots+a_n(t)y^{(0)}(t)=0$$

correspond to the polynomial equation $Q(\lambda)=0$, we show that if $q_n(t)$ is a root of $Q(\lambda)$, then $e^{\int_0^t q_n(\tau)d\tau}$ is a normal mode of the system.

EXAMPLE 16.2.– for $y^{(2)}(t)=0$ which corresponds to:

$$(\lambda+\tfrac{1}{t+k})(\lambda-\tfrac{1}{t+k})\quad \forall k\in\Re=0.$$

A root $\tfrac{1}{t+k}$ provides the normal mode: $e^{\int_0^t 1/(\tau+k)d\tau}=t+k$ (well known).

16.3.3. *Stability*

The solutions of the homogenous equation form a vector space of size smaller or equal to its n degree [AMI 54, ZHU 89].

The general solution is written:

$$y(t)=\sum_i c_i e^{\int_0^t q_i(\tau)d\tau}$$

We infer that the system is stable if:

$$\lim_{t\to\infty}\Re(q_i(t))<0\quad \forall i$$

where $\Re$ means real part.

16.4. Algebra of non-stationary linear systems

A major interest in the transfer function is due to the possibility of easily calculating the transfer function of associated systems, either serially or in parallel.

16.4.1. *Serial systems*

Let Σ_1 and Σ_2 be two systems of transfer functions $H_1(\lambda) = Q_1(\lambda)^{-1} P_1(\lambda)$ and $H_2(\lambda) = Q_2(\lambda)^{-1} P_2(\lambda)$ respectively; the transfer function of the system Σ obtained by serializing Σ_1 and Σ_2 can be calculated as follows:

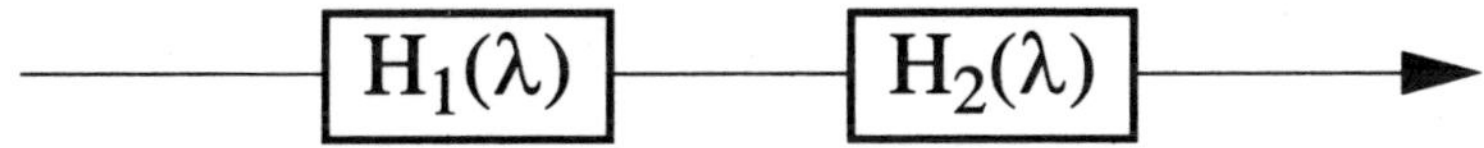

Let $M(\lambda)$ be the RLCM of $P_2(\lambda)$ and of $Q_1(\lambda)$:

$$M(\lambda) = \hat{P}_2(\lambda)P_2(\lambda) = \hat{Q}_1(\lambda)Q_1(\lambda)$$

then we have:

$$H(\lambda) = H_2(\lambda)H_1(\lambda) = Q_2^{-1}(\lambda)P_2(\lambda)Q_1^{-1}(\lambda)P_1(\lambda)$$
$$= Q_2^{-1}(\lambda)\hat{P}_2^{-1}(\lambda)\hat{P}_2(\lambda)P_2(\lambda)Q_1(\lambda)^{-1}P_1(\lambda)$$
$$= Q_2^{-1}(\lambda)\hat{P}_2^{-1}(\lambda)M(\lambda)Q_1(\lambda)^{-1}P_1(\lambda)$$
$$= Q_2^{-1}(\lambda)\hat{P}_2^{-1}(\lambda)\hat{Q}_1(\lambda)Q_1(\lambda)Q_1(\lambda)^{-1}P_1(\lambda)$$

and finally:

$$H(\lambda) = [P_2(\lambda)Q_2(\lambda)]^{-1}\hat{Q}_1(\lambda)P_1(\lambda)$$

EXAMPLE 16.3.

If $\Sigma_1 \dot{y}_1(t) + 1/t\, y_1(t) = \dot{u}(t)$ and $\Sigma_2 \dot{y}_2(t) - 1/t\, y_2(t) = y_1(t)$

$$H_1(\lambda) = Q_1^{-1}(\lambda)P_1(\lambda) = (\lambda + 1/t)^{-1}\lambda \quad H_2(\lambda) = Q_2^{-1}(\lambda)P_2(\lambda) = (\lambda - 1/t)^{-1}$$

we obtain: $M(\lambda) = \hat{P}_2(\lambda)P_2(\lambda) = \hat{Q}_1(\lambda)Q_1(\lambda) = \lambda + 1/t$ and $H(\lambda) = \lambda^{-1}$

so $\Sigma \quad \dot{y}(t) = u(t)$

16.4.2. *Parallel systems*

Let Σ_1 and Σ_2 be two systems of transfer functions $H_1(\lambda) = Q_1(\lambda)^{-1}P_1(\lambda)$ and $H_2(\lambda) = Q_2(\lambda)^{-1}P_2(\lambda)$ respectively; the transfer function of the system Σ obtained by putting Σ_1 and Σ_2 in parallel is provided by:

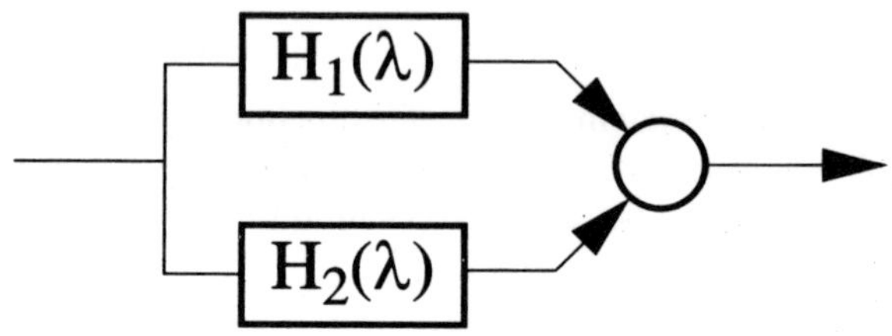

Let $M(\lambda)$ be the RLCM of $Q_1(\lambda)$ and $Q_2(\lambda)$:

$$M(\lambda) = \hat{Q}_1(\lambda)Q_1(\lambda) = \hat{Q}_2(\lambda)Q_2(\lambda)$$

then we simply have:

$$H(\lambda) = H_1(\lambda) + H_2(\lambda) = Q_1^{-1}(\lambda)P_1(\lambda) + Q_2^{-1}(\lambda)P_2(\lambda)$$
$$H(\lambda) = M^{-1}(\lambda)(\hat{Q}_1(\lambda)^{-1}P_1(\lambda) + \hat{Q}_2(\lambda)^{-1}P_2(\lambda))$$

EXAMPLE 16.4.

$$\Sigma_1 \quad \dot{y}_1(t) + 1/t\, y_1(t) = u(t) \text{ and } \Sigma_2 \quad \dot{y}_2(t) - 1/t\, y_2(t) = u(t)$$

$$H_1(\lambda) = Q_1^{-1}(\lambda)P_1(\lambda) = (\lambda + 1/t)^{-1} \; H_2(\lambda) = Q_2^{-1}(\lambda)P_2(\lambda) = (\lambda - 1/t)^{-1}$$

gives: $M(\lambda) = \lambda(\lambda + 1/t) = (\lambda + 2/t)(\lambda - 1/t) = \lambda^2 + 1/t\,\lambda + 2/t$

and $H(\lambda) = (\lambda^2 + 1/t\,\lambda + 2/t)^{-1}2(\lambda + 1/t)$

or: $\Sigma \quad \ddot{y}(t) + \dfrac{1}{t}\dot{y}(t) + \dfrac{2}{t}y(t) = 2(\dot{u}(t) + \dfrac{1}{t}u(t))$

16.5. Applications

In this section, two types of usage of this algebra are presented in the field of modeling and control.

16.5.1. *Modeling*

One of the methods of signal and control processing consists of designing models capable of representing the magnitudes in question. Hence, models MA, AR, ARMA with constant coefficients have been very successful due to their capability to characterize a sufficiently large variety of dynamic behaviors. Their properties were the object of numerous studies and their applications are extremely diversified. However, these models are sometimes insufficient. A current approach is, similarly to the one that facilitated the traditional models MA, AR and ARMA, the research for new models taking into account the highly non-stationary character.

We can illustrate this requirement for the modeling of polynomial phase signals, which we frequently encounter in physics, especially for processing signals coming from radar, sonar, etc. These signals are non-stationary with frequency characteristics that continuously evolve in time with variation speeds that may be significant. The dynamic model of the single-component signal is very easy to establish, whereas that of the multi-component signal is very complex. However, this is indispensable when we deal, for example, with the problem of multiple trajectories due to reflections. The algebra developed here makes it possible to create the complete dynamic model.

Let us consider the following multi-component signal:

$$y(t) = \sum_{i=1}^{n} y_i(t) = \sum_{i=1}^{n} a_i e^{j(\frac{\alpha}{2}t^2 + \beta_i t + \gamma_i)}$$

It corresponds to a signal ("chirp") recorded from a main trajectory to which are associated $n - 1$ reflections.

For each component $y_i(t)$, it is easy to obtain the differential equation that gives it:

$$\ddot{y}_i(t) - \frac{\alpha}{\alpha t + \beta_i} \dot{y}_i(t) + (\alpha t + \beta_i)^2 y_i(t) = 0 \quad \forall i \in [1, n]$$

However, obtaining the one that governs the sum $y(t)$ requires the use of the results presented in the previous section.

Formally, from:

$$\left[\lambda^2 - \frac{\alpha}{\alpha t + \beta_i}\lambda + (\alpha t + \beta_i)^2\right]y_i(t) = [P_i(\lambda)]y_i(t) = 0 \quad \forall i \in [1,n]$$

we obtain for $y(t)$:

$$[M(\lambda)]y(t) = 0 \quad \text{with} \quad M(\lambda) = P.P.C.M.D.(P_i(\lambda))$$

16.5.2. *Pole placement*

Let us consider the system (Σ) written by its transfer function $H(\lambda) = A^{-1}(\lambda)B(\lambda)$, is there a looping defined by $Q^{-1}(\lambda)P(\lambda)$ so that the closed loop transfer has a dynamics set *a priori* by a polynomial $C(\lambda)$?

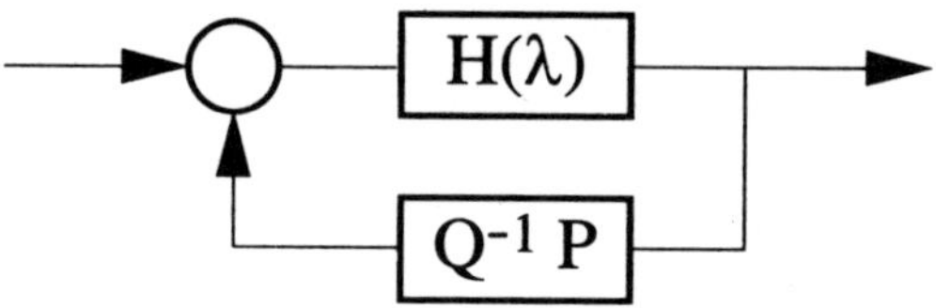

From $A(\lambda)y = B(\lambda)u$, $Q(\lambda)r = P(\lambda)y$ and $u = -r + y_c$ we obtain:

$$A(\lambda)y = B(\lambda)u$$
$$Q(\lambda)u = -P(\lambda)y + Q(\lambda)y_c$$

The solution to this problem goes through an intermediary problem which consists of seeking the controller in a factorized form:

$$\tilde{P}(\lambda)\tilde{Q}^{-1}(\lambda)$$

which is connected to the initial factorization by:

$$Q(\lambda)\tilde{P}(\lambda) = P(\lambda)\tilde{Q}(\lambda)$$

which leads to:

$$(A(\lambda)\tilde{Q}(\lambda) + B(\lambda)\tilde{P}(\lambda))y = A(\lambda)\tilde{Q}(\lambda)y_c$$

In order for the closed loop poles to be given by $C(\lambda)$, it is necessary that:

$$C(\lambda) = A(\lambda)\tilde{Q}(\lambda) + B(\lambda)\tilde{P}(\lambda)$$

with: $P(\lambda)\tilde{Q}(\lambda) = Q(\lambda)\tilde{P}(\lambda)$

Which leads to the following algorithm:

– solution of the Diophantine equation;

– search of RLCM $M(\lambda)$ of $\tilde{Q}(\lambda), \tilde{P}(\lambda)$;

– $P(\lambda)$ and $Q(\lambda)$ are quotient polynomials of $M(\lambda)$ by $\tilde{Q}(\lambda)$ and $\tilde{P}(\lambda)$ respectively.

The resolution of the Diophantine equation is done the same way as in the case where polynomials can be switched [KUC 79].

EXAMPLE 16.5.– let us consider the system Σ described by the differential equation:

$$\ddot{y} + t\dot{y} + y = \dot{u} + u$$

We choose $C(\lambda) = (\lambda + 1)^3$

The previous algorithm applied to:

$$(\lambda^2 + t\lambda + 1)\tilde{Q}(\lambda) + (\lambda + 1)\tilde{P}(\lambda) = (\lambda + 1)^3$$

leads to:

$$\tilde{Q}(\lambda) = \lambda + \frac{t}{t-1} \quad \text{and} \quad \tilde{P}(\lambda) = \frac{t^2 - 3t + 3}{(1-t)}\lambda + \frac{1}{(1-t)^2}$$

And the condition:

$$Q(\lambda)\tilde{P}(\lambda) = P(\lambda)\tilde{Q}(\lambda)$$

finally gives:

$$Q(\lambda) = \lambda - \frac{t^4 - 4t^3 + 6t^2 - 2t + 5}{(1-t)(t^3 - 3t^2 + 3t + 1)}$$

$$P(\lambda) = -\frac{t^2 - 3t + 3}{t^3 - 3t^2 + 3t + 1}\left((t^2 - 3t + 3)\lambda + \frac{t^3 - 6t^2 + 15t + 14}{t^3 - 3t^2 + 3t + 1} \right)$$

which gives the following controller:

$$\dot{u} - \frac{t^4 - 4t^3 + 6t^2 - 2t + 5}{(1-t)(t^3 - 3t^2 + 3t + 1)}u =$$

$$-\frac{t^2 - 3t + 3}{t^3 - 3t^2 + 3t + 1}\left((t^2 - 3t + 3)y + \frac{t^3 - 6t^2 + 15t + 14}{t^3 - 3t^2 + 3t + 1}y \right)$$

16.6. Conclusion

Due to the use of the algebra defined on the non-commutative body of rational fractions, it was shown that, not only could the concept of transfer function of a linear dynamic system with time variable coefficients be extended, but also the traditional operations on these systems had simple solutions, based on simple algebraic operations defined on the body of the related fractions. These mathematical results make it possible to raise and solve traditional control or/and identification problems. Obviously, the complexity of calculations is increased with respect to traditional systems (i.e. with constant coefficients) and, in practice, it is necessary to use formal calculation tools. Finally, the approach presented here was to voluntarily consider the continuous-time systems but an analogous approach can be followed for the discrete-time systems (see, for example, Kamen's works).

16.7. Bibliography

[AMI 54] AMITSUR A.S., "Differential polynomials and division algebras", *Annals of mathematics*, vol. 59, p. 245-278, 1954.

[BAS 93] BASSEVILLE M., NIKIFOROV I., *Detection of abrupt changes, Theory and Application*, Prentice Hall, Information and System Science Series, vol. 1, Englewood Cliffs, New Jersey, 1993.

[CLE 98] CLEMENT A., "An ordinary differential equation for the Green function of time-domain free-surface hydrodynamics", *Journal of Engineering Mathematics,* vol. 33, 1998.

[GUG 01] GUGLIELMI M., NORET E., "Une classe des systèmes auto-similaires et à mémoire longue", *Techniques et sciences informatiques*, vol. 20, no. 9, 2001.

[HUR 65] HURST H.E., BLACK R.P., SINAIKA Y.M., "Long term storage in reservoirs. An experimental study", *Constable*, London, p. 1153-1173, 1965.

[KUC 79] KUCERA V., *Discrete Linear Control: the Polynomial Approach*, John Wiley and sons, 1979.

[ORE 33] ORE O., "Theory of non-commutative polynomials", *Annals of mathematics*, vol. 34, p. 480-508, 1933.

[VAN 88] VAN DER ZIEL, "Unified presentation of 1/f noise in electronic devices: fundamental 1/f noise sources", *Proceedings of IEEE*, vol. 76, no. 3, p 233-258, 1988.

[WIL 95] WILINGER W., TAQQU M.S., LELAND W.E., WILSON V., "Self-similarity in high-speed packed traffic: analysis and modelling of Ethernet traffic measurements", *Statistical Science*, vol. 10, p. 676-685, 1995.

[ZHU 89]] ZHU J., JOHNSON C.D., "New results in the reduction of linear time-varying dynamical systems", *SIAM J. Control & Optimization*, p. 476-494, 1989.

List of Authors

Alain BARRAUD
Laboratoire d'Automatique de Grenoble
Ecole Nationale Supérieure d'Ingénieurs Electriciens de Grenoble ENSIEG
Institut National Polytechnique de Grenoble
France

Dominique BEAUVOIS
Supélec – Service Automatique
Gif-sur-Yvette, France

Patrick BOUCHER
Supélec
Gif-sur-Yvette, France

Philippe CHEVREL
IRCCyN
CNRS – Ecole des Mines de Nantes
France

Martial DEMERLE
Service Automatique, Supélec
Gif-sur-Yvette, France

Gilles DUC
École Supérieure d'Électricité (Supélec)
Service Automatique
Gif-sur-Yvette, France

Didier DUMUR
Supélec
Gif-sur-Yvette, France

Sylvianne GENTIL
Laboratoire d'Automatique de Grenoble
Ecole Nationale Supérieure d'Ingénieurs Electriciens de Grenoble ENSIEG
Institut National Polytechnique de Grenoble
France

Michel GUGLIELMI
IRCCYN
Ecole centrale de Nantes
Nantes, France

Philippe de LARMINAT
IRCCyN
Nantes, France

Eric LE CARPENTIER
IRCCYN
Ecole centrale de Nantes
France

Yann LE GORREC
LAGEP
Villeurbanne, France

Suzanne LESEC
Laboratoire d'Automatique de Grenoble
Ecole Nationale Supérieure d'Ingénieurs Electriciens de Grenoble ENSIEG
Institut National Polytechnique de Grenoble
France

Jean-François MAGNI
ONERA-CERT, Département de commande des systèmes
Toulouse, France

Michel MALABRE
IRCCYN
Ecole centrale de Nantes
France

Houria SIGUERDIDJANE
Service Automatique, Supélec
Gif-sur-Yvette, France

Yves TANGUY
Supélec – Service Automatique
Gif-sur-Yvette, France

Gérard THOMAS
Dept EEA Ecole Centrale de Lyon
Ecully, France

Patrick TURELLE
Supélec
Gif-sur-Yvette, France

Index

error
  dynamic 271, 277, 285, 287
  static 271-274, 284, 289, 295, 301, 304, 308, 311, 314, 317, 320, 376, 482
Euler's method 239, 245, 246

F, I

filter
  Kalman 412, 415
  non-observable 488
Fourier transform 14, 18, 33, 34, 41, 142-145, 147, 149-151
  continuous-time 88, 142-144
  discrete-time 88, 96, 142-144
internal stability 132, 133, 135, 328, 332, 335, 409, 416, 419

J, K, L

Jury criterion 93
Kalman filtering 168
loop
  closed 253-258, 260, 264-267, 269, 280, 281, 283-286, 309, 324, 332, 382, 390, 408, 409, 449, 458-460, 463, 465, 472, 473, 475, 485, 499, 503, 533
  open 253, 254, 256, 258-265, 272, 273, 276, 278, 283-287, 297, 308, 315, 316, 363, 380, 451, 462-465, 470, 486

M

margin
  delay 268, 269, 360, 381
  gain 267, 270, 284, 287, 380
  phase 188, 267-270, 284, 287, 297, 298, 301, 302, 304, 308, 311, 317-320, 323, 360, 380, 414
  stability 185, 187, 267, 314, 380, 381

sufficient phase 284, 323
Markovian system 156
model
  AR 152
  ARMA 150
  ARMAX 153-156
  ARX 153, 154
  behavioral 196, 198, 214
  deterministic
  functional 195
  main 21
  signal 4
  statistical 141
  structural 195

N, O, P

Nyquist criterion 258-260, 263, 265, 486
optimization
  non-linear 214, 217, 221
  quadratic 171, 173, 177, 179, 216
Padé approximant 230, 231, 234, 247
polynomial
  invariant 114,115
  minimal 114
Prony's method 152

R

representation
  external 57, 88
  internal 43, 44, 57, 60, 89
response
  fixed 110, 135, 136
  forced 90, 376, 392
  free 90, 387, 407
  frequency 14, 15, 17, 18, 21, 24, 30, 31, 96, 98, 103, 187, 197, 256-258, 261, 262, 285
  harmonic 204, 267
    closed loop 255

impulse 8-12, 15-18, 21, 23, 89, 94, 96, 98, 105, 150, 151, 153, 215, 220
    nominal 279, 282
    time 98, 204, 257, 456, 476
robustness 185, 242, 354, 356, 370, 380, 381, 395, 400, 401, 414-416, 436, 445, 473, 476, 479-482, 485, 493, 495, 500, 506
Routh criterion 285, 265, 266

S

Shannon condition 101
signal
    continuous-time 4, 81, 100, 142, 144, 148
    deterministic 141, 142, 145
    discrete-time 81, 83, 100, 142-144, 149
    ergodic 145, 146, 148, 149
    stationary 145-150, 153
    white 145, 149
structure
    cascade 382, 385, 386, 389
    RST 353, 370, 379, 380, 385

T

trajectory
    asymptotic 434, 436
    reference 374, 375, 382, 383, 390, 393, 394
transfer function 12-21, 23, 27, 28, 44, 57, 58, 60, 61, 96, 97, 101, 102, 105, 130, 150, 151, 153-155, 186, 187, 197, 202-209, 211, 212, 214, 244, 246-248, 254, 255, 257-261, 263, 269, 270, 278, 280, 281, 284, 285, 293, 297, 299, 303, 306, 307, 310, 313-316, 319, 322, 324, 376, 377, 380, 482, 483, 485, 494, 496, 497, 506, 507, 522, 526, 527, 529, 530, 532

W

white noise 376, 404, 412, 426, 430

Z

zeros
    finite 116, 118, 127, 131, 136
    infinite 116, 119, 127, 129-131, 133-135